I do like CFD, VOL.1

Governing Equations and Exact Solutions

Second Edition

Katate Masatsuka

I do like CFD, VOL.1, second edition
Copyright ©2009, 2013 by Katate Masatsuka
ISBN 978-1-304-82793-7

Author:

Katate Masatsuka
e-mail: info@cfdbooks.com
URL: http://www.cfdbooks.com

Comments and suggestions are appreciated.
Please feel free to send messages to the author.

[Note: Katate Masatsuka is a pen name.]

Published by Katate Masatsuka.

Printed and bound in the United States of America by Lulu.com.

PDF Version:

PDF version of this book is available at http://www.cfdbooks.com.
Extra contents are available only in this hard copy version.

In Memory of Professor Emeritus Haruo Oguro

Contents

Preface

Second Edition (2013)

I still like CFD and found more topics that I like after the publication of the first edition. Also, I received reports on typos and errors from people who like CFD (Thank you so much!). So, I decided to revise the book. The contents remain highly biased, but can still be a good reference book for students studying basics of CFD and researchers developing fundamental CFD algorithms. I hope that the second edition will make them like CFD even more.

Katate Masatsuka
Yorktown, October 2013

First Edition (2009)

CFD: Computational Fluid Dynamics.

I like CFD. I like it so much that I decided to write how much I like CFD. Here, in this volume, I focus on basic notations and formulas, governing equations, and exact solutions used in CFD. I hope that you like them too.

- **Basics:**

  Some basic stuff such as notations, formulas, and theorems.

- **Model Equations:**

  Model equations commonly used in the algorithm development.

- **Euler Equations:**

  The Euler equations and related equations for inviscid flows.

- **Navier-Stokes Equations:**

  The Navier-Stokes equations and related equations for viscous flows.

- **Turbulence Equations:**

  Averaged Navier-Stokes equations and a little about turbulence models.

- **Exact Solutions I:**

  General solutions and techniques for deriving exact solutions.

- **Exact Solutions II:**

  Exact solutions for selected equations which can be used for accuracy study.

Naturally, the content is highly biased. This is everything about what I like in CFD. It does not contain any topics that I don't like. So, this is certainly not a textbook, but can still be a good reference book. In particular, this volume may be useful for students studying basics of CFD and researchers developing fundamental CFD algorithms (I hope that they like CFD too).

Katate Masatsuka
Yorktown, February 2009

Chapter 1

Basics

1.1 Differential Notations

The following notations for derivatives are widely used:

- First derivatives:

$$\frac{\partial u}{\partial x} = \partial_x u = u_x. \tag{1.1.1}$$

- Second derivatives:

$$\frac{\partial^2 u}{\partial x^2} = \partial_{xx} u = u_{xx}. \tag{1.1.2}$$

- n-th derivative:

$$\frac{\partial^n u}{\partial x^n} = \underbrace{\partial_{x \cdots x}}_{n} u = \underbrace{u_{x \cdots x}}_{n}. \tag{1.1.3}$$

It is nice to have more than one way to express the same thing. In particular, I like the subscript notation such as u_x because it is very easy to write and also takes much less vertical space than others. Of course, it will be very inconvenient for very high-order derivatives (because the expression will be too long), but such high-order derivatives do not usually arise in CFD books. pa

- Material/Substantial/Lagrangian derivative:

$$\frac{D\alpha}{Dt} = \frac{\partial \alpha}{\partial t} + u\frac{\partial \alpha}{\partial x} + v\frac{\partial \alpha}{\partial y} + w\frac{\partial \alpha}{\partial z}. \tag{1.1.4}$$

I like this derivative. It is very interesting. It represents the time rate of change of the quantity α when it is being convected in a flow with the velocity (u, v, w). So, when this derivative is zero, $D\alpha/Dt = 0$, the distribution (e.g., a contour plot) of $\alpha(x, y, z)$ in space moves with the velocity (u, v, w), preserving its initial shape. On the other hand, when the time derivative is zero, $\partial_t \alpha = 0$, the distribution of $\alpha(x, y, z)$ does not move and stays fixed in space for all times.

1.2   Vectors and Operators

1.2.1   Vectors and Tensors

Vectors and tensors are usually denoted by boldface letters, rather than by arrows put on top of them. It is simple and very effective. I like it. By the way, in this book, the vector is almost always a column vector and a row vector is expressed as a transpose of a column vector indicated by the superscript t.

$$
\mathbf{a} = \begin{bmatrix} a_1 \\ a_2 \\ a_3 \end{bmatrix}, \quad
\mathbf{b} = \begin{bmatrix} b_1 \\ b_2 \\ b_3 \end{bmatrix}, \quad
\mathbf{a}^t = [a_1, a_2, a_3], \quad \mathbf{b}^t = [b_1, b_2, b_3]. \tag{1.2.1}
$$

Examples of the second-rank tensors:

$$
\mathbf{A} = \begin{bmatrix} A_{11} & A_{12} & A_{13} \\ A_{21} & A_{22} & A_{23} \\ A_{31} & A_{32} & A_{33} \end{bmatrix}, \quad
\mathbf{A}^t = \begin{bmatrix} A_{11} & A_{21} & A_{31} \\ A_{12} & A_{22} & A_{32} \\ A_{13} & A_{23} & A_{33} \end{bmatrix}, \tag{1.2.2}
$$

where $\mathbf{A}^t$ is a transpose of $\mathbf{A}$ that is defined as a matrix with rows and columns interchanged.

1.2.2   Dot/Inner Product and Tensor Product

The dot/inner product yields a scalar quantity by combining a pair of vectors as follows:

$$
\mathbf{a} \cdot \mathbf{b} \;=\; \mathbf{a}^t \mathbf{b} = a_1 b_1 + a_2 b_2 + a_3 b_3, \tag{1.2.3}
$$
$$
\mathbf{a}^2 \;=\; \mathbf{a} \cdot \mathbf{a} = \mathbf{a}^t \mathbf{a}, \tag{1.2.4}
$$

or a pair of tensors as follows:

$$
\begin{aligned}
\mathbf{A} : \mathbf{B} \;=\;& \sum_{\text{all } i,j} A_{ij} B_{ij} \\
=\;& A_{11}B_{11} + A_{12}B_{12} + A_{13}B_{13} + A_{21}B_{21} + A_{22}B_{22} \\
& + A_{23}B_{23} + A_{31}B_{31} + A_{32}B_{32} + A_{33}B_{33}.
\end{aligned} \tag{1.2.5}
$$

The latter is often called the tensor product. Two dots are used for the tensor product because it is summed over two indices, perhaps. In any case, it produces a single real number from vectors/tensors no matter how many components they have. This is very nice. Of course, we have

$$
\mathbf{a} \cdot \mathbf{b} \;=\; \mathbf{b} \cdot \mathbf{a}, \quad \mathbf{A} : \mathbf{B} = \mathbf{B} : \mathbf{A}. \tag{1.2.6}
$$

1.2.3   Vector/Outer Product

The vector/outer product produces a vector:

$$
\mathbf{a} \times \mathbf{b} = \begin{vmatrix} \mathbf{e}_1 & \mathbf{e}_2 & \mathbf{e}_3 \\ a_1 & a_2 & a_3 \\ b_1 & b_2 & b_3 \end{vmatrix} = (a_2 b_3 - a_3 b_2)\mathbf{e}_1 + (a_3 b_1 - a_1 b_3)\mathbf{e}_2 + (a_1 b_2 - a_2 b_1)\mathbf{e}_3 = \begin{bmatrix} a_2 b_3 - a_3 b_2 \\ a_3 b_1 - a_1 b_3 \\ a_1 b_2 - a_2 b_1 \end{bmatrix}, \tag{1.2.7}
$$

where $\mathbf{e}_1$, $\mathbf{e}_2$, $\mathbf{e}_3$ are the unit vectors, i.e., $\mathbf{a} = a_1 \mathbf{e}_1 + a_2 \mathbf{e}_2 + a_3 \mathbf{e}_3$ and $\mathbf{b} = b_1 \mathbf{e}_1 + b_2 \mathbf{e}_2 + b_3 \mathbf{e}_3$. I like the fact that $\mathbf{a} \times \mathbf{b}$ and $\mathbf{b} \times \mathbf{a}$ represent two different vectors, pointing opposite directions with the same magnitude. It is very geometrical and easy to visualize.

1.2.4 Dyadic Tensor

The dyadic tensor, denoted by $\otimes$, is a second-rank tensor made out of two vectors:

$$\mathbf{a}\otimes\mathbf{b} = \mathbf{ab}^t = \begin{bmatrix} a_1 \\ a_2 \\ a_3 \end{bmatrix} [b_1, b_2, b_3] = \begin{bmatrix} a_1b_1 & a_1b_2 & a_1b_3 \\ a_2b_1 & a_2b_2 & a_2b_3 \\ a_3b_1 & a_3b_2 & a_3b_3 \end{bmatrix}, \tag{1.2.8}$$

so, the (i, j)-th component is given by

$$(\mathbf{a}\otimes\mathbf{b})_{ij} = a_i b_j. \tag{1.2.9}$$

In fact, the dyadic tensor is defined, with $\mathbf{c} = [c_1, c_2, c_3]^t$, by

$$(\mathbf{a}\otimes\mathbf{b})\mathbf{c} = \mathbf{a}(\mathbf{b} \cdot \mathbf{c}), \tag{1.2.10}$$

from which you can figure out what $\mathbf{a}\otimes\mathbf{b}$ should be, i.e., write out the right hand side in components and rearrange the result in the form as in the left hand side to find Equation (1.2.8). Of course, we have in general,

$$\mathbf{a}\otimes\mathbf{b} \neq \mathbf{b}\otimes\mathbf{a}. \tag{1.2.11}$$

Unlike the vector product, the geometrical meaning of this is not immediately clear to me. But I like it. It is simple enough.

1.2.5 Scalar Triple Product

I like the scalar triple product defined for three vectors $\mathbf{a}$, $\mathbf{b}$, and $\mathbf{c}$ by

$$\mathbf{a} \cdot (\mathbf{b} \times \mathbf{c}). \tag{1.2.12}$$

It is, in fact, equal to the volume of the parallelepiped formed by the vectors $\mathbf{a}$, $\mathbf{b}$, and $\mathbf{c}$. I like it very much because then I can easily understand various formulas associated with the scalar triple product. For example, we have

$$\mathbf{a} \cdot (\mathbf{b} \times \mathbf{c}) = \mathbf{b} \cdot (\mathbf{c} \times \mathbf{a}) = \mathbf{c} \cdot (\mathbf{a} \times \mathbf{b}), \tag{1.2.13}$$

and of course

$$(\mathbf{b} \times \mathbf{c}) \cdot \mathbf{a} = (\mathbf{c} \times \mathbf{a}) \cdot \mathbf{b} = (\mathbf{a} \times \mathbf{b}) \cdot \mathbf{c}, \tag{1.2.14}$$

which are trivial relations because all represent the same volume. It is also easy to understand that the volume represented by the scalar triple product is a signed volume, meaning that switching the two vectors in the vector product results in the negative of the scalar triple product (because the vector product points the opposite direction):

$$\mathbf{a} \cdot (\mathbf{b} \times \mathbf{c}) = -\mathbf{a} \cdot (\mathbf{c} \times \mathbf{b}), \tag{1.2.15}$$
$$\mathbf{b} \cdot (\mathbf{c} \times \mathbf{a}) = -\mathbf{b} \cdot (\mathbf{a} \times \mathbf{c}), \tag{1.2.16}$$
$$\mathbf{c} \cdot (\mathbf{a} \times \mathbf{b}) = -\mathbf{c} \cdot (\mathbf{b} \times \mathbf{a}). \tag{1.2.17}$$

As you might expect and can easily verify, the scalar triple product can be expressed by the determinant:

$$\mathbf{a} \cdot (\mathbf{b} \times \mathbf{c}) = \begin{vmatrix} a_1 & a_2 & a_3 \\ b_1 & b_2 & b_3 \\ c_1 & c_2 & c_3 \end{vmatrix}. \tag{1.2.18}$$

It is quite easy to see that the scalar triple product vanishes if the two vectors in the vector product are the same (because then the vector product becomes the zero vector). It is also quite easy to understand that the scalar triple product vanishes if any two vectors are identical, e.g.,

$$\mathbf{a} \cdot (\mathbf{a} \times \mathbf{c}) = \mathbf{a} \cdot (\mathbf{b} \times \mathbf{a}) = 0. \tag{1.2.19}$$

It vanishes, of course, because the parallelepiped becomes flat in that case and the volume (= the scalar triple product) becomes zero. I now tell you why I really like the scalar triple product. I like it because the volume V of a tetrahedron formed by the three vectors, $\mathbf{a}$, $\mathbf{b}$, and $\mathbf{c}$, is given by

$$V = \frac{1}{6} \left| \mathbf{a} \cdot (\mathbf{b} \times \mathbf{c}) \right|, \tag{1.2.20}$$

and a tetrahedron is one of my favorite computational cells in CFD. Note that the vectors $\mathbf{a}$, $\mathbf{b}$, and $\mathbf{c}$ are the edge vectors centered at a common vertex of the tetrahedron. The volume is defined with the absolute value of the scalar triple product to keep it positive in case any of the vectors points the opposite direction. Of course, then, it is easy to understand that the scalar triple product does not change at all under rotation of coordinate system:

$$\mathbf{a} \cdot (\mathbf{b} \times \mathbf{c}) = \mathbf{a}' \cdot (\mathbf{b}' \times \mathbf{c}'), \tag{1.2.21}$$

where $\mathbf{a}'$, $\mathbf{b}'$, and $\mathbf{c}'$ are the vectors rotated by a rotation matrix $\mathbf{R}$,

$$\mathbf{a}' = \mathbf{R}\mathbf{a}, \quad \mathbf{b}' = \mathbf{R}\mathbf{b}, \quad \mathbf{c}' = \mathbf{R}\mathbf{c}. \tag{1.2.22}$$

This is natural because rotating a parallelepiped (or a tetrahedron) does not change its volume.

1.2.6   Vector Triple Product

Of course, there is a vector triple product. The vector triple product is defined by

$$\mathbf{a} \times (\mathbf{b} \times \mathbf{c}). \tag{1.2.23}$$

The following formula is known as Lagrange's formula or vector product expansion:

$$\mathbf{a} \times (\mathbf{b} \times \mathbf{c}) = \mathbf{b}(\mathbf{a} \cdot \mathbf{c}) - \mathbf{c}(\mathbf{a} \cdot \mathbf{b}). \tag{1.2.24}$$

It is known also as the BAC-CAB identity. That is useful for memorizing the formula. But that's not the reason that I like the vector triple product. I like it because it is useful in CFD. To see how useful it is, consider the dot product of the vector triple product and $\mathbf{a}$:

$$[\mathbf{a} \times (\mathbf{b} \times \mathbf{c})] \cdot \mathbf{a} = 0. \tag{1.2.25}$$

Yes, it is zero because it is a scalar triple product with two identical vectors: $\mathbf{a}$, $\mathbf{b} \times \mathbf{c}$, and $\mathbf{a}$. It means that the vector triple product of $\mathbf{a}$, $\mathbf{b}$, and $\mathbf{c}$ is perpendicular to $\mathbf{a}$. That is interesting. In CFD, we sometimes need to find vectors tangent to a face of a computational cell from the face normal vector, or more generally, find a vector perpendicular to a given vector. For example, the situation arises in the rotated-hybrid Riemann solvers [112]. Usually in finite-volume methods, a single numerical flux is computed in the direction normal to the face, $\hat{\mathbf{n}}$, where the hat indicates that the vector is a unit vector. The rotated-hybrid method computes the face-normal numerical flux as a combination of two numerical fluxes by breaking $\hat{\mathbf{n}}$ into two orthogonal directions and applying two different numerical fluxes in the two directions. The method applies a robust but dissipative numerical flux in a physically meaningful direction, $\hat{\mathbf{n}}_1$, e.g., normal to a shock wave, and a less robust but much less dissipative numerical flux in the direction $\hat{\mathbf{n}}_2$ that is perpendicular to $\hat{\mathbf{n}}_1$. The resulting flux has been shown to be very robust for strong shocks without adding too much dissipation in smooth regions. Since $\hat{\mathbf{n}}_1$ is determined by a physical consideration, the problem is to find $\hat{\mathbf{n}}_2$ that is perpendicular to a given vector, $\hat{\mathbf{n}}_1$. Then, we can find $\hat{\mathbf{n}}_2$ by the vector triple product [112]:

$$\hat{\mathbf{n}}_2 = \frac{\hat{\mathbf{n}}_1 \times (\hat{\mathbf{n}} \times \hat{\mathbf{n}}_1)}{|\hat{\mathbf{n}}_1 \times (\hat{\mathbf{n}} \times \hat{\mathbf{n}}_1)|}, \tag{1.2.26}$$

where the denominator is the magnitude of the numerator to make $\hat{\mathbf{n}}_2$ a unit vector. Note that the numerator is not necessarily a unit vector. Clearly, we have $\hat{\mathbf{n}}_2 \cdot \hat{\mathbf{n}}_1 = 0$, i.e., $\hat{\mathbf{n}}_2$ is perpendicular to $\hat{\mathbf{n}}_1$ as desired. That is nice. I like the vector triple product very much.

1.2.7   Identity Matrix

This is a matrix whose diagonal entries are all 1 and other entries are zero. Such a matrix is called the identity matrix. For example, the 3×3 identity matrix is given by

$$\mathbf{I} = \begin{bmatrix} 1 & 0 & 0 \\ 0 & 1 & 0 \\ 0 & 0 & 1 \end{bmatrix}. \tag{1.2.27}$$

Note that the $n \times n$ identity matrix has $n^2 - n$ zeroes. This means that the number of zeroes increases quadratically with n. Of course, I like the identity matrix. In particular, I like its alternative notation which is very compact. For example, the 3×3 identity matrix above is denoted by

$$\mathbf{I} = diag(1,1,1). \tag{1.2.28}$$

This is very useful when we write the identity matrix of a very large size. Why is it called the identity matrix? Well, I think it's because we have

$$\mathbf{IA} = \mathbf{AI} = \mathbf{A}, \tag{1.2.29}$$

for any matrix $\mathbf{A}$ in any dimension.

1.3   Orthogonal Vectors

Consider three orthogonal unit vectors, $\mathbf{n}$, $\boldsymbol{\ell}$, $\mathbf{m}$ (see Figure 1.3.1):

$$\mathbf{n} \cdot \mathbf{n} = \boldsymbol{\ell} \cdot \boldsymbol{\ell} = \mathbf{m} \cdot \mathbf{m} = 1, \tag{1.3.1}$$
$$\mathbf{n} \cdot \boldsymbol{\ell} = \boldsymbol{\ell} \cdot \mathbf{m} = \mathbf{m} \cdot \mathbf{n} = 0, \tag{1.3.2}$$
$$\mathbf{n} = \boldsymbol{\ell} \times \mathbf{m}, \tag{1.3.3}$$
$$\boldsymbol{\ell} = \mathbf{m} \times \mathbf{n}, \tag{1.3.4}$$
$$\mathbf{m} = \mathbf{n} \times \boldsymbol{\ell}, \tag{1.3.5}$$
$$\mathbf{n} \otimes \mathbf{n} + \boldsymbol{\ell} \otimes \boldsymbol{\ell} + \mathbf{m} \otimes \mathbf{m} = \mathbf{I}. \tag{1.3.6}$$

Suppose that $\mathbf{n} = [n_x, n_y, n_z]^t$, $\boldsymbol{\ell} = [\ell_x, \ell_y, \ell_z]^t$, $\mathbf{m} = [m_x, m_y, m_z]^t$. Then, the above relations can be expanded as follows:

$$n_x^2 + n_y^2 + n_z^2 = 1, \tag{1.3.7}$$
$$\ell_x^2 + \ell_y^2 + \ell_z^2 = 1, \tag{1.3.8}$$
$$m_x^2 + m_y^2 + m_z^2 = 1, \tag{1.3.9}$$

$$n_x \ell_x + n_y \ell_y + n_z \ell_z = 0, \tag{1.3.10}$$
$$\ell_x m_x + \ell_y m_y + \ell_z m_z = 0, \tag{1.3.11}$$
$$m_x n_x + m_y n_y + m_z n_z = 0, \tag{1.3.12}$$

$$\begin{array}{lll}
\ell_y m_z - \ell_z m_y = n_x, & n_y \ell_z - n_z \ell_y = m_x, & m_y n_z - m_z n_y = \ell_x, \\
\ell_z m_x - \ell_x m_z = n_y, & n_z \ell_x - n_x \ell_z = m_y, & m_z n_x - m_x n_z = \ell_y, \\
\ell_x m_y - \ell_y m_x = n_z, & n_x \ell_y - n_y \ell_x = m_z, & m_x n_y - m_y n_x = \ell_z,
\end{array} \tag{1.3.13}$$

$$n_x^2 + \ell_x^2 + m_x^2 = 1, \tag{1.3.14}$$
$$n_y^2 + \ell_y^2 + m_y^2 = 1, \tag{1.3.15}$$
$$n_z^2 + \ell_z^2 + m_z^2 = 1, \tag{1.3.16}$$
$$n_x n_y + \ell_x \ell_y + m_x m_y = 0, \tag{1.3.17}$$
$$n_x n_z + \ell_x \ell_z + m_x m_z = 0, \tag{1.3.18}$$
$$n_y n_z + \ell_y \ell_z + m_y m_z = 0. \tag{1.3.19}$$

Introduce an arbitrary vector $\mathbf{v} = [u, v, w]^t$, and define the following quantities,

$$q_n = \mathbf{v} \cdot \mathbf{n}, \quad q_\ell = \mathbf{v} \cdot \boldsymbol{\ell}, \quad q_m = \mathbf{v} \cdot \mathbf{m}, \tag{1.3.20}$$

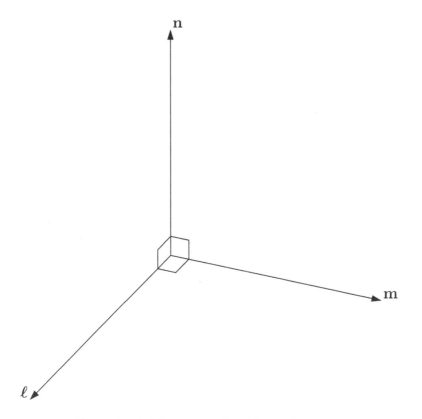

Figure 1.3.1: Three mutually orthogonal vectors.

then we have

$$q_n \mathbf{n} + q_\ell \boldsymbol{\ell} + q_m \mathbf{m} = \mathbf{v}, \tag{1.3.21}$$

$$q_n^2 + q_\ell^2 + q_m^2 = \mathbf{v}^2. \tag{1.3.22}$$

Equation (1.3.21) can be expanded as follows:

$$q_n n_x + q_\ell \ell_x + q_m m_x = u, \tag{1.3.23}$$

$$q_n n_y + q_\ell \ell_y + q_m m_y = v, \tag{1.3.24}$$

$$q_n n_z + q_\ell \ell_z + q_m m_z = w. \tag{1.3.25}$$

These relationships can be differentiated as follows:

$$dq_n n_x + dq_m m_x + dq_\ell \ell_x = du, \tag{1.3.26}$$

$$dq_n n_y + dq_m m_y + dq_\ell \ell_y = dv, \tag{1.3.27}$$

$$dq_n n_z + dq_m m_z + dq_\ell \ell_z = dw. \tag{1.3.28}$$

$$q_n \, dq_n + q_\ell \, dq_\ell + q_m \, dq_m = u \, du + v \, dv + w \, dw. \tag{1.3.29}$$

I like these relations because they can be very useful when I write a three-dimensional CFD code with $\mathbf{v}$ as a velocity vector, and $\mathbf{n}$, $\boldsymbol{\ell}$, $\mathbf{m}$ as a normal vector of a cell face and two tangent vectors respectively. In particular, these relations can be used to eliminate tangent vectors from expressions such as the absolute value of a Jacobian matrix projected along a normal vector which arises, for example, in the Roe solver [128]. See Sections 3.6.1 and 3.13.

Now consider a symmetric second-rank tensor $\boldsymbol{\tau}$:

$$\boldsymbol{\tau} = \begin{bmatrix} \tau_{xx} & \tau_{xy} & \tau_{xz} \\ \tau_{yx} & \tau_{yy} & \tau_{yz} \\ \tau_{zx} & \tau_{zy} & \tau_{zz} \end{bmatrix}, \tag{1.3.30}$$

where $\tau_{yx} = \tau_{xy}$, $\tau_{zx} = \tau_{xz}$, and $\tau_{zy} = \tau_{yz}$. We define the following quantities:

$$\tau_{nx} = \tau_{xx}n_x + \tau_{xy}n_y + \tau_{xz}n_z, \tag{1.3.31}$$
$$\tau_{ny} = \tau_{yx}n_x + \tau_{yy}n_y + \tau_{yz}n_z, \tag{1.3.32}$$
$$\tau_{nz} = \tau_{zx}n_x + \tau_{zy}n_y + \tau_{zz}n_z, \tag{1.3.33}$$
$$\tau_{nn} = \tau_{nx}n_x + \tau_{ny}n_y + \tau_{nz}n_z, \tag{1.3.34}$$
$$\tau_{n\ell} = \tau_{nx}\ell_x + \tau_{ny}\ell_y + \tau_{nz}\ell_z, \tag{1.3.35}$$
$$\tau_{nm} = \tau_{nx}m_x + \tau_{ny}m_y + \tau_{nz}m_z, \tag{1.3.36}$$
$$\tau_{nv} = \tau_{nx}u + \tau_{ny}v + \tau_{nz}w. \tag{1.3.37}$$

Then, we have the following identities:

$$\tau_{nn}n_x + \tau_{nm}m_x + \tau_{n\ell}\ell_x = \tau_{nx}, \tag{1.3.38}$$
$$\tau_{nn}n_y + \tau_{nm}m_y + \tau_{n\ell}\ell_y = \tau_{ny}, \tag{1.3.39}$$
$$\tau_{nn}n_z + \tau_{nm}m_z + \tau_{n\ell}\ell_z = \tau_{nz}, \tag{1.3.40}$$

$$\tau_{nn}d\tau_{nn} + \tau_{nm}d\tau_{nm} + \tau_{n\ell}d\tau_{n\ell} = \tau_{nx}d\tau_{nx} + \tau_{ny}d\tau_{ny} + \tau_{nz}d\tau_{nz}, \tag{1.3.41}$$
$$\tau_{nn}dq_n + \tau_{nm}dq_m + \tau_{n\ell}dq_\ell = \tau_{nx}du + \tau_{ny}dv + \tau_{nz}dw, \tag{1.3.42}$$
$$d\tau_{nn}n_x + d\tau_{nm}m_x + d\tau_{n\ell}\ell_x = d\tau_{nx}, \tag{1.3.43}$$
$$d\tau_{nn}n_y + d\tau_{nm}m_y + d\tau_{n\ell}\ell_y = d\tau_{ny}, \tag{1.3.44}$$
$$d\tau_{nn}n_z + d\tau_{nm}m_z + d\tau_{n\ell}\ell_z = d\tau_{nz}. \tag{1.3.45}$$

I like these identities very much because they can be useful for eliminating tangent vectors from expressions such as the absolute value of a Jacobian matrix projected along a normal vector that arises, for example, in the upwind viscous flux [106].

1.4 Index Notation

I like index notation because it is compact and useful especially for proving vector identities. Also, I can save my time and money because I don't need to write Σ for summations any more.

1.4.1 Einstein's Summation Convention

Einstein's summation convention means that whenever two indices are repeated, summation is taken over that index: for example,

$$a_i b_i = a_1 b_1 + a_2 b_2 + a_3 b_3, \tag{1.4.1}$$

which is nice because the dot product $\mathbf{a} \cdot \mathbf{b}$ can be simply written as $a_i b_i$ without using boldface letters or Σ. The i-th component of the matrix-vector product $\mathbf{Ab}$ can be expressed as

$$(\mathbf{Ab})_i = A_{ij}b_j = A_{i1}b_1 + A_{i2}b_2 + A_{i3}b_3, \tag{1.4.2}$$

where $(\mathbf{Ab})_i$ denotes the i-th component of $\mathbf{Ab}$. Thus, we can write

$$\mathbf{Ab} = \begin{bmatrix} (\mathbf{Ab})_1 \\ (\mathbf{Ab})_2 \\ (\mathbf{Ab})_3 \end{bmatrix} = \begin{bmatrix} A_{11}b_1 + A_{12}b_2 + A_{13}b_3 \\ A_{21}b_1 + A_{22}b_2 + A_{23}b_3 \\ A_{31}b_1 + A_{32}b_2 + A_{33}b_3 \end{bmatrix}. \tag{1.4.3}$$

Of course, the index used for the summation is a dummy, and so we can use any letter we like, e.g.,

$$\mathbf{a} \cdot \mathbf{b} = a_i b_i = a_j b_j = a_k b_k, \tag{1.4.4}$$

which are all equal to $a_1 b_1 + a_2 b_2 + a_3 b_3$.

1.4.2 Kronecker's Delta

Of course, I like the second-rank tensor called Kronecker's delta:

$$\delta_{ij} = \begin{cases} 1 & \text{if} \quad i = j, \\ 0 & \text{if} \quad i \neq j. \end{cases} \tag{1.4.5}$$

This is nothing but the components of the identity matrix:

$$\mathbf{I} = \begin{bmatrix} \delta_{11} & \delta_{12} & \delta_{13} \\ \delta_{21} & \delta_{22} & \delta_{23} \\ \delta_{31} & \delta_{32} & \delta_{33} \end{bmatrix} = \begin{bmatrix} 1 & 0 & 0 \\ 0 & 1 & 0 \\ 0 & 0 & 1 \end{bmatrix}. \tag{1.4.6}$$

I like Kronecker's delta especially because it is sometimes called the substitution operator base on the following property:

$$\delta_{ij} a_j = a_i, \tag{1.4.7}$$

which looks like i is substituted into j. This is nice also because δ_{ij} has disappeared on the right hand side, making the right hand side very simple. By the substitution, we have

$$\delta_{ij} a_i b_j = a_i b_i, \tag{1.4.8}$$

which means

$$\mathbf{I} : (\mathbf{a} \otimes \mathbf{b}) = \mathbf{a}^t \mathbf{I} \mathbf{b} = \mathbf{a} \cdot \mathbf{b}. \tag{1.4.9}$$

In a way, we have just proved a vector identity. It is also easy to show that

$$\delta_{ii} = 3, \tag{1.4.10}$$

which is the trace of the identity matrix, and also that

$$\delta_{ij} \delta_{jk} = \delta_{ik}, \tag{1.4.11}$$

i.e., the product of two identity matrices is the identity matrix:

$$\mathbf{II} = \mathbf{I}. \tag{1.4.12}$$

1.4.3 Eddington's Epsilon

The third-rank tensor called Eddington's epsilon:

$$\varepsilon_{ijk} = \begin{cases} 0 & \text{if any two of } i, j, k \text{ are the same,} \\ 1 & \text{for even permutation,} \\ -1 & \text{for odd permutation,} \end{cases} \tag{1.4.13}$$

reminds me of the Rubik cube that I like. By definition, we have

$$\varepsilon_{ijk} = -\varepsilon_{jik} = -\varepsilon_{ikj}, \tag{1.4.14}$$

$$\varepsilon_{ijk} = \varepsilon_{jki} = \varepsilon_{kij}. \tag{1.4.15}$$

And we have naturally

$$\delta_{ij} \varepsilon_{ijk} = 0, \tag{1.4.16}$$

which is because Kronecker's delta turns ε_{ijk} into ε_{iik} that is zero by definition. I would use $\mathbf{E}$ to denote the third-rank tensor of Eddington's epsilon. Then, I have

$$\mathbf{E}_{1jk} \quad \rightarrow \quad \begin{bmatrix} \varepsilon_{111} & \varepsilon_{112} & \varepsilon_{113} \\ \varepsilon_{121} & \varepsilon_{122} & \varepsilon_{123} \\ \varepsilon_{131} & \varepsilon_{132} & \varepsilon_{133} \end{bmatrix} = \begin{bmatrix} 0 & 0 & 0 \\ 0 & 0 & 1 \\ 0 & -1 & 0 \end{bmatrix}, \tag{1.4.17}$$

$$\mathbf{E}_{2jk} \quad \rightarrow \quad \begin{bmatrix} \varepsilon_{211} & \varepsilon_{212} & \varepsilon_{213} \\ \varepsilon_{221} & \varepsilon_{222} & \varepsilon_{223} \\ \varepsilon_{231} & \varepsilon_{232} & \varepsilon_{233} \end{bmatrix} = \begin{bmatrix} 0 & 0 & -1 \\ 0 & 0 & 0 \\ 1 & 0 & 0 \end{bmatrix}, \tag{1.4.18}$$

$$\mathbf{E}_{3jk} \quad \rightarrow \quad \begin{bmatrix} \varepsilon_{311} & \varepsilon_{312} & \varepsilon_{313} \\ \varepsilon_{321} & \varepsilon_{322} & \varepsilon_{323} \\ \varepsilon_{331} & \varepsilon_{332} & \varepsilon_{333} \end{bmatrix} = \begin{bmatrix} 0 & 1 & 0 \\ -1 & 0 & 0 \\ 0 & 0 & 0 \end{bmatrix}, \tag{1.4.19}$$

which look to me like the slices of the Rubic cube. I like $\mathbf{E}$ especially because I can write the vector product as

$$\mathbf{a} \times \mathbf{b} = \mathbf{E}\left(\mathbf{a} \otimes \mathbf{b}\right).$$ (1.4.20)

As you might expect, then, Eddington's epsilon can be used to express the determinant of $\mathbf{A}$,

$$\det \mathbf{A} = \varepsilon_{ijk} A_{1i} A_{2j} A_{3k},$$ (1.4.21)

or of three vectors,

$$\varepsilon_{ijk} a_i b_j c_k = \begin{vmatrix} a_1 & a_2 & a_3 \\ b_1 & b_2 & b_3 \\ c_1 & c_2 & c_3 \end{vmatrix}.$$ (1.4.22)

Even more interestingly, it can be rewritten as

$$\varepsilon_{ijk} a_i b_j c_k = a_i \begin{vmatrix} \delta_{i1} & \delta_{i2} & \delta_{i3} \\ b_1 & b_2 & b_3 \\ c_1 & c_2 & c_3 \end{vmatrix} = a_i b_j \begin{vmatrix} \delta_{i1} & \delta_{i2} & \delta_{i3} \\ \delta_{j1} & \delta_{j2} & \delta_{j3} \\ c_1 & c_2 & c_3 \end{vmatrix} = a_i b_j c_k \begin{vmatrix} \delta_{i1} & \delta_{i2} & \delta_{i3} \\ \delta_{j1} & \delta_{j2} & \delta_{j3} \\ \delta_{k1} & \delta_{k2} & \delta_{k3} \end{vmatrix},$$ (1.4.23)

which means

$$\varepsilon_{ijk} = \begin{vmatrix} \delta_{i1} & \delta_{i2} & \delta_{i3} \\ \delta_{j1} & \delta_{j2} & \delta_{j3} \\ \delta_{k1} & \delta_{k2} & \delta_{k3} \end{vmatrix}.$$ (1.4.24)

I like this formula because I can use it to derive various identities. Consider

$$\varepsilon_{ijk} \varepsilon_{lmn} = \begin{vmatrix} \delta_{i1} & \delta_{i2} & \delta_{i3} \\ \delta_{j1} & \delta_{j2} & \delta_{j3} \\ \delta_{k1} & \delta_{k2} & \delta_{k3} \end{vmatrix} \begin{vmatrix} \delta_{l1} & \delta_{l2} & \delta_{l3} \\ \delta_{m1} & \delta_{m2} & \delta_{m3} \\ \delta_{n1} & \delta_{n2} & \delta_{n3} \end{vmatrix},$$ (1.4.25)

which can be written as

$$\varepsilon_{ijk} \varepsilon_{lmn} = \begin{vmatrix} \delta_{i1} & \delta_{i2} & \delta_{i3} \\ \delta_{j1} & \delta_{j2} & \delta_{j3} \\ \delta_{k1} & \delta_{k2} & \delta_{k3} \end{vmatrix} \begin{vmatrix} \delta_{l1} & \delta_{m2} & \delta_{n3} \\ \delta_{l1} & \delta_{m2} & \delta_{n3} \\ \delta_{l1} & \delta_{m2} & \delta_{n3} \end{vmatrix} = \begin{vmatrix} \delta_{ir}\delta_{lr} & \delta_{ir}\delta_{mr} & \delta_{ir}\delta_{nr} \\ \delta_{jr}\delta_{lr} & \delta_{jr}\delta_{mr} & \delta_{jr}\delta_{nr} \\ \delta_{kr}\delta_{lr} & \delta_{kr}\delta_{mr} & \delta_{kr}\delta_{nr} \end{vmatrix}.$$ (1.4.26)

Therefore, we obtain by Equation (1.4.11)

$$\varepsilon_{ijk} \varepsilon_{lmn} = \begin{vmatrix} \delta_{il} & \delta_{im} & \delta_{in} \\ \delta_{jl} & \delta_{jm} & \delta_{jn} \\ \delta_{kl} & \delta_{km} & \delta_{kn} \end{vmatrix}.$$ (1.4.27)

This is a very useful formula. For example, with $l = k$, it becomes

$$\begin{aligned}
\varepsilon_{ijk} \varepsilon_{kmn} &= \delta_{kk} \begin{vmatrix} \delta_{im} & \delta_{in} \\ \delta_{jm} & \delta_{jn} \end{vmatrix} - \delta_{jk} \begin{vmatrix} \delta_{im} & \delta_{in} \\ \delta_{km} & \delta_{kn} \end{vmatrix} + \delta_{ik} \begin{vmatrix} \delta_{jm} & \delta_{jn} \\ \delta_{km} & \delta_{kn} \end{vmatrix} \\
&= 3 \begin{vmatrix} \delta_{im} & \delta_{in} \\ \delta_{jm} & \delta_{jn} \end{vmatrix} - \begin{vmatrix} \delta_{im} & \delta_{in} \\ \delta_{jk}\delta_{km} & \delta_{jk}\delta_{kn} \end{vmatrix} + \begin{vmatrix} \delta_{jm} & \delta_{jn} \\ \delta_{ik}\delta_{km} & \delta_{ik}\delta_{kn} \end{vmatrix} \\
&= 3 \begin{vmatrix} \delta_{im} & \delta_{in} \\ \delta_{jm} & \delta_{jn} \end{vmatrix} - \begin{vmatrix} \delta_{im} & \delta_{in} \\ \delta_{jm} & \delta_{jn} \end{vmatrix} + \begin{vmatrix} \delta_{jm} & \delta_{jn} \\ \delta_{im} & \delta_{in} \end{vmatrix} \\
&= 3 \begin{vmatrix} \delta_{im} & \delta_{in} \\ \delta_{jm} & \delta_{jn} \end{vmatrix} - \begin{vmatrix} \delta_{im} & \delta_{in} \\ \delta_{jm} & \delta_{jn} \end{vmatrix} - \begin{vmatrix} \delta_{im} & \delta_{in} \\ \delta_{jm} & \delta_{jn} \end{vmatrix},
\end{aligned}$$ (1.4.28)

and thus

$$\varepsilon_{ijk}\varepsilon_{kmn} = \begin{vmatrix} \delta_{im} & \delta_{in} \\ \delta_{jm} & \delta_{jn} \end{vmatrix} = \delta_{im}\delta_{jn} - \delta_{in}\delta_{jm}, \tag{1.4.29}$$

which is sometimes called the ε-δ identity. Furthermore, with $l = i$ and $m = j$, Equation (1.4.27) becomes

$$\varepsilon_{ijk}\varepsilon_{ijn} = 3\delta_{kn} - \delta_{kn} = 2\delta_{kn}. \tag{1.4.30}$$

Finally, by setting $l = i$, $m = j$, and $n = k$, we obtain

$$\varepsilon_{ijk}\varepsilon_{ijk} = 6. \tag{1.4.31}$$

These relations can be used to prove vector identities. For example, the i-th component of the triple vector product $\mathbf{a} \times (\mathbf{b} \times \mathbf{c})$ can be expressed as

$$(\mathbf{a} \times (\mathbf{b} \times \mathbf{c}))_i = \varepsilon_{ijk}a_j(\varepsilon_{klm}b_l c_m), \tag{1.4.32}$$

which becomes by the formula (1.4.29)

$$\begin{aligned}
(\mathbf{a} \times (\mathbf{b} \times \mathbf{c}))_i &= \varepsilon_{ijk}\varepsilon_{klm}a_j b_l c_m \\
&= \varepsilon_{kij}\varepsilon_{klm}a_j b_l c_m \\
&= (\delta_{il}\delta_{jm} - \delta_{jl}\delta_{im})a_j b_l c_m \\
&= \delta_{il}\delta_{jm}a_j b_l c_m - \delta_{jl}\delta_{im}a_j b_l c_m \\
&= a_j b_i c_j - a_j b_j c_i \\
&= b_i(a_j c_j) - c_i(a_j b_j),
\end{aligned} \tag{1.4.33}$$

and we have thus proved Lagrange's formula,

$$\mathbf{a} \times (\mathbf{b} \times \mathbf{c}) = \mathbf{b}(\mathbf{a} \cdot \mathbf{c}) - \mathbf{c}(\mathbf{a} \cdot \mathbf{b}). \tag{1.4.34}$$

1.5   Div, Grad, Curl

1.5.1   Coordinate Systems

I like vector differential operators, especially in Cartesian coordinates (x, y, z). In cylindrical (r, θ, z) or spherical coordinates (r, θ, ϕ) (see Figures 1.5.1 and 1.5.2), I like them, of course, but they can be a little complicated. Note that these coordinates are related: in cylindrical coordinates,

$$x = r \cos\theta, \tag{1.5.1}$$
$$y = r \sin\theta, \tag{1.5.2}$$
$$z = z, \tag{1.5.3}$$

while in spherical coordinates,

$$x = (r \sin\phi) \cos\theta, \tag{1.5.4}$$
$$y = (r \sin\phi) \sin\theta, \tag{1.5.5}$$
$$z = r \cos\phi. \tag{1.5.6}$$

Now, denote the unit vectors for these coordinates by

$$\mathbf{e}_x, \mathbf{e}_y, \mathbf{e}_z, \quad \text{in Cartesian coordinates,} \tag{1.5.7}$$
$$\mathbf{e}_r, \mathbf{e}_\theta, \mathbf{e}_z, \quad \text{in cylindrical coordinates,} \tag{1.5.8}$$
$$\mathbf{e}_r, \mathbf{e}_\theta, \mathbf{e}_\phi, \quad \text{in spherical coordinates,} \tag{1.5.9}$$

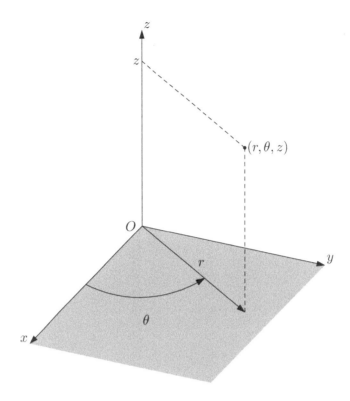

Figure 1.5.1: Cylindrical coordinates.

and vector components as in

$$
\begin{aligned}
\mathbf{a} &= a_1\,\mathbf{e}_x + a_2\,\mathbf{e}_y + a_3\,\mathbf{e}_z, & &\text{in Cartesian coordinates,} & &(1.5.10)\\
\mathbf{a} &= a_r\,\mathbf{e}_r + a_\theta\,\mathbf{e}_\theta + a_z\,\mathbf{e}_z, & &\text{in cylindrical coordinates,} & &(1.5.11)\\
\mathbf{a} &= a_r\,\mathbf{e}_r + a_\theta\,\mathbf{e}_\theta + a_\phi\,\mathbf{e}_\phi, & &\text{in spherical coordinates,} & &(1.5.12)
\end{aligned}
$$

then we can write vector differential operators as in the following sections.

1.5.2   Gradient of Scalar Functions

The gradient of a scalar function $\alpha(x, y, z)$ is defined by

$$
\operatorname{grad}\alpha = \lim_{\Delta V \to 0} \frac{1}{\Delta V} \oint \alpha\mathbf{n}\,dS, \tag{1.5.13}
$$

or simply by

$$
d\alpha = \operatorname{grad}\alpha \cdot d\mathbf{x}, \tag{1.5.14}
$$

where

$$
\begin{aligned}
d\mathbf{x} &= dx\,\mathbf{e}_x + dy\,\mathbf{e}_y + dz\,\mathbf{e}_z, & &\text{in Cartesian coordinates,} & &(1.5.15)\\
d\mathbf{x} &= dr\,\mathbf{e}_r + rd\theta\,\mathbf{e}_\theta + dz\,\mathbf{e}_z, & &\text{in cylindrical coordinates,} & &(1.5.16)\\
d\mathbf{x} &= dr\,\mathbf{e}_r + r\sin\phi\,d\theta\,\mathbf{e}_\theta + r\,d\phi\,\mathbf{e}_\phi, & &\text{in spherical coordinates.} & &(1.5.17)
\end{aligned}
$$

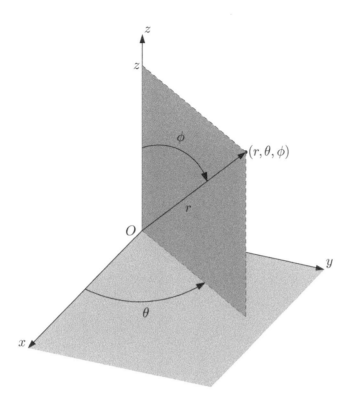

Figure 1.5.2: Spherical coordinates.

The gradient in all three coordinate systems:

$$\operatorname{grad} \alpha = \frac{\partial \alpha}{\partial x}\, \mathbf{e}_x + \frac{\partial \alpha}{\partial y}\, \mathbf{e}_y + \frac{\partial \alpha}{\partial z}\, \mathbf{e}_z, \qquad\qquad \text{in Cartesian coordinates,} \qquad (1.5.18)$$

$$\operatorname{grad} \alpha = \frac{\partial \alpha}{\partial r}\, \mathbf{e}_r + \frac{1}{r}\frac{\partial \alpha}{\partial \theta}\, \mathbf{e}_\theta + \frac{\partial \alpha}{\partial z}\, \mathbf{e}_z, \qquad\qquad \text{in cylindrical coordinates,} \qquad (1.5.19)$$

$$\operatorname{grad} \alpha = \frac{\partial \alpha}{\partial r}\, \mathbf{e}_r + \frac{1}{r \sin \phi}\frac{\partial \alpha}{\partial \theta}\, \mathbf{e}_\theta + \frac{1}{r}\frac{\partial \alpha}{\partial \phi}\, \mathbf{e}_\phi, \qquad\qquad \text{in spherical coordinates.} \qquad (1.5.20)$$

These are relatively intuitive and easy to understand. I like them. By the way, if you want to derive these formulas (and all formulas hereafter), simply apply the definition (1.5.13) to a small volume element, e.g., a small cube in Cartesian coordinates, or simply expand the differential definition (1.5.14).

1.5.3 Divergence of Vector Functions

The divergence of a vector function $\mathbf{a}(x, y, z)$ is defined by

$$\operatorname{div} \mathbf{a} = \lim_{\Delta V \to 0} \frac{1}{\Delta V} \oint \mathbf{a} \cdot \mathbf{n}\, dS, \qquad (1.5.21)$$

and yields

$$\operatorname{div} \mathbf{a} = \frac{\partial a_1}{\partial x} + \frac{\partial a_2}{\partial y} + \frac{\partial a_3}{\partial z} \qquad\qquad \text{in Cartesian coordinates,} \qquad (1.5.22)$$

$$\operatorname{div} \mathbf{a} = \frac{1}{r}\frac{\partial (r a_r)}{\partial r} + \frac{1}{r}\frac{\partial a_\theta}{\partial \theta} + \frac{\partial a_z}{\partial z} \qquad\qquad \text{in cylindrical coordinates,} \qquad (1.5.23)$$

$$\operatorname{div} \mathbf{a} = \frac{1}{r^2}\frac{\partial (r^2 a_r)}{\partial r} + \frac{1}{r \sin \phi}\frac{\partial a_\theta}{\partial \theta} + \frac{1}{r \sin \phi}\frac{\partial (a_\phi \sin \phi)}{\partial \phi} \qquad\qquad \text{in spherical coordinates.} \qquad (1.5.24)$$

1.5.4 Curl of Vector Functions

The curl of a vector function $\mathbf{a}(x, y, z)$ is defined by

$$\text{curl}\,\mathbf{a} = -\lim_{\Delta V \to 0} \frac{1}{\Delta V} \oint \mathbf{a} \times \mathbf{n}\, dS, \tag{1.5.25}$$

which becomes

$$\text{curl}\,\mathbf{a} = \begin{vmatrix} \mathbf{e}_x & \mathbf{e}_y & \mathbf{e}_z \\ \dfrac{\partial}{\partial x} & \dfrac{\partial}{\partial y} & \dfrac{\partial}{\partial z} \\ a_1 & a_2 & a_3 \end{vmatrix} \qquad \text{in Cartesian coordinates,} \tag{1.5.26}$$

$$\text{curl}\,\mathbf{a} = \frac{1}{r}\begin{vmatrix} \mathbf{e}_r & r\mathbf{e}_\theta & \mathbf{e}_z \\ \dfrac{\partial}{\partial r} & \dfrac{\partial}{\partial \theta} & \dfrac{\partial}{\partial z} \\ a_r & ra_\theta & a_z \end{vmatrix} \qquad \text{in cylindrical coordinates,} \tag{1.5.27}$$

$$\text{curl}\,\mathbf{a} = \frac{1}{r^2 \sin\phi}\begin{vmatrix} \mathbf{e}_r & r\mathbf{e}_\phi & r\sin\phi\,\mathbf{e}_\theta \\ \dfrac{\partial}{\partial r} & \dfrac{\partial}{\partial \phi} & \dfrac{\partial}{\partial \theta} \\ a_r & ra_\phi & \sin\phi\,a_\theta \end{vmatrix} \qquad \text{in spherical coordinates,} \tag{1.5.28}$$

or equivalently

$$\text{curl}\,\mathbf{a} = \left(\frac{\partial a_3}{\partial y} - \frac{\partial a_2}{\partial z}\right)\mathbf{e}_x + \left(\frac{\partial a_1}{\partial z} - \frac{\partial a_3}{\partial x}\right)\mathbf{e}_y + \left(\frac{\partial a_2}{\partial x} - \frac{\partial a_1}{\partial y}\right)\mathbf{e}_z, \tag{1.5.29}$$

$$\text{curl}\,\mathbf{a} = \frac{1}{r}\left[\left(\frac{\partial a_z}{\partial \theta} - \frac{\partial (ra_\theta)}{\partial z}\right)\mathbf{e}_r + r\left(\frac{\partial a_r}{\partial z} - \frac{\partial a_z}{\partial r}\right)\mathbf{e}_\theta + \left(\frac{\partial (ra_\theta)}{\partial r} - \frac{\partial a_r}{\partial \theta}\right)\mathbf{e}_z\right], \tag{1.5.30}$$

$$\begin{aligned}
\text{curl}\,\mathbf{a} &= \frac{1}{r^2 \sin\phi}\left[\left(\frac{\partial (ra_\theta \sin\phi)}{\partial \phi} - \frac{\partial (ra_\phi)}{\partial \theta}\right)\mathbf{e}_r + r\sin\theta\left(\frac{\partial (ra_\phi)}{\partial r} - \frac{\partial a_r}{\partial \phi}\right)\mathbf{e}_\theta \right. \\
&\qquad \left. + r\left(\frac{\partial a_r}{\partial \theta} - \frac{\partial (ra_\theta \sin\phi)}{\partial r}\right)\mathbf{e}_\phi\right] \\
&= \frac{1}{r\sin\phi}\left(\frac{\partial (a_\theta \sin\phi)}{\partial \phi} - \frac{\partial a_\phi}{\partial \theta}\right)\mathbf{e}_r + \frac{1}{r}\left(\frac{\partial (ra_\phi)}{\partial r} - \frac{\partial a_r}{\partial \phi}\right)\mathbf{e}_\theta \\
&\qquad + \frac{1}{r}\left(\frac{1}{\sin\phi}\frac{\partial a_r}{\partial \theta} - \frac{\partial (ra_\theta)}{\partial r}\right)\mathbf{e}_\phi.
\end{aligned} \tag{1.5.31}$$

1.5.5 Gradient of Vector Functions

The gradient of a vector function $\mathbf{a}(x, y, z)$ is defined, as in the scalar case, by

$$\text{grad}\,\mathbf{a} = \lim_{\Delta V \to 0} \frac{1}{\Delta V} \oint \mathbf{a} \otimes \mathbf{n}\, dS, \tag{1.5.32}$$

or simply by

$$d\mathbf{a} = (\text{grad}\,\mathbf{a})\, d\mathbf{x}. \tag{1.5.33}$$

In Cartesian coordinates:

$$\operatorname{grad}\mathbf{a} = \begin{bmatrix} \dfrac{\partial a_1}{\partial x} & \dfrac{\partial a_1}{\partial y} & \dfrac{\partial a_1}{\partial z} \\[2mm] \dfrac{\partial a_2}{\partial x} & \dfrac{\partial a_2}{\partial y} & \dfrac{\partial a_2}{\partial z} \\[2mm] \dfrac{\partial a_3}{\partial x} & \dfrac{\partial a_3}{\partial y} & \dfrac{\partial a_3}{\partial z} \end{bmatrix}. \tag{1.5.34}$$

In cylindrical coordinates:

$$\operatorname{grad}\mathbf{a} = \begin{bmatrix} \dfrac{\partial a_r}{\partial r} & \dfrac{1}{r}\dfrac{\partial a_r}{\partial \theta} - \dfrac{a_\theta}{r} & \dfrac{\partial a_r}{\partial z} \\[2mm] \dfrac{\partial a_\theta}{\partial r} & \dfrac{1}{r}\dfrac{\partial a_\theta}{\partial \theta} + \dfrac{a_r}{r} & \dfrac{\partial a_\theta}{\partial z} \\[2mm] \dfrac{\partial a_z}{\partial r} & \dfrac{1}{r}\dfrac{\partial a_z}{\partial \theta} & \dfrac{\partial a_z}{\partial z} \end{bmatrix}. \tag{1.5.35}$$

In spherical coordinates:

$$\operatorname{grad}\mathbf{a} = \begin{bmatrix} \dfrac{\partial a_r}{\partial r} & \dfrac{1}{r\sin\phi}\dfrac{\partial a_r}{\partial \theta} - \dfrac{a_\phi}{r} & \dfrac{1}{r}\dfrac{\partial a_r}{\partial \phi} - \dfrac{a_\phi}{r} \\[2mm] \dfrac{\partial a_\theta}{\partial r} & \dfrac{1}{r\sin\phi}\dfrac{\partial a_\theta}{\partial \theta} + \dfrac{a_r}{r} + \dfrac{a_\phi}{r\tan\phi} & \dfrac{1}{r}\dfrac{\partial a_\theta}{\partial \phi} \\[2mm] \dfrac{\partial a_\phi}{\partial r} & \dfrac{1}{r\sin\phi}\dfrac{\partial a_\phi}{\partial \theta} - \dfrac{a_\theta}{r\tan\phi} & \dfrac{1}{r}\dfrac{\partial a_\phi}{\partial \phi} + \dfrac{a_r}{r} \end{bmatrix}. \tag{1.5.36}$$

I like the gradient of a vector, especially those 'extra' terms arising in cylindrical and spherical coordinates. They make the formulas look very special. These terms come from the fact that some of the unit vectors vary in space; their derivatives are non-zero: in cylindrical coordinates,

$$\frac{\partial \mathbf{e}_r}{\partial \theta} = \mathbf{e}_\theta, \quad \frac{\partial \mathbf{e}_\theta}{\partial \theta} = -\mathbf{e}_r, \tag{1.5.37}$$

and in spherical coordinates,

$$\frac{\partial \mathbf{e}_r}{\partial \theta} = \sin\phi\,\mathbf{e}_\theta, \quad \frac{\partial \mathbf{e}_\theta}{\partial \theta} = -\cos\phi\,\mathbf{e}_\phi - \sin\phi\,\mathbf{e}_r, \quad \frac{\partial \mathbf{e}_\phi}{\partial \theta} = \cos\phi\,\mathbf{e}_\theta, \quad \frac{\partial \mathbf{e}_r}{\partial \phi} = \mathbf{e}_\phi, \quad \frac{\partial \mathbf{e}_\phi}{\partial \phi} = -\mathbf{e}_r. \tag{1.5.38}$$

For example, in cylindrical coordinates, the extra terms arise as follows. Consider the vector,

$$\mathbf{a} = a_r\,\mathbf{e}_r + a_\theta\,\mathbf{e}_\theta + a_z\,\mathbf{e}_z. \tag{1.5.39}$$

Take the differential of $\mathbf{a}$,

$$
\begin{aligned}
d\mathbf{a} \;=\; & \left(\frac{\partial a_r}{\partial r}dr + \frac{\partial a_r}{r\partial\theta}rd\theta + \frac{\partial a_r}{\partial z}dz \right)\mathbf{e}_r + a_r\left(\frac{\partial \mathbf{e}_r}{\partial r}dr + \frac{\partial \mathbf{e}_r}{r\partial\theta}rd\theta + \frac{\partial \mathbf{e}_r}{\partial z}dz \right) \\[2mm]
& + \left(\frac{\partial a_\theta}{\partial r}dr + \frac{\partial a_\theta}{r\partial\theta}rd\theta + \frac{\partial a_\theta}{\partial z}dz \right)\mathbf{e}_\theta + a_\theta\left(\frac{\partial \mathbf{e}_\theta}{\partial r}dr + \frac{\partial \mathbf{e}_\theta}{r\partial\theta}rd\theta + \frac{\partial \mathbf{e}_\theta}{\partial z}dz \right) \\[2mm]
& + \left(\frac{\partial a_z}{\partial r}dr + \frac{\partial a_z}{r\partial\theta}rd\theta + \frac{\partial a_z}{\partial z}dz \right)\mathbf{e}_z + a_z\left(\frac{\partial \mathbf{e}_z}{\partial r}dr + \frac{\partial \mathbf{e}_z}{r\partial\theta}rd\theta + \frac{\partial \mathbf{e}_z}{\partial z}dz \right),
\end{aligned}
$$

which by Equation(1.5.37) becomes

$$
\begin{aligned}
d\mathbf{a} &= \left(\frac{\partial a_r}{\partial r} dr + \frac{\partial a_r}{r\partial\theta} r d\theta + \frac{\partial a_r}{\partial z} dz \right) \mathbf{e}_r + \left(\frac{a_r}{r} \right) r d\theta\, \mathbf{e}_\theta \\[2mm]
&+ \left(\frac{\partial a_\theta}{\partial r} dr + \frac{\partial a_\theta}{r\partial\theta} r d\theta + \frac{\partial a_\theta}{\partial z} dz \right) \mathbf{e}_\theta - \left(\frac{a_\theta}{r} \right) r d\theta\, \mathbf{e}_r \\[2mm]
&+ \left(\frac{\partial a_z}{\partial r} dr + \frac{\partial a_z}{r\partial\theta} r d\theta + \frac{\partial a_z}{\partial z} dz \right) \mathbf{e}_z \\[2mm]
&= \left(\frac{\partial a_r}{\partial r} dr + \left(\frac{\partial a_r}{r\partial\theta} - \frac{a_\theta}{r} \right) r d\theta + \frac{\partial a_r}{\partial z} dz \right) \mathbf{e}_r \\[2mm]
&+ \left(\frac{\partial a_\theta}{\partial r} dr + \left(\frac{\partial a_\theta}{r\partial\theta} + \frac{a_r}{r} \right) r d\theta + \frac{\partial a_\theta}{\partial z} dz \right) \mathbf{e}_\theta \\[2mm]
&+ \left(\frac{\partial a_z}{\partial r} dr + \frac{\partial a_z}{r\partial\theta} r d\theta + \frac{\partial a_z}{\partial z} dz \right) \mathbf{e}_z.
\end{aligned}
\tag{1.5.40}
$$

We find the gradient by writing the result in the form, $d\mathbf{a} = (\operatorname{grad} \mathbf{a})\, d\mathbf{x}$, where $d\mathbf{x} = (dr, r d\theta, dz)$:

$$
d\mathbf{a} =
\begin{bmatrix}
\dfrac{\partial a_r}{\partial r} & \dfrac{1}{r}\dfrac{\partial a_r}{\partial\theta} - \dfrac{a_\theta}{r} & \dfrac{\partial a_r}{\partial z} \\[3mm]
\dfrac{\partial a_\theta}{\partial r} & \dfrac{1}{r}\dfrac{\partial a_\theta}{\partial\theta} + \dfrac{a_r}{r} & \dfrac{\partial a_\theta}{\partial z} \\[3mm]
\dfrac{\partial a_z}{\partial r} & \dfrac{1}{r}\dfrac{\partial a_z}{\partial\theta} & \dfrac{\partial a_z}{\partial z}
\end{bmatrix}
\begin{bmatrix}
dr \\[3mm] r d\theta \\[3mm] dz
\end{bmatrix}.
\tag{1.5.41}
$$

That is, the matrix above is the gradient of $\mathbf{a}$ in cylindrical coordinates. I like deriving, in the same way, the gradient in spherical coordinates also, but I'll let you do it because it is simple and fun. Enjoy.

The divergence of a tensor will also have extra terms and the Laplacian of a vector will also have such in these coordinates as we will see below.

1.5.6   Divergence of Tensor Functions

The divergence of a tensor function $\mathbf{A}(x, y, z)$ is defined, in the same way as in the scalar case, by

$$
\operatorname{div}\mathbf{A} = \lim_{\Delta V \to 0} \frac{1}{\Delta V} \oint \mathbf{A}\mathbf{n}\, dS.
\tag{1.5.42}
$$

In Cartesian coordinates:

$$
\operatorname{div}\mathbf{A} =
\begin{bmatrix}
\dfrac{\partial A_{11}}{\partial x} + \dfrac{\partial A_{12}}{\partial y} + \dfrac{\partial A_{13}}{\partial z} \\[3mm]
\dfrac{\partial A_{21}}{\partial x} + \dfrac{\partial A_{22}}{\partial y} + \dfrac{\partial A_{23}}{\partial z} \\[3mm]
\dfrac{\partial A_{31}}{\partial x} + \dfrac{\partial A_{32}}{\partial y} + \dfrac{\partial A_{33}}{\partial z}
\end{bmatrix}.
\tag{1.5.43}
$$

In cylindrical coordinates:

$$
\mathbf{A} =
\begin{bmatrix}
A_{rr} & A_{r\theta} & A_{rz} \\
A_{\theta r} & A_{\theta\theta} & A_{\theta z} \\
A_{zr} & A_{z\theta} & A_{zz}
\end{bmatrix},
\tag{1.5.44}
$$

$$
\text{div } \mathbf{A} =
\begin{bmatrix}
\dfrac{1}{r}\dfrac{\partial(rA_{rr})}{\partial r} + \dfrac{1}{r}\dfrac{\partial A_{r\theta}}{\partial \theta} + \dfrac{\partial A_{rz}}{\partial z} - \dfrac{A_{\theta\theta}}{r} \\[2.5ex]
\dfrac{1}{r}\dfrac{\partial(rA_{\theta r})}{\partial r} + \dfrac{1}{r}\dfrac{\partial A_{\theta\theta}}{\partial \theta} + \dfrac{\partial A_{\theta z}}{\partial z} + \dfrac{A_{r\theta}}{r} \\[2.5ex]
\dfrac{1}{r}\dfrac{\partial(rA_{zr})}{\partial r} + \dfrac{1}{r}\dfrac{\partial A_{z\theta}}{\partial \theta} + \dfrac{\partial A_{zz}}{\partial z}
\end{bmatrix}.
\tag{1.5.45}
$$

In spherical coordinates:

$$
\mathbf{A} =
\begin{bmatrix}
A_{rr} & A_{r\theta} & A_{r\phi} \\
A_{\theta r} & A_{\theta\theta} & A_{\theta\phi} \\
A_{\phi r} & A_{\phi\theta} & A_{\phi\phi}
\end{bmatrix},
\tag{1.5.46}
$$

$$
\text{div } \mathbf{A} =
\begin{bmatrix}
\dfrac{1}{r^2}\dfrac{\partial(r^2 A_{rr})}{\partial r} + \dfrac{1}{r\sin\phi}\dfrac{\partial A_{r\theta}}{\partial \theta} + \dfrac{1}{r\sin\phi}\dfrac{\partial(A_{r\phi}\sin\phi)}{\partial \phi} - \dfrac{A_{\theta\theta}}{r} - \dfrac{A_{\phi\phi}}{r} \\[2.5ex]
\dfrac{1}{r^2}\dfrac{\partial(r^2 A_{\theta r})}{\partial r} + \dfrac{1}{r\sin\phi}\dfrac{\partial A_{\theta\theta}}{\partial \theta} + \dfrac{1}{r\sin\phi}\dfrac{\partial(A_{\theta\phi}\sin\phi)}{\partial \phi} + \dfrac{A_{r\theta}}{r} + \dfrac{A_{\phi\theta}}{r\tan\phi} \\[2.5ex]
\dfrac{1}{r^2}\dfrac{\partial(r^2 A_{\phi r})}{\partial r} + \dfrac{1}{r\sin\phi}\dfrac{\partial A_{\phi\theta}}{\partial \theta} + \dfrac{1}{r\sin\phi}\dfrac{\partial(A_{\phi\phi}\sin\phi)}{\partial \phi} - \dfrac{A_{\theta\theta}}{r\tan\phi} + \dfrac{A_{r\phi}}{r}
\end{bmatrix}.
\tag{1.5.47}
$$

Again, I like those extra terms arising in cylindrical and spherical coordinates. What is interesting is the way they arise. OK, so let me show you how to derive the divergence of a tensor function in cylindrical coordinates. I would begin with the definition (1.5.42). Consider a small volume defined by the increment in each coordinate direction:

$$
V = dr \times (rd\theta) \times dz = r\,dr\,d\theta\,dz.
\tag{1.5.48}
$$

Since the volume is small, the surface integral on the right hand side of the definition (1.5.42) can be expressed as

$$
\oint \mathbf{A}\,n\,dS = \frac{\partial(\mathbf{A}n_r\mathbf{e}_r dS)}{\partial r}dr + \frac{\partial(\mathbf{A}n_\theta\mathbf{e}_\theta dS)}{\partial \theta}d\theta + \frac{\partial(\mathbf{A}n_z\mathbf{e}_z dS)}{\partial z}dz,
\tag{1.5.49}
$$

where $\mathbf{n} = n_r\mathbf{e}_r + n_\theta\mathbf{e}_\theta + n_z\mathbf{e}_z$. It is actually more convenient to write by $dS_r = n_r dS$, $dS_\theta = n_\theta dS$, $dS_z = n_z dS$,

$$
\oint \mathbf{A}\,n\,dS = \frac{\partial(\mathbf{A}dS_r\mathbf{e}_r)}{\partial r}dr + \frac{\partial(\mathbf{A}dS_\theta\mathbf{e}_\theta)}{\partial \theta}d\theta + \frac{\partial(\mathbf{A}dS_z\mathbf{e}_z)}{\partial z}dz,
\tag{1.5.50}
$$

where

$$
dS_r = r\,d\theta\,dz, \quad dS_\theta = dr\,dz, \quad dS_z = r\,d\theta\,dr.
\tag{1.5.51}
$$

Note that the above expression has been obtained by Taylor expansion between the two opposite surface of the volume and the surface vector has also been expanded together because it can change along a coordinate direction. Each term can be computed in a straightforward manner:

$$
\begin{aligned}
\frac{\partial(\mathbf{A}dS_r\mathbf{e}_r)}{\partial r}dr &= \frac{\partial}{\partial r}\left\{
\begin{bmatrix}
A_{rr} & A_{r\theta} & A_{rz} \\
A_{\theta r} & A_{\theta\theta} & A_{\theta z} \\
A_{zr} & A_{z\theta} & A_{zz}
\end{bmatrix}
\begin{bmatrix}
dS_r \\ 0 \\ 0
\end{bmatrix}
\right\}dr \\[2ex]
&= \frac{\partial}{\partial r}\left\{ dS_r\left[A_{rr}\,\mathbf{e}_r + A_{\theta r}\,\mathbf{e}_\theta + A_{zr}\,\mathbf{e}_z\right]\right\}dr \\[2ex]
&= \frac{\partial}{\partial r}\left\{ r\left[A_{rr}\,\mathbf{e}_r + A_{\theta r}\,\mathbf{e}_\theta + A_{zr}\,\mathbf{e}_z\right]\right\}dr\,d\theta\,dz \\[2ex]
&= \left[\left(\frac{\partial(rA_{rr})}{\partial r}\,\mathbf{e}_r + \frac{\partial(rA_{\theta r})}{\partial r}\,\mathbf{e}_\theta + \frac{\partial(rA_{zr})}{\partial r}\,\mathbf{e}_z\right) + r\left(A_{rr}\cancel{\frac{\partial\mathbf{e}_r}{\partial r}} + A_{\theta r}\cancel{\frac{\partial\mathbf{e}_\theta}{\partial r}} + A_{zr}\cancel{\frac{\partial\mathbf{e}_z}{\partial r}}\right)\right]dr\,d\theta\,dz \\[2ex]
&= \left(\frac{1}{r}\frac{\partial(rA_{rr})}{\partial r}\,\mathbf{e}_r + \frac{1}{r}\frac{\partial(rA_{\theta r})}{\partial r}\,\mathbf{e}_\theta + \frac{1}{r}\frac{\partial(rA_{zr})}{\partial r}\,\mathbf{e}_z\right)V.
\end{aligned}
\tag{1.5.52}
$$

Similarly, we obtain

$$\frac{\partial(\mathbf{A}dS_\theta \mathbf{e}_\theta)}{\partial\theta}d\theta = \frac{\partial}{\partial r}\left\{dS_\theta\left[A_{r\theta}\,\mathbf{e}_r + A_{\theta\theta}\,\mathbf{e}_\theta + A_{z\theta}\,\mathbf{e}_z\right]\right\}d\theta$$

$$= \left\{\left(\frac{\partial A_{r\theta}}{\partial\theta}\,\mathbf{e}_r + \frac{\partial A_{\theta\theta}}{\partial\theta}\,\mathbf{e}_\theta + \frac{\partial A_{z\theta}}{\partial\theta}\,\mathbf{e}_z\right) + \left(A_{r\theta}\frac{\partial\mathbf{e}_r}{\partial\theta} + A_{\theta\theta}\frac{\partial\mathbf{e}_\theta}{\partial\theta} + A_{z\theta}\frac{\partial\mathbf{e}_z}{\partial\theta}\right)\right\}dr\,d\theta\,dz,$$

which becomes by Equation (1.5.37)

$$\frac{\partial(\mathbf{A}dS_\theta \mathbf{e}_\theta)}{\partial\theta}d\theta = \left\{\left(\frac{\partial A_{r\theta}}{\partial\theta}\,\mathbf{e}_r + \frac{\partial A_{\theta\theta}}{\partial\theta}\,\mathbf{e}_\theta + \frac{\partial A_{z\theta}}{\partial\theta}\,\mathbf{e}_z\right) + (A_{r\theta}\mathbf{e}_\theta - A_{\theta\theta}\mathbf{e}_r)\right\}dr\,d\theta\,dz$$

$$= \left[\left(\frac{1}{r}\frac{\partial A_{r\theta}}{\partial\theta} - \frac{A_{\theta\theta}}{r}\right)\mathbf{e}_r + \left(\frac{1}{r}\frac{\partial A_{\theta\theta}}{\partial\theta} + \frac{A_{r\theta}}{r}\right)\mathbf{e}_\theta + \frac{1}{r}\frac{\partial A_{z\theta}}{\partial\theta}\,\mathbf{e}_z\right]V. \tag{1.5.53}$$

The last component is the simplest one:

$$\frac{\partial(\mathbf{A}dS_z \mathbf{e}_z)}{\partial z}dz = \left[\frac{\partial A_{rz}}{\partial z}\,\mathbf{e}_r + \frac{\partial A_{\theta z}}{\partial z}\,\mathbf{e}_\theta + \frac{\partial A_{zz}}{\partial z}\,\mathbf{e}_z\right]V. \tag{1.5.54}$$

Finally, substituting these results into Equation (1.5.50) and subsequently substituting it into the definition (1.5.42), where ΔV is denoted here by the small volume V, we find

$$\text{div}\,\mathbf{A} = \lim_{V\to 0}\frac{1}{V}\oint \mathbf{A}\mathbf{n}\,dS$$

$$= \left(\frac{1}{r}\frac{\partial(rA_{rr})}{\partial r} + \frac{1}{r}\frac{\partial A_{r\theta}}{\partial\theta} - \frac{A_{\theta\theta}}{r} + \frac{\partial A_{rz}}{\partial z}\right)\mathbf{e}_r$$

$$+ \left(\frac{1}{r}\frac{\partial(rA_{\theta r})}{\partial r} + \frac{1}{r}\frac{\partial A_{\theta\theta}}{\partial\theta} + \frac{A_{r\theta}}{r} + \frac{\partial A_{\theta z}}{\partial z}\right)\mathbf{e}_\theta$$

$$+ \left(\frac{1}{r}\frac{\partial(rA_{zr})}{\partial r} + \frac{1}{r}\frac{\partial A_{z\theta}}{\partial\theta} + \frac{\partial A_{zz}}{\partial z}\right)\mathbf{e}_z, \tag{1.5.55}$$

which is the same as Equation (1.5.45). We have thus derived the divergence in cylindrical coordinates. As we have seen above, the extra terms are generated by the derivatives of the unit vectors, which vary in space. I find it very interesting. Do you?

1.5.7 Laplacian of Scalar Functions

The Laplacian of a scalar function $\alpha(x, y, z)$ is defined as the divergence of the gradient:

$$\text{div}\,\text{grad}\,\alpha = \frac{\partial^2\alpha}{\partial x^2} + \frac{\partial^2\alpha}{\partial y^2} + \frac{\partial^2\alpha}{\partial z^2}, \tag{1.5.56}$$

$$\text{div}\,\text{grad}\,\alpha = \frac{1}{r}\frac{\partial}{\partial r}\left(r\frac{\partial\alpha}{\partial r}\right) + \frac{1}{r^2}\frac{\partial^2\alpha}{\partial\theta^2} + \frac{\partial^2\alpha}{\partial z^2}, \tag{1.5.57}$$

$$\text{div}\,\text{grad}\,\alpha = \frac{1}{r^2}\frac{\partial}{\partial r}\left(r^2\frac{\partial\alpha}{\partial r}\right) + \frac{1}{r^2\sin^2\phi}\frac{\partial^2\alpha}{\partial\theta^2} + \frac{1}{r^2\sin\phi}\frac{\partial}{\partial\phi}\left(\sin\phi\frac{\partial\alpha}{\partial\phi}\right), \tag{1.5.58}$$

in Cartesian, cylindrical, and spherical coordinates respectively. Incidentally, the Laplacian operator is often denoted by Δ:

$$\Delta\alpha = \text{div}\,\text{grad}\,\alpha. \tag{1.5.59}$$

This is a very simple and economical notation. I like it.

1.5.8   Laplacian of Vector Functions

The Laplacian of a vector function is defined as the divergence of the gradient just like in the scalar case. But it is a bit more complicated than the scalar case in cylindrical and spherical coordinates.

In Cartesian coordinates:

$$\text{div}\,\text{grad}\,\mathbf{a} = \begin{bmatrix} \dfrac{\partial^2 a_1}{\partial x^2} + \dfrac{\partial^2 a_1}{\partial y^2} + \dfrac{\partial^2 a_1}{\partial z^2} \\[2ex] \dfrac{\partial^2 a_2}{\partial x^2} + \dfrac{\partial^2 a_2}{\partial y^2} + \dfrac{\partial^2 a_2}{\partial z^2} \\[2ex] \dfrac{\partial^2 a_3}{\partial x^2} + \dfrac{\partial^2 a_3}{\partial y^2} + \dfrac{\partial^2 a_3}{\partial z^2} \end{bmatrix}. \tag{1.5.60}$$

In cylindrical coordinates:

$$\text{div}\,\text{grad}\,\mathbf{a} = \begin{bmatrix} \text{div}\,\text{grad}\,a_r - \dfrac{a_r}{r} - \dfrac{2}{r^2}\dfrac{\partial a_\theta}{\partial \theta} \\[2ex] \text{div}\,\text{grad}\,a_\theta + \dfrac{2}{r^2}\dfrac{\partial a_r}{\partial \theta} - \dfrac{a_\theta}{r^2} \\[2ex] \text{div}\,\text{grad}\,a_z \end{bmatrix}, \tag{1.5.61}$$

where the Laplacian for each scalar component is given by Equation (1.5.57).

In spherical coordinates:

$$\text{div}\,\text{grad}\,\mathbf{a} = \begin{bmatrix} \text{div}\,\text{grad}\,a_r - \dfrac{2a_r}{r^2} - \dfrac{2}{r^2}\dfrac{\partial a_\phi}{\partial \phi} - \dfrac{2a_\phi}{r^2 \tan \phi} - \dfrac{2}{r^2 \sin \phi}\dfrac{\partial a_\theta}{\partial \theta} \\[2ex] \text{div}\,\text{grad}\,a_\theta + \dfrac{2}{r^2}\dfrac{\partial a_r}{\partial \phi} - \dfrac{a_\phi}{r^2 \sin^2 \phi} - \dfrac{2}{r^2 \sin \phi}\dfrac{\partial a_\theta}{\partial \theta} \\[2ex] \text{div}\,\text{grad}\,a_\phi - \dfrac{a_\theta}{r^2 \sin^2 \phi} + \dfrac{2}{r^2 \sin^2 \phi}\dfrac{\partial a_r}{\partial \theta} + \dfrac{2\cos \phi}{r^2 \sin^2 \phi}\dfrac{\partial a_\phi}{\partial \theta} \end{bmatrix}, \tag{1.5.62}$$

where the Laplacian for each scalar component is given by Equation (1.5.58).

1.5.9   Material/Substantial/Lagrangian Derivative

The material derivative of a scalar function: in Cartesian coordinates,

$$\frac{D\alpha}{Dt} = \frac{\partial \alpha}{\partial t} + (\text{grad}\,\alpha)\cdot \mathbf{v} = \frac{d\alpha}{dt} + u\frac{d\alpha}{dx} + v\frac{d\alpha}{dy} + w\frac{d\alpha}{dz}, \tag{1.5.63}$$

in cylindrical coordinates with the velocity (v_r, v_θ, v_z),

$$\frac{D\alpha}{Dt} = \frac{\partial \alpha}{\partial t} + (\text{grad}\,\alpha)\cdot \mathbf{v} = \frac{d\alpha}{dt} + v_r\frac{\partial \alpha}{\partial r} + \frac{v_\theta}{r}\frac{\partial \alpha}{\partial \theta} + v_z\frac{\partial \alpha}{\partial z}, \tag{1.5.64}$$

and in spherical coordinates with the velocity (v_r, v_θ, v_ϕ),

$$\frac{D\alpha}{Dt} = \frac{\partial \alpha}{\partial t} + (\text{grad}\,\alpha)\cdot \mathbf{v} = \frac{d\alpha}{dt} + v_r\frac{\partial \alpha}{\partial r} + \frac{v_\theta}{r \sin \phi}\frac{\partial \alpha}{\partial \theta} + \frac{v_\phi}{r}\frac{\partial \alpha}{\partial \phi}. \tag{1.5.65}$$

The material derivative of a vector function $\mathbf{a}(x,y,z,t)$ is simple in Cartesian coordinates:

$$\frac{D\mathbf{a}}{Dt} = \frac{\partial \mathbf{a}}{\partial t} + (\text{grad}\,\mathbf{a})\mathbf{v} = \begin{bmatrix} \dfrac{Da_1}{Dt} \\[2ex] \dfrac{Da_2}{Dt} \\[2ex] \dfrac{Da_3}{Dt} \end{bmatrix}, \tag{1.5.66}$$

but of course it has some 'extra' terms in cylindrical coordinates,

$$\frac{D\mathbf{a}}{Dt} = \frac{\partial \mathbf{a}}{\partial t} + (\mathrm{grad}\,\mathbf{a})\mathbf{v} = \begin{bmatrix} \dfrac{Da_r}{Dt} - \dfrac{a_\theta v_\theta}{r} \\[2mm] \dfrac{Da_\theta}{Dt} + \dfrac{a_r v_\theta}{r} \\[2mm] \dfrac{Da_z}{Dt} \end{bmatrix}, \tag{1.5.67}$$

where the material derivative for each scalar component is given by Equation (1.5.64), and also in spherical coordinates,

$$\frac{D\mathbf{a}}{Dt} = \frac{\partial \mathbf{a}}{\partial t} + (\mathrm{grad}\,\mathbf{a})\mathbf{v} = \begin{bmatrix} \dfrac{Da_r}{Dt} - \dfrac{a_\theta v_\theta + a_\phi v_\phi}{r} \\[2mm] \dfrac{Da_\theta}{Dt} + \dfrac{v_\theta(a_r + a_\phi \cot\phi)}{r} \\[2mm] \dfrac{Da_\phi}{Dt} + \dfrac{a_r v_\phi - a_\theta v_\theta \cot\phi}{r} \end{bmatrix}, \tag{1.5.68}$$

where the material derivative for each scalar component is given by Equation (1.5.65).

1.5.10 Complex Lamellar and Beltrami Flows

A vector field defined by $\mathbf{a}(x,y,z)$ which satisfies

$$\mathbf{a} \cdot (\mathrm{curl}\,\mathbf{a}) = 0, \tag{1.5.69}$$

is called a complex lamellar field. If $\mathbf{a}$ is a fluid velocity, this corresponds to a flow where the velocity and the vorticity (the curl of the velocity) are orthogonal to each other. It is then called a complex lamellar flow. I like it because it sounds like a flow over a lifting wing section. On the other hand, a vector field that satisfies

$$\mathbf{a} \times (\mathrm{curl}\,\mathbf{a}) = 0, \tag{1.5.70}$$

is called a Beltrami field. If $\mathbf{a}$ is a velocity, this corresponds to a flow where the velocity and vorticity are parallel to each other. This is called a Beltrami flow. I like this one also. It reminds me of trailing vortices of an aircraft.

Consult other books such as Ref.[67] to find out more details on the vector differential operators.

1.6 Del Operator

The del operator (which is called nabla as a symbol) in Cartesian coordinates is defined by

$$\nabla = \left[\frac{\partial}{\partial x}, \frac{\partial}{\partial y}, \frac{\partial}{\partial z} \right]. \tag{1.6.1}$$

I like the del operator. But some people don't like it because they say it depends on the coordinate system and doesn't carry much physical meaning while div, grad, and curl are independent of coordinate systems and do carry clear physical meanings. I know what they mean, and actually like it. Nonetheless, it is a useful operator. Div, grad, and curl can be expressed by using the del operator as follows:

$$\mathrm{grad}\,\alpha = \nabla\alpha, \tag{1.6.2}$$

$$\mathrm{div}\,\mathbf{a} = \nabla \cdot \mathbf{a}, \tag{1.6.3}$$

$$\mathrm{curl}\,\mathbf{a} = \nabla \times \mathbf{a}, \tag{1.6.4}$$

$$\mathrm{grad}\,\mathbf{a} = \nabla\mathbf{a}, \tag{1.6.5}$$

$$\mathrm{div}\,\mathbf{A} = \nabla \cdot \mathbf{A}. \tag{1.6.6}$$

What is nice about the del operator is that it can be treated as a vector and thus can be used as follows.

$$\text{div}\,\text{grad}\,\alpha \;=\; \nabla \cdot \nabla \alpha = \nabla^2 \alpha = \Delta \alpha, \tag{1.6.7}$$

$$\text{curl}(\text{grad}\,\alpha) \;=\; \nabla \times \nabla \alpha, \tag{1.6.8}$$

$$\text{grad}(\text{div}\,\mathbf{a}) \;=\; \nabla(\nabla \cdot \mathbf{a}), \tag{1.6.9}$$

$$\text{div}(\text{curl}\,\mathbf{a}) \;=\; \nabla \cdot \nabla \times \mathbf{a}, \tag{1.6.10}$$

$$\text{curl}(\text{curl}\,\mathbf{a}) \;=\; \nabla \times \nabla \times \mathbf{a}, \tag{1.6.11}$$

$$(\text{grad}\,\mathbf{a})\mathbf{b} \;=\; (\mathbf{b} \cdot \nabla)\,\mathbf{a}. \tag{1.6.12}$$

This is a great advantage of ∇. But it must be kept in mind that it is actually an operator. That is, $\mathbf{a} \cdot \mathbf{b} = \mathbf{b} \cdot \mathbf{a}$, but $\mathbf{a} \cdot \nabla \neq \nabla \cdot \mathbf{a}$.

1.7   Vector Identities

Do you like to memorize the following vector identities?

$$\text{curl}\,\mathbf{a} \;=\; \mathbf{E}\,(\text{grad}\,\mathbf{a})^t, \tag{1.7.1}$$

$$\text{div}\,\text{curl}\,\mathbf{a} \;=\; 0, \tag{1.7.2}$$

$$\text{curl}\,\text{grad}\,\alpha \;=\; 0, \tag{1.7.3}$$

$$\text{grad}\,(\alpha\beta) \;=\; \alpha\,\text{grad}\,\beta + \beta\,\text{grad}\,\alpha, \tag{1.7.4}$$

$$\text{grad}\,(\alpha\mathbf{a}) \;=\; \mathbf{a} \otimes \text{grad}\,\alpha + \alpha\,\text{grad}\,\mathbf{a}, \tag{1.7.5}$$

$$\mathbf{a} \times \text{curl}\,\mathbf{b} \;=\; (\text{grad}\,\mathbf{b})^t\mathbf{a} - (\text{grad}\,\mathbf{b})\mathbf{a}, \tag{1.7.6}$$

$$(\text{grad}\,\mathbf{a})\mathbf{a} \;=\; \text{grad}\,\frac{\mathbf{a} \cdot \mathbf{a}}{2} - \mathbf{a} \times \text{curl}\,\mathbf{a}, \tag{1.7.7}$$

$$\mathbf{a} \cdot (\text{grad}\,\mathbf{a})\mathbf{a} \;=\; \mathbf{a} \cdot \text{grad}\,\frac{\mathbf{a} \cdot \mathbf{a}}{2}, \tag{1.7.8}$$

$$(\mathbf{a} \otimes \mathbf{b}) : \text{grad}\,\mathbf{a} \;=\; \mathbf{b} \cdot \text{grad}\,\frac{\mathbf{a} \cdot \mathbf{a}}{2}, \tag{1.7.9}$$

$$\text{grad}\,(\mathbf{a} \cdot \mathbf{b}) \;=\; (\text{grad}\,\mathbf{a})\mathbf{b} + (\text{grad}\,\mathbf{b})\mathbf{a} + \mathbf{a} \times \text{curl}\,\mathbf{b} + \mathbf{b} \times \text{curl}\,\mathbf{a}, \tag{1.7.10}$$

$$\text{div}\,(\alpha\mathbf{a}) \;=\; \alpha\,\text{div}\,\mathbf{a} + \text{grad}\,\alpha \cdot \mathbf{a}, \tag{1.7.11}$$

$$\text{div}\,\text{grad}\,\mathbf{a} \;=\; \text{grad}(\text{div}\,\mathbf{a}) - \text{curl}(\text{curl}\,\mathbf{a}), \tag{1.7.12}$$

$$\text{div}\,(\mathbf{a} \times \mathbf{b}) \;=\; \mathbf{b} \cdot \text{curl}\,\mathbf{a} - \mathbf{a} \cdot \text{curl}\,\mathbf{b}, \tag{1.7.13}$$

$$\text{div}(\mathbf{a} \otimes \mathbf{b}) \;=\; (\text{grad}\,\mathbf{a})\mathbf{b} + \mathbf{a}\,\text{div}\,\mathbf{b}, \tag{1.7.14}$$

$$\mathbf{a} \cdot \text{div}(\mathbf{a} \otimes \mathbf{b}) \;=\; \text{div}\left(\frac{\mathbf{a} \cdot \mathbf{a}}{2}\mathbf{b}\right) + \frac{\mathbf{a} \cdot \mathbf{a}}{2}\,\text{div}\,\mathbf{b}, \tag{1.7.15}$$

$$\mathbf{a} \cdot \text{div}(\mathbf{b} \otimes \mathbf{c}) \;=\; (\mathbf{a} \cdot \mathbf{b})\,\text{div}\,\mathbf{c} + (\mathbf{a} \otimes \mathbf{c}) : \text{grad}\,\mathbf{b}, \tag{1.7.16}$$

$$\text{div}\,(\alpha\mathbf{A}) \;=\; \mathbf{A}\,\text{grad}\,\alpha + \alpha\,\text{div}\,\mathbf{A}, \tag{1.7.17}$$

$$\text{div}\,(\mathbf{A}\mathbf{b}) \;=\; (\text{div}\,\mathbf{A}^t) \cdot \mathbf{b} + \mathbf{A}^t : \text{grad}\,\mathbf{b}, \tag{1.7.18}$$

$$\text{curl}\,(\alpha\mathbf{a}) \;=\; \alpha\,\text{curl}\,\mathbf{a} + \text{grad}\,\alpha \times \mathbf{a}, \tag{1.7.19}$$

$$\text{curl}\,(\mathbf{a} \times \mathbf{b}) \;=\; \mathbf{a}\,\text{div}\,\mathbf{b} + (\text{grad}\,\mathbf{a})\mathbf{b} - (\text{div}\,\mathbf{a})\mathbf{b} - (\text{grad}\,\mathbf{b})\mathbf{a}, \tag{1.7.20}$$

$$\mathbf{a} \cdot (\mathbf{A}\mathbf{b}) \;=\; \mathbf{A} : (\mathbf{a} \otimes \mathbf{b}), \tag{1.7.21}$$

$$\mathbf{a} \cdot (\mathbf{A}\mathbf{b}) \;=\; (\mathbf{A}^t\mathbf{a}) \cdot \mathbf{b}, \tag{1.7.22}$$

$$\mathbf{a} \cdot (\mathbf{A}\mathbf{b}) \;=\; (\mathbf{A}\mathbf{a}) \cdot \mathbf{b} \quad \text{if } \mathbf{A} \text{ is symmetric}, \tag{1.7.23}$$

where $\mathbf{E}$ is Eddington's epsilon; α and β are scalar functions; $\mathbf{a}$, $\mathbf{b}$, and $\mathbf{c}$, are vector functions; and $\mathbf{A}$ is a second-rank tensor function. To me, it is more interesting to prove these identities than to memorize them. In doing so, the component notation is very useful. For example, the (i, j)-th component of div $(\alpha\mathbf{a})$ can be written and expanded as

$$[\text{div}\,(\alpha\mathbf{a})]_{ij} = \frac{\partial(\alpha a_j)}{\partial x_j}$$

$$= \alpha\frac{\partial a_j}{\partial x_j} + \frac{\partial\alpha}{\partial x_j}a_j, \tag{1.7.24}$$

and so, we have just proved

$$\text{div}\,(\alpha\mathbf{a}) = \alpha\,\text{div}\,\mathbf{a} + \text{grad}\,\alpha \cdot \mathbf{a}. \tag{1.7.25}$$

Another example would be $\mathbf{a} \times \text{curl}\,\mathbf{b}$, whose i-th component is

$$(\mathbf{a} \times \text{curl}\,\mathbf{b})_i = \varepsilon_{ijk}a_j\varepsilon_{kmn}\frac{\partial b_n}{\partial x_m}$$

$$= \varepsilon_{ijk}\varepsilon_{kmn}a_j\frac{\partial b_n}{\partial x_m}, \tag{1.7.26}$$

which becomes by the formula (1.4.29),

$$
\begin{aligned}
(\mathbf{a} \times \text{curl}\,\mathbf{b})_i &= (\delta_{im}\delta_{jn} - \delta_{in}\delta_{jm})a_j\frac{\partial b_n}{\partial x_m} \\
&= \delta_{im}\delta_{jn}a_j\frac{\partial b_n}{\partial x_m} - \delta_{in}\delta_{jm}a_j\frac{\partial b_n}{\partial x_m} \\
&= \delta_{im}a_j\frac{\partial b_j}{\partial x_m} - \delta_{in}a_j\frac{\partial b_n}{\partial x_j} \\
&= a_j\frac{\partial b_j}{\partial x_i} - a_j\frac{\partial b_i}{\partial x_j} \\
&= ((\text{grad}\,\mathbf{b})^t\mathbf{a})_i - ((\text{grad}\,\mathbf{b})\mathbf{a})_i. \tag{1.7.27}
\end{aligned}
$$

Thus, Equation (1.7.6) has been proved. Try it for yourself. It's fun.

1.8 Coordinate Transformation of Vectors

I like to switch back and forth among the thee coordinate systems, and actually often need to transform a vector such as a fluid velocity from one coordinate system to another. Let $\mathbf{v}$ be the velocity vector whose components are denoted by

$$\mathbf{v} = [u, v, w]^t \qquad \text{in Cartesian coordinates,} \tag{1.8.1}$$

$$\mathbf{v} = [u_r, u_\theta, u_z]^t \qquad \text{in cylindrical coordinates,} \tag{1.8.2}$$

$$\mathbf{v} = [u_r, u_\theta, u_\phi]^t \qquad \text{in spherical coordinates.} \tag{1.8.3}$$

Then, the velocity vector can be transformed from cylindrical coordinates to the Cartesian coordinates by

$$
\begin{bmatrix} u \\ v \\ w \end{bmatrix} = \begin{bmatrix} \cos\theta & -\sin\theta & 0 \\ \sin\theta & \cos\theta & 0 \\ 0 & 0 & 1 \end{bmatrix} \begin{bmatrix} u_r \\ u_\theta \\ u_z \end{bmatrix}, \tag{1.8.4}
$$

and from the Cartesian coordinates to cylindrical coordinates by

$$
\begin{bmatrix} u_r \\ u_\theta \\ u_z \end{bmatrix} = \begin{bmatrix} \cos\theta & \sin\theta & 0 \\ -\sin\theta & \cos\theta & 0 \\ 0 & 0 & 1 \end{bmatrix} \begin{bmatrix} u \\ v \\ w \end{bmatrix}. \tag{1.8.5}
$$

Similarly, the velocity vector can be transformed from spherical coordinates to the Cartesian coordinates by

$$
\begin{bmatrix} u \\ v \\ w \end{bmatrix} = \begin{bmatrix} \cos\theta\sin\phi & -\sin\theta & \cos\theta\cos\phi \\ \sin\theta\sin\phi & \cos\theta & \sin\theta\cos\phi \\ \cos\phi & 0 & -\sin\phi \end{bmatrix} \begin{bmatrix} u_r \\ u_\theta \\ u_\phi \end{bmatrix}, \tag{1.8.6}
$$

and from the Cartesian coordinates to spherical coordinates by

$$
\begin{bmatrix} u_r \\ u_\theta \\ u_\phi \end{bmatrix} = \begin{bmatrix} \cos\theta\sin\phi & \sin\theta\sin\phi & \cos\phi \\ -\sin\theta & \cos\theta & 0 \\ \cos\theta\cos\phi & \sin\theta\cos\phi & -\sin\phi \end{bmatrix} \begin{bmatrix} u \\ v \\ w \end{bmatrix}. \tag{1.8.7}
$$

1.9 Coordinate Transformation of Partial Derivatives

Sometime I feel like transforming partial derivatives form one coordinate system to another. For cylindrical coordinates, we have

$$
x = r\cos\theta, \quad y = r\sin\theta, \quad z = z, \tag{1.9.1}
$$

and

$$
r = \sqrt{x^2 + y^2}, \tag{1.9.2}
$$

$$
\cos\theta = \frac{x}{\sqrt{x^2 + y^2}}, \tag{1.9.3}
$$

$$
\sin\theta = \frac{y}{\sqrt{x^2 + y^2}}. \tag{1.9.4}
$$

Then, by the chain rule, we have

$$
\frac{\partial}{\partial x} = \frac{\partial r}{\partial x}\frac{\partial}{\partial r} + \frac{\partial\theta}{\partial x}\frac{\partial}{\partial\theta} + \frac{\partial z}{\partial x}\frac{\partial}{\partial z}, \tag{1.9.5}
$$

$$
\frac{\partial}{\partial y} = \frac{\partial r}{\partial y}\frac{\partial}{\partial r} + \frac{\partial\theta}{\partial y}\frac{\partial}{\partial\theta} + \frac{\partial z}{\partial y}\frac{\partial}{\partial z}, \tag{1.9.6}
$$

$$
\frac{\partial}{\partial z} = \frac{\partial r}{\partial z}\frac{\partial}{\partial r} + \frac{\partial\theta}{\partial z}\frac{\partial}{\partial\theta} + \frac{\partial z}{\partial z}\frac{\partial}{\partial z}. \tag{1.9.7}
$$

Of course, we have

$$
\frac{\partial z}{\partial x} = 0, \quad \frac{\partial z}{\partial y} = 0, \quad \frac{\partial r}{\partial z} = 0, \quad \frac{\partial\theta}{\partial z} = 0, \quad \frac{\partial z}{\partial z} = 1, \tag{1.9.8}
$$

and we find from Equations (1.9.2)

$$
\frac{\partial r}{\partial x} = \cos\theta, \quad \frac{\partial r}{\partial y} = \sin\theta, \tag{1.9.9}
$$

and from Equations (1.9.3) and (1.9.4)

$$
\frac{\partial\theta}{\partial x} = -\frac{\sin\theta}{r}, \quad \frac{\partial\theta}{\partial y} = \frac{\cos\theta}{r}. \tag{1.9.10}
$$

Therefore, we have the following transformation:

$$
\begin{bmatrix} \dfrac{\partial}{\partial x} \\[2mm] \dfrac{\partial}{\partial y} \\[2mm] \dfrac{\partial}{\partial z} \end{bmatrix} = \begin{bmatrix} \cos\theta & -\dfrac{\sin\theta}{r} & 0 \\[2mm] \sin\theta & \dfrac{\cos\theta}{r} & 0 \\[2mm] 0 & 0 & 1 \end{bmatrix} \begin{bmatrix} \dfrac{\partial}{\partial r} \\[2mm] \dfrac{\partial}{\partial\theta} \\[2mm] \dfrac{\partial}{\partial z} \end{bmatrix}. \tag{1.9.11}
$$

The inverse transformation is given by

$$
\begin{bmatrix} \dfrac{\partial}{\partial r} \\[2ex] \dfrac{\partial}{\partial \theta} \\[2ex] \dfrac{\partial}{\partial z} \end{bmatrix} = \begin{bmatrix} \cos\theta & \sin\theta & 0 \\[2ex] -r\sin\theta & r\cos\theta & 0 \\[2ex] 0 & 0 & 1 \end{bmatrix} \begin{bmatrix} \dfrac{\partial}{\partial x} \\[2ex] \dfrac{\partial}{\partial y} \\[2ex] \dfrac{\partial}{\partial z} \end{bmatrix}.
\tag{1.9.12}
$$

For spherical coordinates, we have

$$
x = (r\sin\phi)\cos\theta, \quad y = (r\sin\phi)\sin\theta, \quad z = r\cos\phi,
\tag{1.9.13}
$$

and

$$
r = \sqrt{x^2 + y^2 + z^2},
\tag{1.9.14}
$$

$$
\cos\theta = \frac{x}{\sqrt{x^2 + y^2}},
\tag{1.9.15}
$$

$$
\sin\theta = \frac{y}{\sqrt{x^2 + y^2}},
\tag{1.9.16}
$$

$$
\cos\phi = \frac{z}{\sqrt{x^2 + y^2 + z^2}},
\tag{1.9.17}
$$

$$
\cos\phi = \frac{\sqrt{x^2 + y^2}}{\sqrt{x^2 + y^2 + z^2}}.
\tag{1.9.18}
$$

Then, by the chain rule, we have

$$
\frac{\partial}{\partial x} = \frac{\partial r}{\partial x}\frac{\partial}{\partial r} + \frac{\partial\theta}{\partial x}\frac{\partial}{\partial\theta} + \frac{\partial\phi}{\partial x}\frac{\partial}{\partial\phi},
\tag{1.9.19}
$$

$$
\frac{\partial}{\partial y} = \frac{\partial r}{\partial y}\frac{\partial}{\partial r} + \frac{\partial\theta}{\partial y}\frac{\partial}{\partial\theta} + \frac{\partial\phi}{\partial y}\frac{\partial}{\partial\phi},
\tag{1.9.20}
$$

$$
\frac{\partial}{\partial z} = \frac{\partial r}{\partial z}\frac{\partial}{\partial r} + \frac{\partial\theta}{\partial z}\frac{\partial}{\partial\theta} + \frac{\partial\phi}{\partial z}\frac{\partial}{\partial\phi}.
\tag{1.9.21}
$$

We find from Equations (1.9.14)

$$
\frac{\partial r}{\partial x} = \cos\theta\sin\phi, \quad \frac{\partial r}{\partial y} = \sin\theta\sin\phi, \quad \frac{\partial r}{\partial z} = \cos\phi,
\tag{1.9.22}
$$

from Equations (1.9.15) and (1.9.16)

$$
\frac{\partial\theta}{\partial x} = -\frac{\sin\theta}{r\sin\phi}, \quad \frac{\partial\theta}{\partial y} = \frac{\cos\theta}{r\sin\phi}, \quad \frac{\partial\theta}{\partial z} = 0,
\tag{1.9.23}
$$

and Equations (1.9.17) and (1.9.18)

$$
\frac{\partial\theta}{\partial x} = \frac{\cos\theta\cos\phi}{r}, \quad \frac{\partial\theta}{\partial y} = \frac{\sin\theta\cos\phi}{r}, \quad \frac{\partial\theta}{\partial z} = -\frac{\sin\phi}{r}.
\tag{1.9.24}
$$

Therefore, we obtain the following transformation:

$$
\begin{bmatrix} \dfrac{\partial}{\partial x} \\[2ex] \dfrac{\partial}{\partial y} \\[2ex] \dfrac{\partial}{\partial z} \end{bmatrix} = \begin{bmatrix} \sin\phi\cos\theta & -\dfrac{\sin\theta}{r\sin\phi} & \dfrac{\cos\theta\cos\phi}{r} \\[2ex] \sin\phi\sin\theta & \dfrac{\cos\theta}{r\sin\phi} & \dfrac{\sin\theta\cos\phi}{r} \\[2ex] \cos\phi & 0 & -\dfrac{\sin\phi}{r} \end{bmatrix} \begin{bmatrix} \dfrac{\partial}{\partial r} \\[2ex] \dfrac{\partial}{\partial\theta} \\[2ex] \dfrac{\partial}{\partial\phi} \end{bmatrix}.
\tag{1.9.25}
$$

The inverse transformation is given by

$$
\begin{bmatrix} \dfrac{\partial}{\partial r} \\[2mm] \dfrac{\partial}{\partial \theta} \\[2mm] \dfrac{\partial}{\partial \phi} \end{bmatrix} = \begin{bmatrix} \sin\phi\cos\theta & \sin\phi\sin\theta & \cos\phi \\[2mm] -r\sin\phi\sin\theta & r\sin\phi\cos\theta & 0 \\[2mm] r\cos\phi\cos\theta & r\sin\theta\cos\phi & -r\sin\phi \end{bmatrix} \begin{bmatrix} \dfrac{\partial}{\partial x} \\[2mm] \dfrac{\partial}{\partial y} \\[2mm] \dfrac{\partial}{\partial z} \end{bmatrix}. \tag{1.9.26}
$$

If I wish, I can use these transformations to derive the vector operators such as the gradient in cylindrical or spherical coordinates from those in Cartesian coordinates. Anyway, it is always good to understand the relationship between different coordinate systems. So, of course, I like these transformations.

1.10   Eigenvalues and Eigenvectors

A scalar quantity λ is called an eigenvalue of $\mathbf{A}$ if it satisfies

$$(\mathbf{A} - \lambda\mathbf{I})\mathbf{r} = 0, \tag{1.10.1}$$

for a square matrix $\mathbf{A}$ and a non-zero vector $\mathbf{r}$. The vector $\mathbf{r}$ is called the right eigenvector of $\mathbf{A}$ associated with this particular eigenvalue λ. A row vector $\boldsymbol{\ell}^t$ that satisfies,

$$\boldsymbol{\ell}^t(\mathbf{A} - \lambda\mathbf{I}) = 0, \tag{1.10.2}$$

is called the left eigenvector of $\mathbf{A}$, again associated with this particular eigenvalue λ. I like these right and left eigenvectors because they are not uniquely determined and any scalar multiple of an eigenvector is also an eigenvector, i.e., still satisfies the above equation. So, I can scale the eigenvectors in any way I want. This is nice.

Note that there exist n eigenvalues (together with associated eigenvectors) for an $n \times n$ matrix $\mathbf{A}$:

$$\lambda_1, \lambda_2, \lambda_3, \cdots, \lambda_n. \tag{1.10.3}$$

If these values are distinct, their associated eigenvectors are linearly independent. (But even if there are some repeated eigenvalues, their eigenvectors may still be linearly independent.) If the eigenvectors are linearly independent, then we can diagonalize the matrix $\mathbf{A}$:

$$\mathbf{\Lambda} = \mathbf{R}^{-1}\mathbf{A}\mathbf{R}, \tag{1.10.4}$$

where $\mathbf{R}$ is the right eigenvector matrix,

$$
\mathbf{R} = \begin{bmatrix} \mathbf{r}_1 & \mathbf{r}_2 & \mathbf{r}_3 & \cdots & \mathbf{r}_n \end{bmatrix}. \tag{1.10.5}
$$

$\mathbf{\Lambda}$ is a diagonal matrix with the eigenvalues of $\mathbf{A}$ placed along its diagonal,

$$
\mathbf{\Lambda} = \begin{bmatrix} \lambda_1 & 0 & 0 & 0 & \cdots & 0 \\ 0 & \lambda_2 & 0 & 0 & \cdots & 0 \\ 0 & 0 & \lambda_3 & 0 & \cdots & 0 \\ 0 & 0 & 0 & \lambda_4 & \ddots & \vdots \\ \vdots & \vdots & \vdots & \ddots & \ddots & 0 \\ 0 & 0 & 0 & \cdots & 0 & \lambda_n \end{bmatrix}, \tag{1.10.6}
$$

and $\mathbf{L} = \mathbf{R}^{-1}$ is in fact the left eigenvector matrix (whose rows are the left eigenvectors),

$$\mathbf{L} = \begin{bmatrix} \ell_1^t \\ \hline \ell_2^t \\ \hline \ell_3^t \\ \hline \vdots \\ \hline \ell_n^t \end{bmatrix}. \tag{1.10.7}$$

Although $\mathbf{R}$ is not unique (since the eigenvectors cannot be uniquely defined), $\mathbf{L}$ is uniquely defined this way for a given $\mathbf{R}$. By the way, if the matrix $\mathbf{A}$ is symmetric, its eigenvalues are all real and the eigenvectors corresponding to different eigenvalues are orthogonal to one another. This is nice. I like symmetric matrices.

1.11   Similar Matrices

Consider square matrices, $\mathbf{A}$, $\mathbf{B}$, and $\mathbf{M}$ of the same size. If $\mathbf{A}$ and $\mathbf{B}$ are related through $\mathbf{M}$ in the form:

$$\mathbf{B} = \mathbf{M}^{-1}\mathbf{AM}, \tag{1.11.1}$$

then $\mathbf{A}$ and $\mathbf{B}$ are said to be similar and Equation (1.11.1) is called a similarity transformation. If $\mathbf{A}$ and $\mathbf{B}$ are similar, they have the same eigenvalues. Moreover, their eigenvectors, say $\mathbf{r}_A$ and $\mathbf{r}_B$ for a particular eigenvalue, are related by

$$\mathbf{r}_B = \mathbf{M}^{-1}\mathbf{r}_A. \tag{1.11.2}$$

So, once I know the eigenvectors of $\mathbf{A}$, I can find eigenvectors of any matrices similar to $\mathbf{A}$ simply by using this relation. In particular, this is very useful when we analyze conservation laws in various forms because change of variables always results in a similarity transformation of the Jacobian matrices and so it suffices to analyze just one form of the equation. This is very nice. I like the similarity transformation.

1.12   Divergence Theorem

The divergence theorem is given by

$$\int_V \operatorname{div} \mathbf{a} \, dV \;\; = \;\; \oint_S \mathbf{a} \cdot \mathbf{n} \, dS, \tag{1.12.1}$$

where $\mathbf{n}$ is the normal vector to the surface of the volume V. I like it, but sometimes I get confused because it is called the Gauss theorem, or the Green-Gauss theorem, or Green's theorem in two dimensions (not quite right, though; see Section 1.13),

$$\int_V \left(\frac{\partial a_x}{\partial x} + \frac{\partial a_y}{\partial y} \right) dxdy \;\; = \;\; \oint_S (a_x dy - a_y dx), \tag{1.12.2}$$

where (dx, dy) denotes the infinitesimal element along the boundary in the counter-clockwise direction. Also, some people call the following equation the divergence theorem,

$$\int_V \operatorname{grad} \alpha \, dV \;\; = \;\; \oint_S \alpha \mathbf{n} \, dS, \tag{1.12.3}$$

which should be called the gradient theorem, I think. That is probably because each component can be considered as a part of the divergence theorem above. For example, in two dimensions, we have

$$\int_V \frac{\partial \alpha}{\partial x} dxdy \;\; = \;\; \oint_S \alpha \, dy, \tag{1.12.4}$$

which is basically the first term in Equation (1.12.2). Incidentally, the divergence theorem may be thought of as an integration by parts. Look at this. We obtain from Equation (1.7.11),

$$\int_V \alpha \operatorname{div} \mathbf{a}\, dV = \int_V \operatorname{div}(\alpha \mathbf{a})\, dV - \int_V \operatorname{grad}\alpha \cdot \mathbf{a}\, dV$$

$$= \oint_S \alpha \mathbf{a} \cdot \mathbf{n}\, dS - \int_V \operatorname{grad}\alpha \cdot \mathbf{a}\, dV. \tag{1.12.5}$$

Of course, the divergence theorem applies to second-rank tensors (denoted by $\mathbf{A}$):

$$\int_V \operatorname{div} \mathbf{A}\, dV = \oint_S \mathbf{A}\,\mathbf{n}\, dS. \tag{1.12.6}$$

It is also interesting to derive another form [120] by replacing $\mathbf{a}$ by $\mathbf{c} \times \mathbf{b}$, where $\mathbf{c}$ is a constant vector in Equation (1.12.1):

$$\int_V \operatorname{div}(\mathbf{c} \times \mathbf{b})\, dV = \oint_S (\mathbf{c} \times \mathbf{b}) \cdot \mathbf{n}\, dS, \tag{1.12.7}$$

which becomes by Equations (1.7.13) and (1.2.14)

$$\int_V (\mathbf{b} \cdot \operatorname{curl}\mathbf{c} - \mathbf{c} \cdot \operatorname{curl}\mathbf{b})\, dV = \oint_S (\mathbf{b} \times \mathbf{n}) \cdot \mathbf{c}\, dS, \tag{1.12.8}$$

$$\mathbf{c} \cdot \left(-\int_V \operatorname{curl}\mathbf{b}\, dV \right) = \oint_S (\mathbf{b} \times \mathbf{n})\, dS, \tag{1.12.9}$$

and thus

$$\int_V \operatorname{curl}\mathbf{b}\, dV = -\oint_S (\mathbf{b} \times \mathbf{n})\, dS. \tag{1.12.10}$$

Anyway, these formulas are often used in CFD, for example, to evaluate derivatives, to reduce the order of differential equations (e.g., the Galerkin formulation of the Laplace equation), or to simply compute the volume of a computational cell. So, the divergence theorem is very useful in CFD, and I like it.

1.13   Stokes' Theorem

If there is the divergence theorem, there is, of course, the curl theorem. But it is more widely known as Stokes' theorem. It is given, for an arbitrary vector $\mathbf{a}$, by

$$\int_S \operatorname{curl}\mathbf{a} \cdot \mathbf{n}\, dS = \oint_{\partial S} \mathbf{a} \cdot d\mathbf{l}, \tag{1.13.1}$$

where S is a bounded surface, dS and $\mathbf{n}$ are the infinitesimal area and its unit outward vector, respectively, and $d\mathbf{l} = (dx, dy, dz)$ denotes the infinitesimal line-element in the counter-clockwise direction over the boundary of S denoted by ∂S. In Cartesian coordinates, we write $\mathbf{a} = (a_x, a_y, a_z)$ and obtain the curl theorem as

$$\int_S \left[\left(\frac{\partial a_z}{\partial y} - \frac{\partial a_y}{\partial z} \right) dy dz + \left(\frac{\partial a_x}{\partial z} - \frac{\partial a_z}{\partial x} \right) dz dx + \left(\frac{\partial a_y}{\partial x} - \frac{\partial a_x}{\partial y} \right) dx dy \right] = \oint_{\partial S} (a_x dx + a_y dy + a_z dz). \tag{1.13.2}$$

I like it because Green's theorem is immediately obtained by setting $a_z = 0$ and others to be constant in z:

$$\int_S \left(\frac{\partial a_y}{\partial x} - \frac{\partial a_x}{\partial y} \right) dx dy = \oint_{\partial S} (a_x dx + a_y dy). \tag{1.13.3}$$

It is also possible to derive the following from the Stoke's theorem [120]: replace $\mathbf{a}$ by $\mathbf{c} \times \mathbf{b}$, where $\mathbf{c}$ is a constant vector in Equation (1.13.1):

$$\int_S \operatorname{curl}(\mathbf{c} \times \mathbf{b}) \cdot \mathbf{n}\, dS = \oint_{\partial S} (\mathbf{c} \times \mathbf{b}) \cdot d\mathbf{l}, \tag{1.13.4}$$

which becomes by Equations (1.7.20) and (1.2.14)

$$\int_S (\mathbf{c}\,\mathrm{div}\,\mathbf{b} + (\mathrm{grad}\,\mathbf{c})\mathbf{b} - (\mathrm{div}\,\mathbf{c})\mathbf{b} - (\mathrm{grad}\,\mathbf{b})\mathbf{c}) \cdot \mathbf{n}\,dS = \oint_{\partial S} \mathbf{c} \cdot (\mathbf{b} \times d\mathbf{l}), \tag{1.13.5}$$

$$\mathbf{c} \cdot \left(\int_S (\mathbf{n}\,\mathrm{div}\,\mathbf{b} - (\mathrm{grad}\,\mathbf{b})^t\mathbf{n})\,dS = \oint_{\partial S} (\mathbf{b} \times d\mathbf{l}) \right), \tag{1.13.6}$$

and thus

$$\int_S (\mathbf{n}\,\mathrm{div}\,\mathbf{b} - (\mathrm{grad}\,\mathbf{b})^t\mathbf{n})\,dS = \oint_{\partial S} (\mathbf{b} \times d\mathbf{l}). \tag{1.13.7}$$

In fluid dynamics, Stokes' theorem arises naturally in Kelvin's circulation theorem as in Section 3.24 and also in relation to the three-dimensional stream functions as in Section 4.24, for example. In fact, the curl theorem and the divergence theorem are both special cases of Stokes' theorem (also called the Kelvin-Stokes theorem). So, I actually like to call them both Stokes' theorem. See differential geometry textbooks, e.g., Ref.[148], for more details.

1.14 Conservation Laws

Many flow equations come in the form of conservation laws. That is, the time rate of change of a quantity within a control volume is balanced by the net change per unit time brought by the local flow. A typical example of such a quantity is the fluid mass. To express the conservation law mathematically, consider a control volume V with the boundary S. Let $\mathbf{U}$ be a vector of conservative variables (quantities which are conserved), and $\mathbf{F}_n$ be the flux vector normal to the control volume boundary (e.g., 'mass' flow per unit time). Then, we can write the conservation laws as

$$\frac{d}{dt}\int_V \mathbf{U}\,dV + \oint_S \mathbf{F}_n\,dS = 0. \tag{1.14.1}$$

This is called the integral form of conservation laws. It simply states that the integral value of $\mathbf{U}$ changes in time only due to the net effect of the flux across the control volume boundary. In Cartesian coordinates, the normal flux vector can be written as

$$\mathbf{F}_n = \mathbf{F}n_x + \mathbf{G}n_y + \mathbf{H}n_z, \tag{1.14.2}$$

where $\mathbf{n} = (n_x, n_y, n_z)$, and $\mathbf{F}, \mathbf{G}, \mathbf{H}$ are the flux vectors in the directions of x, y, z, respectively. Also, we may write

$$\frac{d}{dt}\int_V \mathbf{U}\,dV + \oint_S \mathcal{F}\mathbf{n}\,dS = 0, \tag{1.14.3}$$

where $\mathcal{F}$ is a second-rank flux tensor defined by $\mathcal{F} = [\mathbf{F}, \mathbf{G}, \mathbf{H}]$. I like this form because I then immediately notice that the surface integral can be written as a volume integral by the divergence theorem:

$$\frac{d}{dt}\int_V \mathbf{U}\,dV + \int_V \mathrm{div}\,\mathcal{F}\,dV = 0, \tag{1.14.4}$$

where, of course, every component of $\mathcal{F}$ must be differentiable within the control volume. A nice thing about this is that as long as the control volume is fixed in space, we can write this as a single integral,

$$\int_V \left(\frac{\partial \mathbf{U}}{\partial t} + \mathrm{div}\,\mathcal{F} \right) dV = 0, \tag{1.14.5}$$

and since this is valid for arbitrary control volumes, we can write

$$\frac{\partial \mathbf{U}}{\partial t} + \mathrm{div}\,\mathcal{F} = 0. \tag{1.14.6}$$

This is called the differential form of conservation laws. In Cartesian coordinates, it is written as

$$\frac{\partial \mathbf{U}}{\partial t} + \frac{\partial \mathbf{F}}{\partial x} + \frac{\partial \mathbf{G}}{\partial y} + \frac{\partial \mathbf{H}}{\partial z} = 0. \tag{1.14.7}$$

I like it, but I prefer the integral form because the integral form allows discontinuous solutions (or weak solutions, more generally) while the differential form does not. The integral form is more fundamental, more general, and thus more useful especially in developing shock-capturing methods. It is also useful even when the control volume deforms in time, i.e., for moving mesh methods [52]. Moreover, the integral form is applicable to control volumes of any shape; it is the basis of the finite-volume method. So, it is indeed very nice.

Having said that, I actually like the differential form (1.14.6) very much because it can be written in a conservative form by expressing the divergence in terms of the surface integral as suggested by the definition (1.5.21):

$$\frac{\partial \mathbf{U}}{\partial t} + \lim_{V \to 0} \frac{1}{V} \oint \mathcal{F} \cdot \mathbf{n} \, dS = 0. \qquad (1.14.8)$$

Of course, it cannot be exact unless $V \to 0$, and so it will be an approximation in a finite (computational) volume:

$$\frac{\partial \mathbf{U}}{\partial t} + \frac{1}{V} \oint \mathcal{F} \cdot \mathbf{n} \, dS \approx 0. \qquad (1.14.9)$$

But it is actually widely used in node-centered finite-volume methods, which can be thought of as solving the differential form at nodes in a conservative manner. It can be a good approximation especially for second-order accurate methods. It is also very useful to construct a conservative scheme. In fact, all advantages mentioned above about the integral form are valid also for the conservative integral formulation of the differential form. If you don't like such an approximation, the exact conversion of the divergence to the surface integral is possible by the primitive function approach. Define $\mathcal{G}$ by

$$\mathcal{F} = \frac{1}{V} \int_V \mathcal{G}(\xi, \eta, \zeta) \, d\xi \, d\eta \, d\zeta, \qquad (1.14.10)$$

where V is the control volume defined in the neighborhood of a point (x, y, z) with a fixed volume V, and (ξ, η, ζ) is a set of dummy variables to perform the integral over the control volume. Then, $\mathcal{F}$ is the volume average of $\mathcal{G}$, or equivalently, $\mathcal{G}$ is the primitive function of $\mathcal{F}$:

$$\mathrm{div}\mathcal{F} = \frac{1}{V} \, \mathrm{div} \int_V \mathcal{G}(\xi, \eta, \zeta) \, d\xi \, d\eta \, d\zeta, = \frac{1}{V} \oint \mathcal{G} \cdot \mathbf{n} \, dS, \qquad (1.14.11)$$

which is exact for a fixed V. Therefore, the differential form (1.14.6) can be written *exactly* as

$$\frac{\partial \mathbf{U}}{\partial t} + \frac{1}{V} \oint \mathcal{G} \cdot \mathbf{n} \, dS = 0. \qquad (1.14.12)$$

Note that the flux needed at the control volume boundary is the primitive function of the physical flux in this case. This form is the basis of high-order finite-difference methods typically developed for regular grids [181].

I like to compare the three forms above on a one-dimensional control volume. Consider a control volume of spacing Δx, which we call j, defined by the left boundary $x_j - \Delta x/2$ and the right boundary $x_j + \Delta x/2$, where x_j is the center of the control volume.

Integral Form

Applying the integral form (1.14.1) over the control volume, we obtain

$$\frac{d\overline{\mathbf{U}}_j}{dt} + \frac{1}{\Delta x} \left[\mathbf{F}_{j+1/2} - \mathbf{F}_{j-1/2} \right] = 0, \qquad (1.14.13)$$

where $\mathbf{F}_{j\pm1/2}$ denotes the flux at $x_j \pm \Delta x/2$, and $\overline{\mathbf{U}}_j$ is the averaged solution over the control volume,

$$\overline{\mathbf{U}}_j = \frac{1}{\Delta x} \int_{x_j - \Delta x/2}^{x_j + \Delta x/2} \mathbf{U} \, dx. \qquad (1.14.14)$$

I like it very much because the fluxes on the left and right boundaries can be computed from the solution at the boundaries, i.e.,

$$\mathbf{F}_{j+1/2} = \mathbf{F}(\mathbf{U}_{j+1/2}), \quad \mathbf{F}_{j-1/2} = \mathbf{F}(\mathbf{U}_{j-1/2}), \qquad (1.14.15)$$

which is nice and simple. Note that the formulation (1.14.13) is exact and therefore higher-order numerical schemes can be developed by improving the accuracy of the interface fluxes (and the quadrature of the flux integral in two and three dimensions). This is a very popular formulation in CFD.

Approximate Conservative Differential Form

Next, applying the conservative approximation to the differential form (1.14.9) at the center $x = x_j$, we get

$$\frac{d\mathbf{U}_j}{dt} + \frac{1}{\Delta x}\left[\mathbf{F}_{j+1/2} - \mathbf{F}_{j-1/2}\right] = 0, \tag{1.14.16}$$

where $\mathbf{U}_j$ is the point value of the solution at $x = x_j$. This is another popular formulation in practical CFD codes. Obviously, it is a second-order accurate approximation to the differential form of the conservation law because it is equivalent to the central difference formula to the derivative $\partial \mathbf{F}/\partial x$. It means that higher-order approximation to the fluxes does not necessarily lead to higher-order accuracy in the whole equation. To achieve higher-order accuracy, a more sophisticated discretization is required for the time derivative, e.g, the Galerkin discretization, rather than the point evaluation as above. If it has source terms, they also need to be discretized carefully for higher-order accuracy [102].

Exact Conservative Differential Form

On the other hand, applying the exact form (1.14.12) at the center $x = x_j$, we obtain

$$\frac{d\mathbf{U}_j}{dt} + \frac{1}{\Delta x}\left[\mathbf{F}^{\mathcal{G}}_{j+1/2} - \mathbf{F}^{\mathcal{G}}_{j-1/2}\right] = 0, \tag{1.14.17}$$

where $\mathbf{F}^{\mathcal{G}}$ is the x-component of $\mathcal{G}$, i.e., the primitive function of $\mathbf{F}$,

$$\mathbf{F}(x) = \frac{1}{\Delta x}\int_{x-\Delta x/2}^{x+\Delta x/2} \mathbf{F}^{\mathcal{G}}(\xi)\,d\xi, \quad \frac{\partial \mathbf{F}}{\partial x} = \frac{1}{\Delta x}\left[\mathbf{F}^{\mathcal{G}}_{j+1/2} - \mathbf{F}^{\mathcal{G}}_{j-1/2}\right]. \tag{1.14.18}$$

This form is exact. Therefore, high-order accuracy in $\mathbf{F}^{\mathcal{G}}_{j+1/2}$ and $\mathbf{F}^{\mathcal{G}}_{j-1/2}$ does yield high-order accuracy in the whole equation, which is very nice. Note that it is the primitive function of the flux, not the flux itself, that is required at the control volume boundaries. Note also that the point value of the flux is now the volume average of the primitive function:

$$\mathbf{F}(x_j) = \frac{1}{\Delta x}\int_{x_j-\Delta x/2}^{x_j+\Delta x/2} \mathbf{F}^{\mathcal{G}}(x)\,dx. \tag{1.14.19}$$

As you might have noticed by now, thee forms are equivalent to one another for second-order accurate methods because the volume-averaged solution differs from the point value by $O(\Delta x^2)$:

$$\overline{\mathbf{U}}_j = \mathbf{U}_j + \mathcal{O}(\Delta x^2). \tag{1.14.20}$$

which can be easily proved by substituting the Taylor expansion of $\mathbf{U}(x)$ around $x = x_j$ into the right hand side of Equation (1.14.14). I like second-order methods because it doesn't really matter what quantity to compute (volume averages or point values). That is, I can say that my second-order CFD code computes volume averages when I set the initial solution by volume averages, or I can also say that it computes point values when I set the initial solution by the point values. I like it very much.

1.15   Conservation Laws in Generalized Coordinates

Consider the coordinate transformation:

$$\xi = \xi(x,y,z), \quad \eta = \eta(x,y,z) \quad \zeta = \zeta(x,y,z), \tag{1.15.1}$$

which defines a map between the physical coordinates (x, y, z) and the computational (or generalized) coordinates (ξ, η, ζ). It maps a curvilinear grid in the physical space to a Cartesian grid in the computational space. Then, partial derivatives are transformed by the chain rule,

$$\frac{\partial}{\partial x} = \xi_x \frac{\partial}{\partial \xi} + \eta_x \frac{\partial}{\partial \eta} + \zeta_x \frac{\partial}{\partial \zeta}, \tag{1.15.2}$$

$$\frac{\partial}{\partial y} = \xi_y \frac{\partial}{\partial \xi} + \eta_y \frac{\partial}{\partial \eta} + \zeta_y \frac{\partial}{\partial \zeta}, \tag{1.15.3}$$

$$\frac{\partial}{\partial z} = \xi_z \frac{\partial}{\partial \xi} + \eta_z \frac{\partial}{\partial \eta} + \zeta_z \frac{\partial}{\partial \zeta}. \tag{1.15.4}$$

Metric coefficients such as ξ_x, η_x, ζ_x, are given by

$$\xi_x = J(y_\eta z_\zeta - y_\zeta z_\eta), \quad \xi_y = J(x_\zeta z_\eta - x_\eta z_\zeta), \quad \xi_z = J(x_\eta y_\zeta - x_\zeta y_\eta), \tag{1.15.5}$$

$$\eta_x = J(y_\zeta z_\xi - y_\xi z_\zeta), \quad \eta_y = J(x_\xi z_\zeta - x_\zeta z_\xi), \quad \eta_z = J(x_\zeta y_\xi - x_\xi y_\zeta), \tag{1.15.6}$$

$$\zeta_x = J(y_\xi z_\eta - y_\eta z_\xi), \quad \zeta_y = J(x_\eta z_\xi - x_\xi z_\eta), \quad \zeta_z = J(x_\xi y_\eta - x_\eta y_\xi), \tag{1.15.7}$$

$$J = \frac{1}{x_\xi(y_\eta z_\zeta - y_\zeta z_\eta) - x_\eta(y_\xi z_\zeta - y_\zeta z_\xi) - x_\zeta(y_\xi z_\eta - y_\eta z_\xi)}, \tag{1.15.8}$$

which can be readily computed once the transformation is given in the following form,

$$x = x(\xi, \eta, \zeta), \quad y = y(\xi, \eta, \zeta), \quad z = z(\xi, \eta, \zeta). \tag{1.15.9}$$

For more details, see Refs.[36, 57, 155]. For example, the differential form of the conservation law (1.14.7) can be written as

$$\frac{\partial \mathbf{U}^\star}{\partial t} + \frac{\partial \mathbf{F}^\star}{\partial \xi} + \frac{\partial \mathbf{G}^\star}{\partial \eta} + \frac{\partial \mathbf{H}^\star}{\partial \zeta} = 0, \tag{1.15.10}$$

where

$$\mathbf{U}^\star = \frac{\mathbf{U}}{J}, \tag{1.15.11}$$

$$\mathbf{F}^\star = \frac{1}{J}\left(\mathbf{F}\xi_x + \mathbf{G}\xi_y + \mathbf{H}\xi_z\right), \tag{1.15.12}$$

$$\mathbf{G}^\star = \frac{1}{J}\left(\mathbf{F}\eta_x + \mathbf{G}\eta_y + \mathbf{H}\eta_z\right), \tag{1.15.13}$$

$$\mathbf{H}^\star = \frac{1}{J}\left(\mathbf{F}\zeta_x + \mathbf{G}\zeta_y + \mathbf{H}\zeta_z\right). \tag{1.15.14}$$

If $\mathbf{F}$, $\mathbf{G}$, $\mathbf{H}$ depend on the derivatives of $\mathbf{U}$, these derivatives are transformed by Equations (1.15.2), (1.15.3), and (1.15.4). I like this form because now I can work with a regular Cartesian grid in the space (ξ, η, ζ), and apply the straightforward finite-difference formulas.

1.16   Classification of Systems of Differential Equations

1.16.1   General System of Equations

Consider the system of n quasi-linear partial differential equations in the form,

$$\mathbf{U}_t + \mathbf{A}\mathbf{U}_x + \mathbf{B}\mathbf{U}_y + \mathbf{C}\mathbf{U}_z = 0, \tag{1.16.1}$$

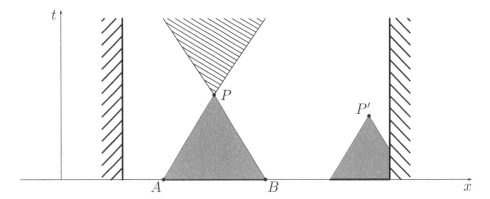

Figure 1.16.1: Domain of dependence (lower triangle) and range of influence (upper triangle) of a point P, for a linear 2×2 hyperbolic system. They are bounded by the characteristic lines (wave paths) of right- and left-running waves. Domain of dependence of a point P' includes the right boundary.

where $\mathbf{U}$ is a set of n solution variables, and $\mathbf{A}$, $\mathbf{B}$, and $\mathbf{C}$ are $n{\times}n$ square matrices. I like such a system because it may have a plane wave solution,

$$\mathbf{U} = f(\mathbf{x} \cdot \mathbf{n} - \lambda_n t)\, \mathbf{U}_0, \qquad (1.16.2)$$

where f is a scalar function, $\mathbf{n} = (n_x, n_y, n_z)$ is a unit vector normal to the plane wave, λ_n is the plane wave speed in the direction of $\mathbf{n}$, and $\mathbf{U}_0$ is a constant vector. In one dimension, this becomes

$$\mathbf{U} = f(x - \lambda t)\, \mathbf{U}_0, \qquad (1.16.3)$$

and the wave travels to the left if $\lambda < 0$ or to the right if $\lambda > 0$. In the space with (x, t) coordinates, the wave is represented by a curve defined by a local slope $dx/dt = \lambda$, along which f is constant. Such a curve is called a characteristic curve, and f is called the Riemann invariant. In two dimensions, we may use the polar angle to indicate the normal direction $\mathbf{n} = (\cos\theta, \sin\theta)$, and write the solution (1.16.2) as

$$\mathbf{U} = f(x\cos\theta + y\sin\theta - \lambda(\theta)t)\, \mathbf{U}_0, \qquad (1.16.4)$$

where $\lambda(\theta)$ denotes the wave speed in the direction of θ. Once $\lambda(\theta)$ is known, we can determine the wave front by constructing an envelope of plane waves. This is discussed in Section 1.21. In two dimensions, the wave may propagate not only in a particular direction as in one dimension but also in all directions at varying speeds. In the space with (x, y, t) coordinates, such a wave is generally represented by a surface. Such a surface is called characteristic surface, and again f is the Riemann invariant.

In order to see if the system (1.16.1) has a plane wave solution, we substitute Equation (1.16.2) into the system to get

$$(\mathbf{A}n_x + \mathbf{B}n_y + \mathbf{C}n_z - \lambda_n \mathbf{I})\, \mathbf{U}_0 = 0. \qquad (1.16.5)$$

This means that the speed of the plane wave λ_n corresponds to an eigenvalue of the projected Jacobian,

$$\mathbf{A}_n = \mathbf{A}n_x + \mathbf{B}n_y + \mathbf{C}n_z, \qquad (1.16.6)$$

and $\mathbf{U}_0$ is the corresponding eigenvector, and that there may be n such plane wave solutions. Therefore, if all eigenvalues of $\mathbf{A}_n$ are real and the corresponding eigenvectors are linearly independent, there will be n independent characteristic surfaces and n Riemann invariants, and thus a solution at a point in space can be completely determined, in principle by solving a set of n equations of constant Riemann invariants. In this case, the system (1.16.1) is called hyperbolic in time. Moreover, if the eigenvalues are distinct, it is called strictly hyperbolic. In that case, the eigenvectors are automatically linearly independent. If all eigenvalues are complex, then the system is called elliptic. Of course, the system will be mixed if it has both real and complex eigenvalues. Also, if there are less than n real eigenvalues (i.e., $\mathbf{A}_n$ is not full rank), then the system is said to be parabolic.

An important implication of the existence of the plane wave solution is that a solution at a point in space depends only on the solution values in the region where the plane wave has traveled before reaching that point. Such a region

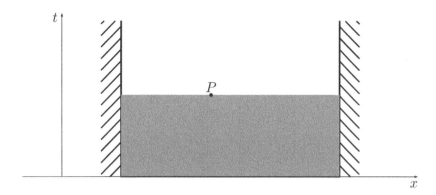

Figure 1.16.2: Domain of dependence of a point P for a parabolic problem.

is called the domain of dependence. In one dimension, a typical 2×2 hyperbolic system has two plane wave solutions: one running to the left and the other to the right. Then, the domain of dependence of a certain point is the union of the left and right intervals where the right- and left-running waves pass through until they reach that point. This grows in time, and will look like a triangle bounded by characteristic lines (or curves for nonlinear waves) in (x, t)-space (see Figure 1.16.1). Note that for a scalar hyperbolic equation the domain of dependence becomes a single point, such as the point A in Figure 1.16.1 for a right-running wave. Also, it is interesting to note that it will take a finite time for a solution at a point P to influence other points in the domain. Of course, it cannot propagate faster than the plane waves (characteristics), and therefore the range of influence (the region influenced by the solution at P) is bounded by the characteristics passing through the point P (see Figure 1.16.1). Since the domain of dependence is bounded, a hyperbolic problem can be solved for a given initial solution if the domain is infinite or periodic (i.e., no boundary). This type of problem is often called an initial-value problem. If the domain is bounded as in Figure 1.16.1, then a part of the solution needs to be specified on the boundary because there might be a characteristic which cannot be traced back to the initial solution (see P' in Figure 1.16.1). Such a problem may be called an initial-boundary value problem. On the other hand, there are no plane wave solutions for elliptic systems. These systems typically arise in steady state problems, i.e., $\mathbf{U}_t = 0$. Since there are no waves whatsoever, the domain of dependence must be the entire domain. In other words, since there are no reasons that a solution at any point depends on a certain bounded region in the domain, it must depend on the entire domain. Elliptic problems, therefore, require solution values to be specified everywhere on the boundary. This type of problem is called a boundary-value problem. A parabolic problem typically requires both initial and boundary values. A solution at a point P depends not only on all solutions at all previous time, but also on the solution at the current time (see Figure 1.16.2). This is another initial-boundary value problem.

By the way, if $\mathbf{A}$, $\mathbf{B}$, $\mathbf{C}$ can be simultaneously symmetrized, e.g., by way of change of variables, the system is guaranteed to be hyperbolic. Of course, symmetrization is not always possible. But usually, physically meaningful systems are symmetrizable: e.g., the system of the Euler equations is symmetrizable [10, 39]. It would be even nicer if the whole system were diagonalizable, i.e., if $\mathbf{A}$, $\mathbf{B}$, $\mathbf{C}$ could be simultaneously diagonalized. This means that the system is transformed into a set of independent scalar advection equations. This is nice. It is certainly possible in one dimension with only one matrix, but it is generally very difficult to achieve in two and three dimensions.

1.16.2 One-Dimensional Equations

Consider the one-dimensional hyperbolic system of the form,

$$\mathbf{U}_t + \mathbf{A}\mathbf{U}_x = 0. \tag{1.16.7}$$

This can be diagonalized by multiplying it by the left-eigenvector matrix $\mathbf{L}$ of $\mathbf{A}$ from the left,

$$\mathbf{L}\mathbf{U}_t + \mathbf{L}\mathbf{A}\mathbf{R}\mathbf{L}\mathbf{U}_x = 0, \tag{1.16.8}$$

$$\mathbf{L}\mathbf{U}_t + \mathbf{\Lambda}\mathbf{L}\mathbf{U}_x = 0, \tag{1.16.9}$$

where $\mathbf{R}$ is the right-eigenvector matrix of $\mathbf{A}$ and $\mathbf{\Lambda}$ is the diagonal eigenvalue matrix of $\mathbf{A}$. For linear hyperbolic systems (i.e., $\mathbf{A}$ is constant, and so $\mathbf{L}$ is constant) we can write

$$(\mathbf{L}\mathbf{U})_t + \mathbf{\Lambda}(\mathbf{L}\mathbf{U})_x = 0. \tag{1.16.10}$$

The quantity $\mathbf{LU}$ is the Riemann invariant or characteristic variable. I like this diagonalized system very much because it is now a set of scalar equations. The k-th component is given by

$$\frac{\partial v_k}{\partial t} + \lambda_k \frac{\partial v_k}{\partial x} = 0, \tag{1.16.11}$$

where v_k is the k-th component of the Riemann invariant,

$$v_k = (\mathbf{LU})_k, \tag{1.16.12}$$

which is simply convected at the speed given by the k-th eigenvalue λ_k of $\mathbf{A}$. This is very nice. A scalar advection equation is so much easier to deal with than the system of equations. On the other hand, for nonlinear systems, the Riemann invariant can be expressed only in the differential form $\mathbf{L}\partial\mathbf{U}$ (difficult to integrate it analytically). Namely, there must be some quantity (i.e., the characteristic variables) $\mathbf{V}$ such that $\mathbf{L}\partial\mathbf{U} = \partial\mathbf{V}$, but it is difficult to find it.

Suppose that the system (1.16.7) involves n equations for n variables. Then, note that we can write

$$\partial\mathbf{U} = \mathbf{R}\partial\mathbf{V} = \mathbf{r}_1\partial V_1 + \mathbf{r}_2\partial V_2 + \mathbf{r}_3\partial V_3 + \cdots + \mathbf{r}_n\partial V_n, \tag{1.16.13}$$

where $\mathbf{V} = [V_1, V_2, V_3, \ldots, V_n]^t$ and $\mathbf{r}_k$ is k-th column of $\mathbf{R}$. The above equation indicates that any change in the solution $\mathbf{U}$ can be decomposed into the sum of the changes in the characteristic variables and each contribution is proportional to the corresponding right eigenvector. This means that the k-th right eigenvector indicates what variables can change due to the change in the k-th characteristic variables (or the k-th wave): for example, one of the eigenvectors of the one-dimensional Euler equations, with the primitive variables $\mathbf{U} = [\rho, u, p]^t$, is given by

$$\mathbf{r} = [1, 0, 0]^t, \tag{1.16.14}$$

which indicates that only the density can change due to this wave. Incidentally, this is called the entropy wave because the corresponding characteristic variable is associated with the entropy. See Section 3.4.2.

I like also to look at Equation (1.16.13) in the following form:

$$\partial\mathbf{U} = \Pi_1\partial\mathbf{U} + \Pi_2\partial\mathbf{U} + \Pi_3\partial\mathbf{U} + \cdots + \Pi_n\partial\mathbf{U}, \tag{1.16.15}$$

where the matrix Π_k, $k = 1, 2, 3, \ldots, n$, is defined by

$$\Pi_k = \mathbf{r}_k\boldsymbol{\ell}_k^t, \tag{1.16.16}$$

where $\boldsymbol{\ell}_k^t$ is the k-th row of $\mathbf{L}$. In this form, we see that the matrix Π_k projects the solution change into the space of the k-th eigenvector. For this reason, it is often called the projection matrix [99, 106]. It arises also in the context of projected spatial operators in two dimensional equations [117]. Of course, we have

$$\sum_{k=1}^{n} \Pi_k = \mathbf{I}. \tag{1.16.17}$$

By using the projection matrix, we can write the decomposition of the Jacobian matrix, $\mathbf{A} = \mathbf{R}\Lambda\mathbf{L}$, as

$$\mathbf{A} = \sum_{k=1}^{n} \lambda_k \Pi_k. \tag{1.16.18}$$

Then, the positive and negative parts of the Jacobian matrix, which are the parts corresponding to positive and negative eigenvalues, respectively, can be easily identified:

$$\mathbf{A}^- = \sum_{\lambda_k \leq 0} \lambda_k \Pi_k, \quad \mathbf{A}^+ = \sum_{\lambda_k > 0} \lambda_k \Pi_k, \tag{1.16.19}$$

where $\mathbf{A}^-$ and $\mathbf{A}^+$ denote the positive and negative parts, respectively. Such a decomposition is often used to construct upwind schemes: one-sided scheme towards the upwind direction. See Ref.[99, 106], for examples that explicitly utilize the projection matrix.

It is also interesting to write the conservation law (1.16.7) as follows,

$$\mathbf{U}_t + \mathbf{RLARLU}_x = 0, \tag{1.16.20}$$

$$\mathbf{U}_t + \mathbf{R}\Lambda\mathbf{V}_x = 0, \tag{1.16.21}$$

$$\mathbf{U}_t + \sum_{k=1}^{n} \lambda_k \frac{\partial v_k}{\partial x} \mathbf{r}_k = 0. \tag{1.16.22}$$

This shows how each subproblem (advection term) affects the time rate of conservative variables, and again, each contribution is proportional to the right eigenvector. It is also very clear from this that each subproblem, i.e., Equation (1.16.11), is obtained simply by multiplying the system by the left eigenvector $\boldsymbol{\ell}_k^t$ from the left.

1.16.3 Two-Dimensional Steady Equations

The classification discussed in Section 1.16.1 can be extended to steady equations. Consider the steady two-dimensional system,

$$\mathbf{A}\mathbf{U}_x + \mathbf{B}\mathbf{U}_y = 0. \tag{1.16.23}$$

This admits a steady plane wave solution of the form,

$$\mathbf{U} = f(x n_x + y n_y)\, \mathbf{U}_0, \tag{1.16.24}$$

provided

$$(\mathbf{A} n_x + \mathbf{B} n_y)\, \mathbf{U}_0 = 0. \tag{1.16.25}$$

That is, if the generalized eigenvalue problem,

$$\det(\mathbf{B} - \lambda \mathbf{A}) = 0, \tag{1.16.26}$$

has a real solution $\lambda = -n_x/n_y$. For a system of n equations, there will be n eigenvalues. If all n eigenvalues are real and the corresponding eigenvectors are linearly independent, the steady system is called hyperbolic. If all are complex, it is called elliptic. It can, of course, be mixed if there are both real and complex eigenvalues. Again, if there are less than n real eigenvalues, the system will be parabolic.

As an example, convert the linear scalar second-order partial differential equation,

$$a\,\phi_{xx} + b\,\phi_{xy} + c\,\phi_{yy} + d\,\phi_x + e\,\phi_y + f = 0, \tag{1.16.27}$$

where a, b, c, d, e, f are functions of (x, y, z), into a system of the form (1.16.23), by introducing new variables,

$$u = \phi_x, \quad v = \phi_y, \tag{1.16.28}$$

leading to

$$\mathbf{A}\mathbf{U}_x + \mathbf{B}\mathbf{U}_y = \mathbf{S}, \tag{1.16.29}$$

where

$$\mathbf{U} = \begin{bmatrix} u \\ v \end{bmatrix}, \quad \mathbf{A} = \begin{bmatrix} a & b/2 \\ 0 & 1 \end{bmatrix}, \quad \mathbf{B} = \begin{bmatrix} b/2 & c \\ -1 & 0 \end{bmatrix}, \quad \mathbf{S} = \begin{bmatrix} -du - ev - f \\ 0 \end{bmatrix}, \tag{1.16.30}$$

and the second equation ($v_x - u_y = 0$) is a necessary condition that comes from the definition of the new variables (1.16.28). Note also that we split $b\phi_{xy}$ into two: $\frac{b}{2}\phi_{xy} + \frac{b}{2}\phi_{xy} = \frac{b}{2} u_y + \frac{b}{2} v_x$. Now, we have

$$\det(\mathbf{B} - \lambda \mathbf{A}) = 0, \tag{1.16.31}$$

i.e.,

$$\det \begin{bmatrix} b/2 - \lambda a & c - b\lambda/2 \\ -1 & -\lambda \end{bmatrix}, \tag{1.16.32}$$

or

$$a\lambda^2 - b\lambda + c = 0, \tag{1.16.33}$$

and therefore we see that the system is hyperbolic if $b^2 - 4ac > 0$ (and the eigenvectors are linearly independent), and elliptic if $b^2 - 4ac < 0$. Of course, it is called parabolic if $b^2 - 4ac = 0$. In this case, there exists only a single (degenerate) characteristic, and it is not sufficient to describe the solution behavior. This type of equation requires low-order terms in $\mathbf{S}$ to describe the solution. For example, the parabolic equation $u_y = u_{xx}$ is basically the heat equation with y interpreted as time. Obviously, its character cannot be described without the first-order term, u_y.

It is interesting to note that if $\mathbf{A}^{-1}$ exists, we can write the two-dimensional system (1.16.23) in the form,

$$\mathbf{U}_x + \mathbf{A}^{-1}\mathbf{B}\mathbf{U}_y = 0. \tag{1.16.34}$$

This looks like a one-dimensional system with x taken as time. It will really behave like a time-dependent system if x-axis is contained entirely in the range of influence of any point (no backward characteristics). If this is true, then, x is called time-like and y is called space-like. The two-dimensional steady Euler system is a good example. See Section 3.17 for details.

By the way, I like adding a time-derivative term to the steady system because then the whole system can be hyperbolic in time: a real eigenvalue indicates a wave traveling in a particular direction while a complex eigenvalue indicates a wave traveling isotropically (all directions). Finally, it is interesting to know that the eigenvalues and eigenvectors of the steady system can be used to decompose the system into hyperbolic and elliptic parts. See Ref.[117] for details.

1.17 Classification of Scalar Differential Equations

Consider, again, the second-order partial differential equation,

$$a\,\phi_{xx} + b\,\phi_{xy} + c\,\phi_{yy} + d\,\phi_x + e\,\phi_y + f = 0, \tag{1.17.1}$$

where a, b, c, d, e, f are functions of (x, y, z). It is really nice that we can easily identify the type of the equation, before attempting to solve it, by

$$b^2 - 4ac < 0, \quad \text{Elliptic},$$
$$b^2 - 4ac = 0, \quad \text{Parabolic}, \tag{1.17.2}$$
$$b^2 - 4ac > 0, \quad \text{Hyperbolic},$$

as shown in the previous subsection. This is very important because the strategy for numerically solving the equation can be quite different for different types. A typical elliptic equation is the Laplace equation,

$$u_{xx} + u_{yy} = 0. \tag{1.17.3}$$

This equation requires a boundary condition everywhere on the domain boundary: specify the solution (Dirichlet condition), specify the solution derivative normal to the boundary (Neumann condition), or specify the combination of the solution and the normal derivative (Robin condition) [155]. The solution inside the domain is smooth. In particular, its maximum/minimum occurs only on the boundary; this is often called the maximum principle.

A typical parabolic equation is the heat equation,

$$u_y - u_{xx} = 0, \tag{1.17.4}$$

where y-axis acts like time. See Figure 1.16.2 with t replaced by y. An initial distribution of u (temperature) along x-axis ($y = 0$) will continue to diffuse along x-direction as y increases. This equation requires boundary conditions at $y = 0$ (initial temperature distribution specified) and along the left and right boundaries (temperature at two end points on x-axis). The solution is again smooth. Even if an initial solution is not smooth, it will be smeared out as time goes on. This is very nice.

A typical hyperbolic equation is the wave equation,

$$u_{xx} - u_{yy} = 0, \tag{1.17.5}$$

where again y may be considered as time. This describes a solution traveling in x with time y (recall that characteristic surfaces exist for hyperbolic systems). It requires an initial solution to be specified at $y = 0$, but the solution at each (x, y) depends only on a part of the initial data. Basically, it only needs to know where it comes from in the initial data. Also, by taking a domain to be periodic, i.e., $u(x_{min}) = u(x_{max})$, we can avoid boundary conditions, thus leading to a pure initial-value problem. The solution does not have to be smooth and can be even discontinuous. But such a solution is not differentiable and will not satisfy the wave equation above. In fact, there is an integral equation associated with the wave equation which allows discontinuous solutions. See Section 2.1 for details.

I like the fact that only the first three terms, i.e., the second derivative terms, determine the character of the equation; we do not need to take into account all the terms to analyze the equations. In fact, generally, many important features of solutions of partial differential equations depend only on the highest-order terms (called the principal part) of the equation [180]. This is really nice. Suppose that you have a partial differential equation in the form (1.16.27) with an extremely complicated source term $f(x, y)$, e.g., those involving terms like $\phi_x \phi_y$ or ϕ_x^2. Then, just look at the three second-order terms only, $a\,\phi_{xx} + b\,\phi_{xy} + c\,\phi_{yy}$, and you will find the type of your equation by Equation (1.17.2), which then guides you to an appropriate numerical method for solving it.

1.18 Pseudo-Transient Equations

I like CFD because partial differential equations are often manipulated for computational advantages or just for convenience. They even change the type. For example, consider the following elliptic equation:

$$u_{xx} + u_{yy} = 0. \tag{1.18.1}$$

It can be solved as a time-dependent equation,

$$u_\tau = u_{xx} + u_{yy}, \tag{1.18.2}$$

where τ is a time-like variable called pseudo time. The solution to the original problem is then obtained as a steady state solution to the pseudo-time-dependent problem. The method is generally called the pseudo-transient method (or continuation). It is very interesting and useful because methods developed for time-dependent initial-value problems can be directly applied to steady problems (those without time derivatives). Note that the type of partial differential equations have been changed from elliptic to parabolic. This is very interesting. The elliptic equation can be solved as a parabolic equation; the solution to the elliptic equation is obtained in the steady state. I like it because then I think I can also define a pseudo-transient equation by introducing the second derivative $u_{\tau\tau}$ instead of the first-derivative u_τ:

$$u_{\tau\tau} = u_{xx} + u_{yy}, \tag{1.18.3}$$

which is the wave equation and thus hyperbolic. In this case, the solution to the elliptic equation is obtained by solving a hyperbolic equation towards the steady state. I'm actually not aware of any work in CFD that actually does so. However, I do know a method that does it in the form of first-order system:

$$u_\tau = \nu(p_x + q_y), \quad p_\tau = \frac{1}{T_r}(u_x - p), \quad q_\tau = \frac{1}{T_r}(u_y - q), \tag{1.18.4}$$

where T_r is a free parameter. This system is hyperbolic, but reduces to the elliptic equation (1.18.1) in the steady state for any positive value of T_r as described in Section 2.5. This is very interesting because we can apply methods developed for hyperbolic systems to solve the elliptic problem. I like it very much. See Section 2.5 and Refs.[98, 99, 103, 104] for details on the method of first-order hyperbolic systems.

The pseudo-transient method is, in a way, a method of manipulated governing equations as mentioned at the beginning. Consider a hyperbolic conservation law in two dimensions:

$$\mathbf{U}_t + \mathbf{F}_x + \mathbf{G}_y = 0, \tag{1.18.5}$$

where $\mathbf{U}$ is a solution vector, and $\mathbf{F}$ and $\mathbf{G}$ are the flux vectors. If we are only interested in the steady state, then the time t can be thought of as pseudo time τ:

$$\mathbf{U}_\tau + \mathbf{F}_x + \mathbf{G}_y = 0. \tag{1.18.6}$$

The system is still hyperbolic, but we can alter the character of wave propagations by multiplying the spatial part by a matrix $\mathbf{P}$:

$$\mathbf{U}_\tau + \mathbf{P}\left(\mathbf{F}_x + \mathbf{G}_y\right) = 0, \tag{1.18.7}$$

where $\mathbf{P}$ may be taken as a positive definite matrix so that the waves do not change the direction of propagation and the same type of boundary conditions apply. The matrix $\mathbf{P}$ can be designed, for example, to optimize the condition number of the equation, the ratio of the fastest wave speed to the slowest wave speed, in order to accelerate the convergence to the steady state (i.e., no waves left behind) [77]. Note that the system reduces to the original system in the steady state as long as $\mathbf{P}$ is non-singular (which is true for a positive definite matrix). That is, the transient solution may be completely wrong, but it converges faster to the steady solution of the original equation. This technique is called the local-preconditioning technique [46, 80, 117, 161, 168]. It has an added benefit of improving the accuracy of numerical schemes in the low-Mach-number limit [45, 163]. Furthermore, some preconditioning matrices can be used to split the governing equations into hyperbolic and elliptic parts [117], enabling optimal discretization [94, 113] as well as optimal multigrid convergence [116]. That is very interesting. I like it. I also find it interesting that the pseudo-compressible formulation [21] for the incompressible Euler and Navier-Stokes equations can also be thought of as a kind of pseudo-transient method.

Another area where the pseudo-transient method is found useful and actually widely applied is the solution method for nonlinear systems, e.g., Newton's method. Newton's method is a very powerful method for solving nonlinear systems, but it can take a long time to achieve a rapid convergence unless the initial solution is close enough to the solution. To enhance the convergence, the pseudo-transient method can be employed in its early stage (typically by gradually increasing the pseudo time step). Such a technique is often employed in practical CFD codes. It is a simple but very effective method and I like it very much.

Finally, time-dependent computations can be performed with the pseudo-transient method by the dual-time approach [59] in which the pseudo-transient method can be used to solve, at each physical time step, the system of equations arising from an implicit-time stepping scheme. I like the fact that a fast steady solver is an essential building block of implicit-time stepping.

1.19 Operator Splitting

In CFD, it is sometimes convenient to solve a partial differential equation term by term. Consider the two-dimensional evolution equation of a solution $u(x, y, t)$ in the form:

$$u_t = (\mathcal{A} + \mathcal{B}) \, u, \tag{1.19.1}$$

where $\mathcal{A}$ and $\mathcal{B}$ are two different linear differential operators. For example, $\mathcal{A}$ may be the advective operator and $\mathcal{B}$ can be the diffusive operator, or $\mathcal{A}$ is an x-derivative operator and $\mathcal{B}$ is a y-derivative operator. I like that $\mathcal{A}$ and $\mathcal{B}$ are linear because then I can easily generate as many time derivatives as I like:

$$u_{tt} = (\mathcal{A} + \mathcal{B}) \, u_t = (\mathcal{A} + \mathcal{B})^2 \, u, \tag{1.19.2}$$

$$u_{ttt} = (\mathcal{A} + \mathcal{B}) \, u_{tt} = (\mathcal{A} + \mathcal{B})^3 \, u, \tag{1.19.3}$$

$$u_{tttt} = (\mathcal{A} + \mathcal{B}) \, u_{ttt} = (\mathcal{A} + \mathcal{B})^4 \, u, \tag{1.19.4}$$

and so on. I would use them to compute the solution at $t + \Delta t$ from the solution u at t as follows:

$$
\begin{aligned}
u(x, y, z, t + \Delta t) &= u + \Delta t \, u_t + \frac{\Delta t^2}{2} u_{tt} + \frac{\Delta t^3}{6} u_{ttt} + \cdots \\[2mm]
&= u + \Delta t (\mathcal{A} + \mathcal{B}) \, u + \frac{\Delta t^2}{2} (\mathcal{A} + \mathcal{B})^2 \, u + \frac{\Delta t^3}{6} (\mathcal{A} + \mathcal{B})^3 \, u + \cdots \\[2mm]
&= \left(1 + \Delta t (\mathcal{A} + \mathcal{B}) + \frac{\Delta t^2}{2} (\mathcal{A} + \mathcal{B})^2 + \frac{\Delta t^3}{6} (\mathcal{A} + \mathcal{B})^3 + \cdots \right) u,
\end{aligned}
\tag{1.19.5}
$$

which can be expressed as

$$u(x, y, t + \Delta t) = \exp\left((\mathcal{A} + \mathcal{B}) \Delta t \right) u. \tag{1.19.6}$$

Note that u and all its time derivatives on the right hand side are understood as values at t. In CFD, it is sometimes easier to solve each operator independently in a sequential manner: e.g., solve

$$u_t^* = \mathcal{A} \, u, \tag{1.19.7}$$

and then solve

$$u_t = \mathcal{B} \, u^*, \tag{1.19.8}$$

to compute the solution at the next time step. This is called the operator splitting method or the fractional step method. The above splitting in particular is often called the Godunov splitting [5]. It is a simple and interesting idea. There are many good reasons that I like it. In this way, a numerical method can be designed specifically for each term, and the coding can be made simple. Also, accuracy and stability of the numerical method may be controlled independently at each step. For example, we may solve the first step by an explicit high-order upwind scheme and the second step by an implicit diffusion scheme to allow large time steps. However, it is not immediately guaranteed that the operator splitting formulation is equivalent to the original formulation. To see this, expand the solutions at the two stages:

$$u^*(x, y, t + \Delta t) = \left(1 + \Delta t \mathcal{A} + \frac{\Delta t^2}{2} \mathcal{A}^2 + \frac{\Delta t^3}{6} \mathcal{A}^3 + \cdots \right) u, \tag{1.19.9}$$

$$u(x, y, t + \Delta t) = \left(1 + \Delta t \mathcal{B} + \frac{\Delta t^2}{2} \mathcal{B}^2 + \frac{\Delta t^3}{6} \mathcal{B}^3 + \cdots \right) u^*(x, y, t + \Delta t), \tag{1.19.10}$$

and therefore,

$$
\begin{aligned}
u(x, y, t + \Delta t) &= \left(1 + \Delta t \mathcal{B} + \frac{\Delta t^2}{2} \mathcal{B}^2 + \frac{\Delta t^3}{6} \mathcal{B}^3 + \cdots \right) \left(1 + \Delta t \mathcal{A} + \frac{\Delta t^2}{2} \mathcal{A}^2 + \frac{\Delta t^3}{6} \mathcal{A}^3 + \cdots \right) u \\[2mm]
&= \left(1 + \Delta t (\mathcal{A} + \mathcal{B}) + \frac{\Delta t^2}{2} (\mathcal{A}^2 + 2\mathcal{B}\mathcal{A} + \mathcal{B}^2) + O(\Delta t^3) \right) u.
\end{aligned}
\tag{1.19.11}
$$

Comparing this with Equation (1.19.5), we find that the solution obtained by the operator splitting method matches the exact one only up to $O(\Delta t)$ term because

$$(\mathcal{A} + \mathcal{B})^2 = \mathcal{A}^2 + \mathcal{A}\mathcal{B} + \mathcal{B}\mathcal{A} + \mathcal{B}^2 \neq \mathcal{A}^2 + 2\mathcal{B}\mathcal{A} + \mathcal{B}^2, \tag{1.19.12}$$

unless

$$\mathcal{B}\mathcal{A} = \mathcal{A}\mathcal{B}, \tag{1.19.13}$$

i.e., unless the operators $\mathcal{A}$ and $\mathcal{B}$ commute. For simple linear operators, the operators often commute straightforwardly, but it is not always the case especially for nonlinear equations of practical interest. If the splitting error remains, the accuracy of the operator-split method would be first-order in time at best no matter how accurate each step is. Of course, the Godunov splitting is not the only splitting method. We have the Strang splitting [150], which recovers second-order accuracy:

$$
\begin{aligned}
u_t^* &= \mathcal{A}\,u && \text{with } \Delta t/2, & (1.19.14)\\
u_t^{**} &= \mathcal{B}\,u^* && \text{with } \Delta t, & (1.19.15)\\
u_t &= \mathcal{A}\,u^{**} && \text{with } \Delta t/2. & (1.19.16)
\end{aligned}
$$

In this case, we find

$$
\begin{aligned}
u(x, y, t + \Delta t) &= \left(1 + \frac{\Delta t}{2}\mathcal{A} + \frac{\Delta t^2}{8}\mathcal{A}^2 + O(\Delta t^3)\right)\left(1 + \Delta t\mathcal{B} + \frac{\Delta t^2}{2}\mathcal{B}^2 + O(\Delta t^3)\right)\left(1 + \frac{\Delta t}{2}\mathcal{A} + \frac{\Delta t^2}{8}\mathcal{A}^2 + O(\Delta t^3)\right) u \\
&= \left(1 + \frac{\Delta t}{2}\mathcal{A} + \Delta t\mathcal{B} + \frac{\Delta t^2}{8}\mathcal{A}^2 + \frac{\Delta t^2}{2}\mathcal{B}^2 + \frac{\Delta t^2}{2}\mathcal{A}\mathcal{B} + O(\Delta t^3)\right)\left(1 + \frac{\Delta t}{2}\mathcal{A} + \frac{\Delta t^2}{8}\mathcal{A}^2 + O(\Delta t^3)\right) u \\
&= \left(1 + \Delta t(\mathcal{A} + \mathcal{B}) + \frac{\Delta t^2}{2}(\mathcal{A}^2 + \mathcal{A}\mathcal{B} + \mathcal{B}\mathcal{A} + \mathcal{B}^2) + O(\Delta t^3)\right) u, \tag{1.19.17}
\end{aligned}
$$

which matches the exact solution up to $O(\Delta t^2)$ and thus leads to second order accuracy. This is nice. Although we have to solve the $\mathcal{A}$-operator problem twice in this method, it may worth paying the price; a good method doesn't come for free. Well, in fact, the Strang splitting can be compounded into a simpler operator and the cost can be reduced [41, 150]. Let $\mathcal{A}_{\Delta t/2}$ denotes the operator that solves the $\mathcal{A}$-operator problem with $\Delta t/2$, and $\mathcal{B}_{\Delta t}$ denotes the operator that solves the $\mathcal{B}$-operator problem with Δt. Then, the Strang splitting over one time step can be expressed as

$$u(x, y, t + \Delta t) = \mathcal{A}_{\Delta t/2}\mathcal{B}_{\Delta t}\mathcal{A}_{\Delta t/2}\, u(x, y, t). \tag{1.19.18}$$

In the next time step, we have

$$u(x, y, t + 2\Delta t) = \mathcal{A}_{\Delta t/2}\mathcal{B}_{\Delta t}\mathcal{A}_{\Delta t/2}\mathcal{A}_{\Delta t/2}\mathcal{B}_{\Delta t}\mathcal{A}_{\Delta t/2}\, u(x, y, t), \tag{1.19.19}$$

which can be simplified by $\mathcal{A}_{\Delta t/2}\mathcal{A}_{\Delta t/2} = \mathcal{A}_{\Delta t} + O(\Delta t^3)$ as

$$u(x, y, t + 2\Delta t) = \mathcal{A}_{\Delta t/2}\mathcal{B}_{\Delta t}\mathcal{A}_{\Delta t}\mathcal{B}_{\Delta t}\mathcal{A}_{\Delta t/2}\, u(x, y, t). \tag{1.19.20}$$

In the same fashion, we have

$$u(x, y, t + 3\Delta t) = \mathcal{A}_{\Delta t/2}\mathcal{B}_{\Delta t}\mathcal{A}_{\Delta t}\mathcal{B}_{\Delta t}\mathcal{A}_{\Delta t}\mathcal{B}_{\Delta t}\mathcal{A}_{\Delta t/2}\, u(x, y, t), \tag{1.19.21}$$

and so on. As you see, two operators are combined into a single operator at each time step. The half-time step operator is required only at the beginning and at the time when the solution is sought. This is nice. Higher-order operator splitting is also possible. See Refs.[14, 178] for examples. Also, you might want to check for yourself if the Strang splitting still gives second-order accuracy for equations with three or more operators, and also if the splitting error vanishes for higher-order terms as well if the operators commute.

I like the operator splitting method. It is simple and systematic. I just would have to be careful for problems with strong interactions among the operators and also steady state problems where the operator-splitting method may not converge to the correct solution (e.g., consider the case in which $(\mathcal{A} + \mathcal{B})u = 0$ with $\mathcal{A} = -\mathcal{B}$).

1.20   Rotational Invariance

Let $\mathbf{T}$ be a matrix of rotation. Then, for the conservation law,

$$\frac{\partial \mathbf{U}}{\partial t} + \frac{\partial \mathbf{F}}{\partial x} + \frac{\partial \mathbf{G}}{\partial y} + \frac{\partial \mathbf{H}}{\partial z} = 0, \qquad (1.20.1)$$

in which we often denote $\mathbf{F}(\mathbf{U})$ to indicate that $\mathbf{F}$ is a function of $\mathbf{U}$, if we have

$$\mathbf{F}_n = n_x \mathbf{F} + n_y \mathbf{G} + n_z \mathbf{H} = \mathbf{T}^{-1} \mathbf{F}(\mathbf{TU}), \qquad (1.20.2)$$

for any non-zero vector $\mathbf{n} = [n_x, n_y, n_z]^t$, then, the conservation law is said to satisfy the rotational invariance property. For example, the Euler equations satisfy the rotational invariance property, with the rotation matrix defined by

$$\mathbf{T} = \begin{pmatrix} 1 & 0 & 0 & 0 & 0 \\ 0 & n_x & n_y & n_z & 0 \\ 0 & \ell_x & \ell_y & \ell_z & 0 \\ 0 & m_x & m_y & m_z & 0 \\ 0 & 0 & 0 & 0 & 1 \end{pmatrix}, \qquad (1.20.3)$$

$$\mathbf{T}^{-1} = \mathbf{T}^t = \begin{pmatrix} 1 & 0 & 0 & 0 & 0 \\ 0 & n_x & \ell_x & m_x & 0 \\ 0 & n_y & \ell_y & m_y & 0 \\ 0 & n_z & \ell_z & m_z & 0 \\ 0 & 0 & 0 & 0 & 1 \end{pmatrix}, \qquad (1.20.4)$$

where $\mathbf{n}, \boldsymbol{\ell}, \mathbf{m}$ are three orthogonal unit vectors. Using this property, we can find a flux vector projected in any direction by using $\mathbf{F}$ only. In particular, by setting $\mathbf{n} = [0, 1, 0]^t$, $\boldsymbol{\ell} = [1, 0, 0]^t$, $\mathbf{m} = [0, 0, 1]^t$, we get

$$\mathbf{G} = \mathbf{T}^{-1} \mathbf{F}(\mathbf{TU}), \quad \mathbf{T} = \mathbf{T}^{-1} = \begin{pmatrix} 1 & 0 & 0 & 0 & 0 \\ 0 & 0 & 1 & 0 & 0 \\ 0 & 1 & 0 & 0 & 0 \\ 0 & 0 & 0 & 1 & 0 \\ 0 & 0 & 0 & 0 & 1 \end{pmatrix}, \qquad (1.20.5)$$

also by setting $\mathbf{n} = [0, 0, 1]^t$, $\boldsymbol{\ell} = [0, 1, 0]^t$, $\mathbf{m} = [1, 0, 0]^t$, we get

$$\mathbf{H} = \mathbf{T}^{-1} \mathbf{F}(\mathbf{TU}), \quad \mathbf{T} = \mathbf{T}^{-1} = \begin{pmatrix} 1 & 0 & 0 & 0 & 0 \\ 0 & 0 & 0 & 1 & 0 \\ 0 & 0 & 1 & 0 & 0 \\ 0 & 1 & 0 & 0 & 0 \\ 0 & 0 & 0 & 0 & 1 \end{pmatrix}. \qquad (1.20.6)$$

I like this; it says we don't need $\mathbf{G}$ and $\mathbf{H}$. See Section 3.8 for more details on the rotational invariance of the Euler system.

The rotational invariance is very useful. Many people like it. For example, in a finite-volume code, some people write just a subroutine to compute an interface flux based on $\mathbf{F}$ only, and then use the same subroutine to compute a flux in the face-normal direction by rotating it by Equation (1.20.2). It is nice and useful.

1.21   Plane Waves of Hyperbolic Equations

The wave front of a wave in hyperbolic equations can be determined by Huygen's principle, i.e., the wave front by a point disturbance is the envelope of all plane wave fronts that pass simultaneously through that point, leading to the formula [79]:

$$\begin{bmatrix} X(\theta) \\ Y(\theta) \end{bmatrix} = \begin{bmatrix} \cos\theta & -\sin\theta \\ \sin\theta & \cos\theta \end{bmatrix} \begin{bmatrix} \lambda(\theta) \\ \lambda'(\theta) \end{bmatrix}, \qquad (1.21.1)$$

where $\lambda(\theta)$ is the plane wave speed in the direction of θ $(0 \leq \theta < 2\pi)$. See Figure 1.20.1. In particular, I like the following examples.

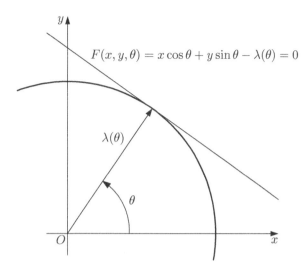

Figure 1.20.1: Envelope of plane waves defined by the equation of the plane wave (the tangent to the envelope) $F(x, y, \theta) = 0$ and the equation of vanishing derivative $\partial F/\partial \theta = 0$.

(a) **Linear Advection:**

Consider the linear scalar hyperbolic equation,

$$u_t + au_x + bu_y = 0, \tag{1.21.2}$$

where a and b are constants. To see if this has a plane wave solution, we substitute a general plane wave solution,

$$u = f(x\cos\theta + y\sin\theta - \lambda(\theta)t), \tag{1.21.3}$$

into Equation (1.21.3) and solve the resulting equation for $\lambda(\theta)$. The solution is

$$\lambda(\theta) = a\cos\theta + b\sin\theta, \tag{1.21.4}$$

which is real, meaning that it has a plane wave solution. We then proceed to find the wave front by

$$\begin{bmatrix} X(\theta) \\ Y(\theta) \end{bmatrix} = \begin{bmatrix} \cos\theta & -\sin\theta \\ \sin\theta & \cos\theta \end{bmatrix} \begin{bmatrix} a\cos\theta + b\sin\theta \\ -a\sin\theta + b\cos\theta \end{bmatrix} = \begin{bmatrix} a \\ b \end{bmatrix}. \tag{1.21.5}$$

Therefore, a point disturbance simply propagates in the direction (a, b).

(b) **Isotropic Advection:**

Consider the following equation,

$$u_t + (R\cos\theta)u_x + (R\sin\theta)u_y = 0, \tag{1.21.6}$$

In this case, we find $\lambda(\theta) = R$ and $\lambda'(\theta) = 0$, and so

$$\begin{bmatrix} X(\theta) \\ Y(\theta) \end{bmatrix} = \begin{bmatrix} \cos\theta & -\sin\theta \\ \sin\theta & \cos\theta \end{bmatrix} \begin{bmatrix} R \\ 0 \end{bmatrix} = \begin{bmatrix} R\cos\theta \\ R\sin\theta \end{bmatrix}. \tag{1.21.7}$$

Therefore, we find that the envelope is a circle defined by

$$X^2 + Y^2 = R^2, \tag{1.21.8}$$

i.e., a disturbance propagates in every direction at the same speed.

(c) **System of Equations:**

Consider a system of hyperbolic equations,

$$\mathbf{U}_t + \mathbf{A}\mathbf{U}_x + \mathbf{B}\mathbf{U}_y = 0. \tag{1.21.9}$$

This has a plane wave solution of the form

$$\mathbf{U} = f(x\cos\theta + y\sin\theta - \lambda(\theta)t)\mathbf{U}_0, \tag{1.21.10}$$

if $\lambda(\theta)$ is an eigenvalue of the matrix,

$$\mathbf{A}_n = \mathbf{A}\cos\theta + \mathbf{B}\sin\theta. \tag{1.21.11}$$

As an example, consider the two-dimensional Euler equations, for which we find

$$\lambda(\theta) = q_n \pm a, \; q_n, \; q_n, \tag{1.21.12}$$

where $q_n = (u,v)\cdot(\cos\theta,\sin\theta)$ is the flow speed in the direction of θ and a is the speed of sound. So, two waves, corresponding to those with the speed q_n, simply propagate in the flow direction at the flow speed $q = \sqrt{u^2 + v^2}$; and the other two, corresponding to those with the speeds $q_n \pm a$, have the envelopes given by

$$\begin{bmatrix} X(\theta) \\ Y(\theta) \end{bmatrix} = \begin{bmatrix} \cos\theta & -\sin\theta \\ \sin\theta & \cos\theta \end{bmatrix} \begin{bmatrix} u\cos\theta + v\sin\theta \pm a \\ -u\sin\theta + v\cos\theta \end{bmatrix} = \begin{bmatrix} u \pm a\cos\theta \\ v \pm a\sin\theta \end{bmatrix}, \tag{1.21.13}$$

i.e., the envelope is a circle centered at (u,v),

$$(X-u)^2 + (Y-v)^2 = a^2. \tag{1.21.14}$$

In general, I like wave fronts because they can be in quite interesting shapes; see Refs.[60, 117], for example, for intriguing magnetohydrodynamic waves.

1.22   Rankine-Hugoniot Relation

Let $\mathbf{n}$ be the unit normal vector of a discontinuous surface (a shock wave), and V_D be the normal speed of the discontinuity. Then, apply the integral form of conservation laws (1.14.1) across the discontinuity, and let $V \to 0$ to obtain the Rankine-Hugoniot relation,

$$[\mathbf{F}_n] = V_D[\mathbf{U}], \tag{1.22.1}$$

where [ ] denotes the jump across the discontinuity. Derive this by yourself or see Refs.[50, 51, 60] for derivations. Suppose that the jump is small, so that we can write

$$d\mathbf{F}_n = V_D\, d\mathbf{U}, \tag{1.22.2}$$

and hence

$$\frac{\partial \mathbf{F}_n}{\partial \mathbf{U}} d\mathbf{U} = V_D\, d\mathbf{U}. \tag{1.22.3}$$

This shows that the shock speed V_D is an eigenvalue of $\dfrac{\partial \mathbf{F}_n}{\partial \mathbf{U}}$ and the jump $d\mathbf{U}$ is the corresponding eigenvector. Note that this is always true for linear equations, but not for nonlinear equations. For nonlinear equations, this is true only for small jumps. I emphasize that the shock speed is NOT an eigenvalue of the Jacobian matrix for large jumps. The one-dimensional Euler equations is a good example: the shock speed associated with one of the acoustic waves is given by (derived from the Rankine-Hugoniot relation)

$$V_D = u_L + a_L\sqrt{1 + \frac{\gamma+1}{2\gamma}\left(\frac{p_R - p_L}{p_L}\right)}, \tag{1.22.4}$$

where $\gamma = 1.4$; u_L, p_L, and a_L are the velocity, pressure, and the speed of sound on one side of the shock; and p_R is the pressure on the other side of the shock. For small change, i.e., $p_R - p_L \approx 0$, it reduces to

$$V_D \approx u_L + a_L, \tag{1.22.5}$$

which is nothing but the eigenvalue of the Jacobian, i.e., the one associated with the same acoustic wave (see Section 3.4.2). This is actually nice. There are two similar eigenvalues in the Euler equations, $u + a$ and $u - a$, which are sometimes called two nonlinear fields associated with the acoustic wave. We then see from Equation (1.22.5) that the shock speed given by Equation (1.22.4) is associated with the field of $u + a$, not the one of $u - a$. Yes, we can identify the field associated with a shock wave by taking the limit $p_R \to p_L$ in the shock speed. In fact, if we look at the other shock speed (again, derived from the Rankine-Hugoniot relation):

$$V_D' = u_L - a_L \sqrt{1 + \frac{\gamma + 1}{2\gamma} \left(\frac{p_R - p_L}{p_L} \right)}, \tag{1.22.6}$$

we find by expanding it for small change $p_R - p_L \approx 0$,

$$V_D' \approx u_L - a_L. \tag{1.22.7}$$

So, V_D' is the shock speed associated with the other field. This is nice, isn't it? I like it very much.

1.23   Linearly Degenerate and Genuinely Nonlinear Fields

Consider a hyperbolic system of the form,

$$\mathbf{U}_t + \mathbf{A}\mathbf{U}_x = 0. \tag{1.23.1}$$

The k-th eigenvalue of the coefficient matrix $\mathbf{A}$, $\lambda_k(\mathbf{U})$ (or the eigenvector $\mathbf{r}_k(\mathbf{U})$) defines a characteristic field called the λ_k-th field (or $\mathbf{r}_k$-th field), or simply k-th characteristic field.

A k-th characteristic field is said to be linearly degenerate if

$$\frac{\partial \lambda_k(\mathbf{U})}{\partial \mathbf{U}} \cdot \mathbf{r}_k(\mathbf{U}) = 0, \quad \text{for all } \mathbf{U}. \tag{1.23.2}$$

In the case of linear hyperbolic systems, all eigenvalues are constant and therefore all fields are linearly degenerate. The contact discontinuity is a discontinuity in a linearly degenerate field: the characteristics are all parallel across such a discontinuity just like the scalar linear advection. On the other hand, a k-th characteristic field is said to be genuinely nonlinear if

$$\frac{\partial \lambda_k(\mathbf{U})}{\partial \mathbf{U}} \cdot \mathbf{r}_k(\mathbf{U}) \neq 0, \quad \text{for all } \mathbf{U}. \tag{1.23.3}$$

I like this genuinely nonlinear one because the characteristic speed can now change across waves, thus creating much more sophisticated waves than the linearly degenerate field. Specifically, there can be nonlinear shocks (converging characteristics) and nonlinear expansions (diverging characteristics) in this case. Note that any nonlinear scalar conservation laws with non-zero characteristic speed is genuinely nonlinear and such scalar conservations laws are said to have a convex flux. See Refs.[81, 159] for details.

The one-dimensional Euler equations in the primitive variables $\mathbf{U} = [\rho, u, p]^t$ can be a good example: the eigenvalues and eigenvectors are given by (see Section 3.4.2)

$$\lambda_1 = u - a, \quad \lambda_2 = u, \quad \lambda_3 = u + a, \tag{1.23.4}$$

$$\mathbf{r}_1 = \begin{bmatrix} -\dfrac{\rho}{a} \\ 1 \\ -\rho a \end{bmatrix}, \quad \mathbf{r}_2 = \begin{bmatrix} 1 \\ 0 \\ 0 \end{bmatrix}, \quad \mathbf{r}_3 = \begin{bmatrix} \dfrac{\rho}{a} \\ 1 \\ \rho a \end{bmatrix}, \tag{1.23.5}$$

where $a = \sqrt{\gamma p / \rho}$, from which we find

$$\frac{\partial \lambda_1}{\partial \mathbf{U}} \cdot \mathbf{r}_1 = \begin{bmatrix} \partial(u-a)/\partial \rho \\ \partial(u-a)/\partial u \\ \partial(u-a)/\partial p \end{bmatrix} \cdot \begin{bmatrix} -\dfrac{\rho}{a} \\ 1 \\ -\rho a \end{bmatrix} = \begin{bmatrix} \dfrac{a}{2\rho} \\ 1 \\ -\dfrac{a}{2p} \end{bmatrix} \cdot \begin{bmatrix} -\dfrac{\rho}{a} \\ 1 \\ -\rho a \end{bmatrix} = \frac{\gamma+1}{2} \neq 0, \tag{1.23.6}$$

$$\frac{\partial \lambda_2}{\partial \mathbf{U}} \cdot \mathbf{r}_2 = \begin{bmatrix} \partial u/\partial \rho \\ \partial u/\partial u \\ \partial u/\partial p \end{bmatrix} \cdot \begin{bmatrix} 1 \\ 0 \\ 0 \end{bmatrix} = \begin{bmatrix} 0 \\ 1 \\ 0 \end{bmatrix} \cdot \begin{bmatrix} 1 \\ 0 \\ 0 \end{bmatrix} = 0, \tag{1.23.7}$$

$$\frac{\partial \lambda_3}{\partial \mathbf{U}} \cdot \mathbf{r}_3 = \begin{bmatrix} \partial(u+a)/\partial \rho \\ \partial(u+a)/\partial u \\ \partial(u+a)/\partial p \end{bmatrix} \cdot \begin{bmatrix} \dfrac{\rho}{a} \\ 1 \\ \rho a \end{bmatrix} = \begin{bmatrix} -\dfrac{a}{2\rho} \\ 1 \\ \dfrac{a}{2p} \end{bmatrix} \cdot \begin{bmatrix} \dfrac{\rho}{a} \\ 1 \\ \rho a \end{bmatrix} = \frac{\gamma+1}{2} \neq 0, \tag{1.23.8}$$

so that the λ_1-field and λ_3-field (acoustic waves) are genuinely nonlinear but the λ_2-field (entropy wave) is linearly degenerate.

Chapter 2

Model Equations

2.1 Linear Advection

I like the linear advection equation:

$$u_t + (\operatorname{grad} u) \cdot \boldsymbol{\lambda} = 0, \tag{2.1.1}$$

or

$$u_t + \operatorname{div}(u\boldsymbol{\lambda}) = 0, \tag{2.1.2}$$

or simply

$$\frac{Du}{Dt} = 0, \tag{2.1.3}$$

where $\boldsymbol{\lambda}$ is a constant advection velocity at which the solution u travels. In Cartesian coordinates, the advection equation becomes, with $\boldsymbol{\lambda} = (a, b, c)$,

$$1\text{D}: \qquad u_t + au_x = 0, \tag{2.1.4}$$
$$2\text{D}: \qquad u_t + au_x + bu_y = 0, \tag{2.1.5}$$
$$3\text{D}: \qquad u_t + au_x + bu_y + cu_z = 0. \tag{2.1.6}$$

The linear advection equation is a model equation for a general quasi-linear system of equations, e.g., in one dimension,

$$\mathbf{U}_t + \mathbf{A}\mathbf{U}_x = 0, \tag{2.1.7}$$

where $\mathbf{U}$ is a vector of solutions and $\mathbf{A}$ is a coefficient matrix having real eigenvalues. This includes the Euler equations of gas dynamics. To develop a method for solving such a system, it is a good idea to first study the model equation which is simpler but retains important features of the general system. Once we develop and study a method for the model equation, we may move on to extend it to the system of equations. For example, since the system (2.1.7) can be diagonalized, i.e., decomposed into a set of linear advection equations (see Section 1.16.2), a method developed for the linear advection equation can be directly applied to each component.

In studying the model equation, I like its integral form,

$$\frac{d}{dt} \int_V u \, dV + \oint_S u\boldsymbol{\lambda} \cdot \mathbf{n} \, dS = 0, \tag{2.1.8}$$

because this allows discontinuous solutions: any initial solutions, whether continuous or discontinuous, will be simply advected at the velocity $\boldsymbol{\lambda}$. This is, in fact, a good model for developing a method for computing flows with shock waves. Many techniques for accurately and stably computing discontinuous solutions have been successfully extended to more general equations such as the Euler equations.

Note that the linear advection equation is closely related to the wave equation in one dimension, but not so much in higher dimensions. In one dimension, the wave equation is

$$u_{tt} - a^2 u_{xx} = 0, \tag{2.1.9}$$

45

which can be written as

$$(\partial_t - a\partial_x)(\partial_t u + a\partial_x u) = 0. \tag{2.1.10}$$

Therefore, the wave equation has two general solutions: one traveling at the speed a and the other at the speed $-a$. The linear advection equation governs only the one traveling at the speed a. On the other hand, in two dimensions, the wave equation is defined by

$$u_{tt} - a^2(u_{xx} + u_{yy}) = 0, \tag{2.1.11}$$

where $a > 0$. This has a plane wave solution in all directions at the same speed a (substituting a function of the form $f(x\cos\theta + y\sin\theta - \lambda t)$ into the wave equation, and solving the resulting equation for the wave speed λ, we obtain $\lambda = \pm a$ which is independent of the direction θ). In contrast, the linear advection equation has a plane wave solution only in a particular direction, i.e., specified by the constant advection vector (a, b). To describe an isotropic wave propagation, the advection vector needs to be a function of space:

$$(a, b) = a(\cos\theta, \sin\theta). \tag{2.1.12}$$

This describes a wave traveling isotropically at speed a (see Section 1.21). The advection equation is then still linear, but it might be nice to stay with constant coefficients. Then, we may use a system description of the isotropic wave:

$$u_t = a(p_x + q_y), \tag{2.1.13}$$
$$p_t = au_x, \tag{2.1.14}$$
$$q_t = au_y. \tag{2.1.15}$$

It is easy to show by straightforward differentiations that this system is equivalent to the wave equation (2.1.11). Note however that this system representation has introduced an additional constraint, $(p_y - q_x)_t = 0$, to the wave equation. This condition implies that any zero or nonzero 'vorticity' is preserved at all later times. This type of model system was used in Ref.[95] for developing vorticity-preserving schemes. In three dimensions, basically the same is true, except that the number of constraints now becomes three; the vorticity has three components in three dimensions.

Incidentally, some people use the term 'convection' instead of 'advection'. Personally, I like 'advection' better simply because I'm used to it. I like not to discuss which is better.

2.2 Circular Advection

The advection equation will still be linear even if $\boldsymbol{\lambda} = (a, b, c)$ is a function of space. For example, the two-dimensional advection equation,

$$u_t + au_x + bu_y = 0, \tag{2.2.1}$$

with the advection velocity,

$$(a, b) = (y, -x), \tag{2.2.2}$$

represents a flow field that rotates clockwise around the origin. This is called the circular advection equation. A three-dimensional circular advection equation would be

$$u_t + au_x + bu_y + cu_z = 0, \tag{2.2.3}$$

with

$$(a, b, c) = (y, -x, 0). \tag{2.2.4}$$

This represents a circular flow field around z-axis. Alternatively, we can take $\boldsymbol{\lambda} = (0, z, -y)$ to create a flow around x-axis, or $\boldsymbol{\lambda} = (z, 0, -x)$ to create a flow around y-axis. Yet, we may take $\boldsymbol{\lambda} = (y, -x, V)$ where V is a constant. I like this one because it describes a more 3D flow: rotating around z-axis and simultaneously moving in the z direction at a constant speed V. Note that the circular advection equation can be written also in the conservative form,

$$u_t + \text{div}(u\boldsymbol{\lambda}) = 0, \tag{2.2.5}$$

$\boldsymbol{\lambda} = (y, -x)$ in two dimensions, and $\boldsymbol{\lambda} = (y, -x, 0)$, $(z, 0, -x)$, $(z, 0, -x)$ or $(y, -x, V)$ in three dimensions.

2.3 Diffusion Equation

I like the diffusion equation:

$$u_t = \text{div}\left(\nu \, \text{grad} \, u\right),$$ (2.3.1)

where ν is a positive diffusion coefficient. As long as ν does not depend on the solution itself, this equation is linear. In Cartesian coordinates, the diffusion equation becomes

1D:	$u_t = (\nu u_x)_x,$	(2.3.2)
2D:	$u_t = (\nu u_x)_x + (\nu u_y)_y,$	(2.3.3)
3D:	$u_t = (\nu u_x)_x + (\nu u_y)_y + (\nu u_z)_z.$	(2.3.4)

The diffusion equation is parabolic in time (prove it!). It requires both an initial solution and boundary conditions. Generally, its solution is very smooth. Even if an initial solution is not smooth, it will be made smooth as time goes on because of the smoothing effect of the second-derivative term on the right hand side, i.e., 'diffusion term'. The solution will reach a steady state, unless we introduce a time-dependent boundary condition and/or source term. Then, the steady solution will be equivalent to that of the Laplace equation (see Section 2.8).

Of course, I like its integral form also,

$$\frac{d}{dt} \int_V u \, dV + \oint_S \nu \, \text{grad} \, u \cdot \mathbf{n} \, dS = 0,$$ (2.3.5)

or

$$\frac{d}{dt} \int_V u \, dV + \oint_S \nu \, \frac{\partial u}{\partial n} \, dS = 0.$$ (2.3.6)

This is a good starting point for developing a finite-volume scheme for the diffusion equation. In general, it is relatively easy to develop a scheme for the diffusion equation in the sense that it only deals with smooth solutions (no discontinuities), especially for second-order methods on regular computational grids. The issue, if any, would be the efficiency of the numerical method. In particular, an explicit scheme can be very inefficient due to a severely restricted time step; implicit schemes are generally preferred. All these (smoothing effect and efficiency) are characteristics of the viscous part of the Navier-Stokes equations. Therefore, the diffusion equation is a typical model for the viscous terms of the Navier-Stokes equations. Numerical methods for the viscous terms are often developed based on the diffusion equations or equivalently on the Laplace equation for spatial discretizations. By the way, the integral form above shows that the integral value of the solution depends only on the normal solution gradient over the control volume boundary, S. This is very interesting. It is basically one-dimensional, and the normal gradient can be easily evaluated on regular grids in which the normal direction often is perfectly aligned with the direction of two solution data points, i.e., a simple finite-difference formula gives the normal gradient. I like it because I think it is one of the reasons that it is relatively easy to develop diffusion schemes on regular grids. For irregular grids, on the other hand, the construction of diffusion schemes is not necessarily easy. Accuracy and convergence can be easily deteriorated unless the normal gradient is carefully evaluated. See, for example, Refs.[101, 108] for diffusion schemes on irregular grids.

Note that the viscous terms of the compressible Navier-Stokes equations are actually more than the diffusion equation because they involve not only the normal gradient but also the gradients tangential to the control volume boundary as described in Section 4.12. Some numerical schemes ignore the tangential gradients, e.g., a thin-layer-approximation [89, 111]. Such schemes are known to be robust, but accuracy is not obtained because they are not consistent schemes. Nevertheless, I like such schemes because they can be useful for constructing an implicit scheme for the compressible Navier-Stokes equations [111]. See Section 4.12.

2.4 Diffusion Equation: Mixed Formulation

Often, the diffusion equation is written and discretized in the following first-order system:

$$u_t = \nu \, \text{div} \, \mathbf{p},$$ (2.4.1)

$$\mathbf{p} = \text{grad} \, u,$$ (2.4.2)

where $\mathbf{p} = (p, q, r)$ is the vector of variables that represents the gradients. In Cartesian coordinates, it reads in one dimension,

$$u_t = \nu p_x, \quad p = u_x, \tag{2.4.3}$$

in two dimensions,

$$u_t = \nu(p_x + q_y), \quad p = u_x, \quad q = u_y, \tag{2.4.4}$$

and in three dimensions,

$$u_t = \nu(p_x + q_y + r_z), \quad p = u_x, \quad q = u_y, \quad r = u_z. \tag{2.4.5}$$

This formulation is known as the mixed formulation in the finite-element-method community [182], and widely used, for example, in high-order methods [25, 66, 115]. I like the mixed formulation because apparently it involves only first-order derivatives, which are easier to discretize, especially on unstructured grids, than second-order derivatives. In fact, a variety of successful diffusion schemes have been developed based on the mixed formulation [176, 182]. Note however that we cannot really avoid the second-derivatives in this formulation. The system is perfectly equivalent to the diffusion equation of second derivatives and the numerical schemes developed for the system is subject to the well-known severe limitation: the explicit time step of $O(h^2)$, where h is a mesh spacing. I actually like it, though, because it shows that the system is indeed equivalent to the diffusion equation. Notice that the first-order system is parabolic in time and thus does not alter the physics of the original diffusion equation. The first-order formulation is used just for the sake of convenience in discretization. I like such an approach very much.

2.5   Diffusion Equation: First-Order Hyperbolic System

There is an interesting first-order hyperbolic system associated with the diffusion equation:

$$u_t = \nu p_x, \quad p_t = \frac{1}{T_r}(u_x - p), \tag{2.5.1}$$

where T_r is a constant. In two and three dimensions, it is given, respectively, by

$$u_t = \nu(p_x + q_y), \quad p_t = \frac{1}{T_r}(u_x - p), \quad q_t = \frac{1}{T_r}(u_y - q), \tag{2.5.2}$$

and

$$u_t = \nu(p_x + q_y + r_z), \quad p_t = \frac{1}{T_r}(u_x - p), \quad q_t = \frac{1}{T_r}(u_y - q), \quad r_t = \frac{1}{T_r}(u_z - r). \tag{2.5.3}$$

This system is equivalent to the diffusion equation in the limit $T_r \to 0$ [20], and often called the hyperbolic heat equations. Consider the vector form:

$$\partial_t \mathbf{U} + \partial_x \mathbf{F} + \partial_y \mathbf{G} + \partial_y \mathbf{H} = \mathbf{S}, \tag{2.5.4}$$

where

$$\mathbf{U} = \begin{bmatrix} u \\ p \\ q \\ r \end{bmatrix}, \quad \mathbf{F} = \begin{bmatrix} -\nu p \\ -u/T_r \\ 0 \\ 0 \end{bmatrix}, \quad \mathbf{G} = \begin{bmatrix} -\nu q \\ 0 \\ -u/T_r \\ 0 \end{bmatrix}, \quad \mathbf{H} = \begin{bmatrix} -\nu r \\ 0 \\ 0 \\ -u/T_r \end{bmatrix}, \quad \mathbf{S} = \begin{bmatrix} 0 \\ -p/T_r \\ -q/T_r \\ -r/T_r \end{bmatrix}. \tag{2.5.5}$$

The flux Jacobians are given by

$$\mathbf{A} = \frac{\partial \mathbf{F}}{\partial \mathbf{U}} = \begin{bmatrix} 0 & -\nu & 0 & 0 \\ -\dfrac{1}{T_r} & 0 & 0 & 0 \\ 0 & 0 & 0 & 0 \\ 0 & 0 & 0 & 0 \end{bmatrix}, \mathbf{B} = \frac{\partial \mathbf{G}}{\partial \mathbf{U}} = \begin{bmatrix} 0 & 0 & -\nu & 0 \\ 0 & 0 & 0 & 0 \\ -\dfrac{1}{T_r} & 0 & 0 & 0 \\ 0 & 0 & 0 & 0 \end{bmatrix}, \mathbf{C} = \frac{\partial \mathbf{H}}{\partial \mathbf{U}} = \begin{bmatrix} 0 & 0 & 0 & -\nu \\ 0 & 0 & 0 & 0 \\ 0 & 0 & 0 & 0 \\ -\dfrac{1}{T_r} & 0 & 0 & 0 \end{bmatrix}. \tag{2.5.6}$$

Consider the projected Jacobian along an arbitrary vector, $\mathbf{n} = [n_x, n_y, n_z]^t$:

$$\mathbf{A}_n = \mathbf{A}n_x + \mathbf{B}n_y + \mathbf{C}n_z = \begin{bmatrix} 0 & -\nu n_x & -\nu n_y & -\nu n_z \\ -n_x/T_r & 0 & 0 & 0 \\ -n_y/T_r & 0 & 0 & 0 \\ -n_z/T_r & 0 & 0 & 0 \end{bmatrix}. \tag{2.5.7}$$

The projected Jacobian has the following eigenvalues:

$$\lambda_1 = -\sqrt{\frac{\nu}{T_r}}, \quad \lambda_2 = \sqrt{\frac{\nu}{T_r}}, \quad \lambda_{3,4} = 0. \tag{2.5.8}$$

The eigenvalues are real, and the associated eigenvectors can be shown to be linearly independent. Hence the system is hyperbolic. Note that the eigenvalues are independent of $\mathbf{n}$, and therefore the system describes a wave propagating isotropically. The third and fourth eigenvalues correspond to the inconsistency damping mode [98], acting on the quantities, $q_x - p_y$ and $r_x - p_z$ that should be zero. Although it has a hyperbolic character, it is hyperbolic only for a very short time of the order T_r. After a long time compared with T_r, it behaves like the diffusion equation as the waves will be damped by the effect of the source term. There have been efforts on solving this system instead of the diffusion equation [129] as it models in part a relaxation system for the Navier-Stokes equations. I like this approach because it avoids the discretization of second derivatives. Also, it allows us to apply numerical schemes well developed for hyperbolic systems to diffusion problems. However, these advantages do not come free. The small relaxation time T_r makes the system 'stiff' and consequently makes it difficult to solve numerically if not impossible.

On the other hand, if we are interested only in the steady state, then, T_r can take *any* positive value [98]. Observe that when time derivatives vanish, the hyperbolic system becomes, for any positive T_r, in one dimension,

$$\begin{array}{rcl} 0 & = & \nu\, p_x, \\ 0 & = & (u_x - p)/T_r, \end{array} \rightarrow \begin{array}{rcl} 0 & = & \nu\, p_x, \\ p & = & u_x, \end{array} \rightarrow \quad 0 \;=\; \nu\, u_{xx}, \tag{2.5.9}$$

in two dimensions,

$$\begin{array}{rcl} 0 & = & \nu\,(p_x + q_y), \\ 0 & = & (u_x - p)/T_r, \\ 0 & = & (u_y - q)/T_r, \end{array} \rightarrow \begin{array}{rcl} 0 & = & \nu\,(p_x + q_y), \\ p & = & u_x, \\ q & = & u_y, \end{array} \rightarrow \quad 0 \;=\; \nu\,(u_{xx} + u_{yy}), \tag{2.5.10}$$

and in three dimensions,

$$\begin{array}{rcl} 0 & = & \nu\,(p_x + q_y + r_z), \\ 0 & = & (u_x - p)/T_r, \\ 0 & = & (u_y - q)/T_r, \\ 0 & = & (u_z - r)/T_r, \end{array} \rightarrow \begin{array}{rcl} 0 & = & \nu\,(p_x + q_y + r_z), \\ p & = & u_x, \\ q & = & u_y, \\ r & = & u_z, \end{array} \rightarrow \quad 0 \;=\; \nu\,(u_{xx} + u_{yy} + u_{zz}). \tag{2.5.11}$$

Therefore, the system is equivalent to the steady diffusion equation (the Laplace equation) for arbitrary T_r (> 0) in contrast to the hyperbolic heat equations that are equivalent to the diffusion equation at any instant of time but require extremely small T_r. To distinguish the two, the hyperbolic system designed to solve steady problems with arbitrary T_r is simply called the *hyperbolic diffusion system*. The equivalence to the Laplace equation emerges naturally also in the energy estimate [104]. Define the energy associated with the hyperbolic diffusion system as

$$E \equiv \frac{u^2 + L_r^2(p^2 + q^2)}{2}. \tag{2.5.12}$$

The governing equation for the energy can be derived by multiplying the hyperbolic system (2.5.4) by the row vector, $\ell^E = (u, L_r^2 p, L_r^2 q)$, from the left:

$$\partial_t E + \partial_x f^E + \partial_y g^E + \partial_y h^E = -\nu(p^2 + q^2 + r^2), \tag{2.5.13}$$

where

$$f^E = -\nu u p, \quad g^E = -\nu u q, \quad h^E = -\nu u r. \tag{2.5.14}$$

By integrating the energy equation over the domain Ω, we obtain

$$\frac{d}{dt} \int_\Omega E \, dV = - \oint_{\partial\Omega} \mathbf{f}^E \cdot \mathbf{n} \, dA - \nu \int_\Omega (p^2 + q^2 + r^2) \, dV, \tag{2.5.15}$$

where $\partial\Omega$ denotes the boundary of the domain, $\mathbf{f}^E = (f^E, g^E, h^E)$, $\mathbf{n}$ is the unit outward normal vector, and dA denotes the infinitesimal boundary area. This integral form shows that the total energy is damped by the magnitude of the solution gradient and changed only by the boundary flux. In the steady state, we have $(p, q, r) = \nabla u$, and the energy estimate reduces to

$$\int_\Omega \nabla u \cdot \nabla u \, dV - \oint_{\partial\Omega} u \frac{\partial u}{\partial n} dA = 0, \tag{2.5.16}$$

which is the well-known energy estimate of the Laplace equation.

I really like that the stiffness is no longer an issue in this approach. The free parameter, T_r, can therefore be chosen to make the system strongly hyperbolic for all times, and can also be defined to accelerate the convergence of numerical methods [98, 99, 104]. In fact, it naturally leads to a very fast method for computing steady state solutions of the diffusion equation with $O(h)$ explicit time step, not $O(h^2)$ [98, 99, 103, 104]. One might have noticed that the first-order hyperbolic system is equivalent to the mixed formulation in the steady state. However, the discretization is very different in that methods for hyperbolic systems, e.g., upwind scheme, are directly applicable to the hyperbolic diffusion system but not to the mixed formulation. This is a great advantage of the hyperbolic formulation because various schemes and techniques developed in the past decades for hyperbolic systems can be applied to the discretization of diffusion. Naturally, it is also possible to develop first-order schemes for diffusion [104]. In addition, in this approach, the solution gradient (p, q, r) can be computed to the same order of accuracy to that of the solution u [98]. Furthermore, the Neumann condition can be implemented as the Dirichlet condition by the extra variables that will be equivalent to the solution gradients in the steady state. Do I like this approach? Oh, yes, I do like this approach very much.

2.6   Advection-Diffusion Equation

I like the advection-diffusion equation:

$$u_t + (\text{grad } u) \cdot \boldsymbol{\lambda} = \text{div} \left(\nu \, \text{grad } u \right), \tag{2.6.1}$$

or in the conservative form,

$$u_t + \text{div} \left(u \boldsymbol{\lambda} - \nu \, \text{grad } u \right) = 0. \tag{2.6.2}$$

In Cartesian coordinates, it is written as

$$\begin{array}{lll} \text{1D:} & u_t + a u_x = (\nu u_x)_x, & (2.6.3) \\ \text{2D:} & u_t + a u_x + b u_y = (\nu u_x)_x + (\nu u_y)_y, & (2.6.4) \\ \text{3D:} & u_t + a u_x + b u_y + c u_z = (\nu u_x)_x + (\nu u_y)_y + (\nu u_z)_z, & (2.6.5) \end{array}$$

or

$$\begin{array}{lll} \text{1D:} & u_t + (au - \nu u_x)_x = 0, & (2.6.6) \\ \text{2D:} & u_t + (au - \nu u_x)_x + (bu - \nu u_y)_y = 0, & (2.6.7) \\ \text{3D:} & u_t + (au - \nu u_x)_x + (bu - \nu u_y)_y + (cu - \nu u_z)_z = 0. & (2.6.8) \end{array}$$

The advection-diffusion equation is a model equation for a general quasi-linear system of equations of the form, e.g., in one dimension,

$$\mathbf{U}_t + \mathbf{A} \mathbf{U}_x = \mathbf{D} \mathbf{U}_{xx}, \tag{2.6.9}$$

where $\mathbf{U}$ is a vector of solutions, $\mathbf{A}$ is a coefficient matrix having real eigenvalues, and $\mathbf{D}$ is a positive coefficient matrix. This includes the Navier-Stokes equations written in the quasi-linear form (see Section 4.16). Basically, it is simply the sum of the advection and diffusion equations. But to solve this numerically, we should be very careful if we want to simply add a diffusion scheme to an advection scheme. Such a scheme may fail in the intermediate case

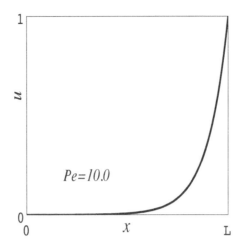

Figure 2.6.1: A typical steady solution of the advection-diffusion equation. $u(0) = 0$ and $u(L) = 1$.

where the advective effect and the diffusive effect are equally important. In fact, some high-order schemes are known to lose the formal accuracy due to incompatible discretizations of the advection and diffusion terms [24, 115, 125]. One important consideration in preserving accuracy is to treat the advection and diffusion terms in a unified manner. For this purpose, the conservative form (2.6.2) may be particularly useful in which the advection and diffusion terms are equally treated in the flux form. Naturally, then, the integral form,

$$\frac{d}{dt} \int_V u \, dV + \oint_S \left(u \boldsymbol{\lambda} \cdot \mathbf{n} - \nu \frac{\partial u}{\partial n} \right) dS = 0, \tag{2.6.10}$$

will be useful also.

Another difficulty, which may arise in advection-diffusion problems, is a rather drastic change in the character of the equation. Consider the one-dimensional advection-diffusion equation,

$$u_t + a u_x = \nu u_{xx}, \tag{2.6.11}$$

where $a > 0$, in $x \in (0, L)$. This is a typical parabolic problem, and therefore we must specify solution values at both boundaries: $u(0)$ and $u(L)$ which must be different for the solution to be non-trivial. However, if $\nu = 0$, this becomes a simple advection equation; we are allowed to specify only one value at the left boundary, i.e., $u(0)$. The solution at the right boundary $u(L)$ must be left unknown and determined by the wave coming all the way from the left boundary. Then, what happens in general ($\nu \neq 0$) is that the wave attempts to propagate the left boundary value $u(0)$ to the right, but cannot continue to the right boundary because $u(0) \neq u(L)$; the diffusion term resolves the discrepancy by smoothing the solution between $u(L)$ and the one transported by the wave from the left (see Figure 2.6.1). Certainly, if the advective effect dominates the diffusive effect, the influence of diffusion will be confined within a narrow region near the right boundary. There is a nice dimensionless parameter that characterizes this balance:

$$Pe = \frac{aL}{\nu}, \tag{2.6.12}$$

which is called the Peclet number. As $Pe \to 0$, the diffusion dominates; thus the narrow region gets broader and broader. On the other hand, as $Pe \to \infty$, the advection dominates; thus the narrow region gets extremely narrow (as if making it look like a pure advection problem by ignoring the right boundary condition $u(L)$). The corresponding parameter in the Navier-Stokes equations is the Reynolds number; the narrow region is observed near a body in a flow and called the boundary layer. Of course, we may as well ignore such a layer by discarding the diffusion term completely (solve the Euler equations instead of the Navier-Stokes equations). But if we really need to resolve the layer (e.g., need to compute the viscous drag or the heat transfer on a body), we will have to retain the diffusion term and confront a variety of issues in computing such a solution in terms of both accuracy and efficiency.

I like the advection-diffusion equation. It is a very simple linear scalar equation, but contains essential difficulties that we will encounter when solving the Navier-Stokes equations for practical problems. It is useful not only for illustrating such difficulties in a simpler setting, but also for developing a new method for the Navier-Stokes equations.

2.7   Advection-Reaction Equation

The advection-reaction equation is given by

$$u_t + (\text{grad } u) \cdot \boldsymbol{\lambda} = -ku. \tag{2.7.1}$$

In Cartesian coordinates, it is written as

$$\begin{array}{llr}
\text{1D:} & u_t + au_x = -ku, & (2.7.2) \\
\text{2D:} & u_t + au_x + bu_y = -ku, & (2.7.3) \\
\text{3D:} & u_t + au_x + bu_y + cu_z = -ku. & (2.7.4)
\end{array}$$

This equation models the transport of a material in a flow with a constant speed $\boldsymbol{\lambda}$, decaying at the rate of k. This is a typical model equation for conservation laws with a source term:

$$u_t + \text{div } \mathbf{f} = s, \tag{2.7.5}$$

where $\mathbf{f} = u\boldsymbol{\lambda}$ and $s = -ku$. Basically I like it, but not with large k because it makes the equation stiff and very difficult to solve numerically.

2.8   Poisson/Laplace Equations

The Poisson equation:

$$\text{div grad } u = f, \tag{2.8.1}$$

where f is a given function of space. If $f = 0$, it is called the Laplace equation. In Cartesian coordinates, the Poisson equation is written as

$$\begin{array}{llr}
\text{1D} & u_{xx} = f(x), & (2.8.2) \\
\text{2D} & u_{xx} + u_{yy} = f(x,y), & (2.8.3) \\
\text{3D} & u_{xx} + u_{yy} + u_{zz} = f(x,y,z). & (2.8.4)
\end{array}$$

I like the Laplace equation because its solution is very smooth. In particular, it is well known that the minimum and the maximum of the solution can occur only on the boundary (maaaximum principle), implying a smooth solution in the interior. On the other hand, solutions of the Poisson equation could be non-smooth, depending on the source term. Anyway, in both cases, the operator is the Laplacian which is very easy to discretize on regular grids: a simple central differencing works fine. Although typically it takes a long time to obtain a numerical solution (expensive direct methods or a large number of iterations), the multigrid technique works extremely well for this type of equation [17]. Well, if the grid is not regular and/or high-order accuracy is required, that is a different story as mentioned in Section 2.3.

2.9   Cauchy-Riemann Equations

2.9.1   Velocity Components

In incompressible irrotational flows, the velocity components (u, v) satisfy the Cauchy-Riemann equations:

$$\begin{array}{rclll}
u_x + v_y & = & 0, & \text{for incompressibility,} & (2.9.1) \\
v_x - u_y & = & 0, & \text{for irrotationality.} & (2.9.2)
\end{array}$$

Write the system in the vector form,

$$\mathbf{A}U_x + \mathbf{B}U_y = 0, \tag{2.9.3}$$

where

$$\mathbf{U} = \begin{bmatrix} u \\ v \end{bmatrix}, \quad \mathbf{A} = \begin{bmatrix} 1 & 0 \\ 0 & -1 \end{bmatrix}, \quad \mathbf{B} = \begin{bmatrix} 0 & 1 \\ 1 & 0 \end{bmatrix}. \tag{2.9.4}$$

Then, $\mathbf{A}$ is invertible, and we find the eigenvalues and eigenvectors of $\mathbf{A}^{-1}\mathbf{B}$:

$$\lambda_1 = i, \quad \mathbf{r}_1 = \begin{bmatrix} 1 \\ i \end{bmatrix}, \quad \lambda_2 = -i, \quad \mathbf{r}_2 = \begin{bmatrix} 1 \\ -i \end{bmatrix}, \tag{2.9.5}$$

where $i = \sqrt{-1}$. Therefore, this system is elliptic. If, in addition, u and v are twice differentiable, they satisfy the Laplace equation independently,

$$u_{xx} + u_{yy} = 0, \tag{2.9.6}$$
$$v_{xx} + v_{yy} = 0. \tag{2.9.7}$$

Another interesting fact is that the contours of u and v are orthogonal to each other. To see this, observe that the dot product of the gradients vanishes by the Cauchy-Riemann equations:

$$(\mathrm{grad}\,u) \cdot (\mathrm{grad}\,v) = u_x v_x + u_y v_y = (-v_y)v_x + (v_x)v_y = 0. \tag{2.9.8}$$

I like the Cauchy-Riemann equations because they are basically the same as a pair of Laplace's equations. Discretize the Cauchy-Riemann equations by any method, you will often find yourself solving the associated Laplace equations in effect [61, 107]. Note that solutions of the Laplace equations do not necessarily satisfy the Cauchy-Riemann equations. For example, the solution, $(u, v) = r(\sin\theta, -\cos\theta)$, which represents a sold body rotation, satisfies the Laplace equations, but does not satisfy the Cauchy-Riemann equations (the vorticity does not vanish). So, if you really want to solve the Cauchy-Riemann equations, you need a numerical method that really solves the Cauchy-Riemann equations [105, 107].

The Cauchy-Riemann system is known as a model for the elliptic subsystem of the Euler equations [93, 123] (the acoustic subsystem for subsonic flows). In fact, it is a model for all elliptic subsystems of a general two-dimensional differential system. That is, any two-dimensional system of partial differential equations can be decomposed into a set of advection equations and/or a set of Cauchy-Riemann systems. This is very interesting. See Ref.[117] for details.

2.9.2 Stream Function and Velocity Potential

The stream function ψ and the velocity potential ϕ are defined by

$$u = \phi_x = \psi_y, \tag{2.9.9}$$
$$v = \phi_y = -\psi_x. \tag{2.9.10}$$

Then, it is obvious that ψ and ϕ satisfy the following Cauchy-Riemann equations,

$$\psi_x + \phi_y = 0, \tag{2.9.11}$$
$$\phi_x - \psi_y = 0. \tag{2.9.12}$$

If, in addition, ψ and ϕ are twice differentiable, they satisfy the Laplace equation independently,

$$\phi_{xx} + \phi_{yy} = 0, \tag{2.9.13}$$
$$\psi_{xx} + \psi_{yy} = 0. \tag{2.9.14}$$

I like the stream function because the boundary condition is made very simple, i.e., constant along a solid body. On the other hand, the velocity potential can be tricky because it may not be continuous for multiply-connected regions such as a flow around a lifting airfoil (see Ref.[68], for example). Of course, the contours of ψ and ϕ, or equivalently the streamlines and the equi-potential lines, are orthogonal at their points of intersections because $(\mathrm{grad}\,\psi) \cdot (\mathrm{grad}\,\phi) = 0$.

2.9.3 Vorticity and Pressure

In Stokes' flows, where the viscous terms balance with the pressure gradient, the vorticity and the pressure satisfy the Cauchy-Riemann equations (see Section 4.25):

$$p_x + \mu\,\omega_y = 0, \tag{2.9.15}$$
$$p_y - \mu\,\omega_x = 0, \tag{2.9.16}$$

where p is the pressure, $\omega = v_x - u_y$ is the vorticity, and μ is the viscosity. I like it very much because the theory of complex variables is directly applicable and a variety of exact solutions can be found for viscous flows just like for potential flows [54].

2.9.4   Time-Dependent Systems

Augment the Cauchy-Riemann equations with time-derivatives as follows:

$$u_t + u_x + v_y = 0, \tag{2.9.17}$$
$$v_t - v_x + u_y = 0, \tag{2.9.18}$$

which may be written in the vector form,

$$\mathbf{U}_t + \mathbf{A}\mathbf{U}_x + \mathbf{B}\mathbf{U}_y = 0, \tag{2.9.19}$$

where

$$\mathbf{U} = \begin{bmatrix} u \\ v \end{bmatrix}, \quad \mathbf{A} = \begin{bmatrix} 1 & 0 \\ 0 & -1 \end{bmatrix}, \quad \mathbf{B} = \begin{bmatrix} 0 & 1 \\ 1 & 0 \end{bmatrix}. \tag{2.9.20}$$

I like this system because it is no longer the Cauchy-Riemann equations, but equivalent to the Cauchy-Riemann equations in the steady state. So, I can compute a solution of the Cauchy-Riemann equations by integrating this in time until the solution stops changing [93, 134]. Note that the time terms have been added in such a way that the resulting system is stable [93]. Also note that the eigenvalues and eigenvectors of the projected Jacobian $\mathbf{A}_n = \mathbf{A}n_x + \mathbf{B}n_y$, where (n_x, n_y) is an arbitrary unit vector, are given by

$$\lambda_1 = 1, \quad \mathbf{r}_1 = \begin{bmatrix} n_x + 1 \\ n_y \end{bmatrix}, \quad \lambda_2 = -1, \quad \mathbf{r}_2 = \begin{bmatrix} n_x - 1 \\ n_y \end{bmatrix}. \tag{2.9.21}$$

Therefore, this system is hyperbolic in time (although elliptic in space). On the other hand, it is interesting that if we add time-derivatives to the Laplace equations,

$$u_t = u_{xx} + u_{yy}, \tag{2.9.22}$$
$$v_t = v_{xx} + v_{yy}, \tag{2.9.23}$$

these are not hyperbolic but parabolic in time. If we wish to have hyperbolic equations, we must introduce the second-derivatives, u_{tt} and v_{tt}, thus resulting in the wave equations:

$$u_{tt} = u_{xx} + u_{yy}, \tag{2.9.24}$$
$$v_{tt} = v_{xx} + v_{yy}. \tag{2.9.25}$$

These wave equations can be derived also from the Cauchy-Riemann system (2.9.17) and (2.9.18). In either case, i.e., parabolic or hyperbolic, these equations are equivalent to the Laplace equations in their steady states. But I definitely like the hyperbolic equations better since a simple explicit scheme will work very well (with a large time step) to quickly reach a steady state.

2.10   Biharmonic Equation

The biharmonic equation is a fourth-order partial differential equation given by

$$\operatorname{div}\operatorname{grad}(\operatorname{div}\operatorname{grad} u) = f, \tag{2.10.1}$$

or equivalently by

$$\Delta^2 u = f, \tag{2.10.2}$$

where f is a given function of space. In Cartesian coordinates, the biharmonic equation is written as

1D	$u_{xxxx} = f(x),$	(2.10.3)
2D	$u_{xxxx} + u_{yyyy} + 2u_{xxyy} = f(x, y),$	(2.10.4)
3D	$u_{xxxx} + u_{yyyy} + u_{zzzz} + 2u_{xxyy} + 2u_{yyzz} + 2u_{zzxx} = f(x, y, z).$	(2.10.5)

I like the biharmonic equation because it does arise in CFD. Stokes' equations, which govern very low Reynolds number flows, formulated by the stream function and the vorticity, involve the biharmonic equation for the stream function. The biharmonic equation is typically solved with the following boundary conditions:

$$u = g_1(x, y, z), \quad \frac{\partial u}{\partial n} = g_2(x, y, z), \tag{2.10.6}$$

where g_1 and g_2 are given functions, and $\partial u / \partial n$ is the derivative in the direction normal to the boundary. For two-dimensional Stokes' flow, u corresponds to the stream function. Then, the first boundary condition may be specified with $g_1 = 0$ to for a solid wall boundary, meaning that the wall is a streamline. The second boundary condition specifies the velocity tangential to the boundary, and thus the no-slip condition can be specified with $g_2 = 0$ at the wall.

Discretization of fourth-order derivatives may be straightforward on a regular computational grid, but not so straightforward on irregular grids. An interesting approach is to write the biharmonic equation as a pair of the Laplace equations [160]:

$$\text{div grad } u = v, \tag{2.10.7}$$
$$\text{div grad } v = f, \tag{2.10.8}$$

where v is a new variable. It is easy to verify that these equations are equivalent to the biharmonic equation for u. Then, the biharmonic equation can be solved by any well-known method for the Laplace equation, e.g., multigrid methods. I like it. It is always nice to reduce the order of partial differential equations. However, the boundary condition for the new variable v is not known and may be difficult to apply in some applications [2]. Nevertheless, the number of boundary conditions required is basically two [160]. For Stokes' flows, the boundary conditions as described above will suffice and the boundary condition is not necessary for v. See Ref.[160] for details.

2.11   Euler-Tricomi Equation

Euler-Tricomi equation,

$$y u_{xx} + u_{yy} + u_{zz} = 0, \tag{2.11.1}$$

is an interesting model equation often used in the study of numerical methods for transonic flows. Some people call it the Tricomi equation, and I don't know exactly why they ignore Euler. Anyway, it is an interesting equation because it changes its type depending on y: it is hyperbolic where $y < 0$ and elliptic where $y > 0$. It thus models transonic flows, which is hyperbolic when the local velocity exceeds the speed of sound and elliptic when the velocity is lower than the speed of sound (see Section 3.17 and also Section 3.20). The Euler-Tricomi equation actually develops steep waves in the hyperbolic region and yields a smooth solution in the elliptic region. This model has been employed for testing a numerical algorithm designed for transonic flow calculations [58, 137]. To deal with the mixed behavior, a mixed numerical algorithm (i.e., central-upwind) is employed in Ref.[58] whereas the equation is made hyperbolic in pseudo time and solved by an upwind method in Ref.[137]. The latter seems to be a popular approach as typically done for the Euler equations, which are physically hyperbolic in time but mixed hyperbolic-elliptic in space as shown in Section 3.17.

2.12   Relaxation Model

I like relaxation models but I don't know much about them. See Refs.[62, 88] for details. As far as I know, the one-dimensional relaxation model is given by

$$u_t + v_x = 0, \tag{2.12.1}$$
$$v_t + a^2 u_x = (cu - v)/\tau, \tag{2.12.2}$$

where $\tau > 0, a^2 > c^2$. For $\tau \ll 1$, this system behaves like the advection-diffusion equation,

$$u_t + c u_x = \tau(a^2 - c^2) u_{xx}. \tag{2.12.3}$$

This can be shown by the Chapman-Enskog expansion [81]. Note that this is a simple model for moment equations (derived from the Boltzmann equation) which behave like the Navier-Stokes equations in the limit of small relaxation time. The two-dimensional version is given by

$$u_t + v_x + w_y = 0, \tag{2.12.4}$$
$$v_t + a^2 u_x = -(v - ru)/\tau, \tag{2.12.5}$$
$$w_t + b^2 u_y = -(w - su)/\tau, \tag{2.12.6}$$
$$\tag{2.12.7}$$

where $\tau > 0, a^2 > r^2, b^2 > s^2$. For $\tau \ll 1$, this system behaves like the following advection-diffusion equation,

$$u_t + ru_x + su_y = \tau \left[(a^2 - r^2)u_{xx} - 2rsu_{xy} + (b^2 - s^2)u_{yy} \right]. \tag{2.12.8}$$

Note that it is more advantageous to solve the first-order relaxation system than the advection-diffusion equations because you can avoid any difficulties in dealing with mixed derivatives [165].

We may generalize the relaxation model, and write a system of nonlinear conservation laws,

$$\mathbf{U}_t + \mathbf{F}_x = 0, \tag{2.12.9}$$

as a *linear* system with a source term, by introducing a new set of variables $\mathbf{V}$,

$$\mathbf{U}_t + \mathbf{V}_x = 0, \tag{2.12.10}$$

$$\mathbf{V}_t + \mathbf{A}\mathbf{U}_x = \frac{\mathbf{F}(\mathbf{U}) - \mathbf{V}}{\tau}, \tag{2.12.11}$$

where $\mathbf{A}$ and τ are constants carefully chosen such that the solution to this system is equivalent to the original conservation law: the eigenvalues of $\mathbf{A}$ must satisfy the so-called sub-characteristic condition and τ must be very small (see Ref.[81] for details). I like to avoid nonlinearity in this way ($\mathbf{A}$ is now constant), but we now have a stiff source term instead (τ must be very small). So, it is difficult anyway. Methods which use this approach to avoid nonlinearity are called relaxation methods. See Ref.[81] for details.

2.13   Burgers' Equation

I like Burgers' equation (also known as inviscid Burgers' equation):

$$u_t + uu_x = 0. \tag{2.13.1}$$

This can be written in the conservative form,

$$u_t + \left(u^2/2 \right)_x = 0. \tag{2.13.2}$$

Then, we can derive a shock speed V_s by integrating this over a discontinuity across two states, u_L and u_R, i.e., from the Rankine-Hugoniot relation (1.22.1),

$$V_s = \frac{u_L + u_R}{2}. \tag{2.13.3}$$

Also, if we multiply Equation (2.13.2) by $2u$, then we get another conservation form,

$$(u^2)_t + \left(\frac{2}{3}u^{3/2} \right)_x = 0, \tag{2.13.4}$$

from which we find

$$V_s = \frac{2}{3} \left(u_L + u_R - \frac{u_L u_R}{u_L + u_R} \right). \tag{2.13.5}$$

So, the two conservation forms share the strong solutions (those that satisfy the differential form), but have different weak solutions (those that satisfy the integral form). In fact, it is natural that they have different weak solutions because the conserved quantities are different: Equation (2.13.2) conserves u while Equation (2.13.4) conserves u^2. I like this particular example because it tells us that it is very important to choose the right conservative variables so that we obtain physically meaningful solutions.

Two-dimensional version is given by

$$u_t + uu_x + vu_y = 0, \tag{2.13.6}$$
$$v_t + uv_x + vv_y = 0. \tag{2.13.7}$$

The following version is also often used,

$$u_t + uu_x + u_y = 0. \tag{2.13.8}$$

The latter is basically the same, at a steady state, as the one-dimensional equation (2.13.1), with y taken as a time-like axis. There is also the three-dimensional version:

$$u_t + uu_x + vu_y + wu_z = 0, \tag{2.13.9}$$
$$v_t + uv_x + vv_y + wv_z = 0, \tag{2.13.10}$$
$$w_t + uw_x + vw_y + ww_z = 0. \tag{2.13.11}$$

Incidentally, some people think that it is Burger's equation. This is wrong. It is Burgers' equation because it is the equation of Burgers, not of Burger.

2.14   Traffic Equations

The traffic equation is given by

$$\rho_t + (\rho v)_x = 0, \tag{2.14.1}$$

where ρ is the density of cars traveling on the road (cars/distance) and v is a local speed of a car. It is reasonable to assume that v is a function of ρ, e.g.,

$$\frac{v}{v_{max}} = 1 - \frac{\rho}{\rho_{max}}, \tag{2.14.2}$$

which means that the speed goes up if there are less cars while the speed goes down if there are more cars. Then, the characteristic speed is given by

$$\frac{\partial f}{\partial \rho} = \frac{\partial(\rho v)}{\partial \rho} = v_{max}\left(1 - \frac{2\rho}{\rho_{max}}\right). \tag{2.14.3}$$

The shock speed V_s can also be found, by the Rankine-Hugoniot relation (1.22.1), as

$$V_s = \frac{f_R - f_L}{\rho_R - \rho_L} = v_{max}\left(1 - \frac{\rho_R + \rho_L}{\rho_{max}}\right). \tag{2.14.4}$$

I like the traffic equation because the relation between the speed and the car density, such as Equation (2.14.2), is required for the traffic equation to be closed but can be determined in any way we want. Various forms may be proposed for modeling a real traffic flow. But remember that we would have to derive the characteristic and shock speeds for each relation we choose. For example, why don't you take this relation,

$$\frac{v}{v_{max}} = 1 - \frac{\rho^2}{\rho_{max}^2}, \tag{2.14.5}$$

and derive the characteristic and shock speeds? See Ref.[81, 170] for more details on the traffic equation.

2.15   Viscous Burgers' Equations

Viscous Burgers' equation is given by

$$u_t + u u_x = \nu u_{xx}, \tag{2.15.1}$$

where ν is a positive constant. I like this equation because it is a simple model for the Navier-Stokes equations. Of course, I like the two-dimensional version also:

$$u_t + u u_x + v u_y = \nu \left(u_{xx} + u_{yy} \right), \tag{2.15.2}$$
$$v_t + u v_x + v v_y = \nu \left(v_{xx} + v_{yy} \right), \tag{2.15.3}$$

which is in fact the two-dimensional incompressible Navier-Stokes equations without the pressure gradient. I like this one also,

$$u_t + u u_x + u_y = \nu u_{xx}, \tag{2.15.4}$$

which will behave like the one-dimensional equation (with time-like y) at a steady state. Finally, of course, I like the three-dimensional version,

$$u_t + u u_x + v u_y + w u_z = \nu \left(u_{xx} + u_{yy} + u_{zz} \right), \tag{2.15.5}$$
$$v_t + u v_x + v v_y + w v_z = \nu \left(v_{xx} + v_{yy} + v_{zz} \right), \tag{2.15.6}$$
$$w_t + u w_x + v w_y + w w_z = \nu \left(w_{xx} + w_{yy} + w_{zz} \right). \tag{2.15.7}$$

One of the most interesting features of the viscous Burgers equation is that its solution can be constructed from a solution of the diffusion equation. This is very interesting, isn't it? See Section 6.4.1 for details.

Chapter 3

Euler Equations

3.1 Thermodynamic Relations

I like gases that obey the ideal gas law (also called the thermal equation of state):

$$p = \rho R T, \tag{3.1.1}$$

where p is the gas pressure, ρ is the gas density, T is the gas temperature, and R is the gas constant which is 287 N·m/kg for the sea-level air. Naturally, a gas that obeys this relation is called an ideal gas (or a thermally perfect gas). The ideal gas law implies that the state of a gas at a point in a flow is determined by any two of the three variables: the pressure p, the density ρ, and the temperature T. These variables are then called state variables: the variables which depend only on the state of gas and not on how it was changed from one state to another. It is interesting that this relation (3.1.1) has nothing to do with time. In a flow, all these variables will generally change in time. But their relation (3.1.1) holds at any instant of time at any location in the flow (no matter how violent the flow is). In other words, the gas is always *in equilibrium*. Of course, this may not be true in reality, but this is the basic assumption of thermodynamics, and in fact, it applies remarkably well to a wide range of flows, from low-speed to supersonic. That is why I like thermodynamics so much. It is very useful for fluid dynamics and CFD.

The internal energy of a gas is defined as the sum of all microscopic energies related to molecular motions (translation, rotation, vibration, etc). In this book, we use e to denote the specific internal energy ('specific' means per-unit-mass). The internal energy is generally a state variable and depends on two other state variables, e.g, T and p. But it is well known that for ideal gases it depends only on the temperature:

$$e = e(T). \tag{3.1.2}$$

Note now that from the equation of state that p/ρ also depends only on T. This makes us want to introduce another quantity called the specific enthalpy, denoted by h:

$$h \equiv e + \frac{p}{\rho}, \tag{3.1.3}$$

so that the enthalpy also depends only on the temperature,

$$h = h(T). \tag{3.1.4}$$

It is interesting that the first law of thermodynamics, which is a simple energy conservation law, can be expressed in terms of both e and h: with the specific volume v defined by $v = 1/\rho$,

$$dq = de + p\,dv, \tag{3.1.5}$$

or

$$dq = dh - \frac{1}{\rho}\,dp, \tag{3.1.6}$$

where dq is an added specific heat. I must mention here that both de and dh depend only on the state while dq depends on the process (how the heat is added), and that the work done, say dw, depends on the process also but this has been replaced by pdv, i.e., a reversible work done solely by the pressure (i.e., no viscous forces). Note that the above equations (3.1.5) and (3.1.6) are true for non-ideal gases also. Now, what is even more interesting is that if we define the constant-volume specific heat c_v and the constant-pressure specific heat c_p as the amounts of heat required to increase the temperature of unit mass by one degree at constant volume and pressure respectively,

$$c_v = \left(\frac{\partial q}{\partial T}\right)_v, \quad c_p = \left(\frac{\partial q}{\partial T}\right)_p, \tag{3.1.7}$$

then, it immediately follows from Equations (3.1.5) and (3.1.6) that

$$c_v = \left(\frac{\partial e}{\partial T}\right)_v, \quad c_p = \left(\frac{\partial h}{\partial T}\right)_p. \tag{3.1.8}$$

This is very interesting. It shows that c_v represents a change in e per unit change in T at constant volume while c_p represents a change in h per unit change in T at constant pressure. It is known from kinetic theory that for molecules with α degrees of freedom for translational and rotational energy, the specific heats are given by

$$c_v = \frac{\alpha}{2}R, \tag{3.1.9}$$

$$c_p = \left(1 + \frac{\alpha}{2}\right)R, \tag{3.1.10}$$

and thus their ratio is independent of R,

$$\gamma = \frac{c_p}{c_v} = \frac{\alpha + 2}{\alpha}, \tag{3.1.11}$$

where γ is called the ratio of specific heats (also called the adiabatic exponent). For monatomic gases, we have $\alpha = 3$ and hence $\gamma = 5/3$; for diatomic gases (e.g., air), we have $\alpha = 5$ and hence $\gamma = 7/5$. These results imply that γ is a constant. But this is not a general result; vibrational energy of molecules has not been taken into account. If the gas temperature is high enough to excite the vibrational energy, then it will generally depend on T.

For ideal gases, both e and h are functions of T only, and therefore, we obtain from Equation (3.1.5)

$$c_v = \left(\frac{\partial q}{\partial T}\right)_v = \frac{\partial e}{\partial T}, \tag{3.1.12}$$

$$c_p = \left(\frac{\partial q}{\partial T}\right)_p = \frac{\partial e}{\partial T} + p\left(\frac{\partial v}{\partial T}\right)_p = c_v + p\left(\frac{\partial v}{\partial T}\right)_p. \tag{3.1.13}$$

Since the last term above becomes R/p by the ideal gas law (3.1.1), we have the important relation:

$$c_p - c_v = R. \tag{3.1.14}$$

This is very nice. Now we can solve this with Equations (3.1.9), (3.1.10), and (3.1.11) for c_v and c_p to get

$$c_v = \frac{1}{\gamma - 1}R, \tag{3.1.15}$$

$$c_p = \frac{\gamma}{\gamma - 1}R. \tag{3.1.16}$$

For high temperature gases, γ may depend on T, and so are c_v and c_p. Then, by integrating Equation (3.1.8), which are now ordinary derivatives, we obtain

$$e(T) = \int c_v \, dT + \text{constant}, \tag{3.1.17}$$

$$h(T) = \int c_p \, dT + \text{constant}. \tag{3.1.18}$$

Yet, for a calorically perfect gas, which is defined as a gas with constant specific heats, we can write

$$e = c_v T = \frac{R}{\gamma - 1}T, \tag{3.1.19}$$

$$h = c_p T = \frac{\gamma R}{\gamma - 1}T. \tag{3.1.20}$$

This is very nice. For example, we can arrange the equation of state into the form,

$$p = (\gamma - 1)\rho e = (\gamma - 1)\left(\rho E - \frac{\rho \mathbf{v}^2}{2}\right), \tag{3.1.21}$$

where $\mathbf{v}$ is the velocity vector and E is the specific total energy,

$$E = \frac{1}{\gamma - 1}\frac{p}{\rho} + \frac{1}{2}\mathbf{v}^2. \tag{3.1.22}$$

Relation (3.1.21) is then called the equation of state for a calorically perfect gas (a calorically perfect gas is also called a polytropic gas [76, 81]). By the way, some people use E and H to denote the total energy and enthalpy: $E = \frac{p}{\gamma-1} + \frac{1}{2}\rho\mathbf{v}^2$ and $H = E + p$. Others use E to denote the specific total energy as in Equation (3.1.22) and H to denote the specific total enthalpy,

$$H = E + \frac{p}{\rho}. \tag{3.1.23}$$

I like the latter, but the former is fine too. They are just different definitions. I like not to discuss which is better. Throughout this book, I will use the latter simply because I like it.

There is another quantity that I like. Consider the following,

$$\frac{dq}{T} = \frac{de}{T} + p\frac{dv}{T}, \tag{3.1.24}$$

which is the first law (3.1.5) divided by the temperature. This can be written as, by the ideal gas law,

$$\frac{dq}{T} = c_v\frac{dT}{T} + \rho R d\left(\frac{1}{\rho}\right), \tag{3.1.25}$$

which further can be written as, for a calorically perfect gas, as

$$\begin{aligned}\frac{dq}{T} &= \frac{d(c_v T)}{T} - \frac{d(R\rho)}{\rho} \\ &= d\left(\ln T^{c_v} - \ln \rho^R\right).\end{aligned} \tag{3.1.26}$$

Therefore, $\frac{dq}{T}$ can be written as an exact differential; it is thus a state variable (although dq is not a state variable). This is defined as an entropy s:

$$ds \equiv \frac{dq}{T} = d\left(\ln T^{c_v} - \ln \rho^R\right). \tag{3.1.27}$$

This is nice but we must recall that this relation is subject to the assumption of reversible process. In general, e.g., for viscous flows with heat transfer, this is not true, and the second law of thermodynamics states that the entropy must increase:

$$ds = \frac{dq}{T} + ds', \tag{3.1.28}$$

where ds' denotes a contribution from irreversible processes which is non-negative ($ds' \geq 0$). It follows from this that for an adiabatic process we have

$$ds = ds' \geq 0, \tag{3.1.29}$$

and for an adiabatic and reversible process we have

$$ds = 0. \tag{3.1.30}$$

That is, the entropy is constant for an adiabatic and reversible process. Such a process is called isentropic. For an isentropic process, some very important relations can be derived. We have from Equation (3.1.27)

$$ds = c_v\frac{dT}{T} - R\frac{d\rho}{\rho}, \tag{3.1.31}$$

and thus by the ideal gas law ($dp/p = d\rho/\rho + dT/T$),

$$
\begin{aligned}
ds &= c_v \frac{dp}{p} - c_p \frac{d\rho}{\rho} \\
&= c_p \frac{dT}{T} - R \frac{dp}{p}.
\end{aligned}
$$

(3.1.32)

For a calorically perfect gas, we can write these as

$$
\begin{aligned}
\frac{\gamma - 1}{R} ds &= \frac{dT}{T} - (\gamma - 1) \frac{d\rho}{\rho} \\
&= \frac{dp}{p} - \gamma \frac{d\rho}{\rho} \\
&= \gamma \frac{dT}{T} - (\gamma - 1) \frac{dp}{p},
\end{aligned}
$$

(3.1.33)

and thus

$$
\begin{aligned}
\frac{\gamma - 1}{R} ds &= d\left(\ln \frac{T}{\rho^{\gamma-1}} \right) \\
&= d\left(\ln \frac{p}{\rho^{\gamma}} \right) \\
&= d\left(\ln \frac{T^{\gamma}}{p^{\gamma-1}} \right).
\end{aligned}
$$

(3.1.34)

Therefore, we obtain the following important relations for an isentropic process ($ds = 0$):

$$
\frac{T}{\rho^{\gamma-1}} = \text{constant,}
$$

(3.1.35)

$$
\frac{p}{\rho^{\gamma}} = \text{constant,}
$$

(3.1.36)

$$
\frac{T^{\gamma}}{p^{\gamma-1}} = \text{constant,}
$$

(3.1.37)

where constants on the right hand sides are, of course, not all the same. These relations are generally called isentropic relations or adiabatic relations ($dq = 0$). They are very useful for a wide range of applications because many flows can be considered as adiabatic and reversible. For example, a supersonic flow over an airfoil can be adiabatic and reversible except inside possible shock waves or inside a thin viscous/thermal boundary layer on the airfoil. Even for a curved shock wave across which the entropy increases by different amount for different streamlines, the flow is isentropic along each streamline (except for the jump across the shock wave). Note that in such a case the constants in these relations differ from one streamline to another.

Overall, I like a calorically perfect gas since it has actually been used quite successfully for a wide range of flows. So, in the rest of this book, I will simply call it a perfect gas.

3.2   Speed of Sound

The speed of sound is a speed at which a disturbance produced by a sound wave travels in a gas. The disturbance is generally very small (a sound wave is not at all as strong as a shock wave), so that the process is considered as isentropic. The speed of sound a is then given by

$$
a^2 = \left(\frac{\partial p}{\partial \rho} \right)_s.
$$

(3.2.1)

For a perfect gas, we have the isentropic relation $p/\rho^{\gamma} = \text{constant}$, and thus

$$
a^2 = \gamma \frac{p}{\rho}.
$$

(3.2.2)

Using the equation of state, $p = \rho R T$, we can write this also in terms of T,

$$a^2 = \gamma R T. \tag{3.2.3}$$

Generally, I like to use a to denote the speed of sound, but sometimes I will instead use c to avoid a possible conflict with other variables. But in the special case that the speed of sound is constant everywhere in a flow field (e.g., isothermal flows), I think I would rather use c than a since it looks like standing for 'constant'.

3.3 Euler Equations

Conservative Form

The conservative form of the Euler equations:

$$\partial_t \rho + \operatorname{div}(\rho \mathbf{v}) = 0, \tag{3.3.1}$$
$$\partial_t (\rho \mathbf{v}) + \operatorname{div}(\rho \mathbf{v} \otimes \mathbf{v}) + \operatorname{grad} p = 0, \tag{3.3.2}$$
$$\partial_t (\rho E) + \operatorname{div}(\rho \mathbf{v} H) = 0. \tag{3.3.3}$$

Some people claim that only the momentum equation should be called the Euler equation because Leonhard Euler actually derived the momentum equation only. But usually all these equations (momentum + continuity + energy) are called the Euler equations. Either way, I like the Euler equations.

Primitive Form

The primitive form of the Euler equations:

$$\frac{D\rho}{Dt} + \rho \operatorname{div} \mathbf{v} = 0, \tag{3.3.4}$$

$$\rho \frac{D\mathbf{v}}{Dt} + \operatorname{grad} p = 0, \tag{3.3.5}$$

$$\frac{Dp}{Dt} + \rho c^2 \operatorname{div} \mathbf{v} = 0. \tag{3.3.6}$$

Other Energy Equations

I like energy equations because the energy equations, (3.3.3) and (3.3.6), may be replaced by any of the followings:

$$\frac{\partial(\rho H)}{\partial t} - \frac{\partial p}{\partial t} + \operatorname{div}(\rho \mathbf{v} H) = 0, \tag{3.3.7}$$

$$\rho \frac{DE}{Dt} + \operatorname{div}(p\mathbf{v}) = 0, \tag{3.3.8}$$

$$\rho \frac{De}{Dt} + p \operatorname{div} \mathbf{v} = 0, \tag{3.3.9}$$

$$\rho \frac{Dh}{Dt} - \frac{Dp}{Dt} = 0, \tag{3.3.10}$$

$$\rho \frac{DH}{Dt} - \frac{\partial p}{\partial t} = 0, \tag{3.3.11}$$

$$\rho \frac{D}{Dt} \left(\frac{\mathbf{v}^2}{2} \right) + (\operatorname{grad} p) \cdot \mathbf{v} = 0, \tag{3.3.12}$$

or even by the entropy equation. From Equation (3.1.27), we have

$$ds = c_v \frac{dT}{T} - R\frac{d\rho}{\rho},$$ (3.3.13)

which can be written for a perfect gas as

$$ds = \frac{de}{T} - \frac{p}{T\rho^2}d\rho.$$ (3.3.14)

Therefore, we obtain from Equations (3.3.9) and (3.3.4)

$$\rho\frac{Ds}{Dt} = 0.$$ (3.3.15)

According to the second law of thermodynamics, this should be interpreted as an inequality in general:

$$\rho\frac{Ds}{Dt} \geq 0.$$ (3.3.16)

3.4 1D Euler Equations

3.4.1 Conservative Form

$$\mathbf{U}_t + \mathbf{F}_x = 0,$$ (3.4.1)

where

$$\mathbf{U} = \begin{bmatrix} \rho \\ \rho u \\ \rho E \end{bmatrix}, \quad \mathbf{F} = \begin{bmatrix} \rho u \\ \rho u^2 + p \\ \rho u H \end{bmatrix}.$$ (3.4.2)

Flux Jacobian

$$\mathbf{A} = \frac{\partial \mathbf{F}}{\partial \mathbf{U}} = \begin{bmatrix} 0 & 1 & 0 \\ (\gamma-3)\frac{u^2}{2} & (3-\gamma)u & \gamma-1 \\ \left(\frac{\gamma-1}{2}u^2 - H\right)u & H+(1-\gamma)u^2 & \gamma u \end{bmatrix}.$$ (3.4.3)

Eigenstructure

$$\mathbf{A} = \mathbf{R}\Lambda\mathbf{L},$$ (3.4.4)

where

$$\Lambda = \begin{bmatrix} u-c & 0 & 0 \\ 0 & u & 0 \\ 0 & 0 & u+c \end{bmatrix},$$ (3.4.5)

$$\mathbf{R} = \begin{bmatrix} 1 & 1 & 1 \\ u-c & u & u+c \\ H-uc & u^2/2 & H+uc \end{bmatrix},$$ (3.4.6)

$$
\mathbf{L} = \begin{bmatrix} \frac{1}{2}\left(\frac{\gamma-1}{2c^2}u^2 + \frac{u}{c}\right) & -\frac{1}{2}\left(\frac{\gamma-1}{c^2}u + \frac{1}{c}\right) & \frac{\gamma-1}{2c^2} \\ 1 - \frac{\gamma-1}{2c^2}u^2 & \frac{\gamma-1}{c^2}u & -\frac{\gamma-1}{c^2} \\ \frac{1}{2}\left(\frac{\gamma-1}{2c^2}u^2 - \frac{u}{c}\right) & -\frac{1}{2}\left(\frac{\gamma-1}{c^2}u - \frac{1}{c}\right) & \frac{\gamma-1}{2c^2} \end{bmatrix},
\tag{3.4.7}
$$

$$
\mathbf{L}d\mathbf{U} = \begin{bmatrix} \dfrac{dp - \rho c\, du}{2c^2} \\ -\dfrac{dp - c^2\, d\rho}{c^2} \\ \dfrac{dp + \rho c\, du}{2c^2} \end{bmatrix}.
\tag{3.4.8}
$$

It is interesting to note that the quantity $\frac{dp-c^2\,d\rho}{c^2}$ corresponds to the entropy change ds and the corresponding right eigenvector indicates (as discussed in Section 1.16.2) how the conservative variables change due to this entropy change,

$$
\begin{bmatrix} \partial\rho \\ \partial(\rho u) \\ \partial(\rho E) \end{bmatrix} \propto \begin{bmatrix} 1 \\ u \\ u^2/2 \end{bmatrix}.
\tag{3.4.9}
$$

This is often called an entropy wave. The associated eigenvalue u gives the speed at which the entropy wave travels: it moves with the local fluid velocity. It is actually more interesting to look at it in the primitive form where we see exactly what variable is affected by the entropy change.

3.4.2 Primitive Form

$$
\mathbf{W}_t + \mathbf{A}^w \mathbf{W}_x = 0,
\tag{3.4.10}
$$

where

$$
\mathbf{W} = \begin{bmatrix} \rho \\ u \\ p \end{bmatrix}, \quad \mathbf{A}^w = \begin{bmatrix} u & \rho & 0 \\ 0 & u & 1/\rho \\ 0 & \rho c^2 & u \end{bmatrix}.
\tag{3.4.11}
$$

Incidentally, I wouldn't call $\mathbf{A}^w$ the primitive Jacobian. This is because $\mathbf{A}^w$ is not a Jacobian [166]: there does not exist a (flux) vector $\mathbf{H}$ such that $\partial\mathbf{H}/\partial\mathbf{W} = \mathbf{A}^w$. So, I simply call $\mathbf{A}^w$ the coefficient matrix of the primitive form of the Euler equations.

Eigenstructure

$$
\mathbf{A}^w = \mathbf{R}^w \mathbf{\Lambda} \mathbf{L}^w,
\tag{3.4.12}
$$

where

$$
\mathbf{\Lambda} = \begin{bmatrix} u-c & 0 & 0 \\ 0 & u & 0 \\ 0 & 0 & u+c \end{bmatrix}, \quad \mathbf{R}^w = \begin{bmatrix} -\dfrac{\rho}{2c} & 1 & \dfrac{\rho}{2c} \\ \dfrac{1}{2} & 0 & \dfrac{1}{2} \\ -\dfrac{\rho c}{2} & 0 & \dfrac{\rho c}{2} \end{bmatrix},
\tag{3.4.13}
$$

$$
\mathbf{L}^w = \begin{bmatrix} 0 & 1 & -\dfrac{1}{\rho c} \\ 1 & 0 & -\dfrac{1}{c^2} \\ 0 & 1 & \dfrac{1}{\rho c} \end{bmatrix}, \quad \mathbf{L}^w d\mathbf{U} = \begin{bmatrix} du - \dfrac{dp}{\rho c} \\ d\rho - \dfrac{dp}{c^2} \\ du + \dfrac{dp}{\rho c} \end{bmatrix}.
\tag{3.4.14}
$$

The second component above represents the entropy wave, and we thus find

$$
\begin{bmatrix} \partial \rho \\ \partial u \\ \partial p \end{bmatrix} \propto \begin{bmatrix} 1 \\ 0 \\ 0 \end{bmatrix}. \tag{3.4.15}
$$

Therefore, it is the density that changes due to the entropy wave; the velocity and the pressure are not affected. A typical example is a contact surface (or contact discontinuity) observed in shock tube problems: a surface across which two different gases are initially in contact moves with the local fluid speed carrying the density jump. In fact, the entropy wave is sometimes called the contact wave.

Change of Variables

$$
d\mathbf{U} = \frac{\partial \mathbf{U}}{\partial \mathbf{W}} d\mathbf{W}, \quad d\mathbf{W} = \frac{\partial \mathbf{W}}{\partial \mathbf{U}} d\mathbf{U}, \tag{3.4.16}
$$

where

$$
\frac{\partial \mathbf{U}}{\partial \mathbf{W}} = \begin{bmatrix} 1 & 0 & 0 \\ u & \rho & 0 \\ u^2/2 & \rho u & 1/(\gamma-1) \end{bmatrix}, \tag{3.4.17}
$$

$$
\frac{\partial \mathbf{W}}{\partial \mathbf{U}} = \left(\frac{\partial \mathbf{U}}{\partial \mathbf{W}} \right)^{-1} = \begin{bmatrix} 1 & 0 & 0 \\ -\dfrac{u}{\rho} & \dfrac{1}{\rho} & 0 \\ (\gamma-1)u^2/2 & -(\gamma-1)u & \gamma-1 \end{bmatrix}. \tag{3.4.18}
$$

Note that the coefficient matrix $\mathbf{A}^w$ is similar to the conservative Jacobian $\mathbf{A}$,

$$
\mathbf{A}^w = \frac{\partial \mathbf{W}}{\partial \mathbf{U}} \mathbf{A} \frac{\partial \mathbf{U}}{\partial \mathbf{W}}. \tag{3.4.19}
$$

Hence, the eigenvectors are related by

$$
\mathbf{L} = \mathbf{L}^w \frac{\partial \mathbf{W}}{\partial \mathbf{U}}, \quad \mathbf{R} = \frac{\partial \mathbf{U}}{\partial \mathbf{W}} \mathbf{R}^w, \tag{3.4.20}
$$

and the eigenvalues are the same as those of the conservative Jacobian because similar matrices share the same eigenvalues (see Section 1.11).

3.4.3 Characteristic Variables

Multiply the primitive form (3.4.10) by $\mathbf{L}^w$ from the left to get

$$
\mathbf{L}^w \mathbf{W}_t + \mathbf{L}^w \mathbf{A}^w \mathbf{W}_x = 0, \tag{3.4.21}
$$

$$
\mathbf{L}^w \mathbf{W}_t + \mathbf{L}^w \mathbf{A}^w \mathbf{R}^w \mathbf{L}^w \mathbf{W}_x = 0, \tag{3.4.22}
$$

$$
\mathbf{W}^{\mathbf{c}}{}_t + \mathbf{\Lambda} \mathbf{W}^{\mathbf{c}}{}_x = 0, \tag{3.4.23}
$$

where the new variables defined by

$$
\partial \mathbf{W}^{\mathbf{c}} = \mathbf{L}^w \partial \mathbf{W} = \begin{bmatrix} \partial u - \dfrac{1}{\rho c} \partial p \\ \partial \rho - \dfrac{1}{c^2} \partial p \\ \partial u + \dfrac{1}{\rho c} \partial p \end{bmatrix}, \tag{3.4.24}
$$

are called the characteristic variables, and the equation (3.4.23) is called the characteristic form. It is very nice that the system (3.4.23) is a completely diagonalized form of the Euler equations, so that we have a set of independent scalar nonlinear advection equations:

$$\left(\frac{\partial u}{\partial t} - \frac{1}{\rho c}\frac{\partial p}{\partial t}\right) + (u - c)\left(\frac{\partial u}{\partial x} - \frac{1}{\rho c}\frac{\partial p}{\partial x}\right) = 0, \tag{3.4.25}$$

$$\left(\frac{\partial \rho}{\partial t} - \frac{1}{c^2}\frac{\partial p}{\partial t}\right) + u\left(\frac{\partial \rho}{\partial x} - \frac{1}{c^2}\frac{\partial p}{\partial x}\right) = 0, \tag{3.4.26}$$

$$\left(\frac{\partial u}{\partial t} + \frac{1}{\rho c}\frac{\partial p}{\partial t}\right) + (u + c)\left(\frac{\partial u}{\partial x} + \frac{1}{\rho c}\frac{\partial p}{\partial x}\right) = 0. \tag{3.4.27}$$

The second equation corresponds to an entropy advection, and the other two correspond to an acoustic wave propagation. This shows that the entropy is simply convected along a streamline and not affected by the acoustic waves. In other words, the entropy is constant through the acoustic waves (unless they are strong enough to become shock waves). For adiabatic (isentropic) flows, the characteristic variables can be integrated:

$$\partial \mathbf{W}^c = \begin{bmatrix} \partial u - \dfrac{1}{\rho c}\partial p \\[2mm] \partial \rho - \dfrac{1}{c^2}\partial p \\[2mm] \partial u + \dfrac{1}{\rho c}\partial p \end{bmatrix} = \begin{bmatrix} \partial\left(u - \dfrac{2}{\gamma - 1}c\right) \\[2mm] \partial s \\[2mm] \partial\left(u + \dfrac{2}{\gamma - 1}c\right) \end{bmatrix}, \tag{3.4.28}$$

where the entropy s has been scaled by $-(\gamma - 1)\rho/(\gamma R)$. These integrated quantities are called Riemann invariants. We thus have

$$\frac{\partial}{\partial t}\left(u - \frac{2}{\gamma - 1}c\right) + (u - c)\frac{\partial}{\partial x}\left(u - \frac{2}{\gamma - 1}c\right) = 0, \tag{3.4.29}$$

$$\frac{\partial s}{\partial t} + u\frac{\partial s}{\partial x} = 0, \tag{3.4.30}$$

$$\frac{\partial}{\partial t}\left(u + \frac{2}{\gamma - 1}c\right) + (u + c)\frac{\partial}{\partial x}\left(u + \frac{2}{\gamma - 1}c\right) = 0. \tag{3.4.31}$$

This system is often used to implement a characteristic boundary condition in CFD [51]. Now, if we assume $\gamma = 3$, this system reduces to a set of Burgers' equations.

$$\frac{\partial}{\partial t}(u - c) + (u - c)\frac{\partial}{\partial x}(u - c) = 0, \tag{3.4.32}$$

$$\frac{\partial s}{\partial t} + u\frac{\partial s}{\partial x} = 0, \tag{3.4.33}$$

$$\frac{\partial}{\partial t}(u + c) + (u + c)\frac{\partial}{\partial x}(u + c) = 0. \tag{3.4.34}$$

On the other hand, if we assume a simple wave, i.e., for example,

$$u - \frac{2}{\gamma - 1}c = \text{constant}, \quad s = \text{constant}, \tag{3.4.35}$$

everywhere (alternatively, $u + \frac{2}{\gamma - 1}c = \text{constant}$ may be assumed), then it can be shown that the system of Equations (3.4.29), (3.4.30), and (3.4.31) reduces to the following single Burgers' equation:

$$\frac{\partial}{\partial t}(u + c) + (u + c)\frac{\partial}{\partial x}(u + c) = 0. \tag{3.4.36}$$

To show this, let

$$R_\infty^- = u - \frac{2}{\gamma - 1}c = \text{constant}, \tag{3.4.37}$$

and solve for c,

$$c = \frac{\gamma - 1}{2}(u - R_\infty^-). \tag{3.4.38}$$

Then, we find

$$R^+ = u + \frac{2}{\gamma - 1}c = 2u - R_\infty^-, \tag{3.4.39}$$

$$u + c = \frac{\gamma + 1}{2}u - \frac{\gamma - 1}{2}R_\infty^-, \tag{3.4.40}$$

and derive the relation,

$$R^+ = \frac{4}{\gamma + 1}\left[(u + c) + \frac{\gamma - 3}{4}R_\infty^-\right]. \tag{3.4.41}$$

Finally, substitute this into Equation (3.4.31) and multiply the result by $\frac{\gamma+1}{4}$ to arrive at Equation (3.4.36).

After all, I like the one-dimensional Euler system because it can be diagonalized completely and thus written as a set of independent scalar advection equations as shown above. This is very nice but difficult in higher dimensions.

3.4.4 Parameter Vector

I like the parameter vector $\mathbf{Z}$ [128],

$$\mathbf{Z} = \sqrt{\rho}\begin{bmatrix} 1 \\ u \\ H \end{bmatrix} = \begin{bmatrix} z_1 \\ z_2 \\ z_3 \end{bmatrix}, \tag{3.4.42}$$

because every component of conservative variables and fluxes can be expressed as quadratics in the components of $\mathbf{Z}$,

$$\mathbf{U} = \begin{bmatrix} \rho \\ \rho u \\ \rho E \end{bmatrix} = \begin{bmatrix} z_1^2 \\ z_1 z_2 \\ \dfrac{z_1 z_3}{\gamma} + \dfrac{\gamma - 1}{\gamma}\dfrac{z_2^2}{2} \end{bmatrix}, \tag{3.4.43}$$

$$\mathbf{F} = \begin{bmatrix} \rho u \\ \rho u^2 + p \\ \rho u H \end{bmatrix} = \begin{bmatrix} z_1 z_2 \\ z_2^2 + p \\ z_2 z_3 \end{bmatrix}, \tag{3.4.44}$$

$$p = \frac{\gamma - 1}{\gamma}\left(z_1 z_3 - \frac{z_2^2}{2}\right). \tag{3.4.45}$$

This is very useful, for instance, for linearizing the Euler equations [128].

3.5 2D Euler Equations

3.5.1 Conservative Form

$$\mathbf{U}_t + \mathbf{F}_x + \mathbf{G}_y = 0, \tag{3.5.1}$$

where

$$\mathbf{U} = \begin{bmatrix} \rho \\ \rho u \\ \rho v \\ \rho E \end{bmatrix}, \quad \mathbf{F} = \begin{bmatrix} \rho u \\ \rho u^2 + p \\ \rho u v \\ \rho u H \end{bmatrix}, \quad \mathbf{G} = \begin{bmatrix} \rho v \\ \rho u v \\ \rho v^2 + p \\ \rho v H \end{bmatrix}. \tag{3.5.2}$$

Normal Flux

Projection of the flux in the direction of $\mathbf{n} = [n_x, n_y]^t$:

$$\mathbf{F}_n = [\mathbf{F}, \mathbf{G}] \cdot \mathbf{n} = \mathbf{F}n_x + \mathbf{G}n_y = \begin{bmatrix} \rho q_n \\ \rho q_n u + p n_x \\ \rho q_n v + p n_y \\ \rho q_n H \end{bmatrix}, \tag{3.5.3}$$

where

$$q_n = u n_x + v n_y. \tag{3.5.4}$$

Jacobians

$$\mathbf{A} = \frac{\partial \mathbf{F}}{\partial \mathbf{U}} = \begin{bmatrix} 0 & 1 & 0 & 0 \\ (\gamma-1)\frac{q^2}{2} - u^2 & (3-\gamma)u & (1-\gamma)v & \gamma-1 \\ -uv & v & u & 0 \\ \left(\frac{\gamma-1}{2}q^2 - H\right)u & H + (1-\gamma)u^2 & (1-\gamma)uv & \gamma u \end{bmatrix}, \tag{3.5.5}$$

$$\mathbf{B} = \frac{\partial \mathbf{G}}{\partial \mathbf{U}} = \begin{bmatrix} 0 & 0 & 1 & 0 \\ -uv & v & u & 0 \\ (\gamma-1)\frac{q^2}{2} - v^2 & (1-\gamma)u & (3-\gamma)v & \gamma-1 \\ \left(\frac{\gamma-1}{2}q^2 - H\right)v & (1-\gamma)uv & H + (1-\gamma)v^2 & \gamma v \end{bmatrix}, \tag{3.5.6}$$

$$\begin{aligned}
\mathbf{A}_n &= \frac{\partial \mathbf{F}_n}{\partial \mathbf{U}} = \mathbf{A}n_x + \mathbf{B}n_y \\[2mm]
&= \begin{bmatrix} 0 & n_x & n_y & 0 \\ (\gamma-1)\frac{q^2}{2}n_x - uq_n & un_x - (\gamma-1)un_x + q_n & un_y - (\gamma-1)vn_x & (\gamma-1)n_x \\ (\gamma-1)\frac{q^2}{2}n_y - vq_n & vn_x - (\gamma-1)un_y & vn_y - (\gamma-1)vn_y + q_n & (\gamma-1)n_y \\ \left(\frac{\gamma-1}{2}q^2 - H\right)q_n & Hn_x - (\gamma-1)uq_n & Hn_y - (\gamma-1)vq_n & \gamma q_n \end{bmatrix} \\[2mm]
&= \begin{bmatrix} 0 & \mathbf{n}^t & 0 \\ (\gamma-1)\frac{q^2}{2}\mathbf{n} - q_n\mathbf{v} & \mathbf{v}\otimes\mathbf{n} + (1-\gamma)\mathbf{n}\otimes\mathbf{v} + q_n\mathbf{I} & (\gamma-1)\mathbf{n} \\ \left(\frac{\gamma-1}{2}q^2 - H\right)q_n & H\mathbf{n}^t - (1-\gamma)\mathbf{v}^t q_n & \gamma q_n \end{bmatrix},
\end{aligned} \tag{3.5.7}$$

where $q^2 = u^2 + v^2$, $\mathbf{v} = [u,v]^t$, and $\mathbf{I}$ is the 2×2 identify matrix. I like $\mathbf{A}_n$ very much because it is very convenient: simply by taking $\mathbf{n} = [1,0]^t$ and $[0,1]^t$, we obtain $\mathbf{A}$ and $\mathbf{B}$ respectively.

Eigenstructure

$$\mathbf{A}_n = \mathbf{R}_n \mathbf{\Lambda}_n \mathbf{L}_n, \tag{3.5.8}$$

where

$$
\mathbf{\Lambda}_n \;=\; \begin{bmatrix} q_n - c & 0 & 0 & 0 \\ 0 & q_n & 0 & 0 \\ 0 & 0 & q_n + c & 0 \\ 0 & 0 & 0 & q_n \end{bmatrix},
\tag{3.5.9}
$$

$$
\mathbf{R}_n \;=\; \begin{bmatrix} 1 & 1 & 1 & 0 \\ u - cn_x & u & u + cn_x & \ell_x \\ v - cn_y & v & v + cn_y & \ell_y \\ H - q_n c & q^2/2 & H + q_n c & q_\ell \end{bmatrix},
\tag{3.5.10}
$$

$$
\mathbf{L}_n = \begin{bmatrix}
\dfrac{1}{2}\left(\dfrac{\gamma-1}{2c^2}q^2 + \dfrac{q_n}{c}\right) & -\dfrac{1}{2}\left(\dfrac{\gamma-1}{c^2}u + \dfrac{n_x}{c}\right) & -\dfrac{1}{2}\left(\dfrac{\gamma-1}{c^2}v + \dfrac{n_y}{c}\right) & \dfrac{\gamma-1}{2c^2} \\[2.5ex]
1 - \dfrac{\gamma-1}{2c^2}q^2 & \dfrac{\gamma-1}{c^2}u & \dfrac{\gamma-1}{c^2}v & -\dfrac{\gamma-1}{c^2} \\[2.5ex]
\dfrac{1}{2}\left(\dfrac{\gamma-1}{2c^2}q^2 - \dfrac{q_n}{c}\right) & -\dfrac{1}{2}\left(\dfrac{\gamma-1}{c^2}u - \dfrac{n_x}{c}\right) & -\dfrac{1}{2}\left(\dfrac{\gamma-1}{c^2}v - \dfrac{n_y}{c}\right) & \dfrac{\gamma-1}{2c^2} \\[2.5ex]
-q_\ell & \ell_x & \ell_y & 0
\end{bmatrix},
\tag{3.5.11}
$$

$$
\mathbf{L}_n d\mathbf{U} \;=\; \begin{bmatrix} \dfrac{dp - \rho c\, dq_n}{2c^2} \\[2ex] -\dfrac{dp - c^2\, d\rho}{c^2} \\[2ex] \dfrac{dp + \rho c\, dq_n}{2c^2} \\[2ex] \rho\, dq_\ell \end{bmatrix},
\tag{3.5.12}
$$

where $[\ell_x, \ell_y]^t$ is a tangent vector (perpendicular to $\mathbf{n}$) such as

$$
\begin{bmatrix} \ell_x \\ \ell_y \end{bmatrix} = \begin{bmatrix} -n_y \\ n_x \end{bmatrix},
\tag{3.5.13}
$$

and q_ℓ is the velocity component in that direction,

$$
q_\ell \;=\; u\ell_x + v\ell_y.
\tag{3.5.14}
$$

I like $\mathbf{R}_n$ and $\mathbf{L}_n$ very much. Again, they are very convenient: simply by taking $\mathbf{n} = [1,0]^t$ and $[0,1]^t$, we obtain eigenvectors of $\mathbf{A}$ and $\mathbf{B}$. It is also nice that the normal Jacobian $\mathbf{A}_n$ can be diagonalized as

$$
\mathbf{\Lambda}_n = \mathbf{L}_n \mathbf{A}_n \mathbf{R}_n.
\tag{3.5.15}
$$

It would be even nicer if $\mathbf{A}$ and $\mathbf{B}$ could be diagonalized simultaneously so that the Euler equations could be written as a set of scalar advection equations. This was easy for the one-dimensional Euler equations, but it seems very difficult in two (and also three) dimensions. But it is still nice that the *steady* two-dimensional Euler equations can be fully diagonalized (see Section 3.17).

3.5.2 Primitive Form

$$
\mathbf{W}_t + \mathbf{A}^w \mathbf{W}_x + \mathbf{B}^w \mathbf{W}_y = 0,
\tag{3.5.16}
$$

where

$$
\mathbf{W} = \begin{bmatrix} \rho \\ u \\ v \\ p \end{bmatrix}, \quad \mathbf{A}^w = \begin{bmatrix} u & \rho & 0 & 0 \\ 0 & u & 0 & 1/\rho \\ 0 & 0 & u & 0 \\ 0 & \rho c^2 & 0 & u \end{bmatrix}, \quad \mathbf{B}^w = \begin{bmatrix} v & 0 & \rho & 0 \\ 0 & v & 0 & 0 \\ 0 & 0 & v & 1/\rho \\ 0 & 0 & \rho c^2 & v \end{bmatrix}.
\tag{3.5.17}
$$

Of course, as in Section 3.4.2, I wouldn't call $\mathbf{A}^w$ and $\mathbf{B}^w$ the primitive Jacobians because they are not Jacobians. I do like to call them the coefficient matrices of the primitive form of the Euler equations.

Eigenstructure

$$\mathbf{A}_n^w = \mathbf{R}_n^w \mathbf{\Lambda}_n \mathbf{L}_n^w, \tag{3.5.18}$$

where

$$\mathbf{A}_n^w = \begin{bmatrix} q_n & \rho n_x & \rho n_y & 0 \\ 0 & q_n & 0 & n_x/\rho \\ 0 & 0 & q_n & n_y/\rho \\ 0 & \rho c^2 n_x & \rho c^2 n_y & q_n \end{bmatrix}, \tag{3.5.19}$$

$$\mathbf{\Lambda}_n = \begin{bmatrix} q_n - c & 0 & 0 & 0 \\ 0 & q_n & 0 & 0 \\ 0 & 0 & q_n + c & 0 \\ 0 & 0 & 0 & q_n \end{bmatrix}, \tag{3.5.20}$$

$$\mathbf{R}_n^w = \begin{bmatrix} 1 & 1 & 1 & 0 \\ -\dfrac{n_x c}{\rho} & 0 & \dfrac{n_x c}{\rho} & \dfrac{\ell_x}{\rho} \\ -\dfrac{n_y c}{\rho} & 0 & \dfrac{n_y c}{\rho} & \dfrac{\ell_y}{\rho} \\ c^2 & 0 & c^2 & 0 \end{bmatrix}, \quad \mathbf{L}_n^w = \begin{bmatrix} 0 & -\dfrac{n_x \rho}{2c} & -\dfrac{n_y \rho}{2c} & \dfrac{1}{2c^2} \\ 1 & 0 & 0 & -\dfrac{1}{c^2} \\ 0 & \dfrac{n_x \rho}{2c} & \dfrac{n_y \rho}{2c} & \dfrac{1}{2c^2} \\ 0 & \rho\ell_x & \rho\ell_y & 0 \end{bmatrix}, \tag{3.5.21}$$

$$\mathbf{L}_n^w d\mathbf{W} = \begin{bmatrix} \dfrac{dp - \rho c\, dq_n}{2c^2} \\[2mm] -\dfrac{dp - c^2\, d\rho}{c^2} \\[2mm] \dfrac{dp + \rho c\, dq_n}{2c^2} \\[2mm] \rho\, dq_\ell \end{bmatrix}. \tag{3.5.22}$$

Note that the fourth component represents a wave which brings a change in the tangential velocity component. This is often called a shear wave or a vorticity wave (because such a velocity jump can be represented by a vortex sheet).

Change of Variables

$$d\mathbf{U} = \frac{\partial \mathbf{U}}{\partial \mathbf{W}} d\mathbf{W}, \quad d\mathbf{W} = \frac{\partial \mathbf{W}}{\partial \mathbf{U}} d\mathbf{U}, \tag{3.5.23}$$

where

$$\frac{\partial \mathbf{U}}{\partial \mathbf{W}} = \begin{bmatrix} 1 & 0 & 0 & 0 \\ u & \rho & 0 & 0 \\ v & 0 & \rho & 0 \\ q^2/2 & \rho u & \rho v & 1/(\gamma - 1) \end{bmatrix}, \tag{3.5.24}$$

$$\frac{\partial \mathbf{W}}{\partial \mathbf{U}} = \left(\frac{\partial \mathbf{U}}{\partial \mathbf{W}}\right)^{-1} = \begin{bmatrix} 1 & 0 & 0 & 0 \\ -\dfrac{u}{\rho} & \dfrac{1}{\rho} & 0 & 0 \\ -\dfrac{v}{\rho} & 0 & \dfrac{1}{\rho} & 0 \\ (\gamma-1)q^2/2 & -(\gamma-1)u & -(\gamma-1)v & \gamma-1 \end{bmatrix}, \tag{3.5.25}$$

$$\mathbf{A}_n^w = \frac{\partial \mathbf{W}}{\partial \mathbf{U}} \mathbf{A}_n \frac{\partial \mathbf{U}}{\partial \mathbf{W}}, \quad \mathbf{L}_n = \mathbf{L}_n^w \frac{\partial \mathbf{W}}{\partial \mathbf{U}}, \quad \mathbf{R}_n = \frac{\partial \mathbf{U}}{\partial \mathbf{W}} \mathbf{R}_n^w. \tag{3.5.26}$$

3.5.3   Symmetric Form

I like the symmetric form of the Euler equations because the coefficient matrices become very simple.

$$\mathbf{V}_t + \mathbf{A}^v \mathbf{V}_x + \mathbf{B}^v \mathbf{V}_y = 0, \tag{3.5.27}$$

where

$$d\mathbf{V} = \begin{bmatrix} \dfrac{dp}{\rho c} \\[2mm] du \\[2mm] dv \\[2mm] dp - a^2 d\rho \end{bmatrix}, \quad \mathbf{A}^v = \begin{bmatrix} u & c & 0 & 0 \\ c & u & 0 & 0 \\ 0 & 0 & u & 0 \\ 0 & 0 & 0 & u \end{bmatrix}, \quad \mathbf{B}^v = \begin{bmatrix} v & 0 & c & 0 \\ 0 & v & 0 & 0 \\ c & 0 & v & 0 \\ 0 & 0 & 0 & v \end{bmatrix}, \tag{3.5.28}$$

and $d\mathbf{V}$ is called symmetric variables (symmetric variables are not unique; see Refs.[10, 40] for other choices). Look at the matrices above. They are very sparse and simple. Moreover, I like the fourth equation which clearly shows how the entropy evolves:

$$s_t + u s_x + v s_y = 0, \tag{3.5.29}$$

where the entropy s has been scaled by $(\gamma - 1)p/R$. That is, it is convected along a streamline, thus implying the isentropic flow.

Change of Variables

The symmetric variables and the conservative variables are transformed into each other as follows:

$$d\mathbf{U} = \frac{\partial \mathbf{U}}{\partial \mathbf{V}} d\mathbf{V}, \quad d\mathbf{V} = \frac{\partial \mathbf{V}}{\partial \mathbf{U}} d\mathbf{U}, \tag{3.5.30}$$

where the transformation matrices are given by

$$\frac{\partial \mathbf{U}}{\partial \mathbf{V}} = \begin{bmatrix} \dfrac{\rho}{c} & 0 & 0 & -\dfrac{1}{c^2} \\[3mm] \dfrac{\rho u}{c} & \rho & 0 & -\dfrac{u}{c^2} \\[3mm] \dfrac{\rho v}{c} & 0 & \rho & -\dfrac{v}{c^2} \\[3mm] \rho c\left(\dfrac{q^2}{2c^2} + \dfrac{1}{\gamma - 1}\right) & \rho u & \rho v & -\dfrac{q^2}{2c^2} \end{bmatrix}, \tag{3.5.31}$$

$$\frac{\partial \mathbf{V}}{\partial \mathbf{U}} = \left(\frac{\partial \mathbf{U}}{\partial \mathbf{V}}\right)^{-1} = \begin{bmatrix} \dfrac{\gamma - 1}{2}\dfrac{q^2}{\rho c} & -(\gamma - 1)\dfrac{u}{\rho c} & -(\gamma - 1)\dfrac{v}{\rho c} & \dfrac{\gamma - 1}{\rho c} \\[3mm] -\dfrac{u}{\rho} & \dfrac{1}{\rho} & 0 & 0 \\[3mm] -\dfrac{v}{\rho} & 0 & \dfrac{1}{\rho} & 0 \\[3mm] \dfrac{\gamma - 1}{2}q^2 - c^2 & -(\gamma - 1)u & -(\gamma - 1)v & \gamma - 1 \end{bmatrix}. \tag{3.5.32}$$

Eigenstructure

$$\mathbf{A}_n^v = \mathbf{R}_n^v \mathbf{\Lambda}_n \mathbf{L}_n^v, \tag{3.5.33}$$

where

$$\mathbf{A}_n^v = \begin{bmatrix} q_n & c\,n_x & c\,n_y & 0 \\ c\,n_x & q_n & 0 & 0 \\ c\,n_y & 0 & q_n & 0 \\ 0 & 0 & 0 & q_n \end{bmatrix}, \tag{3.5.34}$$

$$\mathbf{\Lambda}_n^v = \begin{bmatrix} q_n & 0 & 0 & 0 \\ 0 & q_n - c & 0 & 0 \\ 0 & 0 & q_n + c & 0 \\ 0 & 0 & 0 & q_n \end{bmatrix}, \tag{3.5.35}$$

$$\mathbf{R}_n^v = \begin{bmatrix} 0 & 1 & 1 & 0 \\ 0 & -n_x & n_x & \ell_x \\ 0 & -n_y & n_y & \ell_y \\ 1 & 0 & 0 & 0 \end{bmatrix}, \tag{3.5.36}$$

$$\mathbf{L}_n^w = \begin{bmatrix} 0 & 0 & 0 & 1 \\ \frac{1}{2} & -\frac{1}{2}n_x & -\frac{1}{2}n_y & 0 \\ \frac{1}{2} & \frac{1}{2}n_x & \frac{1}{2}n_y & 0 \\ 0 & \ell_x & \ell_y & 0 \end{bmatrix}. \tag{3.5.37}$$

It is interesting to note that the change in the symmetric variables due to the acoustic and shear waves are given by

$$\begin{bmatrix} \dfrac{\partial p}{\rho c} \\ \partial u \\ \partial v \\ \partial p - a^2 \partial \rho \end{bmatrix} \propto \begin{bmatrix} 1 \\ -n_x \\ -n_y \\ 0 \end{bmatrix}, \begin{bmatrix} 1 \\ n_x \\ n_y \\ 0 \end{bmatrix}, \begin{bmatrix} 0 \\ \ell_x \\ \ell_y \\ 0 \end{bmatrix}, \tag{3.5.38}$$

which shows that the entropy is constant through the acoustic and shear waves. Of course, the entropy may increase across a shock wave (a strong acoustic wave); this change is given by the Rankine-Hugoniot relation.

Streamline/Natural Coordinates

I like the streamline coordinates, (ξ, η) where ξ taken along streamlines and η perpendicular to streamlines. The symmetric form (3.5.27) becomes even simpler in this coordinate system:

$$\tilde{\mathbf{V}}_t + \tilde{\mathbf{A}}^v \tilde{\mathbf{V}}_\xi + \tilde{\mathbf{B}}^v \tilde{\mathbf{V}}_\eta = 0, \tag{3.5.39}$$

where

$$d\tilde{\mathbf{V}} = \begin{bmatrix} \dfrac{dp}{\rho c} \\ dq \\ dq_\ell \\ dp - c^2 d\rho \end{bmatrix}, \quad \tilde{\mathbf{A}}^v = \begin{bmatrix} q & c & 0 & 0 \\ c & q & 0 & 0 \\ 0 & 0 & q & 0 \\ 0 & 0 & 0 & q \end{bmatrix}, \quad \tilde{\mathbf{B}}^v = \begin{bmatrix} 0 & 0 & c & 0 \\ 0 & 0 & 0 & 0 \\ c & 0 & 0 & 0 \\ 0 & 0 & 0 & 0 \end{bmatrix}, \tag{3.5.40}$$

$$q = u\cos\theta + v\sin\theta, \tag{3.5.41}$$
$$q_\ell = -u\sin\theta + v\cos\theta, \tag{3.5.42}$$
$$\tilde{\mathbf{A}}^v = \mathbf{A}^v\cos\theta + \mathbf{B}^v\sin\theta, \tag{3.5.43}$$
$$\tilde{\mathbf{B}}^v = -\mathbf{A}^v\sin\theta + \mathbf{B}^v\cos\theta, \tag{3.5.44}$$

$$d\tilde{\mathbf{V}} = \frac{\partial\tilde{\mathbf{V}}}{\partial\mathbf{V}}d\mathbf{V} = \begin{bmatrix} 1 & 0 & 0 & 0 \\ 0 & \cos\theta & \sin\theta & 0 \\ 0 & -\sin\theta & \cos\theta & 0 \\ 0 & 0 & 0 & 1 \end{bmatrix} d\mathbf{V}, \tag{3.5.45}$$

and θ is a local flow angle defined by $\theta = \tan^{-1}\left(\frac{v}{u}\right)$. Look at $\tilde{\mathbf{B}}^v$. It is extremely simple.

We have obtained this form from Equation (3.5.27) by setting x axis to be a streamline direction and y axis to be normal to the streamline. Or more formally, we invert this change of coordinates,

$$\frac{\partial}{\partial \xi} = \cos\theta\frac{\partial}{\partial x} + \sin\theta\frac{\partial}{\partial y}, \tag{3.5.46}$$

$$\frac{\partial}{\partial \eta} = -\sin\theta\frac{\partial}{\partial x} + \cos\theta\frac{\partial}{\partial y}, \tag{3.5.47}$$

to get

$$\frac{\partial}{\partial x} = \cos\theta\frac{\partial}{\partial \xi} - \sin\theta\frac{\partial}{\partial \eta}, \tag{3.5.48}$$

$$\frac{\partial}{\partial y} = \sin\theta\frac{\partial}{\partial \xi} + \cos\theta\frac{\partial}{\partial \eta}, \tag{3.5.49}$$

and substitute these into Equation (3.5.27).

Note that we always have $q_\ell = 0$ by definition (zero velocity across streamlines), but its rate of change is not necessarily zero along the streamlines.

3.5.4   Parameter Vector

The parameter vector is defined by

$$\mathbf{Z} = \sqrt{\rho}\begin{bmatrix} 1 \\ u \\ v \\ H \end{bmatrix} = \begin{bmatrix} z_1 \\ z_2 \\ z_3 \\ z_4 \end{bmatrix}. \tag{3.5.50}$$

I like it because every component of conservative variables and fluxes can be expressed as a quadratic in the components of $\mathbf{Z}$ [128].

$$\mathbf{U} = \begin{bmatrix} \rho \\ \rho u \\ \rho v \\ \rho E \end{bmatrix} = \begin{bmatrix} z_1^2 \\ z_1 z_2 \\ z_1 z_3 \\ \dfrac{z_1 z_4}{\gamma} + \dfrac{\gamma-1}{\gamma}\dfrac{z_2^2 + z_3^2}{2} \end{bmatrix}, \tag{3.5.51}$$

$$\mathbf{F} = \begin{bmatrix} \rho u \\ \rho u^2 + p \\ \rho uv \\ \rho uH \end{bmatrix} = \begin{bmatrix} z_1 z_2 \\ z_2^2 + p \\ z_2 z_3 \\ z_2 z_4 \end{bmatrix}, \tag{3.5.52}$$

$$\mathbf{G} = \begin{bmatrix} \rho v \\ \rho uv \\ \rho v^2 + p \\ \rho vH \end{bmatrix} = \begin{bmatrix} z_1 z_3 \\ z_2 z_3 \\ z_3^2 + p \\ z_3 z_4 \end{bmatrix}, \tag{3.5.53}$$

$$p = \frac{\gamma-1}{\gamma}\left(z_1 z_4 - \frac{z_2^2 + z_3^2}{2}\right). \tag{3.5.54}$$

We can write the conservative form of the Euler equations in terms of the parameter vector variables:

$$\mathbf{U}_t + \mathbf{A}^z\mathbf{Z}_x + \mathbf{B}^z\mathbf{Z}_y = 0, \tag{3.5.55}$$

$$\mathbf{A}^z = \frac{\partial \mathbf{F}}{\partial \mathbf{Z}} = \begin{bmatrix} z_2 & z_1 & 0 & 0 \\ \dfrac{\gamma-1}{\gamma} z_4 & \dfrac{\gamma+1}{\gamma} z_2 & -\dfrac{\gamma-1}{\gamma} z_3 & \dfrac{\gamma-1}{\gamma} z_1 \\ 0 & z_3 & z_2 & 0 \\ 0 & z_4 & 0 & z_2 \end{bmatrix}, \tag{3.5.56}$$

$$\mathbf{B}^z = \frac{\partial \mathbf{G}}{\partial \mathbf{Z}} = \begin{bmatrix} z_3 & 0 & z_1 & 0 \\ 0 & z_3 & z_2 & 0 \\ \dfrac{\gamma-1}{\gamma} z_4 & -\dfrac{\gamma-1}{\gamma} z_2 & \dfrac{\gamma+1}{\gamma} z_3 & \dfrac{\gamma-1}{\gamma} z_1 \\ 0 & 0 & z_4 & z_3 \end{bmatrix}. \tag{3.5.57}$$

Note that the Jacobian matrices are linear in the parameter vector variables, which can be useful to linearize the Euler equations over a computational cell [31].

Perform change of variables as follows:

$$d\mathbf{Z} = \frac{\partial \mathbf{Z}}{\partial \mathbf{U}} d\mathbf{U}, \quad d\mathbf{U} = \frac{\partial \mathbf{U}}{\partial \mathbf{Z}} d\mathbf{Z}, \tag{3.5.58}$$

where

$$\frac{\partial \mathbf{U}}{\partial \mathbf{Z}} = \begin{bmatrix} 2z_1 & 0 & 0 & 0 \\ z_2 & z_1 & 0 & 0 \\ z_3 & 0 & z_1 & 0 \\ \dfrac{z_4}{\gamma} & \dfrac{\gamma-1}{\gamma} z_2 & \dfrac{\gamma-1}{\gamma} z_3 & \dfrac{z_1}{\gamma} \end{bmatrix}, \tag{3.5.59}$$

$$\frac{\partial \mathbf{Z}}{\partial \mathbf{U}} = \frac{1}{z_1} \begin{bmatrix} \dfrac{1}{2} & 0 & 0 & 0 \\ -\dfrac{1}{2}\dfrac{z_2}{z_1} & 1 & 0 & 0 \\ -\dfrac{1}{2}\dfrac{z_3}{z_1} & 0 & 1 & 0 \\ \dfrac{\gamma-1}{2}\dfrac{z_2^2 + z_3^2}{z_1^2} - \dfrac{1}{2}\dfrac{z_4}{z_1} & -(\gamma-1)\dfrac{z_2}{z_1} & -(\gamma-1)\dfrac{z_3}{z_1} & \gamma \end{bmatrix},$$

$$= \frac{1}{\sqrt{\rho}} \begin{bmatrix} \dfrac{1}{2} & 0 & 0 & 0 \\ -\dfrac{1}{2}u & 1 & 0 & 0 \\ -\dfrac{1}{2}v & 0 & 1 & 0 \\ \dfrac{\gamma-1}{2}(u^2 + v^2) - \dfrac{1}{2}H & -(\gamma-1)u & -(\gamma-1)v & \gamma \end{bmatrix}, \tag{3.5.60}$$

from which we find also

$$\mathbf{Z} = 2\frac{\partial \mathbf{Z}}{\partial \mathbf{U}} \mathbf{U}, \quad \mathbf{U} = \frac{1}{2}\frac{\partial \mathbf{U}}{\partial \mathbf{Z}} \mathbf{Z}. \tag{3.5.61}$$

Compare these with Equation (3.5.58). It is very interesting, isn't it? Moreover, since we have

$$\mathbf{F} = \mathbf{A}\mathbf{U}, \quad \mathbf{G} = \mathbf{B}\mathbf{U}, \tag{3.5.62}$$

for the Euler equations (see Section 3.10), we find from Equation (3.5.61) that

$$\mathbf{F} \;=\; \mathbf{AU} = \frac{1}{2}\frac{\partial \mathbf{F}}{\partial \mathbf{U}}\frac{\partial \mathbf{U}}{\partial \mathbf{Z}}\mathbf{Z} = \frac{1}{2}\mathbf{A}^z\mathbf{Z}, \tag{3.5.63}$$

$$\mathbf{G} \;=\; \mathbf{BU} = \frac{1}{2}\frac{\partial \mathbf{G}}{\partial \mathbf{U}}\frac{\partial \mathbf{U}}{\partial \mathbf{Z}}\mathbf{Z} = \frac{1}{2}\mathbf{B}^z\mathbf{Z}. \tag{3.5.64}$$

These show how we can compute the fluxes by the parameter vector. I like these formulas because they would be useful if we carry the parameter vector as working variables in a CFD code (i.e., we can directly compute the fluxes without converting the parameter vector into the primitive variables). I like them also because they show that the physical fluxes are quadratic in $\mathbf{Z}$.

3.6 3D Euler Equations

3.6.1 Conservative Form

$$\mathbf{U}_t + \mathbf{F}_x + \mathbf{G}_y + \mathbf{H}_z = 0, \tag{3.6.1}$$

where

$$\mathbf{U} = \begin{bmatrix} \rho \\ \rho u \\ \rho v \\ \rho w \\ \rho E \end{bmatrix}, \quad \mathbf{F} = \begin{bmatrix} \rho u \\ \rho u^2 + p \\ \rho uv \\ \rho uw \\ \rho uH \end{bmatrix}, \quad \mathbf{G} = \begin{bmatrix} \rho v \\ \rho vu \\ \rho v^2 + p \\ \rho vw \\ \rho vH \end{bmatrix}, \quad \mathbf{H} = \begin{bmatrix} \rho w \\ \rho wu \\ \rho wv \\ \rho w^2 + p \\ \rho wH \end{bmatrix}. \tag{3.6.2}$$

Normal Flux

$$\mathbf{F}_n = [\mathbf{F},\mathbf{G},\mathbf{H}]\cdot\mathbf{n} = \mathbf{F}n_x + \mathbf{G}n_y + \mathbf{H}n_z = \begin{bmatrix} \rho q_n \\ \rho q_n u + p n_x \\ \rho q_n v + p n_y \\ \rho q_n w + p n_z \\ \rho q_n H \end{bmatrix}, \tag{3.6.3}$$

where $\mathbf{n} = [n_x, n_y, n_z]^t$ and

$$q_n = u n_x + v n_y + w n_z. \tag{3.6.4}$$

Jacobians

$$\mathbf{A} = \begin{bmatrix} 0 & 1 & 0 & 0 & 0 \\ \dfrac{\gamma-1}{2}q^2 - u^2 & (3-\gamma)u & (1-\gamma)v & (1-\gamma)w & \gamma-1 \\ -uv & v & u & 0 & 0 \\ -uw & w & 0 & u & 0 \\ \left(\dfrac{\gamma-1}{2}q^2 - H\right)u & H+(1-\gamma)u^2 & (1-\gamma)uv & (1-\gamma)uw & \gamma u \end{bmatrix}, \tag{3.6.5}$$

$$
\mathbf{B} = \begin{bmatrix}
0 & 0 & 1 & 0 & 0 \\
-vu & v & u & 0 & 0 \\
\dfrac{\gamma-1}{2}q^2 - v^2 & (1-\gamma)v & (3-\gamma)u & (1-\gamma)w & \gamma-1 \\
-vw & 0 & w & v & 0 \\
\left(\dfrac{\gamma-1}{2}q^2 - H\right)v & (1-\gamma)uv & H+(1-\gamma)v^2 & (1-\gamma)vw & \gamma v
\end{bmatrix},
\tag{3.6.6}
$$

$$
\mathbf{C} = \begin{bmatrix}
0 & 0 & 0 & 1 & 0 \\
-uw & w & 0 & u & 0 \\
-vw & 0 & w & v & 0 \\
\dfrac{\gamma-1}{2}q^2 - w^2 & (1-\gamma)u & (1-\gamma)v & (3-\gamma)w & \gamma-1 \\
\left(\dfrac{\gamma-1}{2}q^2 - H\right)w & (1-\gamma)uw & (1-\gamma)vw & H+(1-\gamma)w^2 & \gamma w
\end{bmatrix},
\tag{3.6.7}
$$

where $\mathbf{A} = \frac{\partial \mathbf{F}}{\partial \mathbf{U}}$, $\mathbf{B} = \frac{\partial \mathbf{G}}{\partial \mathbf{U}}$, $\mathbf{C} = \frac{\partial \mathbf{H}}{\partial \mathbf{U}}$, and $q^2 = u^2 + v^2 + w^2$. Also, I like this one:

$$
\mathbf{A}_n = \mathbf{A}n_x + \mathbf{B}n_y + \mathbf{C}n_z
$$

$$
= \begin{bmatrix}
0 & n_x & n_y & n_z & 0 \\
\dfrac{K}{2}q^2 n_x - uq_n & un_x - Kun_x + q_n & un_y - Kvn_x & un_z - Kwn_x & Kn_x \\
\dfrac{K}{2}q^2 n_y - vq_n & vn_x - Kun_y & vn_y - Kvn_y + q_n & vn_z - Kwn_y & Kn_y \\
\dfrac{K}{2}q^2 n_z - wq_n & wn_x - Kun_z & wn_y - Kvn_z & wn_z - Kwn_z + q_n & Kn_z \\
\left(\dfrac{K}{2}q^2 - H\right)q_n & Hn_x - Kuq_n & Hn_y - Kvq_n & Hn_z - Kwq_n & \gamma q_n
\end{bmatrix}
$$

$$
= \begin{bmatrix}
0 & \mathbf{n}^t & 0 \\
\dfrac{\gamma-1}{2}q^2\mathbf{n} - \mathbf{v}q_n & \mathbf{v}\otimes\mathbf{n} - (\gamma-1)\mathbf{n}\otimes\mathbf{v} + q_n\mathbf{I} & (\gamma-1)\mathbf{n} \\
\left(\dfrac{\gamma-1}{2}q^2 - H\right)q_n & H\mathbf{n}^t - (\gamma-1)\mathbf{v}^t q_n & \gamma q_n
\end{bmatrix},
\tag{3.6.8}
$$

where $K = \gamma - 1$, $\mathbf{v} = [u, v, w]^t$, and $\mathbf{I}$ is the 3×3 identity matrix. We may take $\mathbf{n} = [1, 0, 0]^t$, $[0, 1, 0]^t$, or $[0, 0, 1]^t$, to obtain $\mathbf{A}$, $\mathbf{B}$, or $\mathbf{C}$, respectively.

Eigenstructure

$$
\mathbf{A}_n = \mathbf{R}_n \mathbf{\Lambda}_n \mathbf{L}_n,
\tag{3.6.9}
$$

where

$$\mathbf{\Lambda}_n = \begin{bmatrix} q_n - c & 0 & 0 & 0 & 0 \\ 0 & q_n & 0 & 0 & 0 \\ 0 & 0 & q_n + c & 0 & 0 \\ 0 & 0 & 0 & q_n & 0 \\ 0 & 0 & 0 & 0 & q_n \end{bmatrix}, \tag{3.6.10}$$

$$\mathbf{R}_n = \begin{bmatrix} 1 & 1 & 1 & 0 & 0 \\ u - c\,n_x & u & u + c\,n_x & \ell_x & m_x \\ v - c\,n_y & v & v + c\,n_y & \ell_y & m_y \\ w - c\,n_z & w & w + c\,n_z & \ell_z & m_z \\ H - q_n c & q^2/2 & H + q_n c & q_\ell & q_m \end{bmatrix}, \tag{3.6.11}$$

$$\mathbf{L}_n = \begin{bmatrix} \dfrac{Kq^2}{4c^2} + \dfrac{q_n}{2c} & -\left(\dfrac{K}{2c^2}u + \dfrac{n_x}{2c}\right) & -\left(\dfrac{K}{2c^2}v + \dfrac{n_y}{2c}\right) & -\left(\dfrac{K}{2c^2}w + \dfrac{n_z}{2c}\right) & \dfrac{K}{2c^2} \\[2mm] 1 - \dfrac{Kq^2}{2c^2} & \dfrac{Ku}{c^2} & \dfrac{Kv}{c^2} & \dfrac{Kw}{c^2} & -\dfrac{K}{c^2} \\[2mm] \dfrac{Kq^2}{4c^2} - \dfrac{q_n}{2c} & -\left(\dfrac{K}{2c^2}u - \dfrac{n_x}{2c}\right) & -\left(\dfrac{K}{2c^2}v - \dfrac{n_y}{2c}\right) & -\left(\dfrac{K}{2c^2}w - \dfrac{n_z}{2c}\right) & \dfrac{K}{2c^2} \\[2mm] -q_\ell & \ell_x & \ell_y & \ell_z & 0 \\[2mm] -q_m & m_x & m_y & m_z & 0 \end{bmatrix}, \tag{3.6.12}$$

$$\mathbf{L}_n d\mathbf{U} = \begin{bmatrix} \dfrac{dp - \rho c\,dq_n}{2c^2} \\[2mm] -\dfrac{dp - c^2\,d\rho}{c^2} \\[2mm] \dfrac{dp + \rho c\,dq_n}{2c^2} \\[2mm] \rho\,dq_\ell \\[2mm] \rho\,dq_m \end{bmatrix}, \tag{3.6.13}$$

where $\boldsymbol{\ell} = [\ell_x, \ell_y, \ell_z]^t$, $\mathbf{m} = [m_x, m_y, m_z]^t$, and $\mathbf{n}$ are mutually orthogonal unit vectors, and

$$q_\ell = u\ell_x + v\ell_y + w\ell_z, \quad q_m = um_x + vm_y + wm_z. \tag{3.6.14}$$

See Section 1.3 for formulas and properties of mutually orthogonal unit vectors. Again, it is nice that the normal Jacobian $\mathbf{A}_n$ can be diagonalized as

$$\mathbf{\Lambda}_n = \mathbf{L}_n \mathbf{A}_n \mathbf{R}_n, \tag{3.6.15}$$

but it is very difficult to diagonalize $\mathbf{A}$, $\mathbf{B}$, and $\mathbf{C}$ simultaneously. In three dimensions, even the steady equations cannot be diagonalized for the same reason that it is hard to diagonalize the two-dimensional time-dependent Euler system.

Absolute Normal Jacobian

Incidentally, I like to show that the absolute value of the normal Jacobian defined by

$$|\mathbf{A}_n| \equiv \mathbf{R}_n |\mathbf{\Lambda}_n| \mathbf{L}_n, \tag{3.6.16}$$

where

$$\mathbf{\Lambda}_n = \operatorname{diag}(|\lambda_1|, |\lambda_2|, |\lambda_3|, |\lambda_4|, |\lambda_5|) \qquad (3.6.17)$$

$$= \begin{bmatrix} |q_n - c| & 0 & 0 & 0 & 0 \\ 0 & |q_n| & 0 & 0 & 0 \\ 0 & 0 & |q_n + c| & 0 & 0 \\ 0 & 0 & 0 & |q_n| & 0 \\ 0 & 0 & 0 & 0 & |q_n| \end{bmatrix}, \qquad (3.6.18)$$

can be expressed without ambiguous tangent vectors, $\boldsymbol{\ell}$ and $\boldsymbol{m}$. First, denote the k-th column of $\mathbf{R}_n$ by $\mathbf{r}_k$ and the k-th row of $\mathbf{L}_n$ by $\boldsymbol{\ell}_k^t$, so that we can write

$$|\mathbf{A}_n| \equiv \mathbf{R}_n |\mathbf{\Lambda}_n| \mathbf{L}_n$$

$$= \sum_{k=1}^{5} |\lambda_k| \, \mathbf{r}_k \boldsymbol{\ell}_k^t. \qquad (3.6.19)$$

(Note that $\boldsymbol{\ell}$ is a tangent vector and $\boldsymbol{\ell}_k^t$ is the k-th left eigenvector.) Now, we have

$$\mathbf{r}_4 \boldsymbol{\ell}_4^t = \begin{bmatrix} 0 & \mathbf{0}^t & 0 \\ -q_\ell \boldsymbol{\ell} & \boldsymbol{\ell} \otimes \boldsymbol{\ell} & 0 \\ -q_\ell^2 & q_\ell \boldsymbol{\ell}^t & 0 \end{bmatrix}, \qquad (3.6.20)$$

where $\mathbf{0}^t = [0, 0, 0]$, and also

$$\mathbf{r}_5 \boldsymbol{\ell}_5^t = \begin{bmatrix} 0 & \mathbf{0}^t & 0 \\ -q_m \boldsymbol{m} & \boldsymbol{m} \otimes \boldsymbol{m} & 0 \\ -q_m^2 & q_m \boldsymbol{m}^t & 0 \end{bmatrix}. \qquad (3.6.21)$$

Then, we can eliminate the tangent vectors by adding these two and using the results from Section 1.3:

$$\mathbf{r}_4 \boldsymbol{\ell}_4^t + \mathbf{r}_5 \boldsymbol{\ell}_5^t = \begin{bmatrix} 0 & \mathbf{0}^t & 0 \\ q_n \mathbf{n} - \mathbf{v} & \mathbf{I} - \mathbf{n} \otimes \mathbf{n} & 0 \\ q_n^2 - q^2 & \mathbf{v}^t - q_n \mathbf{n}^t & 0 \end{bmatrix}, \qquad (3.6.22)$$

where $\mathbf{I}$ is the 3x3 identity matrix. Here, we notice that $\lambda_2 = \lambda_4 = \lambda_5 = |q_n|$ and $\mathbf{r}_2 \boldsymbol{\ell}_2^t$ does not depend on tangent vectors. Then, we would want to add $\mathbf{r}_2 \boldsymbol{\ell}_2^t$,

$$\mathbf{r}_2 \boldsymbol{\ell}_2^t = \begin{bmatrix} 1 - \dfrac{\gamma - 1}{2} M^2 & \dfrac{\gamma - 1}{c^2} \mathbf{v}^t & -\dfrac{\gamma - 1}{c^2} \\ -\dfrac{\gamma - 1}{2} M^2 \mathbf{v} & \dfrac{\gamma - 1}{c^2} \mathbf{v} \otimes \mathbf{v} & -\dfrac{\gamma - 1}{c^2} \mathbf{v} \\ \dfrac{q^2}{2} \left(1 - \dfrac{\gamma - 1}{2} M^2 \right) & \dfrac{\gamma - 1}{2} M^2 \mathbf{v}^t & -\dfrac{\gamma - 1}{2} M^2 \end{bmatrix}, \qquad (3.6.23)$$

where $M^2 = q^2 / c^2$, to Equation (3.6.22) to get

$$|\lambda_2| \mathbf{r}_2 \boldsymbol{\ell}_2^t + \sum_{k=4}^{5} |\lambda_k| \mathbf{r}_k \boldsymbol{\ell}_k^t = |q_n| \left(\mathbf{r}_2 \boldsymbol{\ell}_2^t + \mathbf{r}_4 \boldsymbol{\ell}_4^t + \mathbf{r}_5 \boldsymbol{\ell}_5^t \right)$$

$$= |q_n| \begin{bmatrix} 1 - \dfrac{\gamma - 1}{2} M^2 & \dfrac{\gamma - 1}{c^2} \mathbf{v}^t & -\dfrac{\gamma - 1}{c^2} \\ -\left(1 + \dfrac{\gamma - 1}{2} M^2 \right) \mathbf{v} + q_n \mathbf{n} & \dfrac{\gamma - 1}{c^2} \mathbf{v} \otimes \mathbf{v} + \mathbf{I} - \mathbf{n} \otimes \mathbf{n} & -\dfrac{\gamma - 1}{c^2} \mathbf{v} \\ q_n^2 - \dfrac{q^2}{2} \left(1 + \dfrac{\gamma - 1}{2} M^2 \right) & \left(1 + \dfrac{\gamma - 1}{2} M^2 \right) \mathbf{v}^t - q_n \mathbf{n}^t & -\dfrac{\gamma - 1}{2} M^2 \end{bmatrix}. \quad (3.6.24)$$

Next, consider $\mathbf{r}_1\boldsymbol{\ell}_1^t$ and $\mathbf{r}_3\boldsymbol{\ell}_3^t$:

$$\mathbf{r}_1\boldsymbol{\ell}_1^t = \begin{bmatrix} \dfrac{\gamma-1}{4}M^2 + \dfrac{M_n}{2} & -\dfrac{\gamma-1}{2c^2}\mathbf{v} - \dfrac{\mathbf{n}}{2c} & \dfrac{\gamma-1}{2c^2} \\[2ex] (\mathbf{v}-c\,\mathbf{n})\left(\dfrac{\gamma-1}{4}M^2 + \dfrac{M_n}{2}\right) & (\mathbf{v}-c\,\mathbf{n})\otimes\left(-\dfrac{\gamma-1}{2c^2}\mathbf{v} - \dfrac{\mathbf{n}}{2c}\right) & (\mathbf{v}-c\,\mathbf{n})\dfrac{\gamma-1}{2c^2} \\[2ex] (H-q_nc)\left(\dfrac{\gamma-1}{4}M^2 + \dfrac{M_n}{2}\right) & (H-q_nc)\left(-\dfrac{\gamma-1}{2c^2}\mathbf{v} - \dfrac{\mathbf{n}}{2c}\right) & (H-q_nc)\dfrac{\gamma-1}{2c^2} \end{bmatrix}, \quad (3.6.25)$$

$$\mathbf{r}_3\boldsymbol{\ell}_3^t = \begin{bmatrix} \dfrac{\gamma-1}{4}M^2 - \dfrac{M_n}{2} & -\dfrac{\gamma-1}{2c^2}\mathbf{v} + \dfrac{\mathbf{n}}{2c} & \dfrac{\gamma-1}{2c^2} \\[2ex] (\mathbf{v}+c\,\mathbf{n})\left(\dfrac{\gamma-1}{4}M^2 - \dfrac{M_n}{2}\right) & (\mathbf{v}+c\,\mathbf{n})\otimes\left(-\dfrac{\gamma-1}{2c^2}\mathbf{v} + \dfrac{\mathbf{n}}{2c}\right) & (\mathbf{v}+c\,\mathbf{n})\dfrac{\gamma-1}{2c^2} \\[2ex] (H+q_nc)\left(\dfrac{\gamma-1}{4}M^2 - \dfrac{M_n}{2}\right) & (H+q_nc)\left(-\dfrac{\gamma-1}{2c^2}\mathbf{v} + \dfrac{\mathbf{n}}{2c}\right) & (H+q_nc)\dfrac{\gamma-1}{2c^2} \end{bmatrix}, \quad (3.6.26)$$

where $M_n = q_n/c$. It may be possible to simplify the sum, $|\lambda_1|\,\mathbf{r}_1\boldsymbol{\ell}_1^t + |\lambda_3|\,\mathbf{r}_3\boldsymbol{\ell}_3^t$, but I stop here because they are already independent of the tangent vectors. I like this very much because I can now compute the dissipation term (3.6.19), using the normal vector $\mathbf{n}$ only. This is very useful because tangent vectors are not uniquely defined for a given normal vector $\mathbf{n}$, especially in three dimensions. By the way, the matrix $|\mathbf{A}_n|$ is required, for example, for implementing a class of residual-distribution schemes [15, 19, 30, 164].

I also like to show that the product of the absolute normal Jacobian and the differential of the conservative variables,

$$|\mathbf{A}_n|\,d\mathbf{U} \equiv \mathbf{R}_n|\mathbf{\Lambda}_n|\mathbf{L}_n d\mathbf{U}, \quad (3.6.27)$$

is independent of the tangent vectors. It goes like this:

$$|\mathbf{A}_n|\,d\mathbf{U} = \mathbf{R}_n|\mathbf{\Lambda}_n|\mathbf{L}_n d\mathbf{U}$$

$$= \sum_{k=1}^{5} |\lambda_k|\,\mathbf{r}_k\boldsymbol{\ell}_k^t d\mathbf{U}$$

$$= \sum_{k=1}^{3} |\lambda_k|\,\mathbf{r}_k\boldsymbol{\ell}_k^t d\mathbf{U} + |q_n| \begin{bmatrix} 0 \\ \rho(\ell_x dq_\ell + m_x dq_m) \\ \rho(\ell_y dq_\ell + m_y dq_m) \\ \rho(\ell_z dq_\ell + m_z dq_m) \\ \rho(q_\ell dq_\ell + q_m dq_m) \end{bmatrix}$$

$$= \sum_{k=1}^{3} |\lambda_k|\,\mathbf{r}_k\boldsymbol{\ell}_k^t d\mathbf{U} + |q_n| \begin{bmatrix} 0 \\ \rho(du - dq_n n_x) \\ \rho(dv - dq_n n_y) \\ \rho(dw - dq_n n_z) \\ \rho(u\,du + v\,dv + w\,dw - q_n dq_n) \end{bmatrix}, \quad (3.6.28)$$

where the results in Section 1.3 have been used in the last step. This is very nice. I like it. Basically, the tangent vectors should not appear at all because obviously $|\mathbf{A}_n|$ depends only on the normal vector. It should always be possible to eliminate tangent vectors, $\boldsymbol{\ell}$ and $\mathbf{m}$, from $|\mathbf{A}_n|$ for any equations. By the way, the vector $|\mathbf{A}_n|\,d\mathbf{U}$ is often used, for example, for implementing Roe's approximate Riemann solver [128].

Expanded Form

Sometimes, it is convenient to expand the conservative form of the Euler equations in terms of the derivatives of the primitive variables:

$$
\mathbf{F}_x + \mathbf{G}_y + \mathbf{H}_z = \begin{bmatrix} (\rho u)_x + (\rho v)_y + (\rho w)_z \\[4pt] (\rho u^2)_x + (\rho uv)_y + (\rho uw)_z + p_x \\[4pt] (\rho vu)_x + (\rho v^2)_y + (\rho vw)_z + p_y \\[4pt] (\rho wu)_x + (\rho wv)_y + (\rho w^2)_z + p_z \\[4pt] (\rho uH)_x + (\rho vH)_y + (\rho wH)_x \end{bmatrix}
$$

$$
= \begin{bmatrix} \rho(u_x + v_y + w_z) + u\rho_x + v\rho_y + w\rho_z \\[4pt] u\left\{\rho(u_x + v_y + w_z) + u\rho_x + v\rho_y + w\rho_z\right\} + \rho u u_x + \rho v u_y + \rho w u_z + p_x \\[4pt] v\left\{\rho(u_x + v_y + w_z) + u\rho_x + v\rho_y + w\rho_z\right\} + \rho u v_x + \rho v v_y + \rho w v_z + p_y \\[4pt] w\left\{\rho(u_x + v_y + w_z) + u\rho_x + v\rho_y + w\rho_z\right\} + \rho u w_x + \rho v w_y + \rho w w_z + p_z \\[4pt] \rho H(u_x + v_y + w_z) + u(\rho H)_x + v(\rho H)_y + w(\rho H)_z \end{bmatrix}, (3.6.29)
$$

where

$$
(\rho H)_x = \frac{\gamma}{\gamma - 1}p_x + \frac{1}{2}\rho_x(u^2 + v^2 + w^2) + \rho(uu_x + vv_x + ww_x), \tag{3.6.30}
$$

$$
(\rho H)_y = \frac{\gamma}{\gamma - 1}p_y + \frac{1}{2}\rho_y(u^2 + v^2 + w^2) + \rho(uu_y + vv_y + ww_y), \tag{3.6.31}
$$

$$
(\rho H)_z = \frac{\gamma}{\gamma - 1}p_z + \frac{1}{2}\rho_z(u^2 + v^2 + w^2) + \rho(uu_z + vv_z + ww_z). \tag{3.6.32}
$$

I like it because this form can be useful for numerically computing a source term arising from the method of manufactured solutions where we define simple functions for the primitive variables and substitute them into the Euler equations to derive a source term, so that the simple functions are the exact solutions for the Euler equations with the source term.

3.6.2 Primitive Form

$$
\mathbf{W}_t + \mathbf{A}^w \mathbf{W}_x + \mathbf{B}^w \mathbf{W}_y + \mathbf{C}^w \mathbf{W}_z = 0, \tag{3.6.33}
$$

where

$$
\mathbf{W} = \begin{bmatrix} \rho \\ u \\ v \\ w \\ p \end{bmatrix}, \quad \mathbf{A}^w = \begin{bmatrix} u & \rho & 0 & 0 & 0 \\ 0 & u & 0 & 0 & 1/\rho \\ 0 & 0 & u & 0 & 0 \\ 0 & 0 & 0 & u & 0 \\ 0 & \rho c^2 & 0 & 0 & u \end{bmatrix}, \tag{3.6.34}
$$

$$
\mathbf{B}^w = \begin{bmatrix} v & 0 & \rho & 0 & 0 \\ 0 & v & 0 & 0 & 0 \\ 0 & 0 & v & 0 & 1/\rho \\ 0 & 0 & 0 & v & 0 \\ 0 & 0 & \rho c^2 & 0 & v \end{bmatrix}, \quad \mathbf{C}^w = \begin{bmatrix} w & 0 & 0 & \rho & 0 \\ 0 & w & 0 & 0 & 0 \\ 0 & 0 & w & 0 & 0 \\ 0 & 0 & 0 & w & 1/\rho \\ 0 & 0 & 0 & \rho c^2 & w \end{bmatrix}. \tag{3.6.35}
$$

Eigenstructure

$$\mathbf{A}_n^w = \mathbf{R}_n^w \mathbf{\Lambda}_n \mathbf{L}_n^w, \tag{3.6.36}$$

where

$$\mathbf{A}_n^w = \begin{bmatrix} q_n & \rho n_x & \rho n_y & \rho n_z & 0 \\ 0 & q_n & 0 & 0 & n_x/\rho \\ 0 & 0 & q_n & 0 & n_y/\rho \\ 0 & 0 & 0 & q_n & n_z/\rho \\ 0 & \rho c^2 n_x & \rho c^2 n_y & \rho c^2 n_z & q_n \end{bmatrix}, \tag{3.6.37}$$

$$\mathbf{\Lambda}_n = \begin{bmatrix} q_n - c & 0 & 0 & 0 & 0 \\ 0 & q_n & 0 & 0 & 0 \\ 0 & 0 & q_n + c & 0 & 0 \\ 0 & 0 & 0 & q_n & 0 \\ 0 & 0 & 0 & 0 & q_n \end{bmatrix}, \tag{3.6.38}$$

$$\mathbf{R}_n^w = \begin{bmatrix} 1 & 1 & 1 & 0 & 0 \\ -\dfrac{n_x c}{\rho} & 0 & \dfrac{n_x c}{\rho} & \dfrac{\ell_x}{\rho} & \dfrac{m_x}{\rho} \\ -\dfrac{n_y c}{\rho} & 0 & \dfrac{n_y c}{\rho} & \dfrac{\ell_y}{\rho} & \dfrac{m_y}{\rho} \\ -\dfrac{n_z c}{\rho} & 0 & \dfrac{n_z c}{\rho} & \dfrac{\ell_z}{\rho} & \dfrac{m_z}{\rho} \\ c^2 & 0 & c^2 & 0 & 0 \end{bmatrix}, \tag{3.6.39}$$

$$\mathbf{L}_n^w = \begin{bmatrix} 0 & -\dfrac{n_x \rho}{2c} & -\dfrac{n_y \rho}{2c} & -\dfrac{n_z \rho}{2c} & \dfrac{1}{2c^2} \\ 1 & 0 & 0 & 0 & -\dfrac{1}{c^2} \\ 0 & \dfrac{n_x \rho}{2c} & \dfrac{n_y \rho}{2c} & \dfrac{n_z \rho}{2c} & \dfrac{1}{2c^2} \\ 0 & \rho \ell_x & \rho \ell_y & \rho \ell_z & 0 \\ 0 & \rho m_x & \rho m_y & \rho m_z & 0 \end{bmatrix}. \tag{3.6.40}$$

Again, as in Section 3.4.2, there are no such things as primitive Jacobians. Note that we have two shear/vorticity waves (fourth and fifth components above) in three dimensions.

Derivatives of Conservative Fluxes

We can write the spatial part of the conservative form of the Euler equations in terms of the primitive variables:

$$\mathbf{U}_t + \frac{\partial \mathbf{F}}{\partial \mathbf{W}} \mathbf{W}_x + \frac{\partial \mathbf{G}}{\partial \mathbf{W}} \mathbf{W}_y + \frac{\partial \mathbf{H}}{\partial \mathbf{W}} \mathbf{W}_z = 0, \tag{3.6.41}$$

$$\frac{\partial \mathbf{F}}{\partial \mathbf{W}} = \begin{bmatrix} u & \rho & 0 & 0 & 0 \\ u^2 & 2\rho u & 0 & 0 & 1 \\ uv & \rho v & \rho u & 0 & 0 \\ uw & \rho w & 0 & \rho u & 0 \\ \frac{1}{2}uq^2 & \rho H + \rho u^2 & \rho uv & \rho uw & \frac{\gamma}{\gamma-1}u \end{bmatrix}, \tag{3.6.42}$$

$$\frac{\partial \mathbf{G}}{\partial \mathbf{W}} = \begin{bmatrix} v & 0 & \rho & 0 & 0 \\ vu & \rho v & \rho u & 0 & 0 \\ v^2 & 0 & 2\rho v & 0 & 1 \\ vw & 0 & \rho w & \rho v & 0 \\ \frac{1}{2}vq^2 & \rho vu & \rho H + \rho v^2 & \rho vw & \frac{\gamma}{\gamma-1}v \end{bmatrix}, \tag{3.6.43}$$

$$\frac{\partial \mathbf{H}}{\partial \mathbf{W}} = \begin{bmatrix} w & 0 & 0 & \rho & 0 \\ wu & \rho w & 0 & \rho u & 0 \\ wv & 0 & \rho w & \rho v & 0 \\ w^2 & 0 & 0 & 2\rho w & 1 \\ \frac{1}{2}wq^2 & \rho wu & \rho wv & \rho H + \rho w^2 & \frac{\gamma}{\gamma-1}w \end{bmatrix}. \tag{3.6.44}$$

I like these derivatives of the conservative fluxes because they can be useful in computing the gradient of the fluxes in terms of the gradient of the primitive variables:

$$\frac{\partial \mathbf{F}}{\partial x} = \frac{\partial \mathbf{F}}{\partial \mathbf{W}}\frac{\partial \mathbf{W}}{\partial x}, \quad \frac{\partial \mathbf{F}}{\partial y} = \frac{\partial \mathbf{F}}{\partial \mathbf{W}}\frac{\partial \mathbf{W}}{\partial y}, \quad \frac{\partial \mathbf{F}}{\partial z} = \frac{\partial \mathbf{F}}{\partial \mathbf{W}}\frac{\partial \mathbf{W}}{\partial z}, \tag{3.6.45}$$

and similarly for $\mathbf{G}$ and $\mathbf{H}$. Also, it is possible in the same way to compute the gradient of the normal flux:

$$\frac{\partial \mathbf{F}_n}{\partial x} = \frac{\partial \mathbf{F}_n}{\partial \mathbf{W}}\frac{\partial \mathbf{W}}{\partial x}, \quad \frac{\partial \mathbf{F}_n}{\partial y} = \frac{\partial \mathbf{F}_n}{\partial \mathbf{W}}\frac{\partial \mathbf{W}}{\partial y}, \quad \frac{\partial \mathbf{F}_n}{\partial z} = \frac{\partial \mathbf{F}_n}{\partial \mathbf{W}}\frac{\partial \mathbf{W}}{\partial z}, \tag{3.6.46}$$

where

$$\begin{aligned} \frac{\partial \mathbf{F}_n}{\partial \mathbf{W}} &= \frac{\partial \mathbf{F}}{\partial \mathbf{W}}n_x + \frac{\partial \mathbf{G}}{\partial \mathbf{W}}n_y + \frac{\partial \mathbf{H}}{\partial \mathbf{W}}n_z \\[2mm] &= \begin{bmatrix} q_n & \rho n_x & \rho n_y & \rho n_z & 0 \\ q_n u & \rho(q_n + un_x) & \rho un_y & \rho un_z & n_x \\ q_n v & \rho vn_x & \rho(q_n + vn_y) & \rho vn_z & n_y \\ q_n w & \rho wn_x & \rho wn_y & \rho(q_n + wn_z) & n_z \\ \frac{1}{2}q_n q^2 & \rho Hn_x + \rho q_n u & \rho Hn_y + \rho q_n v & \rho Hn_z + \rho q_n w & \frac{\gamma}{\gamma-1}q_n \end{bmatrix}. \end{aligned} \tag{3.6.47}$$

Change of Variables

$$d\mathbf{U} = \frac{\partial \mathbf{U}}{\partial \mathbf{W}}d\mathbf{W}, \quad d\mathbf{W} = \frac{\partial \mathbf{W}}{\partial \mathbf{U}}d\mathbf{U}, \tag{3.6.48}$$

where

$$\frac{\partial \mathbf{U}}{\partial \mathbf{W}} = \begin{bmatrix} 1 & 0 & 0 & 0 & 0 \\ u & \rho & 0 & 0 & 0 \\ v & 0 & \rho & 0 & 0 \\ w & 0 & 0 & \rho & 0 \\ q^2/2 & \rho u & \rho v & \rho w & 1/(\gamma-1) \end{bmatrix}, \tag{3.6.49}$$

$$\frac{\partial \mathbf{W}}{\partial \mathbf{U}} = \begin{bmatrix} 1 & 0 & 0 & 0 & 0 \\ -u/\rho & \frac{1}{\rho} & 0 & 0 & 0 \\ -v/\rho & 0 & \frac{1}{\rho} & 0 & 0 \\ -w/\rho & 0 & 0 & \frac{1}{\rho} & 0 \\ (\gamma-1)q^2/2 & -(\gamma-1)u & -(\gamma-1)v & -(\gamma-1)w & \gamma-1 \end{bmatrix}, \tag{3.6.50}$$

$$\frac{\partial \mathbf{W}}{\partial \mathbf{U}} = \left(\frac{\partial \mathbf{U}}{\partial \mathbf{W}} \right)^{-1}, \quad \mathbf{A}_n^w = \frac{\partial \mathbf{W}}{\partial \mathbf{U}} \mathbf{A}_n \frac{\partial \mathbf{U}}{\partial \mathbf{W}}, \quad \mathbf{L}_n = \mathbf{L}_n^w \frac{\partial \mathbf{W}}{\partial \mathbf{U}}, \quad \mathbf{R}_n = \frac{\partial \mathbf{U}}{\partial \mathbf{W}} \mathbf{R}_n^w. \tag{3.6.51}$$

3.6.3   Symmetric Form

$$\mathbf{V}_t + \mathbf{A}^v \mathbf{V}_x + \mathbf{B}^v \mathbf{V}_y + \mathbf{C}^v \mathbf{V}_z = 0, \tag{3.6.52}$$

where

$$d\mathbf{V} = \begin{bmatrix} \dfrac{dp}{\rho c} \\ du \\ dv \\ dw \\ dp - a^2 d\rho \end{bmatrix}, \quad \mathbf{A}^v = \begin{bmatrix} u & c & 0 & 0 & 0 \\ c & u & 0 & 0 & 0 \\ 0 & 0 & u & 0 & 0 \\ 0 & 0 & 0 & u & 0 \\ 0 & 0 & 0 & 0 & u \end{bmatrix}, \tag{3.6.53}$$

$$\mathbf{B}^v = \begin{bmatrix} v & 0 & c & 0 & 0 \\ 0 & v & 0 & 0 & 0 \\ c & 0 & v & 0 & 0 \\ 0 & 0 & 0 & v & 0 \\ 0 & 0 & 0 & 0 & v \end{bmatrix}, \quad \mathbf{C}^v = \begin{bmatrix} w & 0 & 0 & c & 0 \\ 0 & w & 0 & 0 & 0 \\ 0 & 0 & w & 0 & 0 \\ c & 0 & 0 & w & 0 \\ 0 & 0 & 0 & 0 & w \end{bmatrix}. \tag{3.6.54}$$

Note that the fifth equation is an advection equation for the entropy:

$$s_t + u s_x + v s_y + w s_z = 0, \tag{3.6.55}$$

where the entropy s has been scaled by $(\gamma-1)p/R$. So, in all dimensions, basically the Euler equations describe isentropic flows. It is the Rankine-Hugoniot relation (the integral form of the Euler equations) that describes non-isentropic flows (shock waves).

Change of Variables

$$d\mathbf{U} = \frac{\partial \mathbf{U}}{\partial \mathbf{V}} d\mathbf{V}, \quad d\mathbf{V} = \frac{\partial \mathbf{V}}{\partial \mathbf{U}} d\mathbf{U}, \tag{3.6.56}$$

where

$$\frac{\partial \mathbf{U}}{\partial \mathbf{V}} = \begin{bmatrix} \dfrac{\rho}{c} & 0 & 0 & 0 & -\dfrac{1}{c^2} \\ \dfrac{\rho u}{c} & \rho & 0 & 0 & -\dfrac{u}{c^2} \\ \dfrac{\rho v}{c} & 0 & \rho & 0 & -\dfrac{v}{c^2} \\ \dfrac{\rho w}{c} & 0 & 0 & \rho & -\dfrac{w}{c^2} \\ \rho c \left(\dfrac{M^2}{2} + \dfrac{1}{\gamma-1} \right) & \rho u & \rho v & \rho w & -\dfrac{M^2}{2} \end{bmatrix}, \tag{3.6.57}$$

$$\frac{\partial \mathbf{V}}{\partial \mathbf{U}} = \left(\frac{\partial \mathbf{U}}{\partial \mathbf{V}}\right)^{-1} = \begin{bmatrix} \dfrac{\gamma-1}{2}\dfrac{q^2}{\rho c} & -(\gamma-1)\dfrac{u}{\rho c} & -(\gamma-1)\dfrac{v}{\rho c} & -(\gamma-1)\dfrac{w}{\rho c} & \dfrac{\gamma-1}{\rho c} \\ -\dfrac{u}{\rho} & \dfrac{1}{\rho} & 0 & 0 & 0 \\ -\dfrac{v}{\rho} & 0 & \dfrac{1}{\rho} & 0 & 0 \\ -\dfrac{w}{\rho} & 0 & 0 & \dfrac{1}{\rho} & 0 \\ \dfrac{\gamma-1}{2}q^2 - c^2 & -(\gamma-1)u & -(\gamma-1)v & -(\gamma-1)w & \gamma-1 \end{bmatrix}. \quad (3.6.58)$$

Eigenstructure

$$\mathbf{A}_n^v = \mathbf{R}_n^v \mathbf{\Lambda}_n \mathbf{L}_n^v, \quad (3.6.59)$$

where

$$\mathbf{A}_n^v = \begin{bmatrix} q_n & c\,n_x & c\,n_y & c\,n_z & 0 \\ c\,n_x & q_n & 0 & 0 & 0 \\ c\,n_y & 0 & q_n & 0 & 0 \\ c\,n_z & 0 & 0 & q_n & 0 \\ 0 & 0 & 0 & 0 & q_n \end{bmatrix}, \quad (3.6.60)$$

$$\mathbf{\Lambda}_n^v = \begin{bmatrix} q_n & 0 & 0 & 0 & 0 \\ 0 & q_n - c & 0 & 0 & 0 \\ 0 & 0 & q_n + c & 0 & 0 \\ 0 & 0 & 0 & q_n & 0 \\ 0 & 0 & 0 & 0 & q_n \end{bmatrix}, \quad (3.6.61)$$

$$\mathbf{R}_n^v = \begin{bmatrix} 0 & 1 & 1 & 0 & 0 \\ 0 & -n_x & n_x & \ell_x & m_x \\ 0 & -n_y & n_y & \ell_y & m_y \\ 0 & -n_z & n_z & \ell_z & m_z \\ 1 & 0 & 0 & 0 & 0 \end{bmatrix}, \quad (3.6.62)$$

$$\mathbf{L}_n^v = \begin{bmatrix} 0 & 0 & 0 & 0 & 1 \\ \frac{1}{2} & -\frac{1}{2}n_x & -\frac{1}{2}n_y & -\frac{1}{2}n_z & 0 \\ \frac{1}{2} & \frac{1}{2}n_x & \frac{1}{2}n_y & \frac{1}{2}n_z & 0 \\ 0 & \ell_x & \ell_y & \ell_z & 0 \\ 0 & m_x & m_y & m_z & 0 \end{bmatrix}. \quad (3.6.63)$$

Again, as in two dimensions, it can be seen from the right eigenvectors that the acoustic and shear waves do not affect the entropy.

Absolute Normal Jacobian

I like the absolute value of the normal Jacobian very very much for the symmetric form,

$$|\mathbf{A}_n^v| \equiv \mathbf{R}_n^v |\mathbf{\Lambda}_n^v| \mathbf{L}_n^v, \quad (3.6.64)$$

because it can be much simpler than that of the conservative form (see Section 3.6.1). If we denote the k-th column of $\mathbf{R}_n^v$ by $\mathbf{r}_k$ and the k-th row of $\mathbf{L}_n^v$ by $\boldsymbol{\ell}_k^t$, then we can write

$$|\mathbf{A}_n^v| = \sum_{k=1}^{5} |\lambda_k|\, \mathbf{r}_k \boldsymbol{\ell}_k^t. \tag{3.6.65}$$

First, consider the following components:

$$\mathbf{r}_1 \boldsymbol{\ell}_1^t = \begin{bmatrix} 0 & 0 & 0 & 0 & 0 \\ 0 & 0 & 0 & 0 & 0 \\ 0 & 0 & 0 & 0 & 0 \\ 0 & 0 & 0 & 0 & 0 \\ 0 & 0 & 0 & 0 & 1 \end{bmatrix}, \tag{3.6.66}$$

$$\mathbf{r}_4 \boldsymbol{\ell}_4^t = \begin{bmatrix} 0 & \mathbf{0}^t & 0 \\ \mathbf{0} & \boldsymbol{\ell}\otimes\boldsymbol{\ell} & \mathbf{0} \\ 0 & \mathbf{0}^t & 0 \end{bmatrix}, \tag{3.6.67}$$

$$\mathbf{r}_5 \boldsymbol{\ell}_5^t = \begin{bmatrix} 0 & \mathbf{0}^t & 0 \\ \mathbf{0} & \mathbf{m}\otimes\mathbf{m} & \mathbf{0} \\ 0 & \mathbf{0}^t & 0 \end{bmatrix}, \tag{3.6.68}$$

where $\mathbf{0}^t = [0,0,0]$, and so

$$|\lambda_1|\, \mathbf{r}_1\boldsymbol{\ell}_1^t + |\lambda_4|\, \mathbf{r}_4\boldsymbol{\ell}_4^t + |\lambda_5|\, \mathbf{r}_5\boldsymbol{\ell}_5^t = |q_n| \begin{bmatrix} 0 & \mathbf{0}^t & 0 \\ \mathbf{0} & \boldsymbol{\ell}\otimes\boldsymbol{\ell} + \mathbf{m}\otimes\mathbf{m} & \mathbf{0} \\ 0 & \mathbf{0}^t & 1 \end{bmatrix}$$

$$= |q_n| \begin{bmatrix} 0 & \mathbf{0}^t & 0 \\ \mathbf{0} & \mathbf{I} - \mathbf{n}\otimes\mathbf{n} & \mathbf{0} \\ 0 & \mathbf{0}^t & 1 \end{bmatrix}, \tag{3.6.69}$$

where $\mathbf{I}$ is the 3x3 identity matrix. Second, consider the other two components:

$$\mathbf{r}_2 \boldsymbol{\ell}_2^t = \frac{1}{2} \begin{bmatrix} 1 & -\mathbf{n}^t & 0 \\ -\mathbf{n} & \mathbf{n}\otimes\mathbf{n} & \mathbf{0} \\ 0 & \mathbf{0}^t & 0 \end{bmatrix}, \tag{3.6.70}$$

$$\mathbf{r}_3 \boldsymbol{\ell}_3^t = \frac{1}{2} \begin{bmatrix} 1 & \mathbf{n}^t & 0 \\ \mathbf{n} & \mathbf{n}\otimes\mathbf{n} & \mathbf{0} \\ 0 & \mathbf{0}^t & 0 \end{bmatrix}, \tag{3.6.71}$$

and so,

$$|\lambda_2|\, \mathbf{r}_2\boldsymbol{\ell}_2^t + |\lambda_3|\, \mathbf{r}_3\boldsymbol{\ell}_3^t = \begin{bmatrix} \overline{|\lambda|} & \widetilde{|\lambda|}\,\mathbf{n}^t & 0 \\ \widetilde{|\lambda|}\,\mathbf{n} & \overline{|\lambda|}\,\mathbf{n}\otimes\mathbf{n} & \mathbf{0} \\ 0 & \mathbf{0}^t & 0 \end{bmatrix}, \tag{3.6.72}$$

where

$$\overline{|\lambda|} = \frac{|\lambda_3| + |\lambda_2|}{2} = \frac{|q_n + c| + |q_n - c|}{2}, \tag{3.6.73}$$

$$\widetilde{|\lambda|} = \frac{|\lambda_3| - |\lambda_2|}{2} = \frac{|q_n + c| - |q_n - c|}{2}. \tag{3.6.74}$$

Finally, we obtain

$$|\mathbf{A}_n^v| = \sum_{k=1}^{5} |\lambda_k| \, \mathbf{r}_k \boldsymbol{\ell}_k^t = \begin{bmatrix} \overline{|\lambda|} & \widetilde{|\lambda|}\, \mathbf{n}^t & 0 \\ \widetilde{|\lambda|}\, \mathbf{n} & |q_n|\,\mathbf{I} + (\overline{|\lambda|} - |q_n|)\, \mathbf{n} \otimes \mathbf{n} & \mathbf{0} \\ 0 & \mathbf{0}^t & |q_n| \end{bmatrix}. \tag{3.6.75}$$

Comparing this with $|\mathbf{A}_n|$ in Section 3.6.1, we immediately see that this is so much simpler. I really like it. Note also that

$$|\mathbf{A}_n| = \mathbf{R}_n |\boldsymbol{\Lambda}_n| \mathbf{L}_n = \frac{\partial \mathbf{U}}{\partial \mathbf{V}} \mathbf{R}_n^v |\boldsymbol{\Lambda}_n| \mathbf{L}_n^v \frac{\partial \mathbf{V}}{\partial \mathbf{U}} = \frac{\partial \mathbf{U}}{\partial \mathbf{V}} |\mathbf{A}_n^v| \frac{\partial \mathbf{V}}{\partial \mathbf{U}}, \tag{3.6.76}$$

which is possible because the conservative form and the symmetric form are related through a similarity transformation. Hence, the absolute Jacobian $|\mathbf{A}_n|$ can be calculated via a much simpler matrix (3.6.75). This is very nice.

Streamline Coordinates

Take x-axis as a streamline direction denoted by ξ, and y-axis and z-axis as mutually orthogonal normal directions denoted by η_ℓ and η_m, then I get the following.

$$\tilde{\mathbf{V}}_t + \tilde{\mathbf{A}}^v \tilde{\mathbf{V}}_\xi + \tilde{\mathbf{B}}^v \tilde{\mathbf{V}}_{\eta_\ell} + \tilde{\mathbf{C}}^v \tilde{\mathbf{V}}_{\eta_m} = 0, \tag{3.6.77}$$

where

$$d\tilde{\mathbf{V}} = \begin{bmatrix} \dfrac{dp}{\rho c} \\ dq \\ dq_\ell \\ dq_m \\ dp - a^2 d\rho \end{bmatrix}, \quad \tilde{\mathbf{A}}^v = \begin{bmatrix} q & c & 0 & 0 & 0 \\ c & q & 0 & 0 & 0 \\ 0 & 0 & q & 0 & 0 \\ 0 & 0 & 0 & q & 0 \\ 0 & 0 & 0 & 0 & q \end{bmatrix}, \tag{3.6.78}$$

$$\tilde{\mathbf{B}}^v = \begin{bmatrix} 0 & 0 & c & 0 & 0 \\ 0 & 0 & 0 & 0 & 0 \\ c & 0 & 0 & 0 & 0 \\ 0 & 0 & 0 & 0 & 0 \\ 0 & 0 & 0 & 0 & 0 \end{bmatrix}, \quad \tilde{\mathbf{C}}^v = \begin{bmatrix} 0 & 0 & 0 & c & 0 \\ 0 & 0 & 0 & 0 & 0 \\ 0 & 0 & 0 & 0 & 0 \\ c & 0 & 0 & 0 & 0 \\ 0 & 0 & 0 & 0 & 0 \end{bmatrix}, \tag{3.6.79}$$

and we have set the unit vector in ξ direction to be $\mathbf{n} = \mathbf{u}/q$, and the unit vectors in the other two directions, η_ℓ and η_m, to be $\boldsymbol{\ell}$ and $\mathbf{m}$, and defined the followings:

$$q = un_x + vn_y + wn_z, \tag{3.6.80}$$
$$q_\ell = u\ell_x + v\ell_y + w\ell_z, \tag{3.6.81}$$
$$q_m = um_x + vm_y + wm_z, \tag{3.6.82}$$
$$\tilde{\mathbf{A}}^v = \mathbf{A}^v n_x + \mathbf{B}^v n_y + \mathbf{C}^v n_z, \tag{3.6.83}$$
$$\tilde{\mathbf{B}}^v = \mathbf{A}^v \ell_x + \mathbf{B}^v \ell_y + \mathbf{C}^v \ell_z, \tag{3.6.84}$$
$$\tilde{\mathbf{C}}^v = \mathbf{A}^v m_x + \mathbf{B}^v m_y + \mathbf{C}^v m_z, \tag{3.6.85}$$

$$d\tilde{\mathbf{V}} = \frac{\partial \tilde{\mathbf{V}}}{\partial \mathbf{V}} d\mathbf{V} = \begin{bmatrix} 1 & 0 & 0 & 0 & 0 \\ 0 & n_x & n_y & n_z & 0 \\ 0 & \ell_x & \ell_y & \ell_z & 0 \\ 0 & m_x & m_y & m_z & 0 \\ 0 & 0 & 0 & 0 & 1 \end{bmatrix} d\mathbf{V}. \tag{3.6.86}$$

3.6.4 Parameter Vector

The parameter vector is defined by

$$
\mathbf{Z} \;=\; \sqrt{\rho}\begin{bmatrix} 1 \\ u \\ v \\ w \\ H \end{bmatrix} = \begin{bmatrix} z_1 \\ z_2 \\ z_3 \\ z_4 \\ z_5 \end{bmatrix}.
\tag{3.6.87}
$$

I like it as in two dimensions because again every component of conservative variables and fluxes can be expressed as a quadratic in the components of $\mathbf{Z}$ [128].

$$
\mathbf{U} \;=\; \begin{bmatrix} \rho \\ \rho u \\ \rho v \\ \rho w \\ \rho E \end{bmatrix} = \begin{bmatrix} z_1^2 \\ z_1 z_2 \\ z_1 z_3 \\ z_1 z_4 \\ \dfrac{z_1 z_5}{\gamma} + \dfrac{\gamma - 1}{\gamma}\dfrac{z_2^2 + z_3^2 + z_4^2}{2} \end{bmatrix},
\tag{3.6.88}
$$

$$
\mathbf{F} \;=\; \begin{bmatrix} \rho u \\ \rho u^2 + p \\ \rho u v \\ \rho u w \\ \rho u H \end{bmatrix} = \begin{bmatrix} z_1 z_2 \\ z_2^2 + p \\ z_2 z_3 \\ z_2 z_4 \\ z_2 z_5 \end{bmatrix},
\tag{3.6.89}
$$

$$
\mathbf{G} \;=\; \begin{bmatrix} \rho v \\ \rho v u \\ \rho v^2 + p \\ \rho v w \\ \rho v H \end{bmatrix} = \begin{bmatrix} z_1 z_3 \\ z_3 z_2 \\ z_3^2 + p \\ z_3 z_4 \\ z_3 z_5 \end{bmatrix},
\tag{3.6.90}
$$

$$
\mathbf{H} \;=\; \begin{bmatrix} \rho w \\ \rho w u \\ \rho w v \\ \rho w^2 + p \\ \rho w H \end{bmatrix} = \begin{bmatrix} z_1 z_4 \\ z_4 z_2 \\ z_4 z_3 \\ z_4^2 + p \\ z_4 z_5 \end{bmatrix},
\tag{3.6.91}
$$

$$
p \;=\; \frac{\gamma - 1}{\gamma}\left(z_1 z_5 - \frac{z_2^2 + z_3^2 + z_4^2}{2} \right).
\tag{3.6.92}
$$

We can write the conservative form of the Euler equations in terms of the parameter vector variables:

$$
\mathbf{U}_t + \mathbf{A}^z \mathbf{Z}_x + \mathbf{B}^z \mathbf{Z}_y + \mathbf{C}^z \mathbf{Z}_z = 0,
\tag{3.6.93}
$$

$$\mathbf{A}^z = \frac{\partial \mathbf{F}}{\partial \mathbf{Z}} = \begin{bmatrix} z_2 & z_1 & 0 & 0 & 0 \\ \frac{\gamma-1}{\gamma}z_5 & \frac{\gamma+1}{\gamma}z_2 & -\frac{\gamma-1}{\gamma}z_3 & -\frac{\gamma-1}{\gamma}z_4 & \frac{\gamma-1}{\gamma}z_1 \\ 0 & z_3 & z_2 & 0 & 0 \\ 0 & z_4 & 0 & z_2 & 0 \\ 0 & z_5 & 0 & 0 & z_2 \end{bmatrix}, \tag{3.6.94}$$

$$\mathbf{B}^z = \frac{\partial \mathbf{G}}{\partial \mathbf{Z}} = \begin{bmatrix} z_3 & 0 & z_1 & 0 & 0 \\ 0 & z_3 & z_2 & 0 & 0 \\ \frac{\gamma-1}{\gamma}z_5 & -\frac{\gamma-1}{\gamma}z_2 & \frac{\gamma+1}{\gamma}z_3 & -\frac{\gamma-1}{\gamma}z_4 & \frac{\gamma-1}{\gamma}z_1 \\ 0 & 0 & z_4 & z_3 & 0 \\ 0 & 0 & z_5 & 0 & z_3 \end{bmatrix}, \tag{3.6.95}$$

$$\mathbf{C}^z = \frac{\partial \mathbf{H}}{\partial \mathbf{Z}} = \begin{bmatrix} z_4 & 0 & 0 & z_1 & 0 \\ 0 & z_4 & 0 & z_2 & 0 \\ 0 & 0 & z_4 & z_3 & 0 \\ \frac{\gamma-1}{\gamma}z_5 & -\frac{\gamma-1}{\gamma}z_2 & -\frac{\gamma-1}{\gamma}z_3 & \frac{\gamma+1}{\gamma}z_4 & \frac{\gamma-1}{\gamma}z_1 \\ 0 & 0 & 0 & z_5 & z_4 \end{bmatrix}. \tag{3.6.96}$$

Again, the Jacobian matrices are linear in the parameter vector variables, and this can be useful to linearize the Euler equations over a computational cell [31]. Of course, I like to perform change of variables:

$$d\mathbf{Z} = \frac{\partial \mathbf{Z}}{\partial \mathbf{U}}d\mathbf{U}, \quad d\mathbf{U} = \frac{\partial \mathbf{U}}{\partial \mathbf{Z}}d\mathbf{Z}, \tag{3.6.97}$$

where

$$\frac{\partial \mathbf{U}}{\partial \mathbf{Z}} = \begin{bmatrix} 2z_1 & 0 & 0 & 0 & 0 \\ z_2 & z_1 & 0 & 0 & 0 \\ z_3 & 0 & z_1 & 0 & 0 \\ z_4 & 0 & 0 & z_1 & 0 \\ \frac{z_5}{\gamma} & \frac{\gamma-1}{\gamma}z_2 & \frac{\gamma-1}{\gamma}z_3 & \frac{\gamma-1}{\gamma}z_4 & \frac{z_1}{\gamma} \end{bmatrix}, \tag{3.6.98}$$

$$\frac{\partial \mathbf{Z}}{\partial \mathbf{U}} = \frac{1}{z_1}\begin{bmatrix} \frac{1}{2} & 0 & 0 & 0 & 0 \\ -\frac{1}{2}\frac{z_2}{z_1} & 1 & 0 & 0 & 0 \\ -\frac{1}{2}\frac{z_3}{z_1} & 0 & 1 & 0 & 0 \\ -\frac{1}{2}\frac{z_4}{z_1} & 0 & 0 & 1 & 0 \\ \frac{\gamma-1}{2}\frac{z_2^2+z_3^2+z_4^2}{z_1^2}-\frac{1}{2}\frac{z_5}{z_1} & -\frac{(\gamma-1)z_2}{z_1} & -\frac{(\gamma-1)z_3}{z_1} & -\frac{(\gamma-1)z_4}{z_1} & \gamma \end{bmatrix} \tag{3.6.99}$$

$$= \frac{1}{\sqrt{\rho}} \begin{bmatrix} \frac{1}{2} & 0 & 0 & 0 & 0 \\ -\frac{u}{2} & 1 & 0 & 0 & 0 \\ -\frac{v}{2} & 0 & 1 & 0 & 0 \\ -\frac{w}{2} & 0 & 0 & 1 & 0 \\ \frac{(\gamma-1)q^2}{2} - \frac{H}{2} & -(\gamma-1)u & -(\gamma-1)v & -(\gamma-1)w & \gamma \end{bmatrix}, \qquad (3.6.100)$$

from which we find

$$\mathbf{Z} = 2\frac{\partial \mathbf{Z}}{\partial \mathbf{U}}\mathbf{U}, \quad \mathbf{U} = \frac{1}{2}\frac{\partial \mathbf{U}}{\partial \mathbf{Z}}\mathbf{Z}. \qquad (3.6.101)$$

As in the two-dimensional case, I find it very interesting. Of course, since we have

$$\mathbf{F} = \mathbf{A}\mathbf{U}, \quad \mathbf{G} = \mathbf{B}\mathbf{U}, \quad \mathbf{H} = \mathbf{C}\mathbf{U}, \qquad (3.6.102)$$

for the Euler equations even in three dimensions (see Section 3.10), we find from Equation (3.6.101) as in two dimensions that

$$\mathbf{F} = \mathbf{A}\mathbf{U} = \frac{1}{2}\frac{\partial \mathbf{F}}{\partial \mathbf{U}}\frac{\partial \mathbf{U}}{\partial \mathbf{Z}}\mathbf{Z} = \frac{1}{2}\mathbf{A}^z\mathbf{Z}, \qquad (3.6.103)$$

$$\mathbf{G} = \mathbf{B}\mathbf{U} = \frac{1}{2}\frac{\partial \mathbf{G}}{\partial \mathbf{U}}\frac{\partial \mathbf{U}}{\partial \mathbf{Z}}\mathbf{Z} = \frac{1}{2}\mathbf{B}^z\mathbf{Z}, \qquad (3.6.104)$$

$$\mathbf{H} = \mathbf{C}\mathbf{U} = \frac{1}{2}\frac{\partial \mathbf{H}}{\partial \mathbf{U}}\frac{\partial \mathbf{U}}{\partial \mathbf{Z}}\mathbf{Z} = \frac{1}{2}\mathbf{C}^z\mathbf{Z}. \qquad (3.6.105)$$

Yes, we can directly compute the fluxes by the parameter vector variables without converting them into the primitive variables. I really like the fact that the physical fluxes are quadratic in $\mathbf{Z}$ as seen in the above equations.

3.6.5   Derivatives of Eigenvalues

Sometimes, it is necessary to differentiate a numerical flux function, for example, for constructing implicit schemes. Then, we would need to differentiate various quantities, e.g., the eigenvalues with respect to the conservative variables. I have done it before, and I like it. Suppose we wish to compute $\partial q_n/\partial \mathbf{U}$. It can be computed as follows.

$$\frac{\partial q_n}{\partial \mathbf{U}} = \frac{\partial q_n}{\partial \mathbf{W}}\frac{\partial \mathbf{W}}{\partial \mathbf{U}}$$

$$= [0, n_x, n_y, n_z, 0] \begin{bmatrix} 1 & 0 & 0 & 0 & 0 \\ -\frac{u}{\rho} & \frac{1}{\rho} & 0 & 0 & 0 \\ -\frac{v}{\rho} & 0 & \frac{1}{\rho} & 0 & 0 \\ -\frac{w}{\rho} & 0 & 0 & \frac{1}{\rho} & 0 \\ \frac{\gamma-1}{2}q^2 & -(\gamma-1)u & -(\gamma-1)v & -(\gamma-1)w & \gamma-1 \end{bmatrix}$$

$$= \left[-\frac{q_n}{\rho}, \frac{n_x}{\rho}, \frac{n_y}{\rho}, \frac{n_z}{\rho}, 0 \right], \qquad (3.6.106)$$

which is the derivative of the eigenvalue, q_n. How about the speed of sound? In the same way, we obtain

$$
\frac{\partial c}{\partial \mathbf{U}} = \frac{\partial c}{\partial \mathbf{W}} \frac{\partial \mathbf{W}}{\partial \mathbf{U}}
$$

$$
= \left[-\frac{c}{2\rho}, 0, 0, 0, \frac{\gamma}{2\rho c} \right]
\begin{bmatrix}
1 & 0 & 0 & 0 & 0 \\
-\dfrac{u}{\rho} & \dfrac{1}{\rho} & 0 & 0 & 0 \\
-\dfrac{v}{\rho} & 0 & \dfrac{1}{\rho} & 0 & 0 \\
-\dfrac{w}{\rho} & 0 & 0 & \dfrac{1}{\rho} & 0 \\
\dfrac{\gamma - 1}{2} q^2 & -(\gamma - 1)u & -(\gamma - 1)v & -(\gamma - 1)w & \gamma - 1
\end{bmatrix}
$$

$$
= \left[-\frac{c}{2\rho}\left(1 - \frac{\gamma(\gamma - 1)}{2}\frac{q^2}{c^2}\right), \frac{-\gamma(\gamma - 1)u}{2\rho c}, \frac{-\gamma(\gamma - 1)v}{2\rho c}, \frac{-\gamma(\gamma - 1)w}{2\rho c}, \frac{\gamma(\gamma - 1)}{2\rho c} \right]. \tag{3.6.107}
$$

So, it is now easy to get the derivatives of the remaining eigenvalues, $q_n - c$ and $q_n + c$.

$$
\frac{\partial (q_n - c)}{\partial \mathbf{U}} = \frac{\partial q_n}{\partial \mathbf{U}} - \frac{\partial c}{\partial \mathbf{U}} =
\begin{bmatrix}
-\dfrac{q_n}{\rho} + \dfrac{c}{2\rho}\left(1 - \dfrac{\gamma(\gamma - 1)}{2}\dfrac{q^2}{c^2}\right) \\[3mm]
\dfrac{n_x}{\rho} + \dfrac{\gamma(\gamma - 1)u}{2\rho c} \\[3mm]
\dfrac{n_y}{\rho} + \dfrac{\gamma(\gamma - 1)v}{2\rho c} \\[3mm]
\dfrac{n_z}{\rho} + \dfrac{\gamma(\gamma - 1)w}{2\rho c} \\[3mm]
-\dfrac{\gamma(\gamma - 1)}{2\rho c}
\end{bmatrix}^{t}, \tag{3.6.108}
$$

$$
\frac{\partial (q_n + c)}{\partial \mathbf{U}} = \frac{\partial q_n}{\partial \mathbf{U}} + \frac{\partial c}{\partial \mathbf{U}} =
\begin{bmatrix}
-\dfrac{q_n}{\rho} - \dfrac{c}{2\rho}\left(1 - \dfrac{\gamma(\gamma - 1)}{2}\dfrac{q^2}{c^2}\right) \\[3mm]
\dfrac{n_x}{\rho} - \dfrac{\gamma(\gamma - 1)u}{2\rho c} \\[3mm]
\dfrac{n_y}{\rho} - \dfrac{\gamma(\gamma - 1)v}{2\rho c} \\[3mm]
\dfrac{n_z}{\rho} - \dfrac{\gamma(\gamma - 1)w}{2\rho c} \\[3mm]
\dfrac{\gamma(\gamma - 1)}{2\rho c}
\end{bmatrix}^{t}. \tag{3.6.109}
$$

Some numerical flux functions involve the maximum eigenvalue, $|q_n| + c$. Its derivative can be obtained in the same way. First compute

$$\frac{\partial |q_n|}{\partial \mathbf{U}} = \frac{\partial |q_n|}{\partial \mathbf{W}} \frac{\partial \mathbf{W}}{\partial \mathbf{U}}$$

$$= [0, \text{sign}(q_n)\, n_x, \text{sign}(q_n)\, n_y, \text{sign}(q_n)\, n_z, 0] \begin{bmatrix} 1 & 0 & 0 & 0 & 0 \\ -\dfrac{u}{\rho} & \dfrac{1}{\rho} & 0 & 0 & 0 \\ -\dfrac{v}{\rho} & 0 & \dfrac{1}{\rho} & 0 & 0 \\ -\dfrac{w}{\rho} & 0 & 0 & \dfrac{1}{\rho} & 0 \\ \dfrac{\gamma-1}{2}q^2 & -(\gamma-1)u & -(\gamma-1)v & -(\gamma-1)w & \gamma-1 \end{bmatrix}$$

$$= \left[-\frac{|q_n|}{\rho}, \frac{\text{sign}(q_n)\, n_x}{\rho}, \frac{\text{sign}(q_n)\, n_y}{\rho}, \frac{\text{sign}(q_n)\, n_z}{\rho}, 0 \right], \tag{3.6.110}$$

and then we obtain

$$\frac{\partial(|q_n| + c)}{\partial \mathbf{U}} = \begin{bmatrix} -\dfrac{|q_n|}{\rho} - \dfrac{c}{2\rho}\left(1 - \dfrac{\gamma(\gamma-1)}{2}\dfrac{q^2}{c^2}\right) \\[2ex] \dfrac{\text{sign}(q_n)\, n_x}{\rho} - \dfrac{\gamma(\gamma-1)u}{2\rho c} \\[2ex] \dfrac{\text{sign}(q_n)\, n_y}{\rho} - \dfrac{\gamma(\gamma-1)v}{2\rho c} \\[2ex] \dfrac{\text{sign}(q_n)\, n_z}{\rho} - \dfrac{\gamma(\gamma-1)w}{2\rho c} \\[2ex] \dfrac{\gamma(\gamma-1)}{2\rho c} \end{bmatrix}^t . \tag{3.6.111}$$

If we wish to differentiate Roe's approximate Riemann solver [128], which is a very popular numerical flux function for the Euler equations, we need to differentiate the eigenvalues evaluated at the Roe-averaged quantities:

$$\hat{\rho} = \sqrt{\rho_L \rho_R}, \tag{3.6.112}$$

$$\hat{u} = \frac{\sqrt{\rho_L}\, u_L + \sqrt{\rho_R}\, u_R}{\sqrt{\rho_L} + \sqrt{\rho_R}}, \tag{3.6.113}$$

$$\hat{v} = \frac{\sqrt{\rho_L}\, v_L + \sqrt{\rho_R}\, v_R}{\sqrt{\rho_L} + \sqrt{\rho_R}}, \tag{3.6.114}$$

$$\hat{w} = \frac{\sqrt{\rho_L}\, w_L + \sqrt{\rho_R}\, w_R}{\sqrt{\rho_L} + \sqrt{\rho_R}}, \tag{3.6.115}$$

$$\hat{H} = \frac{\sqrt{\rho_L} H_L + \sqrt{\rho_R} H_R}{\sqrt{\rho_L} + \sqrt{\rho_R}}, \tag{3.6.116}$$

$$\hat{c} = \sqrt{(\gamma - 1)\left(\hat{H} - \frac{\hat{q}^2}{2}\right)}, \tag{3.6.117}$$

$$\hat{q}^2 = \hat{u}^2 + \hat{v}^2 + \hat{w}^2, \tag{3.6.118}$$

where the subscripts L and R indicate the two states that are averaged. See Ref.[128] for details and Ref.[109] for the Roe-averaged density, $\hat{\rho}$. Here, I'm interested to compute the derivatives of the eigenvalues, $\hat{q}_n$, $\hat{q}_n + \hat{c}$, and $\hat{q}_n - \hat{c}$, where $\hat{q}_n = \hat{u} n_x + \hat{v} n_y + \hat{w} n_z$. We can compute the derivative of $\hat{q}_n$ as follows.

$$\frac{\partial \hat{q}_n}{\partial \mathbf{U}} = \frac{\partial \hat{q}_n}{\partial \hat{\mathbf{W}}^H} \frac{\partial \hat{\mathbf{W}}^H}{\partial \mathbf{W}^H} \frac{\partial \mathbf{W}^H}{\partial \mathbf{W}} \frac{\partial \mathbf{W}}{\partial \mathbf{U}}, \tag{3.6.119}$$

where

$$\hat{\mathbf{W}}^H = \left[\hat{\rho}, \hat{u}, \hat{v}, \hat{w}, \hat{H}\right]^t, \quad \mathbf{W}^H = [\rho, u, v, w, H]^t. \tag{3.6.120}$$

Note that we are interested in $\partial \hat{q}_n / \partial \mathbf{U}_L$ and $\partial \hat{q}_n / \partial \mathbf{U}_R$, but the subscripts L and R have been dropped because the above expression is valid for both L and R. From now on, the quantities with no subscripts are understood as either the state L or the state R. The first derivative on the right hand side, i.e., $\partial \hat{q}_n / \partial \hat{\mathbf{W}}^H$, is given by

$$\frac{\partial \hat{q}_n}{\partial \hat{\mathbf{W}}^H} = [0, n_x, n_y, n_z, 0]. \tag{3.6.121}$$

The transformation matrix $\partial \hat{\mathbf{W}}^H / \partial \mathbf{W}^H$ is given by

$$\frac{\partial \hat{\mathbf{W}}^H}{\partial \mathbf{W}^H} = \begin{bmatrix} \dfrac{\hat{\rho}}{2\rho} & 0 & 0 & 0 & 0 \\[2mm] \dfrac{u - \hat{u}}{2(\rho + \hat{\rho})} & \dfrac{\rho}{\rho + \hat{\rho}} & 0 & 0 & 0 \\[2mm] \dfrac{v - \hat{v}}{2(\rho + \hat{\rho})} & 0 & \dfrac{\rho}{\rho + \hat{\rho}} & 0 & 0 \\[2mm] \dfrac{w - \hat{w}}{2(\rho + \hat{\rho})} & 0 & 0 & \dfrac{\rho}{\rho + \hat{\rho}} & 0 \\[2mm] \dfrac{H - \hat{H}}{2(\rho + \hat{\rho})} & 0 & 0 & 0 & \dfrac{\rho}{\rho + \hat{\rho}} \end{bmatrix}. \tag{3.6.122}$$

I like this matrix very much because it has been arranged such that it can be used to compute both $\partial \hat{\mathbf{W}}^H / \partial \mathbf{W}_L^H$ and $\partial \hat{\mathbf{W}}^H / \partial \mathbf{W}_R^H$. That is,

$$\frac{\partial \hat{\mathbf{W}}^H}{\partial \mathbf{W}_L^H} = \begin{bmatrix} \dfrac{\hat{\rho}}{2\rho_L} & 0 & 0 & 0 & 0 \\[2mm] \dfrac{u_L - \hat{u}}{2(\rho_L + \hat{\rho})} & \dfrac{\rho_L}{\rho_L + \hat{\rho}} & 0 & 0 & 0 \\[2mm] \dfrac{v_L - \hat{v}}{2(\rho_L + \hat{\rho})} & 0 & \dfrac{\rho_L}{\rho_L + \hat{\rho}} & 0 & 0 \\[2mm] \dfrac{w_L - \hat{w}}{2(\rho_L + \hat{\rho})} & 0 & 0 & \dfrac{\rho_L}{\rho_L + \hat{\rho}} & 0 \\[2mm] \dfrac{H_L - \hat{H}}{2(\rho_L + \hat{\rho})} & 0 & 0 & 0 & \dfrac{\rho_L}{\rho_L + \hat{\rho}} \end{bmatrix}, \tag{3.6.123}$$

and

$$\frac{\partial \hat{\mathbf{W}}^H}{\partial \mathbf{W}_R^H} = \begin{bmatrix} \dfrac{\hat{\rho}}{2\rho_R} & 0 & 0 & 0 & 0 \\[2ex] \dfrac{u_R - \hat{u}}{2(\rho_R + \hat{\rho})} & \dfrac{\rho_R}{\rho_R + \hat{\rho}} & 0 & 0 & 0 \\[2ex] \dfrac{v_R - \hat{v}}{2(\rho_R + \hat{\rho})} & 0 & \dfrac{\rho_R}{\rho_R + \hat{\rho}} & 0 & 0 \\[2ex] \dfrac{w_R - \hat{w}}{2(\rho_R + \hat{\rho})} & 0 & 0 & \dfrac{\rho_R}{\rho_R + \hat{\rho}} & 0 \\[2ex] \dfrac{H_R - \hat{H}}{2(\rho_R + \hat{\rho})} & 0 & 0 & 0 & \dfrac{\rho_R}{\rho_R + \hat{\rho}} \end{bmatrix} . \tag{3.6.124}$$

It is nice. I like it. So, I continue to drop the subscripts below. Next, the matrix $\partial \mathbf{W}^H / \partial \mathbf{W}$ is given by

$$\frac{\partial \mathbf{W}^H}{\partial \mathbf{W}} = \begin{bmatrix} 1 & 0 & 0 & 0 & 0 \\[1ex] 0 & 1 & 0 & 0 & 0 \\[1ex] 0 & 0 & 1 & 0 & 0 \\[1ex] 0 & 0 & 0 & 1 & 0 \\[1ex] -\dfrac{a^2}{\rho(\gamma - 1)} & u & v & w & \dfrac{\gamma}{\rho(\gamma - 1)} \end{bmatrix} , \tag{3.6.125}$$

which also can be used to compute both $\partial \hat{\mathbf{W}}_L^H / \partial \mathbf{W}_L^H$ and $\partial \hat{\mathbf{W}}_R^H / \partial \mathbf{W}_R^H$. We are now ready to compute the derivative. First compute

$$\frac{\partial \hat{\mathbf{W}}^H}{\partial \mathbf{W}} = \frac{\partial \hat{\mathbf{W}}^H}{\partial \mathbf{W}^H} \frac{\partial \mathbf{W}^H}{\partial \mathbf{W}}$$

$$= \begin{bmatrix} \dfrac{\hat{\rho}}{2\rho} & 0 & 0 & 0 & 0 \\[2ex] \dfrac{u - \hat{u}}{2(\rho + \hat{\rho})} & \dfrac{\rho}{\rho + \hat{\rho}} & 0 & 0 & 0 \\[2ex] \dfrac{v - \hat{v}}{2(\rho + \hat{\rho})} & 0 & \dfrac{\rho}{\rho + \hat{\rho}} & 0 & 0 \\[2ex] \dfrac{w - \hat{w}}{2(\rho + \hat{\rho})} & 0 & 0 & \dfrac{\rho}{\rho + \hat{\rho}} & 0 \\[2ex] \dfrac{H - \hat{H}}{2(\rho + \hat{\rho})} & 0 & 0 & 0 & \dfrac{\rho}{\rho + \hat{\rho}} \end{bmatrix} \begin{bmatrix} 1 & 0 & 0 & 0 & 0 \\[1ex] 0 & 1 & 0 & 0 & 0 \\[1ex] 0 & 0 & 1 & 0 & 0 \\[1ex] 0 & 0 & 0 & 1 & 0 \\[1ex] -\dfrac{a^2}{\rho(\gamma - 1)} & u & v & w & \dfrac{\gamma}{\rho(\gamma - 1)} \end{bmatrix}$$

$$= \begin{bmatrix} \dfrac{\hat{\rho}}{2\rho} & 0 & 0 & 0 & 0 \\[2ex] \dfrac{u - \hat{u}}{2(\rho + \hat{\rho})} & \dfrac{\rho}{\rho + \hat{\rho}} & 0 & 0 & 0 \\[2ex] \dfrac{v - \hat{v}}{2(\rho + \hat{\rho})} & 0 & \dfrac{\rho}{\rho + \hat{\rho}} & 0 & 0 \\[2ex] \dfrac{w - \hat{w}}{2(\rho + \hat{\rho})} & 0 & 0 & \dfrac{\rho}{\rho + \hat{\rho}} & 0 \\[2ex] \dfrac{H - \hat{H}}{2(\rho + \hat{\rho})} - \dfrac{a^2}{(\gamma - 1)(\rho + \hat{\rho})} & \dfrac{\rho u}{\rho + \hat{\rho}} & \dfrac{\rho v}{\rho + \hat{\rho}} & \dfrac{\rho w}{\rho + \hat{\rho}} & \dfrac{\gamma}{(\gamma - 1)(\rho + \hat{\rho})} \end{bmatrix} , \tag{3.6.126}$$

and then

$$\frac{\partial \hat{\mathbf{W}}^H}{\partial \mathbf{U}} = \frac{\partial \hat{\mathbf{W}}^H}{\partial \mathbf{W}} \frac{\partial \mathbf{W}}{\partial \mathbf{U}}$$

$$= \frac{1}{\rho + \hat{\rho}} \begin{bmatrix} \dfrac{\hat{\rho}(\rho + \hat{\rho})}{2\rho} & 0 & 0 & 0 & 0 \\ \dfrac{u - \hat{u}}{2} & \rho & 0 & 0 & 0 \\ \dfrac{v - \hat{v}}{2} & 0 & \rho & 0 & 0 \\ \dfrac{w - \hat{w}}{2} & 0 & 0 & \rho & 0 \\ \dfrac{H - \hat{H}}{2} - \dfrac{a^2}{\gamma - 1} & \rho u & \rho v & \rho w & \dfrac{\gamma}{\gamma - 1} \end{bmatrix} \begin{bmatrix} 1 & 0 & 0 & 0 & 0 \\ \dfrac{-u}{\rho} & \dfrac{1}{\rho} & 0 & 0 & 0 \\ \dfrac{-v}{\rho} & 0 & \dfrac{1}{\rho} & 0 & 0 \\ \dfrac{-w}{\rho} & 0 & 0 & \dfrac{1}{\rho} & 0 \\ \dfrac{(\gamma - 1)q^2}{2} & (1 - \gamma)u & (1 - \gamma)v & (1 - \gamma)w & \gamma - 1 \end{bmatrix}$$

$$= \frac{1}{\rho + \hat{\rho}} \begin{bmatrix} \dfrac{\hat{\rho}(\rho + \hat{\rho})}{2\rho} & 0 & 0 & 0 & 0 \\ -\dfrac{u + \hat{u}}{2} & 1 & 0 & 0 & 0 \\ -\dfrac{v + \hat{v}}{2} & 0 & 1 & 0 & 0 \\ -\dfrac{w + \hat{w}}{2} & 0 & 0 & 1 & 0 \\ \dfrac{H - \hat{H}}{2} - \dfrac{a^2}{\gamma - 1} + \dfrac{(\gamma - 2)q^2}{2} & (1 - \gamma)u & (1 - \gamma)v & (1 - \gamma)w & \gamma \end{bmatrix}, \qquad (3.6.12$$

and we finally obtain

$$\frac{\partial \hat{q}_n}{\partial \mathbf{U}} = \frac{\partial \hat{q}_n}{\partial \hat{\mathbf{W}}^H} \frac{\partial \hat{\mathbf{W}}^H}{\partial \mathbf{W}^H} \frac{\partial \mathbf{W}^H}{\partial \mathbf{W}} \frac{\partial \mathbf{W}}{\partial \mathbf{U}}$$

$$= \frac{\partial \hat{q}_n}{\partial \hat{\mathbf{W}}^H} \frac{\partial \hat{\mathbf{W}}^H}{\partial \mathbf{W}} \frac{\partial \mathbf{W}}{\partial \mathbf{U}}$$

$$= \frac{\partial \hat{q}_n}{\partial \hat{\mathbf{W}}^H} \frac{\partial \hat{\mathbf{W}}^H}{\partial \mathbf{U}}$$

$$= \frac{1}{\rho + \hat{\rho}} [0, n_x, n_y, n_z, 0] \begin{bmatrix} \dfrac{\hat{\rho}(\rho + \hat{\rho})}{2\rho} & 0 & 0 & 0 & 0 \\ -\dfrac{u + \hat{u}}{2} & 1 & 0 & 0 & 0 \\ -\dfrac{v + \hat{v}}{2} & 0 & 1 & 0 & 0 \\ -\dfrac{w + \hat{w}}{2} & 0 & 0 & 1 & 0 \\ \dfrac{H - \hat{H}}{2} - \dfrac{a^2}{\gamma - 1} + \dfrac{(\gamma - 2)q^2}{2} & (1 - \gamma)u & (1 - \gamma)v & (1 - \gamma)w & \gamma \end{bmatrix}$$

$$= \frac{1}{\rho + \hat{\rho}} \left[-\frac{q_n + \hat{q}_n}{2} \quad n_x, \quad n_y, \quad n_z, \quad 0 \right]. \qquad (3.6.128)$$

In the same way, we can derive also

$$\frac{\partial |\hat{q}_n|}{\partial \mathbf{U}} = \frac{1}{\rho + \hat{\rho}} \left[-\frac{\text{sign}(\hat{q}_n)(q_n + \hat{q}_n)}{2}, \quad \text{sign}(\hat{q}_n)n_x, \quad \text{sign}(\hat{q}_n)n_y, \quad \text{sign}(\hat{q}_n)n_z, \quad 0 \right]. \qquad (3.6.129)$$

Of course, the derivative of the speed of sound can be obtained in the same way. First compute,

$$\frac{\partial \hat{c}}{\partial \hat{\mathbf{W}}^H} = \left[0, -\frac{(\gamma-1)\hat{u}}{2\hat{c}}, -\frac{(\gamma-1)\hat{v}}{2\hat{c}}, -\frac{(\gamma-1)\hat{w}}{2\hat{c}}, \frac{\gamma-1}{2\hat{c}} \right], \tag{3.6.130}$$

and then

$$\frac{\partial \hat{c}}{\partial \mathbf{U}} = \frac{\partial \hat{c}}{\partial \hat{\mathbf{W}}^H} \frac{\partial \hat{\mathbf{W}}^H}{\partial \mathbf{U}} = \frac{1}{\rho + \hat{\rho}} \begin{bmatrix} \frac{\gamma-1}{2\hat{c}}\left(\frac{\hat{\mathbf{v}}^2 + \hat{\mathbf{v}}\cdot\mathbf{v}}{2} + \frac{H-\hat{H}}{2} - \frac{c^2}{\gamma-1} + \frac{(\gamma-2)q^2}{2} \right) \\ -\frac{(\gamma-1)(\hat{u}+(\gamma-1)u)}{2\hat{c}} \\ -\frac{(\gamma-1)(\hat{v}+(\gamma-1)v)}{2\hat{c}} \\ -\frac{(\gamma-1)(\hat{w}+(\gamma-1)w)}{2\hat{c}} \\ \frac{\gamma(\gamma-1)}{2\hat{c}} \end{bmatrix}^t, \tag{3.6.131}$$

where $\hat{\mathbf{v}} = [\hat{u}, \hat{v}, \hat{w}]^t$ and $\mathbf{v} = [u, v, w]^t$. Now we can easily compute the following derivatives.

$$\frac{\partial(\hat{q}_n - \hat{c})}{\partial \mathbf{U}} = \frac{1}{\rho + \hat{\rho}} \begin{bmatrix} -\frac{q_n + \hat{q}_n}{2} - \frac{\gamma-1}{2\hat{c}}\left(\frac{\hat{\mathbf{v}}^2 + \hat{\mathbf{v}}\cdot\mathbf{v}}{2} + \frac{H-\hat{H}}{2} - \frac{c^2}{\gamma-1} + \frac{(\gamma-2)q^2}{2} \right) \\ n_x + \frac{(\gamma-1)(\hat{u}+(\gamma-1)u)}{2\hat{c}} \\ n_y + \frac{(\gamma-1)(\hat{v}+(\gamma-1)v)}{2\hat{c}} \\ n_z + \frac{(\gamma-1)(\hat{w}+(\gamma-1)w)}{2\hat{c}} \\ -\frac{\gamma(\gamma-1)}{2\hat{c}} \end{bmatrix}^t, \tag{3.6.132}$$

$$\frac{\partial(\hat{q}_n + \hat{c})}{\partial \mathbf{U}} = \frac{1}{\rho + \hat{\rho}} \begin{bmatrix} -\frac{q_n + \hat{q}_n}{2} + \frac{\gamma-1}{2\hat{c}}\left(\frac{\hat{\mathbf{v}}^2 + \hat{\mathbf{v}}\cdot\mathbf{v}}{2} + \frac{H-\hat{H}}{2} - \frac{c^2}{\gamma-1} + \frac{(\gamma-2)q^2}{2} \right) \\ n_x - \frac{(\gamma-1)(\hat{u}+(\gamma-1)u)}{2\hat{c}} \\ n_y - \frac{(\gamma-1)(\hat{v}+(\gamma-1)v)}{2\hat{c}} \\ n_z - \frac{(\gamma-1)(\hat{w}+(\gamma-1)w)}{2\hat{c}} \\ \frac{\gamma(\gamma-1)}{2\hat{c}} \end{bmatrix}^t, \tag{3.6.133}$$

$$\frac{\partial(|\hat{q}_n| + \hat{c})}{\partial \mathbf{U}} = \frac{1}{\rho + \hat{\rho}} \begin{bmatrix} -\frac{\operatorname{sign}(\hat{q}_n)(q_n + \hat{q}_n)}{2} + \frac{\gamma-1}{2\hat{c}}\left(\frac{\hat{\mathbf{v}}^2 + \hat{\mathbf{v}}\cdot\mathbf{v}}{2} + \frac{H-\hat{H}}{2} - \frac{c^2}{\gamma-1} + \frac{(\gamma-2)q^2}{2} \right) \\ \operatorname{sign}(\hat{q}_n)n_x - \frac{(\gamma-1)(\hat{u}+(\gamma-1)u)}{2\hat{c}} \\ \operatorname{sign}(\hat{q}_n)n_y - \frac{(\gamma-1)(\hat{v}+(\gamma-1)v)}{2\hat{c}} \\ \operatorname{sign}(\hat{q}_n)n_z - \frac{(\gamma-1)(\hat{w}+(\gamma-1)w)}{2\hat{c}} \\ \frac{\gamma(\gamma-1)}{2\hat{c}} \end{bmatrix}^t. \tag{3.6.134}$$

Other derivatives can be obtained in the same way. It is so easy because we have $\partial \hat{\mathbf{W}}^{\mathbf{H}}/\partial \mathbf{U}$. That is, we first differentiate any Roe-averaged quantity with respect to $\hat{\mathbf{W}}^{\mathbf{H}}$ and then multiply it by the matrix $\partial \hat{\mathbf{W}}^{\mathbf{H}}/\partial \mathbf{U}$ from the right. I like it. I hope you like it, too. Note that we have already obtained the following derivatives from the matrix (3.6.127):

$$
\begin{bmatrix} \dfrac{\partial \hat{\rho}}{\partial \mathbf{U}} \\[2mm] \dfrac{\partial \hat{u}}{\partial \mathbf{U}} \\[2mm] \dfrac{\partial \hat{v}}{\partial \mathbf{U}} \\[2mm] \dfrac{\partial \hat{w}}{\partial \mathbf{U}} \\[2mm] \dfrac{\partial \hat{H}}{\partial \mathbf{U}} \end{bmatrix} = \frac{1}{\rho + \hat{\rho}} \begin{bmatrix} \dfrac{\hat{\rho}(\rho + \hat{\rho})}{2\rho} & 0 & 0 & 0 & 0 \\[3mm] -\dfrac{u + \hat{u}}{2} & 1 & 0 & 0 & 0 \\[3mm] -\dfrac{v + \hat{v}}{2} & 0 & 1 & 0 & 0 \\[3mm] -\dfrac{w + \hat{w}}{2} & 0 & 0 & 1 & 0 \\[3mm] \dfrac{H - \hat{H}}{2} - \dfrac{a^2}{\gamma - 1} + \dfrac{(\gamma - 2)q^2}{2} & (1-\gamma)u & (1-\gamma)v & (1-\gamma)w & \gamma \end{bmatrix}. \quad (3.6.135)
$$

It is, of course, possible to make use of these derivatives for differentiating Roe-averaged quantities. For example, we can express $\partial \hat{c}/\partial \mathbf{U}$ as a combination of $\partial \hat{u}/\partial \mathbf{U}$, $\partial \hat{v}/\partial \mathbf{U}$, $\partial \hat{w}/\partial \mathbf{U}$, and $\partial \hat{H}/\partial \mathbf{U}$, and then evaluate each term by the corresponding row vector in the above matrix. Finally, I say again that all the expressions above are valid for both L and R.

$$
\begin{bmatrix} \dfrac{\partial \hat{\rho}}{\partial \mathbf{U}_L} \\[2mm] \dfrac{\partial \hat{u}}{\partial \mathbf{U}_L} \\[2mm] \dfrac{\partial \hat{v}}{\partial \mathbf{U}_L} \\[2mm] \dfrac{\partial \hat{w}}{\partial \mathbf{U}_L} \\[2mm] \dfrac{\partial \hat{H}}{\partial \mathbf{U}_L} \end{bmatrix} = \frac{1}{\rho_L + \hat{\rho}} \begin{bmatrix} \dfrac{\hat{\rho}(\rho_L + \hat{\rho})}{2\rho_L} & 0 & 0 & 0 & 0 \\[3mm] -\dfrac{u_L + \hat{u}}{2} & 1 & 0 & 0 & 0 \\[3mm] -\dfrac{v_L + \hat{v}}{2} & 0 & 1 & 0 & 0 \\[3mm] -\dfrac{w_L + \hat{w}}{2} & 0 & 0 & 1 & 0 \\[3mm] \dfrac{H_L - \hat{H}}{2} - \dfrac{a_L^2}{\gamma - 1} + \dfrac{(\gamma - 2)q_L^2}{2} & (1-\gamma)u_L & (1-\gamma)v_L & (1-\gamma)w_L & \gamma \end{bmatrix}, \quad (3.6.136)
$$

$$
\begin{bmatrix} \dfrac{\partial \hat{\rho}}{\partial \mathbf{U}_R} \\[2mm] \dfrac{\partial \hat{u}}{\partial \mathbf{U}_R} \\[2mm] \dfrac{\partial \hat{v}}{\partial \mathbf{U}_R} \\[2mm] \dfrac{\partial \hat{w}}{\partial \mathbf{U}_R} \\[2mm] \dfrac{\partial \hat{H}}{\partial \mathbf{U}_R} \end{bmatrix} = \frac{1}{\rho_R + \hat{\rho}} \begin{bmatrix} \dfrac{\hat{\rho}(\rho_R + \hat{\rho})}{2\rho_R} & 0 & 0 & 0 & 0 \\[3mm] -\dfrac{u_R + \hat{u}}{2} & 1 & 0 & 0 & 0 \\[3mm] -\dfrac{v_R + \hat{v}}{2} & 0 & 1 & 0 & 0 \\[3mm] -\dfrac{w_R + \hat{w}}{2} & 0 & 0 & 1 & 0 \\[3mm] \dfrac{H_R - \hat{H}}{2} - \dfrac{a_R^2}{\gamma - 1} + \dfrac{(\gamma - 2)q_R^2}{2} & (1-\gamma)u_R & (1-\gamma)v_R & (1-\gamma)w_R & \gamma \end{bmatrix}. \quad (3.6.137)
$$

To illustrate how these formulas are used in an implicit solver, I wrote a 2D implicit Euler solver and have made it available at http://www.cfdbooks.com. I hope you like it.

3.7 Axisymmetric Euler Equations

I like the axisymmetric Euler equations very much. These equations formally two-dimensional equations but describe three-dimensional flows. That is very efficient. In fact, the axisymmetric Euler equations have been often employed in CFD, e.g., Refs.[145, 173, 179].

Consider cylindrical coordinates in Figure 1.5.1, where the velocity is denoted by $\mathbf{v} = [u_r, u_\theta, u_z]^t$. An axisymmetric flow is defined as a flow with no variation in θ with the z-axis as the axis of symmetry. Hence, $u_\theta = 0$, and one of the momentum equations drops out. The Euler equations are then given by

$$
\frac{\partial}{\partial t}
\begin{bmatrix} \rho \\ \rho u_z \\ \rho u_r \\ \rho E \end{bmatrix}
+
\begin{bmatrix}
\dfrac{1}{r}\dfrac{\partial(r\rho u_r)}{\partial r} + \dfrac{\partial \rho u_z}{\partial z} \\[2mm]
\dfrac{1}{r}\dfrac{\partial(r\rho u_z u_r)}{\partial r} + \dfrac{\partial(\rho u_z^2)}{\partial z} + \dfrac{\partial p}{\partial z} \\[2mm]
\dfrac{1}{r}\dfrac{\partial(r\rho u_r^2)}{\partial r} + \dfrac{\partial(\rho u_r u_z)}{\partial z} + \dfrac{\partial p}{\partial r} \\[2mm]
\dfrac{1}{r}\dfrac{\partial(r\rho u_r H)}{\partial r} + \dfrac{\partial \rho u_z H}{\partial z}
\end{bmatrix}
= 0.
\tag{3.7.1}
$$

Note that the continuity equation comes from Equation (3.3.1) with Equation (1.5.23), the momentum equations from Equations (3.3.2) with Equation (1.5.45), and the energy equation from Equation (3.3.3) with Equation (1.5.23). The system can be written in the conservative form (multiply the both sides by r):

$$
\frac{\partial}{\partial t}
\begin{bmatrix} r\rho \\ r\rho u_z \\ r\rho u_r \\ r\rho E \end{bmatrix}
+
\begin{bmatrix}
\dfrac{\partial(r\rho u_r)}{\partial r} + \dfrac{\partial(r\rho u_z)}{\partial z} \\[2mm]
\dfrac{\partial(r\rho u_z u_r)}{\partial r} + \dfrac{\partial(r\rho u_z^2)}{\partial z} + \dfrac{\partial(rp)}{\partial z} \\[2mm]
\dfrac{\partial(r\rho u_r^2)}{\partial r} + \dfrac{\partial(r\rho u_r u_z)}{\partial z} + \dfrac{\partial(rp)}{\partial r} \\[2mm]
\dfrac{\partial(r\rho u_r H)}{\partial r} + \dfrac{\partial(r\rho u_z H)}{\partial z}
\end{bmatrix}
=
\begin{bmatrix} 0 \\ 0 \\ p \\ 0 \end{bmatrix}.
\tag{3.7.2}
$$

The system is very similar to the two-dimensional Euler system. To see this more clearly, let us replace (z, r) by (x, y) and (u_z, u_r) by (u, v) to get

$$
\frac{\partial}{\partial t}
\begin{bmatrix} y\rho \\ y\rho u \\ y\rho v \\ y\rho E \end{bmatrix}
+
\begin{bmatrix}
\dfrac{\partial(y\rho u)}{\partial x} + \dfrac{\partial(y\rho v)}{\partial y} \\[2mm]
\dfrac{\partial(y\rho u^2)}{\partial x} + \dfrac{\partial(y\rho u v)}{\partial y} + \dfrac{\partial(yp)}{\partial x} \\[2mm]
\dfrac{\partial(y\rho u v)}{\partial x} + \dfrac{\partial(y\rho v^2)}{\partial y} + \dfrac{\partial(yp)}{\partial y} \\[2mm]
\dfrac{\partial(y\rho u H)}{\partial x} + \dfrac{\partial(y\rho v H)}{\partial y}
\end{bmatrix}
=
\begin{bmatrix} 0 \\ 0 \\ p \\ 0 \end{bmatrix}.
\tag{3.7.3}
$$

In the vector form, it can be written as

$$
(y\mathbf{U})_t + (y\mathbf{F})_x + (y\mathbf{G})_y = \mathbf{S},
\tag{3.7.4}
$$

where

$$
\mathbf{U} = \begin{bmatrix} \rho \\ \rho u \\ \rho v \\ \rho E \end{bmatrix}, \quad
\mathbf{F} = \begin{bmatrix} \rho u \\ \rho u^2 + p \\ \rho u v \\ \rho u H \end{bmatrix}, \quad
\mathbf{G} = \begin{bmatrix} \rho v \\ \rho u v \\ \rho v^2 + p \\ \rho v H \end{bmatrix}, \quad
\mathbf{S} = \begin{bmatrix} 0 \\ 0 \\ p \\ 0 \end{bmatrix}.
\tag{3.7.5}
$$

It is clear now that the system differs from the two-dimensional Euler system only by the factor y multiplied by the conservative variables and the flux vectors, and the pressure source term. In Ref.[173], the axisymmetric form and the two-dimensional form are expressed in a single form with a logical switch, $\overline{\omega} = 0$ for the two-dimensional form and $\overline{\omega} = 1$ for the axisymmetric form:

$$
(\overline{\omega}_a \mathbf{U})_t + (\overline{\omega}_a \mathbf{F})_x + (\overline{\omega}_a \mathbf{G})_y = \overline{\omega}\, \mathbf{S},
\tag{3.7.6}
$$

where

$$
\overline{\omega}_a = (1 - \overline{\omega}) + \overline{\omega}\, y.
\tag{3.7.7}
$$

This is nice. If you have a two-dimensional Euler code, you almost have an axisymmetric Euler code, which is exciting because the axisymmetric code is essentially a three-dimensional CFD code. I like it very much.

3.8   Rotational Invariance

The Euler equations have the rotational invariance property (see Section 1.20)

$$\mathbf{F}_n = n_x \mathbf{F} + n_y \mathbf{G} + n_z \mathbf{H} = \mathbf{T}^{-1} \mathbf{F}(\mathbf{T}\mathbf{U}), \tag{3.8.1}$$

where

$$\mathbf{T} = \begin{bmatrix} 1 & 0 & 0 & 0 & 0 \\ 0 & n_x & n_y & n_z & 0 \\ 0 & \ell_x & \ell_y & \ell_z & 0 \\ 0 & m_x & m_y & m_z & 0 \\ 0 & 0 & 0 & 0 & 1 \end{bmatrix}, \tag{3.8.2}$$

$$\mathbf{T}^{-1} = \mathbf{T}^t = \begin{bmatrix} 1 & 0 & 0 & 0 & 0 \\ 0 & n_x & \ell_x & m_x & 0 \\ 0 & n_y & \ell_y & m_y & 0 \\ 0 & n_z & \ell_z & m_z & 0 \\ 0 & 0 & 0 & 0 & 1 \end{bmatrix}, \tag{3.8.3}$$

and $\mathbf{n}$, $\boldsymbol{\ell}$, $\mathbf{m}$ are three orthogonal unit vectors (see Section 1.3). So, we can do the following. First compute $\mathbf{T}\mathbf{U}$,

$$\mathbf{T}\mathbf{U} = \begin{bmatrix} \rho \\ \rho q_n \\ \rho q_\ell \\ \rho q_m \\ \rho E \end{bmatrix}, \tag{3.8.4}$$

and evaluate $\mathbf{F}(\mathbf{T}\mathbf{U})$,

$$\mathbf{F}(\mathbf{T}\mathbf{U}) = \begin{bmatrix} \rho q_n \\ \rho q_n^2 + p \\ \rho q_n q_\ell \\ \rho q_n q_m \\ \rho q_n H \end{bmatrix}, \tag{3.8.5}$$

and then rotate it back to obtain a flux vector projected in an arbitrary direction $\mathbf{n}$,

$$\mathbf{F}_n = \mathbf{T}^{-1} \mathbf{F}(\mathbf{T}\mathbf{U}) = \begin{bmatrix} \rho q_n \\ \rho q_n u + p n_x \\ \rho q_n v + p n_y \\ \rho q_n w + p n_z \\ \rho q_n H \end{bmatrix}, \tag{3.8.6}$$

where the relation (1.3.21) has been used to get u, v, and w.

So, basically, we need only $\mathbf{F}$ and don't really need $\mathbf{G}$ and $\mathbf{H}$. In fact, some people take advantage of this property when they write a finite-volume code: set up one subroutine to evaluate the finite-volume interface flux based on $\mathbf{F}$ only, and then perform the rotation as described above to get a numerical flux in arbitrary directions. I like this way of computing a numerical flux, but I also like to use the form (3.8.6) directly to implement a numerical flux since it seems more intuitive.

3.9   Convenient Ordering

I like to order the components as follows:

$$\mathbf{U} = \begin{bmatrix} \rho \\ \rho E \\ \rho u \\ \rho v \\ \rho w \end{bmatrix}, \quad \mathbf{F} = \begin{bmatrix} \rho u \\ \rho u H \\ \rho u^2 + p \\ \rho u v \\ \rho u w \end{bmatrix}, \quad \mathbf{G} = \begin{bmatrix} \rho v \\ \rho v H \\ \rho v u \\ \rho v^2 + p \\ \rho v w \end{bmatrix}, \quad \mathbf{H} = \begin{bmatrix} \rho w \\ \rho w H \\ \rho w u \\ \rho w v \\ \rho w^2 + p \end{bmatrix}. \tag{3.9.1}$$

Then, simply by ignoring the fifth row and $\mathbf{H}$, we obtain the two-dimensional Euler equations, and by ignoring further the fourth row and $\mathbf{G}$, we obtain the one-dimensional Euler equations. The same is true for the flux, Jacobians, and eigenvectors:

Normal Flux

$$\mathbf{F}_n = \mathbf{F}n_x + \mathbf{G}n_y + \mathbf{H}n_z = \begin{bmatrix} \rho q_n \\ \rho q_n H \\ \rho q_n u + p n_x \\ \rho q_n v + p n_y \\ \rho q_n w + p n_z \end{bmatrix}. \tag{3.9.2}$$

Jacobians

$$\mathbf{A}_n = \begin{bmatrix} 0 & 0 & \mathbf{n}^t \\ \left(\frac{\gamma-1}{2}q^2 - H\right)q_n & \gamma q_n & H\mathbf{n}^t - (\gamma-1)\mathbf{v}^t q_n \\ (\gamma-1)\frac{q^2}{2}\mathbf{n} - \mathbf{v}q_n & (\gamma-1)\mathbf{n} & \mathbf{v}\otimes\mathbf{n} - (\gamma-1)\mathbf{n}\otimes\mathbf{v} + q_n\mathbf{I} \end{bmatrix}. \tag{3.9.3}$$

Note that by taking $\mathbf{n} = [1,0,0]^t$, $[0,1,0]^t$, or $[0,0,1]^t$, we obtain $\mathbf{A}$, $\mathbf{B}$, or $\mathbf{C}$.

Right-Eigenvector Matrix

$$\mathbf{R}_n = \begin{bmatrix} 1 & 1 & 1 & 0 & 0 \\ H - q_n c & q^2/2 & H + q_n c & q_\ell & q_m \\ u - c n_x & u & u + c n_x & \ell_x & m_x \\ v - c n_y & v & v + c n_y & \ell_y & m_y \\ w - c n_z & w & w + c n_z & \ell_z & m_z \end{bmatrix}. \tag{3.9.4}$$

This is convenient in the case that we want to perform one- or two-dimensional calculations with a 3D code (simply limit the size of arrays). I like this technique. It's really simple and useful. The CFD code called EulFS [15] actually takes advantage of this ordering. Of course, alternatively, if we have a 3D code, we can perform one-dimensional calculations by using a string of 3D cells aligned in x-axis. You can also do two-dimensional calculations by using a two-dimensional grid with a unit width added in the third direction (this creates a 3D grid of prismatic cells). But I think that the ordering technique is still useful for other purposes also. For example, it can be used also for extended systems such as magnetohydrodynamic equations: add magnetic field components to the end of the hydrodynamic state vector so that one can run purely hydrodynamic cases with an MHD code simply by limiting the size of the solution array.

3.10   Homogeneity Property of the Euler Fluxes

If we assume that p/ρ is a function of e only (i.e., if $p = \rho f(e)$), then the Euler fluxes are homogeneous of degree 1 in terms of the conservative variables. That is, for any scalar λ, we have

$$\mathbf{F}(\lambda\mathbf{U}) = \lambda\mathbf{F}(\mathbf{U}), \quad \mathbf{G}(\lambda\mathbf{U}) = \lambda\mathbf{G}(\mathbf{U}), \quad \mathbf{H}(\lambda\mathbf{U}) = \lambda\mathbf{H}(\mathbf{U}). \tag{3.10.1}$$

Differentiating both sides with respect to λ, and then setting $\lambda = 1$, we obtain

$$\frac{\partial\mathbf{F}}{\partial\mathbf{U}}\mathbf{U} = \mathbf{A}\mathbf{U} = \mathbf{F}, \quad \frac{\partial\mathbf{G}}{\partial\mathbf{U}}\mathbf{U} = \mathbf{B}\mathbf{U} = \mathbf{G}, \quad \frac{\partial\mathbf{H}}{\partial\mathbf{U}}\mathbf{U} = \mathbf{C}\mathbf{U} = \mathbf{H}. \tag{3.10.2}$$

Therefore, the conservative form,

$$\mathbf{U}_t + \mathbf{F}_x + \mathbf{G}_y + \mathbf{H}_z = 0, \tag{3.10.3}$$

can be written not only as

$$\mathbf{U}_t + \mathbf{A}\mathbf{U}_x + \mathbf{B}\mathbf{U}_y + \mathbf{C}\mathbf{U}_z = 0, \tag{3.10.4}$$

but also as

$$\mathbf{U}_t + (\mathbf{AU})_x + (\mathbf{BU})_y + (\mathbf{CU})_z = 0. \tag{3.10.5}$$

I like this because the Jacobian matrices look like as if they are constant even though they are actually not. In fact, this can be useful. The Steger-Warming flux vector splitting [149] takes advantage of this property.

Also, note that we have

$$d\mathbf{F} = d(\mathbf{AU}) = \mathbf{A}d\mathbf{U} + d\mathbf{A}\,\mathbf{U}, \tag{3.10.6}$$
$$d\mathbf{F} = \mathbf{A}d\mathbf{U}, \tag{3.10.7}$$

from which we find

$$d\mathbf{A}\,\mathbf{U} = 0. \tag{3.10.8}$$

I think that I like it. I'm not sure, but it must be useful.

3.11 Nondimensionalization

3.11.1 Nondimensionalized Euler Equations

Nondimensionalized quantities are nice. No matter how large the actual free stream speed is (which depends on units also), we can set it equal to 1 by scaling the velocity by the free stream speed. So, a CFD code does not have to deal with too large (or too small) numbers. To be even nicer, the Euler equations stay in exactly the same form for various nondimensionalizations. Consider the following nondimensionalization:

$$\rho^* = \frac{\rho}{\rho_r}, \quad p^* = \frac{p}{p_r}, \quad \mathbf{v}^* = \frac{\mathbf{v}}{V_r}, \quad \mathbf{x}^* = \frac{\mathbf{x}}{L}, \quad t^* = \frac{t}{t_r}, \tag{3.11.1}$$

where L is a reference length, the variables with $*$ are the nondimensional variables, and those with the subscript r are some reference values. Substituting these into the continuity equation (3.3.1), we obtain

$$\frac{\partial \rho^*}{\partial t^*} + \frac{t_r V_r}{L} \operatorname{div}^*(\rho^* \mathbf{v}^*) = 0, \tag{3.11.2}$$

where div^* is the divergence operator based on the nondimensionalized coordinates. So, by setting

$$t_r = \frac{L}{V_r}, \tag{3.11.3}$$

we find that the nondimensional variables satisfy exactly the same continuity equation as that for the dimensional variables:

$$\frac{\partial \rho^*}{\partial t^*} + \operatorname{div}^*(\rho^* \mathbf{v}^*) = 0. \tag{3.11.4}$$

For the momentum equations (3.3.2), we find

$$\frac{\partial (\rho^* \mathbf{v}^*)}{\partial t^*} + \operatorname{div}^*(\rho^* \mathbf{v}^* \otimes \mathbf{v}^*) + \frac{p_r}{\rho_r V_r^2} \operatorname{grad}^* p^* = 0, \tag{3.11.5}$$

which suggests that we set

$$p_r = \rho_r V_r^2, \tag{3.11.6}$$

and therefore the momentum equations are also identical to those for dimensional variables:

$$\frac{\partial (\rho^* \mathbf{v}^*)}{\partial t^*} + \operatorname{div}^*(\rho^* \mathbf{v}^* \otimes \mathbf{v}^*) + \operatorname{grad}^* p^* = 0. \tag{3.11.7}$$

For the energy equation (3.3.3), we first nondimensionalize the internal energy by V_r^2:

$$e^* = \frac{e}{V_r^2}, \tag{3.11.8}$$

so that we have

$$E^* = \frac{E}{V_r^2} = e^* + \frac{\mathbf{v}^{*2}}{2}, \tag{3.11.9}$$

and also

$$H^* = \frac{H}{V_r^2} = e^* + \frac{\mathbf{v}^{*2}}{2} + \frac{p^*}{\rho^*}. \tag{3.11.10}$$

Hence, the total specific energy and enthalpy can be calculated in the same forms for the nondimensional variables. Furthermore, it follows that the equation of state for a calorically perfect gas (3.1.21) also holds for the nondimensional variables:

$$p^* = (\gamma - 1)\rho^* e^*. \tag{3.11.11}$$

Finally, with the above nondimensionalization, the energy equation (3.3.3) becomes

$$\frac{\partial (\rho^* E^*)}{\partial t^*} + \text{div}^*(\rho^* \mathbf{v}^* H^*) = 0, \tag{3.11.12}$$

which is again the same form as that for the dimensional variables. So, all equations are invariant for the nondimensionalization. This is very nice. So, I'm happy to summarize the result below.

Euler Equations for Nondimensionalized Variables

$$\frac{\partial \rho^*}{\partial t^*} + \text{div}^*(\rho^* \mathbf{v}^*) = 0, \tag{3.11.13}$$

$$\frac{\partial (\rho^* \mathbf{v}^*)}{\partial t^*} + \text{div}^*(\rho^* \mathbf{v}^* \otimes \mathbf{v}^*) + \text{grad}^* p^* = 0, \tag{3.11.14}$$

$$\frac{\partial (\rho^* E^*)}{\partial t^*} + \text{div}^*(\rho^* \mathbf{v}^* H^*) = 0, \tag{3.11.15}$$

where $E^* = e^* + \frac{\mathbf{v}^{*2}}{2}$, $H^* = E^* + \frac{p^*}{\rho^*}$, and

$$\rho^* = \frac{\rho}{\rho_r}, \quad p^* = \frac{p}{\rho_r V_r^2}, \quad e^* = \frac{e}{V_r^2}, \quad \mathbf{v}^* = \frac{\mathbf{v}}{V_r}, \quad \mathbf{x}^* = \frac{\mathbf{x}}{L}, \quad t^* = \frac{t}{L/V_r}. \tag{3.11.16}$$

The reference length can be taken as anything you like. For example, if the diameter of a cylinder is chosen as the reference length, you generate a computational grid around the cylinder with the diameter 1. If you are given a grid, whatever having the length 1 can be thought of as the part chosen as the reference length. Then, it is now a matter of what values to choose for the reference density and velocity. There can be various choices. I especially like those I describe below.

3.11.2   Free Stream Values

We may use the free stream values (indicated by the subscript ∞) as reference, i.e., set $\rho_r = \rho_\infty$ and $V_r = V_\infty$, which give the following nondimensionalization:

$$\rho^* = \frac{\rho}{\rho_\infty}, \quad p^* = \frac{p}{\rho_\infty V_\infty^2}, \quad e^* = \frac{e}{V_\infty^2}, \quad \mathbf{v}^* = \frac{\mathbf{v}}{V_\infty}, \quad \mathbf{x}^* = \frac{\mathbf{x}}{L}, \quad t^* = \frac{t}{L/V_\infty}. \tag{3.11.17}$$

Then, we can set the free stream values as follows:

$$\rho_\infty^* = 1, \quad p_\infty^* = \frac{1}{\gamma M_\infty^2}, \quad e^* = \frac{p_\infty^*}{\rho_\infty^*(\gamma - 1)} = \frac{M_\infty^2}{\gamma(\gamma - 1)}, \quad \mathbf{v}_\infty^* = \frac{\mathbf{v}_\infty}{V_\infty} = \mathbf{n}_\infty, \tag{3.11.18}$$

where $\mathbf{n}_\infty$ is a unit vector in the direction of the free stream. This means that the free stream speed is 1 and only the direction, i.e., $\mathbf{n}_\infty$, needs to be specified at a free stream inflow. Yes, also the free stream Mach number M_∞ needs to be given, in order to set the free stream pressure. Note that the pressure is inversely proportional to M_∞^2 and therefore it will become extremely large for flows with a very small Mach number. In fact, this is not a popular choice for compressible flow simulations (popular for incompressible flow simulations, though).

3.11.3 Speed of Sound

We may use the free stream speed of sound a_∞ in place of the free stream velocity V_∞. This set of quantities is widely used in compressible flow simulations:

$$\rho^* = \frac{\rho}{\rho_\infty}, \quad p^* = \frac{p}{\rho_\infty a_\infty^2}, \quad e^* = \frac{e}{a_\infty^2}, \quad \mathbf{v}^* = \frac{\mathbf{v}}{a_\infty}, \quad \mathbf{x}^* = \frac{\mathbf{x}}{L}, \quad t^* = \frac{t}{L/a_\infty}. \tag{3.11.19}$$

Then, we can set the free stream values as follows:

$$\rho_\infty^* = 1, \quad p_\infty^* = \frac{1}{\gamma}, \quad e^* = \frac{1}{\gamma(\gamma-1)}, \quad \mathbf{v}_\infty^* = \frac{\mathbf{v}_\infty}{a_\infty} = M_\infty \mathbf{n}_\infty. \tag{3.11.20}$$

Again, we only need to specify the Mach number and its direction at the free stream inflow. I like this nondimensionalization particularly because the velocity becomes equivalent to the Mach number at the free stream inflow and also that the free stream pressure is constant (independent of M_∞). This is a very popular choice for compressible flow simulations.

3.11.4 Stagnation Values

We can also use the stagnation values:

$$\rho^* = \frac{\rho}{\rho_0} = \left[1 + \frac{\gamma-1}{2}M^2\right]^{-\frac{1}{\gamma-1}}, \quad p^* = \frac{p}{\rho_0 a_0^2} = \frac{p}{\gamma p_0} = \frac{1}{\gamma}\left[1 + \frac{\gamma-1}{2}M^2\right]^{-\frac{\gamma}{\gamma-1}}, \tag{3.11.21}$$

(you can derive these relations from Bernoulli's equation for compressible flows in Section 3.18), and

$$e^* = \frac{e}{a_0^2}, \quad \mathbf{v}^* = \frac{\mathbf{v}}{a_0} = \frac{\mathbf{v}}{a}\frac{a}{a_0} = \frac{\mathbf{v}}{a}\sqrt{\gamma\frac{p^*}{\rho^*}} = \frac{\mathbf{v}}{a}a^*, \quad \mathbf{x}^* = \frac{\mathbf{x}}{L}, \quad t^* = \frac{t}{L/a_0}. \tag{3.11.22}$$

Then, we can set the free stream values as follows:

$$\rho_\infty^* = \left[1 + \frac{\gamma-1}{2}M_\infty^2\right]^{-\frac{1}{\gamma-1}}, \quad p_\infty^* = \frac{1}{\gamma}\left[1 + \frac{\gamma-1}{2}M_\infty^2\right]^{-\frac{\gamma}{\gamma-1}}, \tag{3.11.23}$$

$$e^* = \frac{p_\infty^*}{\rho_\infty^*(\gamma-1)}, \quad \mathbf{v}_\infty^* = \frac{\mathbf{v}_\infty}{a_\infty}a_\infty^* = M_\infty \mathbf{n}_\infty \sqrt{\gamma\frac{p_\infty^*}{\rho_\infty^*}}. \tag{3.11.24}$$

Again, only the Mach number and the free stream direction need to be specified at the free stream inflow. This will be convenient when you want to compare a computed pressure at a stagnation point with a known value. See Section 7.13.7 for example.

3.12 Change of Variables

I like converting variables, from conservative to primitive, for example. In particular, I like those I describe in this section since these are valid for any set of nondimensionalized variables discussed in Section 3.11.

3.12.1 One Dimension

Let $\mathbf{U} = [\rho, \rho u, \rho E]^t = [u_1, u_2, u_3]^t$, $\mathbf{W} = [\rho, u, p]^t = [w_1, w_2, w_3]^t$, and $\mathbf{Z} = \sqrt{\rho}\,[1, u, H]^t = [z_1, z_2, z_3]^t$. Then we have the following relations.

From conservative variables to others

$$\rho = u_1, \tag{3.12.1}$$

$$u = u_2/u_1, \tag{3.12.2}$$

$$p = (\gamma-1)\left[u_3 - \frac{1}{2}\frac{u_2^2}{u_1}\right], \tag{3.12.3}$$

$$z_1 = \sqrt{u_1}, \tag{3.12.4}$$

$$z_2 = u_2/\sqrt{u_1}, \tag{3.12.5}$$

$$z_3 = \sqrt{u_1}\left[\gamma\frac{u_3}{u_1} - \frac{\gamma-1}{2}\left(\frac{u_2}{u_1}\right)^2\right], \tag{3.12.6}$$

$$H = \gamma\frac{u_3}{u_1} - \frac{\gamma-1}{2}\left(\frac{u_2}{u_1}\right)^2. \tag{3.12.7}$$

From primitive variables to others

$$\rho = w_1, \tag{3.12.8}$$

$$\rho u = w_1 w_2, \tag{3.12.9}$$

$$\rho E = \frac{w_3}{\gamma-1} + \frac{w_1}{2}w_2^2, \tag{3.12.10}$$

$$z_1 = \sqrt{w_1}, \tag{3.12.11}$$

$$z_2 = \sqrt{w_1}w_2, \tag{3.12.12}$$

$$z_3 = \frac{\gamma}{\gamma-1}\frac{w_3}{\sqrt{w_1}} + \frac{\sqrt{w_1}}{2}w_2^2, \tag{3.12.13}$$

$$H = \frac{\gamma}{\gamma-1}\frac{w_3}{w_1} + \frac{1}{2}w_2^2. \tag{3.12.14}$$

From parameter vector variables to others

$$\rho = z_1^2, \tag{3.12.15}$$

$$u = z_2/z_1, \tag{3.12.16}$$

$$p = \frac{\gamma-1}{\gamma}\left(z_1 z_3 - \frac{z_2^2}{2}\right), \tag{3.12.17}$$

$$\rho u = z_1 z_2, \tag{3.12.18}$$

$$\rho E = \frac{z_1 z_3}{\gamma} + \frac{\gamma-1}{2\gamma}z_2^2, \tag{3.12.19}$$

$$H = z_3/z_1. \tag{3.12.20}$$

3.12.2 Two Dimensions

Let $\mathbf{U} = [\rho, \rho u, \rho v, \rho E]^t = [u_1, u_2, u_3, u_4]^t$, $\mathbf{W} = [\rho, u, v, p]^t = [w_1, w_2, w_3, w_4]^t$, and $\mathbf{Z} = \sqrt{\rho}[1, u, v, H]^t = [z_1, z_2, z_3, z_4]^t$. Then we have the following relations.

From conservative variables to others

$$\rho = u_1, \tag{3.12.21}$$

$$u = u_2/u_1, \tag{3.12.22}$$

$$v = u_3/u_1, \tag{3.12.23}$$

$$p = (\gamma-1)\left[u_4 - \frac{1}{2}\frac{u_2^2 + u_3^2}{u_1}\right], \tag{3.12.24}$$

$$z_1 = \sqrt{u_1}, \tag{3.12.25}$$

$$z_2 = u_2/\sqrt{u_1}, \tag{3.12.26}$$

$$z_3 = u_3/\sqrt{u_1}, \tag{3.12.27}$$

$$z_4 = \sqrt{u_1}\left[\gamma\frac{u_4}{u_1} - \frac{\gamma-1}{2}\frac{u_2^2+u_3^2}{u_1^2}\right], \tag{3.12.28}$$

$$H = \gamma\frac{u_4}{u_1} - \frac{\gamma-1}{2}\frac{u_2^2+u_3^2}{u_1^2}. \tag{3.12.29}$$

From primitive variables to others

$$\rho = w_1, \tag{3.12.30}$$

$$\rho u = w_1 w_2, \tag{3.12.31}$$

$$\rho v = w_1 w_3, \tag{3.12.32}$$

$$\rho E = \frac{w_4}{\gamma-1} + \frac{w_1}{2}\left(w_2^2+w_3^2\right), \tag{3.12.33}$$

$$z_1 = \sqrt{w_1}, \tag{3.12.34}$$

$$z_2 = \sqrt{w_1}w_2, \tag{3.12.35}$$

$$z_3 = \sqrt{w_1}w_3, \tag{3.12.36}$$

$$z_4 = \frac{\gamma}{\gamma-1}\frac{w_4}{\sqrt{w_1}} + \frac{\sqrt{w_1}}{2}\left(w_2^2+w_3^2\right), \tag{3.12.37}$$

$$H = \frac{\gamma}{\gamma-1}\frac{w_4}{w_1} + \frac{1}{2}(w_2^2+w_3^2). \tag{3.12.38}$$

From parameter vector variables to others

$$\rho = z_1^2, \tag{3.12.39}$$

$$u = z_2/z_1, \tag{3.12.40}$$

$$v = z_3/z_1, \tag{3.12.41}$$

$$p = \frac{\gamma-1}{\gamma}\left(z_1 z_4 - \frac{z_2^2+z_3^2}{2}\right), \tag{3.12.42}$$

$$\rho u = z_1 z_2, \tag{3.12.43}$$

$$\rho v = z_1 z_3, \tag{3.12.44}$$

$$\rho E = \frac{z_1 z_4}{\gamma} + \frac{\gamma-1}{2\gamma}\left(z_2^2+z_3^2\right), \tag{3.12.45}$$

$$H = z_4/z_1. \tag{3.12.46}$$

3.12.3 Three Dimensions

Let $\mathbf{U} = [\rho, \rho u, \rho v, \rho w, \rho E]^t = [u_1, u_2, u_3, u_4, u_5]^t$, $\mathbf{W} = [\rho, u, v, w, p]^t = [w_1, w_2, w_3, w_4, w_5]^t$, and $\mathbf{Z} = \sqrt{\rho}\,[1, u, v, w, H]^t = [z_1, z_2, z_3, z_4, z_5]^t$. Then we have the following relations.

From conservative variables to others

$$\rho = u_1, \tag{3.12.47}$$

$$u = u_2/u_1, \tag{3.12.48}$$

$$v = u_3/u_1, \tag{3.12.49}$$

$$w = u_4/u_1, \tag{3.12.50}$$

$$p = (\gamma - 1) \left[u_5 - \frac{1}{2} \frac{u_2^2 + u_3^2 + u_4^2}{u_1} \right], \tag{3.12.51}$$

$$z_1 = \sqrt{u_1}, \tag{3.12.52}$$

$$z_2 = u_2/\sqrt{u_1}, \tag{3.12.53}$$

$$z_3 = u_3/\sqrt{u_1}, \tag{3.12.54}$$

$$z_4 = u_4/\sqrt{u_1}, \tag{3.12.55}$$

$$z_5 = \sqrt{u_1} \left[\gamma \frac{u_5}{u_1} - \frac{\gamma - 1}{2} \frac{u_2^2 + u_3^2 + u_4^2}{u_1^2} \right], \tag{3.12.56}$$

$$H = \gamma \frac{u_5}{u_1} - \frac{\gamma - 1}{2} \frac{u_2^2 + u_3^2 + u_4^2}{u_1^2}. \tag{3.12.57}$$

From primitive variables to others

$$\rho = w_1, \tag{3.12.58}$$

$$\rho u = w_1 w_2, \tag{3.12.59}$$

$$\rho v = w_1 w_3, \tag{3.12.60}$$

$$\rho w = w_1 w_4, \tag{3.12.61}$$

$$\rho E = \frac{w_5}{\gamma - 1} + \frac{w_1}{2} \left(w_2^2 + w_3^2 + w_4^2 \right), \tag{3.12.62}$$

$$z_1 = \sqrt{w_1}, \tag{3.12.63}$$

$$z_2 = \sqrt{w_1} w_2, \tag{3.12.64}$$

$$z_3 = \sqrt{w_1} w_3, \tag{3.12.65}$$

$$z_4 = \sqrt{w_1} w_4, \tag{3.12.66}$$

$$z_5 = \frac{\gamma}{\gamma - 1} \frac{w_5}{\sqrt{w_1}} + \frac{\sqrt{w_1}}{2} \left(w_2^2 + w_3^2 + w_4^2 \right), \tag{3.12.67}$$

$$H = \frac{\gamma}{\gamma - 1} \frac{w_5}{w_1} + \frac{1}{2} (w_2^2 + w_3^2 + w_4^2). \tag{3.12.68}$$

From parameter vector variables to others

$$\rho = z_1^2, \tag{3.12.69}$$

$$u = z_2/z_1, \tag{3.12.70}$$

$$v = z_3/z_1, \tag{3.12.71}$$

$$w = z_4/z_1, \tag{3.12.72}$$

$$p = \frac{\gamma - 1}{\gamma} \left(z_1 z_5 - \frac{z_2^2 + z_3^2 + z_4^2}{2} \right), \tag{3.12.73}$$

$$\rho u = z_1 z_2, \tag{3.12.74}$$

$$\rho v = z_1 z_3, \tag{3.12.75}$$

$$\rho w = z_1 z_4, \tag{3.12.76}$$

$$\rho E = \frac{z_1 z_5}{\gamma} + \frac{\gamma - 1}{2\gamma} \left(z_2^2 + z_3^2 + z_4^2 \right), \tag{3.12.77}$$

$$H = z_5/z_1. \tag{3.12.78}$$

3.13 Incompressible/Pseudo-Compressible Euler Equations

Incompressible fluids are defined as fluids with negligible density variation (e.g., water). This is equivalent to negligible Mach number, and in fact, a flow with a Mach number below 0.3 may be considered as incompressible. If we simplify the Euler equations by using the constant density assumption, $\rho =$ constant, the results will be called the incompressible Euler equations.

Conservative Form of the Incompressible Euler Equations:

$$\text{div}\,(\rho \mathbf{v}) = 0, \tag{3.13.1}$$

$$\partial_t(\rho \mathbf{v}) + \text{div}(\rho \mathbf{v} \otimes \mathbf{v}) + \text{grad}\,p = 0. \tag{3.13.2}$$

Primitive Form of the Incompressible Euler Equations:

$$\text{div}\,\mathbf{v} = 0, \tag{3.13.3}$$

$$\rho \frac{D\mathbf{v}}{Dt} + \text{grad}\,p = 0. \tag{3.13.4}$$

I like these incompressible Euler equations because they are closed without the energy equation (which may be solved *a posteriori*) and also because there is no time derivative in the continuity equation even for unsteady flows (because the density is constant). But in fact the latter can be a disadvantage: a simple time-marching method is not applicable because the whole system cannot be written as a time-dependent system. Now, Chorin proposed a pseudo-compressible method [21], in which a constant artificial speed of sound a^* is introduced to define an artificial density variable ρ^*,

$$\rho^* = p/a^{*2}, \tag{3.13.5}$$

and a time derivative is brought back to the continuity equation in the form,

$$\frac{\partial \rho^*}{\partial t^*} + \text{div}\,(\rho \mathbf{v}) = 0, \tag{3.13.6}$$

i.e.,

$$\frac{\partial P}{\partial t^*} + \text{div}\,(a^{*2}\mathbf{v}) = 0, \tag{3.13.7}$$

where t^* is an artificial time and $P = p/\rho$ which is called the kinematic pressure. This helps the incompressible Euler equations recover the hyperbolic character, so that a time-marching method is now directly applicable. The resulting equations are called the pseudo-compressible Euler equations. This is very interesting. I like it. Note however that only when ρ^* reaches a steady state in the artificial time, is the original continuity equation ($\text{div}\,\mathbf{v} = 0$) satisfied. That is, the solution will be accurate only at a steady state. The value of the artificial speed of sound a^* depends on the problem and it is usually assigned a value between 0.1 and 10.0. See Refs.[21, 75, 97, 131, 132] for details. Note that some people formulate the pseudo-compressible equation with a slightly different notation for the artificial parameter, e.g., with $\beta = \rho a^{*2}$ [97, 131, 132]. I like both notations, but in this book I will use a^* throughout.

The pseudo-compressible Euler equations:

$$\frac{\partial P}{\partial t^*} + \text{div}\,(a^{*2}\mathbf{v}) = 0, \tag{3.13.8}$$

$$\frac{\partial \mathbf{v}}{\partial t^*} + \text{div}(\mathbf{v} \otimes \mathbf{v} + P\mathbf{I}) = 0. \tag{3.13.9}$$

In Cartesian coordinates:

$$\mathbf{U}_{t^*} + \mathbf{F}_x + \mathbf{G}_y + \mathbf{H}_z = 0, \tag{3.13.10}$$

$$\mathbf{U} = \begin{bmatrix} P \\ u \\ v \\ w \end{bmatrix}, \quad \mathbf{F} = \begin{bmatrix} a^{*2}u \\ u^2 + P \\ uv \\ uw \end{bmatrix}, \quad \mathbf{G} = \begin{bmatrix} a^{*2}v \\ vu \\ v^2 + P \\ vw \end{bmatrix}, \quad \mathbf{H} = \begin{bmatrix} a^{*2}w \\ wu \\ wv \\ w^2 + P \end{bmatrix}. \tag{3.13.11}$$

The normal flux:

$$\mathbf{F}_n = \mathbf{F}n_x + \mathbf{G}n_y + \mathbf{H}n_z = \begin{bmatrix} a^{*2}q_n \\ q_n u + Pn_x \\ q_n v + Pn_y \\ q_n w + Pn_z \end{bmatrix}. \tag{3.13.12}$$

The Jacobians:

$$\mathbf{A} = \frac{\partial \mathbf{F}}{\partial \mathbf{U}} = \begin{bmatrix} 0 & a^{*2} & 0 & 0 \\ 1 & 2u & 0 & 0 \\ 0 & v & u & 0 \\ 0 & w & 0 & u \end{bmatrix}, \tag{3.13.13}$$

$$\mathbf{B} = \frac{\partial \mathbf{G}}{\partial \mathbf{U}} = \begin{bmatrix} 0 & 0 & a^{*2} & 0 \\ 0 & v & u & 0 \\ 1 & 0 & 2v & 0 \\ 0 & 0 & w & v \end{bmatrix}, \tag{3.13.14}$$

$$\mathbf{C} = \frac{\partial \mathbf{H}}{\partial \mathbf{U}} = \begin{bmatrix} 0 & 0 & 0 & a^{*2} \\ 0 & w & 0 & u \\ 0 & 0 & w & v \\ 1 & 0 & 0 & 2w \end{bmatrix}, \tag{3.13.15}$$

$$\mathbf{A}_n = \mathbf{A}n_x + \mathbf{B}n_y + \mathbf{C}n_z = \begin{bmatrix} 0 & a^{*2}n_x & a^{*2}n_y & a^{*2}n_z \\ n_x & q_n + un_x & un_y & un_z \\ n_y & vn_x & q_n + vn_y & vn_z \\ n_y & wn_x & wn_y & q_n + wn_z \end{bmatrix}. \tag{3.13.16}$$

The eigenstructure of the Jacobian matrix $\mathbf{A}_n$:

$$\mathbf{A}_n = \mathbf{R}_n \mathbf{\Lambda}_n \mathbf{L}_n, \tag{3.13.17}$$

where

$$\mathbf{\Lambda}_n = \begin{bmatrix} q_n - c & 0 & 0 & 0 \\ 0 & q_n & 0 & 0 \\ 0 & 0 & q_n & 0 \\ 0 & 0 & 0 & q_n + c \end{bmatrix}, \tag{3.13.18}$$

$$c = \sqrt{q_n^2 + a^{*2}}, \tag{3.13.19}$$

$$\mathbf{R}_n = \begin{bmatrix} (q_n + c)c & 0 & 0 & -(q_n - c)c \\ u - (q_n + c)n_x & \ell_x & m_x & u - (q_n - c)n_x \\ v - (q_n + c)n_y & \ell_y & m_y & v - (q_n - c)n_y \\ w - (q_n + c)n_z & \ell_z & m_z & w - (q_n - c)n_z \end{bmatrix}, \tag{3.13.20}$$

$$\mathbf{L}_n = \begin{bmatrix} \dfrac{1}{2c^2} & \dfrac{(q_n - c)n_x}{2c^2} & \dfrac{(q_n - c)n_y}{2c^2} & \dfrac{(q_n - c)n_z}{2c^2} \\[2mm] -\dfrac{q_\ell}{c^2} & \ell_x - \dfrac{q_n q_\ell}{c^2}n_x & \ell_y - \dfrac{q_n q_\ell}{c^2}n_y & \ell_z - \dfrac{q_n q_\ell}{c^2}n_z \\[2mm] -\dfrac{q_m}{c^2} & m_x - \dfrac{q_n q_m}{c^2}n_x & m_y - \dfrac{q_n q_m}{c^2}n_y & m_z - \dfrac{q_n q_m}{c^2}n_z \\[2mm] \dfrac{1}{2c^2} & \dfrac{(q_n + c)n_x}{2c^2} & \dfrac{(q_n + c)n_y}{2c^2} & \dfrac{(q_n + c)n_z}{2c^2} \end{bmatrix}, \tag{3.13.21}$$

$\boldsymbol{\ell} = [\ell_x, \ell_y, \ell_z]^t$, $\mathbf{m} = [m_x, m_y, m_z]^t$, $\mathbf{n}$ are mutually orthogonal unit vectors, and

$$q_\ell = u\ell_x + v\ell_y + w\ell_z, \quad q_m = um_x + vm_y + wm_z. \tag{3.13.22}$$

The characteristic variables (amplitudes of the waves) are given by

$$\mathbf{L}_n d\mathbf{U} = \begin{bmatrix} \dfrac{dP + (q_n - c)dq_n}{2c^2} \\ dq_\ell - \dfrac{q_\ell(dP + q_n dq_n)}{c^2} \\ dq_m - \dfrac{q_m(dP + q_n dq_n)}{c^2} \\ \dfrac{dP + (q_n + c)dq_n}{2c^2} \end{bmatrix}. \tag{3.13.23}$$

The product of the absolute normal Jacobian and the differential of the variables,

$$|\mathbf{A}_n|\, d\mathbf{U} \equiv \mathbf{R}_n |\mathbf{\Lambda}_n| \mathbf{L}_n d\mathbf{U}, \tag{3.13.24}$$

which can be used to construct the upwind flux, is, of course, independent of the tangent vectors as it should:

$$
\begin{aligned}
|\mathbf{A}_n|\, d\mathbf{U} &= \mathbf{R}_n |\mathbf{\Lambda}_n| \mathbf{L}_n d\mathbf{U} \\
&= \sum_{k=1}^{4} |\lambda_k| \mathbf{r}_k \boldsymbol{\ell}_k^t d\mathbf{U} \\
&= |\lambda_1| \mathbf{r}_1 \boldsymbol{\ell}_1^t d\mathbf{U} + |\lambda_4| \mathbf{r}_4 \boldsymbol{\ell}_4^t d\mathbf{U} + |q_n| \begin{bmatrix} 0 \\ (\ell_x dq_\ell + m_x dq_m) - \dfrac{(q_\ell \ell_x + q_m m_x)(dP + q_n dq_n)}{c^2} \\ (\ell_y dq_\ell + m_y dq_m) - \dfrac{(q_\ell \ell_y + q_m m_y)(dP + q_n dq_n)}{c^2} \\ (\ell_z dq_\ell + m_z dq_m) - \dfrac{(q_\ell \ell_z + q_m m_z)(dP + q_n dq_n)}{c^2} \end{bmatrix} \\
&= |\lambda_1| \mathbf{r}_1 \boldsymbol{\ell}_1^t d\mathbf{U} + |\lambda_4| \mathbf{r}_4 \boldsymbol{\ell}_4^t d\mathbf{U} + |q_n| \begin{bmatrix} 0 \\ (du - dq_n n_x) - \dfrac{(u - q_n n_x)(dP + q_n dq_n)}{c^2} \\ (dv - dq_n n_y) - \dfrac{(v - q_n n_y)(dP + q_n dq_n)}{c^2} \\ (dw - dq_n n_z) - \dfrac{(w - q_n n_z)(dP + q_n dq_n)}{c^2} \end{bmatrix},
\end{aligned} \tag{3.13.25}
$$

where the k-th column of $\mathbf{R}_n$ is denoted by $\mathbf{r}_k$, the k-th row of $\mathbf{L}_n$ is denoted by $\boldsymbol{\ell}_k^t$, and the identities in Section 1.3 have been used in the last step.

Note that this system is in conservative form and it is equivalent to the primitive form only in the steady state since the continuity equation will be satisfied only in the steady state. Note also that the time derivative in the momentum equation is actually real but has been written as artificial in the above. This is because it is meaningless to be real if the momentum equation is solved simultaneously with the pseudo-compressible continuity equation.

3.14 Homentropic/Isothermal Euler Equations

As we have seen in the previous sections, in all dimensions, the Euler equations involve a scalar advection equation for the entropy s:

$$s_t + us_x + vs_y + ws_z = 0, \tag{3.14.1}$$

where (u, v, w) is a local flow velocity. This means that the entropy is constant along a streamline. But at the same time, it means also that the entropy is not necessarily uniform for the entire flow field and can differ from one streamline to another. Such a flow is called an isentropic flow. So, the Euler equations basically describe isentropic flows, with a capability of accounting for non-isentropic features such as shocks by the Rankine-Hugoniot relation (the integral form). A typical example of the isentropic flow is a flow behind a curved shock where the entropy is constant along each streamline, but not uniform because its increase across the shock depends on the shock intensity which varies in space. On the other hand, it would be very nice if the entropy were constant everywhere. It would be the case, for example, when the entropy is initially constant everywhere and no shock waves are created anywhere. Such a flow is called a homentropic flow (a special case of an isentropic flow). What is nice about a homentropic flow is that the Euler equations close without the energy equation. For homentropic flows, the adiabatic relation,

$$p/\rho^\gamma = K, \tag{3.14.2}$$

is valid everywhere, i.e., K is a global constant (not only along a streamline), say $K = p_\infty/\rho_\infty^\gamma$. Hence, it follows from the equation of state,

$$\rho E = \frac{p}{\gamma - 1} + \frac{1}{2}\rho \mathbf{v}^2 = \frac{K\rho^\gamma}{\gamma - 1} + \frac{1}{2}\rho \mathbf{v}^2, \tag{3.14.3}$$

that the total energy can be explicitly computed everywhere once the density and the velocity are determined by the continuity and momentum equations, and therefore the energy equation is no longer required. This implies also that more generally, if the pressure is a function of the density only, i.e., $p = p(\rho)$ (such a flow or a process is called barotropic [18]), the Euler equations will close without the energy equation. Then, the speed of sound c is given by

$$c = \sqrt{p'(\rho)}, \tag{3.14.4}$$

which becomes, in homentropic flows,

$$c = \sqrt{\gamma K \rho^{\gamma - 1}}. \tag{3.14.5}$$

On the other hand, a flow with $\gamma = 1$ is called an isothermal flow. Here, by Equation (3.1.11), this is equivalent to a gas with infinite degrees of freedom. Also, from Equations (3.1.9) and (3.1.10), we see that the specific heats become infinite. This means that it is extremely difficult to change the temperature of the gas, thus leading to the term 'isothermal'. I like isothermal flows because in this case also the Euler equations close without the energy equation. In particular, the speed of sound is given by

$$c^2 = RT = p/\rho, \tag{3.14.6}$$

which is now a constant. Isothermal flows are mysterious because $\gamma = 1$ is artificial. There are no such gases in reality.

In either homentropic or isothermal flow, the Euler equations close without the energy equation, and the pressure and the speed of sound are given by

$$\begin{cases} p = K\rho^\gamma, \quad c = \sqrt{\gamma K \rho^{\gamma - 1}}, & \text{for homentropic flows,} \\ p = c^2\rho, \quad c = \sqrt{RT} = \text{constant}, & \text{for isothermal flows.} \end{cases} \tag{3.14.7}$$

In what follows, I write the Euler equations without the energy equation. Note that p and c in these equations are given by the above formulas, depending on the type of flows.

3.14.1   1D Homentropic/Isothermal Euler Equations

$$\mathbf{U}_t + \mathbf{F}_x = 0, \tag{3.14.8}$$

where

$$\mathbf{U} = \begin{bmatrix} \rho \\ \rho u \end{bmatrix}, \mathbf{F} = \begin{bmatrix} \rho u \\ p + \rho u^2 \end{bmatrix}. \tag{3.14.9}$$

Jacobian

$$\mathbf{A} = \frac{\partial \mathbf{F}}{\partial \mathbf{U}} = \begin{bmatrix} 0 & 1 \\ c^2 - u^2 & 2u \end{bmatrix}. \tag{3.14.10}$$

Eigenstructure

$$\mathbf{A} = \mathbf{R}\mathbf{\Lambda}\mathbf{L}, \tag{3.14.11}$$

where

$$\mathbf{\Lambda} = \begin{bmatrix} u - c & 0 \\ 0 & u + c \end{bmatrix}, \tag{3.14.12}$$

$$\mathbf{R} = \begin{bmatrix} 1 & 1 \\ u - c & u + c \end{bmatrix}, \quad \mathbf{L} = \begin{bmatrix} \dfrac{u+c}{2c} & -\dfrac{1}{2c} \\ -\dfrac{u-c}{2c} & \dfrac{1}{2c} \end{bmatrix}, \quad \mathbf{L}d\mathbf{U} = \begin{bmatrix} \dfrac{1}{2}d\rho - \dfrac{\rho}{2c}du \\ \dfrac{1}{2}d\rho + \dfrac{\rho}{2c}du \end{bmatrix}. \tag{3.14.13}$$

Of course, this system is hyperbolic only if c is real, which is true if $p'(\rho) > 0$. Hence, it then follows from Equation (3.14.7) that it is hyperbolic for both homentropic and isothermal flows.

3.14.2 2D Homentropic/Isothermal Euler Equations

$$\mathbf{U}_t + \mathbf{F}_x + \mathbf{G}_y = 0, \tag{3.14.14}$$

where

$$\mathbf{U} = \begin{bmatrix} \rho \\ \rho u \\ \rho v \end{bmatrix}, \quad \mathbf{F} = \begin{bmatrix} \rho u \\ p + \rho u^2 \\ \rho uv \end{bmatrix}, \quad \mathbf{G} = \begin{bmatrix} \rho v \\ \rho uv \\ p + \rho v^2 \end{bmatrix}. \tag{3.14.15}$$

Jacobians

$$\mathbf{A} = \frac{\partial \mathbf{F}}{\partial \mathbf{U}} = \begin{bmatrix} 0 & 1 & 0 \\ c^2 - u^2 & 2u & 0 \\ -uv & v & u \end{bmatrix}, \quad \mathbf{B} = \frac{\partial \mathbf{G}}{\partial \mathbf{U}} = \begin{bmatrix} 0 & 0 & 1 \\ -uv & v & u \\ c^2 - v^2 & 0 & 2v \end{bmatrix}, \tag{3.14.16}$$

$$\mathbf{A}_n = \mathbf{A}n_x + \mathbf{B}n_y = \begin{bmatrix} 0 & n_x & n_y \\ c^2 n_x - uq_n & q_n + un_x & un_y \\ c^2 n_y - vq_n & vn_x & q_n + vn_y \end{bmatrix}, \tag{3.14.17}$$

where

$$q_n = un_x + vn_y. \tag{3.14.18}$$

Eigenstructure

$$\mathbf{A}_n = \mathbf{R}_n\mathbf{\Lambda}_n\mathbf{L}_n, \tag{3.14.19}$$

where

$$\mathbf{\Lambda}_n = \begin{bmatrix} q_n - c & 0 & 0 \\ 0 & q_n & 0 \\ 0 & 0 & q_n + c \end{bmatrix}, \tag{3.14.20}$$

$$\mathbf{R}_n = \begin{bmatrix} 1 & 0 & 1 \\ u - c\,n_x & -n_y & u + c\,n_x \\ v - c\,n_y & n_x & v + c\,n_y \end{bmatrix}, \quad \mathbf{L}_n = \begin{bmatrix} \dfrac{c + q_n}{2c} & -\dfrac{n_x}{2c} & -\dfrac{n_y}{2c} \\ -q_\ell & -n_y & n_x \\ \dfrac{c - q_n}{2c} & \dfrac{n_x}{2c} & \dfrac{n_y}{2c} \end{bmatrix}, \tag{3.14.21}$$

and

$$q_\ell = -un_y + vn_x. \tag{3.14.22}$$

3.14.3   3D Homentropic/Isothermal Euler Equations

$$\mathbf{U}_t + \mathbf{F}_x + \mathbf{G}_y + \mathbf{H}_z = 0, \tag{3.14.23}$$

where

$$\mathbf{U} = \begin{bmatrix} \rho \\ \rho u \\ \rho v \\ \rho w \end{bmatrix}, \ \mathbf{F} = \begin{bmatrix} \rho u \\ p + \rho u^2 \\ \rho u v \\ \rho u w \end{bmatrix}, \ \mathbf{G} = \begin{bmatrix} \rho v \\ \rho v u \\ p + \rho v^2 \\ \rho v w \end{bmatrix}, \ \mathbf{H} = \begin{bmatrix} \rho w \\ \rho w u \\ \rho w v \\ p + \rho w^2 \end{bmatrix}. \tag{3.14.24}$$

Jacobians

$$\mathbf{A} \ = \ \frac{\partial \mathbf{F}}{\partial \mathbf{U}} = \begin{bmatrix} 0 & 1 & 0 & 0 \\ c^2 - u^2 & 2u & 0 & 0 \\ -uv & v & u & 0 \\ -uw & w & 0 & u \end{bmatrix}, \tag{3.14.25}$$

$$\mathbf{B} \ = \ \frac{\partial \mathbf{G}}{\partial \mathbf{U}} = \begin{bmatrix} 0 & 0 & 1 & 0 \\ -vu & v & u & 0 \\ c^2 - v^2 & 0 & 2v & 0 \\ -vw & 0 & w & v \end{bmatrix}, \tag{3.14.26}$$

$$\mathbf{C} \ = \ \frac{\partial \mathbf{H}}{\partial \mathbf{U}} = \begin{bmatrix} 0 & 0 & 0 & 1 \\ -wu & w & 0 & u \\ -wv & 0 & w & v \\ c^2 - w^2 & 0 & 0 & 2w \end{bmatrix}, \tag{3.14.27}$$

$$\mathbf{A}_n = \mathbf{A}n_x + \mathbf{B}n_y + \mathbf{C}n_z = \begin{bmatrix} 0 & n_x & n_y & n_z \\ c^2 n_x - u q_n & u n_x + q_n & u n_y & u n_z \\ c^2 n_y - v q_n & v n_x & v n_y + q_n & v n_z \\ c^2 n_z - w q_n & w n_x & w n_y & w n_z + q_n \end{bmatrix}. \tag{3.14.28}$$

Eigenstructure

$$\mathbf{A}_n = \mathbf{R}_n \mathbf{\Lambda}_n \mathbf{L}_n, \tag{3.14.29}$$

where

$$\mathbf{\Lambda}_n = \begin{bmatrix} q_n - c & 0 & 0 & 0 \\ 0 & q_n & 0 & 0 \\ 0 & 0 & q_n & 0 \\ 0 & 0 & 0 & q_n + c \end{bmatrix}, \tag{3.14.30}$$

$$\mathbf{R}_n = \begin{bmatrix} 1 & 0 & 0 & 1 \\ u - c n_x & \ell_x & m_x & u + c n_x \\ v - c n_y & \ell_y & m_y & v + c n_x \\ w - c n_z & \ell_z & m_z & w + c n_y \end{bmatrix}, \quad \mathbf{L}_n = \begin{bmatrix} \dfrac{c + q_n}{2c} & -\dfrac{n_x}{2c} & -\dfrac{n_y}{2c} & -\dfrac{n_z}{2c} \\ -q_\ell & \ell_x & \ell_y & \ell_z \\ -q_m & m_x & m_y & m_z \\ \dfrac{c - q_n}{2c} & \dfrac{n_x}{2c} & \dfrac{n_y}{2c} & \dfrac{n_z}{2c} \end{bmatrix}, \tag{3.14.31}$$

$\boldsymbol{\ell} = [\ell_x, \ell_y, \ell_z]^t$, $\mathbf{m} = [m_x, m_y, m_z]^t$, $\mathbf{n}$ are mutually orthogonal unit vectors, and

$$q_\ell = u\ell_x + v\ell_y + w\ell_z, \quad q_m = um_x + vm_y + wm_z. \tag{3.14.32}$$

3.15 Linear Acoustics Equations (Linearized Euler Equations)

Decompose variables as follows:

$$\rho = \rho_0 + \rho', \tag{3.15.1}$$

$$p = p_0 + p', \tag{3.15.2}$$

$$(u, v, w) = (u_0, v_0, w_0) + (u', v', w'), \tag{3.15.3}$$

where the variables with the subscript 0 are uniform stream values and those with a prime are small perturbations. Substitute these into the Euler equations, and make approximations (ignoring products of perturbations), then we obtain the so-called linear acoustics equations.

Flows with small perturbations can be considered as homentropic (because shock waves will not appear). Then, the equation of state can be written as

$$p = P(\rho), \tag{3.15.4}$$

e.g., $p = K\rho^\gamma$, from which we get

$$p' = \left(\frac{dP}{d\rho}\right)\bigg|_{\rho=\rho_0} \rho' = c_0^2 \rho', \tag{3.15.5}$$

where c_0 is the speed of sound which is a constant. This will close the linear acoustics equations. Note that

$$c_0 = \sqrt{\left(\frac{dP}{d\rho}\right)\bigg|_{\rho=\rho_0}}, \tag{3.15.6}$$

and therefore $\frac{dP}{d\rho}$ must be positive. Don't worry. It is positive in homentropic flows:

$$c_0 = \sqrt{\gamma K \rho_0^{\gamma-1}} = \sqrt{\gamma \left(\frac{p_0}{\rho_0^\gamma}\right) \rho_0^{\gamma-1}} = \sqrt{\gamma \frac{p_0}{\rho_0}} > 0, \tag{3.15.7}$$

for physically realistic pressure and density.

So, basically, the linear acoustics equations are valid only for flows with small perturbations, such as a sound wave propagation or a flow over a slender body. The governing equations are now linear, and therefore nonlinear waves such as shocks or expansions all reduce to linear waves. Also, simple wave solutions can be superposed to produce more general solutions. This can be very useful. I like it.

3.15.1 1D Linear Acoustics Equations

$$\mathbf{W}'_t + \mathbf{A}^{w'} \mathbf{W}'_x = 0, \tag{3.15.8}$$

where

$$\mathbf{W}' = \begin{bmatrix} p' \\ u' \end{bmatrix}, \quad \mathbf{A}^{w'} = \begin{bmatrix} u_0 & \rho_0 c_0^2 \\ 1/\rho_0 & u_0 \end{bmatrix}. \tag{3.15.9}$$

Eigenstructure

$$\mathbf{A}^{w'} = \mathbf{R}^{w'} \mathbf{\Lambda} \mathbf{L}^{w'}, \tag{3.15.10}$$

where

$$\mathbf{\Lambda}^{w'} = \begin{bmatrix} u_0 - c & 0 \\ 0 & u_0 + c \end{bmatrix}, \quad \mathbf{R}^{w'} = \begin{bmatrix} \rho_0 c_0^2 & \rho_0 c_0^2 \\ -c_0 & c_0 \end{bmatrix}, \quad \mathbf{L}^{w'} = \begin{bmatrix} \dfrac{1}{2\rho_0 c_0^2} & -\dfrac{1}{2c_0} \\ \dfrac{1}{2\rho_0 c_0^2} & \dfrac{1}{2c_0} \end{bmatrix}. \tag{3.15.11}$$

3.15.2   2D Linear Acoustics Equations

$$\mathbf{W}'_t + \mathbf{A}^{w'}\mathbf{W}'_x + \mathbf{B}^{w'}\mathbf{W}'_y = 0, \tag{3.15.12}$$

where

$$\mathbf{W}' = \begin{bmatrix} p' \\ u' \\ v' \end{bmatrix}, \quad \mathbf{A}^{w'} = \begin{bmatrix} u_0 & \rho_0 c_0^2 & 0 \\ 1/\rho_0 & u_0 & 0 \\ 0 & 0 & u_0 \end{bmatrix}, \quad \mathbf{B}^w = \begin{bmatrix} v_0 & 0 & \rho_0 c_0^2 \\ 0 & v_0 & 0 \\ 1/\rho_0 & 0 & v_0 \end{bmatrix}. \tag{3.15.13}$$

Of course, these matrices are not Jacobians (see Section 3.4.2).

Eigenstructure

$$\mathbf{A}_n^{w'} = \mathbf{R}_n^{w'}\mathbf{\Lambda}_n\mathbf{L}_n^{w'}, \tag{3.15.14}$$

where

$$\mathbf{A}_n^{w'} = \begin{bmatrix} q_{n0} & \rho_0 c_0^2 n_x & \rho_0 c_0^2 n_y \\ n_x/\rho_0 & q_{n0} & 0 \\ n_y/\rho_0 & 0 & q_{n0} \end{bmatrix}, \quad \mathbf{\Lambda}_n^{w'} = \begin{bmatrix} q_{n0}-c & 0 & 0 \\ 0 & q_{n0} & 0 \\ 0 & 0 & q_{n0}+c \end{bmatrix}, \tag{3.15.15}$$

$$\mathbf{R}_n^{w'} = \begin{bmatrix} \rho_0 c_0^2 & 0 & \rho_0 c_0^2 \\ -c_0 n_x & \ell_x & c_0 n_x \\ -c_0 n_y & \ell_y & c_0 n_y \end{bmatrix}, \quad \mathbf{L}_n^{w'} = \begin{bmatrix} \dfrac{1}{2\rho_0 c_0^2} & -\dfrac{n_x}{2c_0} & -\dfrac{n_y}{2c_0} \\ 0 & \ell_x & \ell_y \\ \dfrac{1}{2\rho_0 c_0^2} & \dfrac{n_x}{2c_0} & \dfrac{n_y}{2c_0} \end{bmatrix}, \tag{3.15.16}$$

and $(\ell_x, \ell_y) = (-n_y, n_x)$.

3.15.3   3D Linear Acoustics Equations

$$\mathbf{W}'_t + \mathbf{A}^{w'}\mathbf{W}'_x + \mathbf{B}^{w'}\mathbf{W}'_y + \mathbf{C}^{w'}\mathbf{W}'_z = 0, \tag{3.15.17}$$

where

$$\mathbf{W}' = \begin{bmatrix} p' \\ u' \\ v' \\ w' \end{bmatrix}, \quad \mathbf{A}^{w'} = \begin{bmatrix} u_0 & \rho_0 c_0^2 & 0 & 0 \\ 1/\rho_0 & u_0 & 0 & 0 \\ 0 & 0 & u_0 & 0 \\ 0 & 0 & 0 & u_0 \end{bmatrix}, \tag{3.15.18}$$

$$\mathbf{B}^w = \begin{bmatrix} v_0 & 0 & \rho_0 c_0^2 & 0 \\ 0 & v_0 & 0 & 0 \\ 1/\rho_0 & 0 & v_0 & 0 \\ 0 & 0 & 0 & v_0 \end{bmatrix}, \quad \mathbf{C}^w = \begin{bmatrix} w_0 & 0 & 0 & \rho_0 c_0^2 \\ 0 & w_0 & 0 & 0 \\ 0 & 0 & w_0 & 0 \\ 1/\rho_0 & 0 & 0 & w_0 \end{bmatrix}. \tag{3.15.19}$$

Eigenstructure

$$\mathbf{A}_n^{w'} = \mathbf{R}_n^{w'}\mathbf{\Lambda}_n\mathbf{L}_n^{w'}, \tag{3.15.20}$$

where

$$
\mathbf{A}_n^{w'} = \begin{bmatrix} q_{n0} & \rho_0 c_0^2 n_x & \rho_0 c_0^2 n_y & \rho_0 c_0^2 n_z \\ n_x/\rho_0 & q_{n0} & 0 & 0 \\ n_y/\rho_0 & 0 & q_{n0} & 0 \\ n_z/\rho_0 & 0 & 0 & q_{n0} \end{bmatrix},
\tag{3.15.21}
$$

$$
\mathbf{\Lambda}_n^{w'} = \begin{bmatrix} q_{n0} - c & 0 & 0 & 0 \\ 0 & q_{n0} & 0 & 0 \\ 0 & 0 & q_{n0} & 0 \\ 0 & 0 & 0 & q_{n0} + c \end{bmatrix},
\tag{3.15.22}
$$

$$
\mathbf{R}_n^{w'} = \begin{bmatrix} \rho_0 c_0^2 & 0 & 0 & \rho_0 c_0^2 \\ -c_0 n_x & \ell_x & m_x & c_0 n_x \\ -c_0 n_y & \ell_y & m_y & c_0 n_y \\ -c_0 n_z & \ell_z & m_z & c_0 n_z \end{bmatrix},
\tag{3.15.23}
$$

$$
\mathbf{L}_n^{w'} = \begin{bmatrix} \dfrac{1}{2\rho_0 c_0^2} & -\dfrac{n_x}{2c_0} & -\dfrac{n_y}{2c_0} & -\dfrac{n_z}{2c_0} \\ 0 & \ell_x & \ell_y & \ell_z \\ 0 & m_x & m_y & m_z \\ \dfrac{1}{2\rho_0 c_0^2} & \dfrac{n_x}{2c_0} & \dfrac{n_y}{2c_0} & \dfrac{n_z}{2c_0} \end{bmatrix},
\tag{3.15.24}
$$

$\boldsymbol{\ell} = [\ell_x, \ell_y, \ell_z]^t$, $\mathbf{m} = [m_x, m_y, m_z]^t$, $\mathbf{n}$ are mutually orthogonal unit vectors, and

$$
q_\ell = u\ell_x + v\ell_y + w\ell_z, \quad q_m = um_x + vm_y + wm_z.
\tag{3.15.25}
$$

3.16 Quasi-1D Euler Equations

The quasi-1D Euler system is a model for a nozzle flow. The nozzle itself is two-dimensional, i.e., its section area $A(x)$ varies along the axial direction x, but flow variables are assumed to depend only on x. Here is the quasi-1D Euler system:

$$
\frac{\partial \mathbf{U}}{\partial t} + \frac{\partial \mathbf{F}}{\partial x} = \mathbf{S},
\tag{3.16.1}
$$

where

$$
\mathbf{U} = \begin{bmatrix} \rho A \\ \rho u A \\ \rho E A \end{bmatrix}, \quad \mathbf{F} = \begin{bmatrix} \rho u A \\ \rho u^2 A \\ \rho u(E + p/\rho)A \end{bmatrix}, \quad \mathbf{S} = \begin{bmatrix} 0 \\ -A\frac{\partial p}{\partial x} \\ 0 \end{bmatrix},
\tag{3.16.2}
$$

and $E = e + u^2/2$. I like this system because it can be written in the following form ($A(x)$ only in the source term),

$$
\mathbf{U} = \begin{bmatrix} \rho \\ \rho u \\ \rho E \end{bmatrix}, \quad \mathbf{F} = \begin{bmatrix} \rho u \\ \rho u^2 + p \\ \rho u(E + p/\rho) \end{bmatrix}, \quad \mathbf{S} = -\frac{1}{A}\frac{DA}{Dt}\begin{bmatrix} \rho \\ \rho u \\ \rho E + p \end{bmatrix},
\tag{3.16.3}
$$

and also in the following form (no pressure gradient in the source term),

$$
\mathbf{U} = \begin{bmatrix} \rho A \\ \rho u A \\ \rho E A \end{bmatrix}, \quad \mathbf{F} = \begin{bmatrix} \rho u A \\ (\rho u^2 + p)A \\ \rho u(E + p/\rho)A \end{bmatrix}, \quad \mathbf{S} = \begin{bmatrix} 0 \\ p\frac{\partial A}{\partial x} \\ 0 \end{bmatrix}.
\tag{3.16.4}
$$

See Ref.[57] for derivation, and Ref.[159] for the case that A changes in time. A really nice thing about the quasi-1D Euler system is that if I set $\rho A \to \rho$, then it is in the same form as the 1D Euler system (except for the source term), and therefore the eigenstructure is the same as the 1D Euler system (which we already know). This is very nice.

On the other hand, if I set $A(x) = 2\pi x$, then I obtain the Euler equations for cylindrically symmetric wave motion, i.e., the Euler equations in cylindrical coordinates (r, θ, z) with no variations in the z direction. For example, I obtain from Equation (3.16.3),

$$\frac{\partial \mathbf{U}}{\partial t} + \frac{\partial \mathbf{F}}{\partial r} = \mathbf{S}, \tag{3.16.5}$$

where

$$\mathbf{U} = \begin{bmatrix} \rho \\ \rho u \\ \rho E \end{bmatrix}, \quad \mathbf{F} = \begin{bmatrix} \rho u \\ \rho u^2 + p \\ \rho u(E + p/\rho) \end{bmatrix}, \quad \mathbf{S} = -\frac{u}{r} \begin{bmatrix} \rho \\ \rho u \\ \rho E + p \end{bmatrix}, \tag{3.16.6}$$

and u is now the velocity in the radial direction. Note also that I have replaced x by r to emphasize the cylindrical symmetry. Similarly, if I set $A(x) = 4\pi x^2$, then I obtain the Euler equations for spherically symmetric wave motion:

$$\mathbf{U} = \begin{bmatrix} \rho \\ \rho u \\ \rho E \end{bmatrix}, \quad \mathbf{F} = \begin{bmatrix} \rho u \\ \rho u^2 + p \\ \rho u(E + p/\rho) \end{bmatrix}, \quad \mathbf{S} = -\frac{2u}{r} \begin{bmatrix} \rho \\ \rho u \\ \rho E + p \end{bmatrix}. \tag{3.16.7}$$

This is very interesting, isn't it?

3.17 2D Steady Euler Equations

The 2D steady Euler equations are obtained by removing the time derivative from the 2D Euler equations (3.5.16),

$$\mathbf{A}^w \mathbf{W}_x + \mathbf{B}^w \mathbf{W}_y = 0, \tag{3.17.1}$$

where $\mathbf{W} = [\rho, u, v, p]^t$. Since $\mathbf{A}^w$ is invertible (for $u^2 \neq c^2$), we can write this as

$$\mathbf{W}_x + (\mathbf{A}^w)^{-1}\mathbf{B}^w \mathbf{W}_y = 0, \tag{3.17.2}$$

where

$$(\mathbf{A}^w)^{-1}\mathbf{B}^w = \frac{1}{u^2 - c^2} \begin{bmatrix} \frac{v}{u}(u^2 - c^2) & -\rho v & \rho u & \frac{v}{u} \\ 0 & uv & -c^2 & -\frac{v}{\rho} \\ 0 & 0 & \frac{v}{u}(u^2 - c^2) & \frac{u^2 - c^2}{\rho u} \\ 0 & -\rho v c^2 & \rho u c^2 & uv \end{bmatrix}. \tag{3.17.3}$$

Therefore, x-axis may be time-like (depending on flow directions). The eigenvalues and eigenvectors of $(\mathbf{A}^w)^{-1}\mathbf{B}^w$ are given by

$$\lambda_{1,2} = \frac{v}{u}, \tag{3.17.4}$$

$$\lambda_3 = \frac{uv - c\sqrt{q^2 - c^2}}{u^2 - c^2} = \frac{v\beta - u}{u\beta + v}, \tag{3.17.5}$$

$$\lambda_4 = \frac{uv + c\sqrt{q^2 - c^2}}{u^2 - c^2} = \frac{u + v\beta}{u\beta - v}, \tag{3.17.6}$$

$$\mathbf{R}_w = \begin{bmatrix} 1 & 0 & -\dfrac{\rho}{u\beta + v} & \dfrac{\rho}{u\beta - v} \\[2ex] 0 & \dfrac{u}{\rho q^2} & -\dfrac{v\beta - u}{M^2(u\beta + v)} & -\dfrac{v\beta + u}{M^2(u\beta - v)} \\[2ex] 0 & \dfrac{v}{\rho q^2} & \dfrac{1}{M^2} & \dfrac{1}{M^2} \\[2ex] 0 & 0 & \dfrac{-\rho c^2}{u\beta + v} & \dfrac{\rho c^2}{u\beta - v} \end{bmatrix}, \tag{3.17.7}$$

$$\mathbf{L}_w = \begin{bmatrix} 1 & 0 & 0 & -\dfrac{1}{c^2} \\[2ex] 0 & \rho u & \rho v & 1 \\[2ex] 0 & -\dfrac{v}{2c^2\beta}(u\beta + v) & \dfrac{u}{2c^2\beta}(u\beta + v) & -\dfrac{1}{2\rho c^2}(u\beta + v) \\[2ex] 0 & -\dfrac{v}{2c^2\beta}(u\beta - v) & \dfrac{u}{2c^2\beta}(u\beta - v) & \dfrac{1}{2\rho c^2}(u\beta - v) \end{bmatrix}, \tag{3.17.8}$$

where $\beta = \sqrt{M^2 - 1} = \sqrt{q^2/c^2 - 1} = \sqrt{(u^2 + v^2)/c^2 - 1}$ and $c = \sqrt{\gamma p/\rho}$. Clearly, the character of the 2D Steady Euler Equations depends on the Mach number:

$$\begin{array}{lllll} \text{Supersonic } M > 1 & \rightarrow & \text{Real eigenvalues} & \rightarrow & \text{Hyperbolic,} \\ \text{Subsonic } M < 1 & \rightarrow & \text{Complex eigenvalues} & \rightarrow & \text{Elliptic.} \end{array} \tag{3.17.9}$$

Of course, it can be mixed: hyperbolic (supersonic) in some region while elliptic (subsonic) in other regions. In particular, I like supersonic flows because then the system (3.17.2) can be fully diagonalized, with the characteristic variables defined by $\partial\mathbf{V} = \mathbf{L}_w \partial\mathbf{W}$:

$$\frac{\partial v_1}{\partial x} + \frac{u}{v}\frac{\partial v_1}{\partial y} = 0, \tag{3.17.10}$$

$$\frac{\partial v_2}{\partial x} + \frac{u}{v}\frac{\partial v_2}{\partial y} = 0, \tag{3.17.11}$$

$$\frac{\partial v_3}{\partial x} + \frac{v\beta - u}{u\beta + v}\frac{\partial v_3}{\partial y} = 0, \tag{3.17.12}$$

$$\frac{\partial v_4}{\partial x} + \frac{u + v\beta}{u\beta - v}\frac{\partial v_4}{\partial y} = 0, \tag{3.17.13}$$

where

$$\partial\mathbf{V} = \begin{bmatrix} \partial v_1 \\[1ex] \partial v_2 \\[1ex] \partial v_3 \\[1ex] \partial v_4 \end{bmatrix} = \begin{bmatrix} \partial\rho - \dfrac{\partial p}{c^2} \\[2ex] \rho(u\partial u + v\partial v) + \partial p \\[2ex] \dfrac{u\beta + v}{2\rho c^2\beta}\{\rho(u\partial v - v\partial u) - \beta\partial p\} \\[2ex] \dfrac{u\beta - v}{2\rho c^2\beta}\{\rho(u\partial v - v\partial u) + \beta\partial p\} \end{bmatrix}. \tag{3.17.14}$$

Note that the eigenvalues represent the slope dy/dx of the characteristics. Therefore, they can be expressed as vectors:

$$\vec{\lambda}_{1,2} = [u, v]^t, \quad \vec{\lambda}_3 = [u\beta + v, v\beta - u]^t, \quad \vec{\lambda}_4 = [u\beta - v, v\beta + u]^t. \tag{3.17.15}$$

Then, we find

$$\text{div }\vec{\lambda}_3 = \text{div}(\beta\mathbf{v}) + \zeta, \quad \text{div }\vec{\lambda}_4 = \text{div}(\beta\mathbf{v}) - \zeta, \tag{3.17.16}$$

where $\zeta = v_x - u_y$. The left hand side (the divergence of the characteristic speed vector) represents the divergence/convergence of characteristics, implying the existence of expansion/shock waves. I like these relations very much because they indicate connections between the existence of the nonlinear waves and the physical quantities such as the vorticity and divergence of the velocity field. Interestingly, it is not just the vorticity or the divergence but their combinations (the sum or the difference) that tell us whether a shock or expansion wave exists (and also of which family it is). So, if the sum $\text{div}(\beta\mathbf{v}) + \zeta$ is negative/positive, there is a shock/expansion from 3-family; if the difference $\text{div}(\beta\mathbf{v}) - \zeta$ is negative/positive, there is a shock/expansion from 4-family. Or you might want to combine them into one,

$$\left(\text{div}\,\vec{\lambda}_3\right)^2 + \left(\text{div}\,\vec{\lambda}_4\right)^2 = \{\text{div}(\beta\mathbf{v})\}^2 + \zeta^2. \tag{3.17.17}$$

This can also be a useful scalar (non-negative) quantity for detecting nonlinear waves (a shocks and/or expansions of any family). Actually, these relations were successfully used for adaptive Euler schemes in Ref.[100].

I like the steady Euler equations because they are of mixed type. There have been attempts to construct optimal discretization schemes for the Euler equations, i.e., central scheme for the elliptic part and upwind scheme for the hyperbolic part, (see Refs.[94, 113] and references therein). That sounds like an optimal strategy, but typically in most cases numerical schemes for the Euler equations are constructed based on the unsteady character, i.e., hyperbolic character. Yes, the Euler equations are hyperbolic in time, and can never be elliptic in time, which in fact makes it easier to construct a numerical scheme. So, I actually like both approaches.

3.18   Bernoulli's Equation

Bernoulli's equation is basically an integral of the Euler equation (the momentum equation),

$$\rho\frac{\partial\mathbf{v}}{\partial t} + \rho(\text{grad}\mathbf{v})\mathbf{v} + \text{grad}\,p = 0, \tag{3.18.1}$$

which can be written, by Equation (1.7.7), as

$$\rho\frac{\partial\mathbf{v}}{\partial t} + \rho\,\text{grad}\left(\frac{\mathbf{v}^2}{2}\right) - \rho\mathbf{v}\times\text{curl}\mathbf{v} + \text{grad}\,p = 0. \tag{3.18.2}$$

This can be integrated for various flows; thus there are various forms of Bernoulli's equation.

Unsteady, Incompressible, Irrotational Flows

For incompressible irrotational flows, we have $\rho = $ constant and $\text{curl}\mathbf{v} = 0$. The latter implies the existence of the velocity potential ϕ: $\mathbf{v} = \text{grad}\phi$. Then, the Euler equations (3.18.2) can be written as

$$\rho\,\text{grad}\left(\frac{\partial\phi}{\partial t} + \frac{\mathbf{v}^2}{2} + \frac{p}{\rho}\right) = 0. \tag{3.18.3}$$

This means that the quantity in the parenthesis is spatially constant but may change in time. We therefore obtain

$$\frac{\partial\phi}{\partial t} + \frac{\mathbf{v}^2}{2} + \frac{p}{\rho} = f(t), \tag{3.18.4}$$

where $f(t)$ is an arbitrary function of time. This is called the unsteady Bernoulli's equation and its steady version, $\mathbf{v}^2/2 + p/\rho = $ constant, is the famous Bernoulli's equation obtained by Daniel Bernoulli.

Steady, Incompressible, Rotational Flows

For steady incompressible rotational flows, the Euler equations (3.18.2) becomes

$$\rho\,\text{grad}\left(\frac{\mathbf{v}^2}{2} + \frac{p}{\rho}\right) - \rho\mathbf{v}\times\text{curl}\mathbf{v} = 0. \tag{3.18.5}$$

Because of the second term, there can be no spatially constant quantities. However, noting that the vector $(\mathbf{v}\times\text{curl}\mathbf{v})$ is perpendicular to the velocity vector $\mathbf{v}$, we take the dot product of this equation and the velocity,

$$\rho\mathbf{v}\cdot\text{grad}\left(\frac{\mathbf{v}^2}{2} + \frac{p}{\rho}\right) - \rho\mathbf{v}\cdot(\mathbf{v}\times\text{curl}\mathbf{v}) = 0, \tag{3.18.6}$$

so that the second term vanishes and we obtain

$$\rho \mathbf{v} \cdot \operatorname{grad}\left(\frac{\mathbf{v}^2}{2} + \frac{p}{\rho}\right) = 0. \tag{3.18.7}$$

This shows that we do not have a global constant quantity but a streamwise constant quantity:

$$\frac{\mathbf{v}^2}{2} + \frac{p}{\rho} = \text{constant}, \tag{3.18.8}$$

along a streamline, i.e., the stagnation pressure $p_0 = \rho \mathbf{v}^2/2 + p$ takes different constant values for different streamlines. To see how it changes across streamlines, we take the dot product of Equation (3.18.5) and a unit vector $\mathbf{n}$ in the direction normal to the streamline,

$$\rho \mathbf{n} \cdot \operatorname{grad}\left(\frac{\mathbf{v}^2}{2} + \frac{p}{\rho}\right) - \rho \mathbf{n} \cdot (\mathbf{v} \times \operatorname{curl}\mathbf{v}) = 0, \tag{3.18.9}$$

which gives the spatial rate of change of the stagnation pressure normal to the streamline,

$$\frac{\partial p_0}{\partial n} = \frac{\partial}{\partial n}\left(p + \frac{\rho \mathbf{v}^2}{2}\right) = \rho \mathbf{n} \cdot (\mathbf{v} \times \operatorname{curl}\mathbf{v}). \tag{3.18.10}$$

This shows that the vorticity ($\operatorname{curl}\mathbf{v}$) is the source of the spatial variation of p_0. However, it is also clear that in order for p_0 to be constant everywhere, the vorticity does not have to be zero, or equivalently the flow does not have to be irrotational. If the flow satisfies

$$\mathbf{v} \times \operatorname{curl}\mathbf{v} = 0, \tag{3.18.11}$$

everywhere, meaning that it is a Beltrami flow (see Section 1.5), p_0 will be a global constant. This is nice. I also like two-dimensional flows in which we always have $\mathbf{v} \perp \operatorname{curl}\mathbf{v}$, and so by choosing $\mathbf{n}$ in the same direction as ($\mathbf{v} \times \operatorname{curl}\mathbf{v}$), we can write

$$\frac{\partial}{\partial n}\left(p + \frac{\rho \mathbf{v}^2}{2}\right) = \rho q \omega, \tag{3.18.12}$$

q is the flow speed and ω is the magnitude of the vorticity.

Unsteady, Compressible, Irrotational Flows

For compressible flows, the density is not constant. This makes it non-trivial to integrate the pressure gradient term $\frac{1}{\rho}\operatorname{grad} p$ in the Euler equation. But it can still be integrated if we can express it as a gradient of some quantity. This is possible for an adiabatic flow of a calorically perfect gas:

$$\frac{dp}{\rho} = dh - T ds = dh = \frac{\gamma R}{\gamma - 1} dT = \frac{1}{\gamma - 1} d(a^2), \tag{3.18.13}$$

where s is the entropy which is uniform for adiabatic flows and a is the speed of sound. This gives

$$\frac{1}{\rho}\operatorname{grad} p = \frac{1}{\gamma - 1}\operatorname{grad}(a^2). \tag{3.18.14}$$

Now, introducing the velocity potential ϕ, we can write the the Euler equations (3.18.2) as

$$\rho \operatorname{grad}\left(\frac{\partial \phi}{\partial t} + \frac{\mathbf{v}^2}{2} + \frac{a^2}{\gamma - 1}\right) = 0. \tag{3.18.15}$$

Therefore we obtain the Bernoulli's equation for unsteady compressible irrotational flows:

$$\frac{\partial \phi}{\partial t} + \frac{\mathbf{v}^2}{2} + \frac{a^2}{\gamma - 1} = g(t), \tag{3.18.16}$$

where $g(t)$ is an arbitrary function of time.

Steady, Compressible, Rotational Flows

In the same way as in the previous case, we can derive Bernoulli's equations for steady compressible rotational flows. Basically, we just need to replace p/ρ by $\frac{a^2}{\gamma-1}$. Along a streamline, we have

$$\frac{\mathbf{v}^2}{2} + \frac{a^2}{\gamma-1} = \text{constant.} \tag{3.18.17}$$

Normal to the streamline, we have

$$\frac{\partial}{\partial n}\left(\frac{\mathbf{v}^2}{2} + \frac{a^2}{\gamma-1}\right) = \mathbf{n}\cdot(\mathbf{v}\times\text{curl}\,\mathbf{v}), \tag{3.18.18}$$

where $\mathbf{n}$ is a unit vector normal to the streamline.

3.19   Gas Dynamics Equation (Nonlinear Potential Equation)

Assume that the vorticity is everywhere zero: $\text{curl}\,\mathbf{v} = 0$. Then, we can define the velocity potential ϕ as

$$\mathbf{v} = \text{grad}\,\phi, \tag{3.19.1}$$

and substitute this into the continuity equation to get the so-called potential equation:

$$\text{div}\,(\rho\,\text{grad}\,\phi) = 0. \tag{3.19.2}$$

In Cartesian coordinates, the potential is defined as

$$u = \phi_x, \quad v = \phi_y, \quad w = \phi_z, \tag{3.19.3}$$

and the potential equation becomes

$$(\rho\phi_x)_x + (\rho\phi_y)_y + (\rho\phi_z)_z = 0. \tag{3.19.4}$$

This nonlinear equation requires an additional equation for the density which can be obtained from the energy equation (Bernoulli's equation) for an adiabatic flow of a calorically perfect gas (see Section 3.18):

$$\frac{a^2}{\gamma-1} + \frac{u^2+v^2+w^2}{2} = H = \text{constant}, \tag{3.19.5}$$

where $a = \sqrt{\gamma p/\rho}$. The density equation is then derived from

$$\frac{a^2}{\gamma-1} + \frac{u^2+v^2+w^2}{2} = \frac{a_\infty^2}{\gamma-1} + \frac{u_\infty^2+v_\infty^2+w_\infty^2}{2}, \tag{3.19.6}$$

and the adiabatic relation, $p/\rho^\gamma = \text{constant}$, as

$$\frac{\rho}{\rho_\infty} = \left[1 - \frac{\gamma-1}{2}M_\infty^2\left(\frac{u^2+v^2+w^2}{u_\infty^2+v_\infty^2+w_\infty^2} - 1\right)\right]^{1/(\gamma-1)}. \tag{3.19.7}$$

This formulation is valid for a steady adiabatic irrotational flow. It is also worth noting that in a steady adiabatic irrotational flow the entropy is constant everywhere, i.e., the flow is homentropic (see Section 3.23). It follows then that it is not valid for flows with shocks. But this conservative form is still approximately valid if the Mach number is close to 1; it has actually been used successfully for various such flows. See Refs.[51, 155] for more details.

In two dimensions, the nonlinear potential flow can be described by the following Cauchy-Riemann type system:

$$\frac{\partial(\rho u)}{\partial x} + \frac{\partial(\rho v)}{\partial y} = 0, \tag{3.19.8}$$

$$\frac{\partial v}{\partial x} - \frac{\partial u}{\partial y} = 0. \tag{3.19.9}$$

See Ref.[134] for numerical methods for solving this system. I like the Cauchy-Riemann system, but in this case I like the scalar potential equation (3.19.14) better simply because it is a scalar equation. It is so much easier to develop a numerical scheme for a scalar equation than for a system.

I also like the potential equation written in a nonconservative form. To derive it, expand the continuity equation,

$$\rho \operatorname{div} \mathbf{v} + \mathbf{v} \cdot \operatorname{grad} \rho = 0, \tag{3.19.10}$$

and take a dot product of the velocity vector and the momentum equation,

$$\mathbf{v} \cdot [(\operatorname{grad} \mathbf{v})\mathbf{v}] + \frac{a^2}{\rho} \mathbf{v} \cdot \operatorname{grad} \rho = 0, \tag{3.19.11}$$

where the pressure gradient has been replaced by the density gradient by using the relation, $\operatorname{grad} p = a^2 \operatorname{grad} \rho$. Now it is clear that this and the continuity equation can be combined to give

$$\operatorname{div} \mathbf{v} - \frac{1}{a^2} \mathbf{v} \cdot [(\operatorname{grad} \mathbf{v})\mathbf{v}] = 0. \tag{3.19.12}$$

This equation is known as the gas dynamics equation. I like it very much, especially the name, 'Gas Dynamics Equation'. It sounds exciting. In Cartesian coordinates, the gas dynamics equation is written as

$$\left(1 - \frac{u^2}{a^2}\right) u_x + \left(1 - \frac{v^2}{a^2}\right) v_y + \left(1 - \frac{w^2}{a^2}\right) w_z - \frac{uv}{a^2}(u_y + v_x) - \frac{vw}{a^2}(v_z + w_y) - \frac{wu}{a^2}(w_x + u_z) = 0. \tag{3.19.13}$$

Of course, this cannot be solved for the three velocity components (u, v, w) because it is a scalar equation. But if the flow is irrotational and thus the velocity is given by the velocity potential ϕ: $u = \phi_x$, $v = \phi_y$, $w = \phi_z$, then we can write the gas dynamics equation in terms of ϕ:

$$\left(1 - \frac{\phi_x^2}{a^2}\right)\phi_{xx} + \left(1 - \frac{\phi_y^2}{a^2}\right)\phi_{yy} + \left(1 - \frac{\phi_z^2}{a^2}\right)\phi_{zz} - \frac{2\phi_x\phi_y}{a^2}\phi_{xy} - \frac{2\phi_y\phi_z}{a^2}\phi_{yz} - \frac{2\phi_z\phi_x}{a^2}\phi_{zx} = 0. \tag{3.19.14}$$

This is a scalar equation for ϕ, and so can be solved for ϕ with Equation (3.19.7). This equation is perhaps more widely known as the full potential equation rather than as the gas dynamics equation.

Incidentally, the velocity potential exists even in viscous flows as long as the vorticity is zero. The viscous terms actually vanish from the governing equations for irrotational flows (curl $\mathbf{v} = 0$), but the viscous stresses do not necessarily vanish and can be expressed in terms of the velocity potential. See Ref.[65] for an interesting theory of viscous potential flows. Also, the potential formulation can be extended to rotational flows by decomposing the velocity into the irrotational (potential) part and the rotational part. See Refs.[43, 44, 96] for details.

3.20 Linear Potential Equation

In general, I like to linearize nonlinear equations. I would linearize the nonlinear potential equation as follows. In two dimensions, a flow over a slender body with the free stream velocity U_∞ in x direction generates perturbations u', v' ($u', v' << U_\infty$):

$$u = U_\infty + u', \quad v = v'. \tag{3.20.1}$$

Then, introducing the velocity potential and its perturbation as

$$u = \phi_x = U_\infty + \phi_x', \quad v = \phi_y = \phi_y', \tag{3.20.2}$$

we find from Equation (3.19.14) that, for two-dimensional flows,

$$(1 - M_\infty^2)\phi_{xx}' + \phi_{yy}' = M_\infty^2(\gamma + 1)\frac{\phi_x'\phi_{xx}'}{U_\infty}. \tag{3.20.3}$$

Note that the right hand side must be retained for transonic flows $M_\infty \approx 1$ to keep the equation two-dimensional: the x-derivative term $(1 - M_\infty^2)\phi_{xx}'$ will be very small for transonic flows and the equation will become nearly one-dimensional without the right hand side. However, the right hand side can be ignored for subsonic and supersonic flows, resulting in

$$(1 - M_\infty^2)\phi_{xx}' + \phi_{yy}' = 0. \tag{3.20.4}$$

This is the two-dimensional linear potential equation. In three dimensions, we introduce a perturbation velocity w' in z direction as

$$w = w' = \phi_z = \phi'_z << U_\infty,$$
(3.20.5)

and obtain, from Equation (3.19.14), the following three-dimensional linear potential equation:

$$(1 - M_\infty^2)\phi'_{xx} + \phi'_{yy} + \phi'_{zz} = 0.$$
(3.20.6)

In either two or three dimensions, once the perturbation potential ϕ' is determined, the pressure can be computed by

$$C_p = \frac{p - p_\infty}{\frac{1}{2}\rho U_\infty^2} = \frac{2}{\gamma M_\infty^2}(p/p_\infty - 1) = -\left[\frac{2u'}{U_\infty} + (1 - M_\infty^2)\frac{u'^2}{U_\infty^2} + \frac{v'^2 + w'^2}{U_\infty^2}\right].$$
(3.20.7)

I like the linear potential equation because it can be solved easily for small Mach number (it is almost the Laplace equation), but not so much for a large Mach number (it is now hyperbolic. See Section 1.17 for classification). See Refs.[53, 84] for more details on the linearized potential equations, and Refs.[51, 155] for numerical methods for solving them.

3.21 Small Perturbation Equations

We can write the linear potential equation in terms of the velocity components, which are called the small perturbation equations (or the Prandtl-Glauert equations). For supersonic flows ($M_\infty > 1$), these equations are written as

$$\beta^2 u_x - v_y = 0,$$
(3.21.1)

$$v_x - u_y = 0,$$
(3.21.2)

where $\beta^2 = M_\infty^2 - 1 > 0$. It can be written as a vector form:

$$\mathbf{w}_x + \mathbf{A}\mathbf{w}_y = 0,$$
(3.21.3)

where

$$\mathbf{w} = \begin{bmatrix} u \\ v \end{bmatrix}, \quad \mathbf{A} = \begin{bmatrix} 0 & -\dfrac{1}{\beta^2} \\ -1 & 0 \end{bmatrix}.$$
(3.21.4)

Eigenstructure

$$\mathbf{A} = \mathbf{R}\mathbf{\Lambda}\mathbf{L},$$
(3.21.5)

where

$$\mathbf{\Lambda} = \begin{bmatrix} \dfrac{1}{\beta} & 0 \\ 0 & -\dfrac{1}{\beta} \end{bmatrix}, \quad \mathbf{R} = \begin{bmatrix} \dfrac{1}{2\beta} & \dfrac{1}{2\beta} \\ -\dfrac{1}{2} & \dfrac{1}{2} \end{bmatrix}, \quad \mathbf{L} = \begin{bmatrix} \beta & -1 \\ \beta & 1 \end{bmatrix}.$$
(3.21.6)

Characteristic Form

$$\mathbf{L}\mathbf{w}_x + \mathbf{L}\mathbf{A}\mathbf{w}_y = 0,$$
(3.21.7)

$$(\mathbf{L}\mathbf{w})_x + \mathbf{\Lambda}(\mathbf{L}\mathbf{w})_y = 0,$$
(3.21.8)

that is,

$$\frac{\partial}{\partial x}(\beta u - v) + \frac{1}{\beta}\frac{\partial}{\partial y}(\beta u - v) = 0,$$
(3.21.9)

$$\frac{\partial}{\partial x}(\beta u + v) - \frac{1}{\beta}\frac{\partial}{\partial y}(\beta u + v) = 0.$$
(3.21.10)

This shows that $\beta u - v$ is constant along $dy/dx = 1/\beta$ while $\beta u + v$ is constant along $dy/dx = -1/\beta$. Note that this diagonalization is valid only for supersonic flows ($M_\infty > 1$). In particular, I like a flow with $M_\infty = \sqrt{2}$ because then $\beta = 1$ and everything looks very simple.

For subsonic flows ($M_\infty < 1$), we have

$$(1 - M_\infty^2)\, u_x + v_y = 0, \tag{3.21.11}$$

$$v_x - u_y = 0. \tag{3.21.12}$$

Then, the eigenvalues become complex, and therefore the system is elliptic. As for the Mach number, I like to set $M_\infty = 0$ (incompressible flows) because the system reduces to the Cauchy-Riemann system. As a matter of fact, I would use the Prandtl-Glauert transformation:

$$\xi = x, \quad \eta = y\sqrt{1 - M_\infty^2}, \quad U = \sqrt{1 - M_\infty^2}\,u, \quad V = v, \tag{3.21.13}$$

so that the system, (3.21.11) and (3.21.12), becomes the Cauchy-Riemann system,

$$U_\xi + V_\eta = 0, \tag{3.21.14}$$

$$V_\xi - U_\eta = 0, \tag{3.21.15}$$

for any Mach number ($|M_\infty| < 1$). This is very nice. Any subsonic small perturbation flow can be transformed into an incompressible flow. See Refs.[74, 84] for more details on the Prandtl-Glauert transformation.

3.22 Incompressible Potential Equation

As we have seen in the previous sections, the velocity potential ϕ is very useful in computing steady compressible irrotational flows. Here, we consider steady incompressible flows. For steady incompressible flows, the continuity equation is $\operatorname{div} \mathbf{v} = 0$. Therefore, by substituting the velocity potential (3.19.1) into the continuity equation, we obtain

$$\operatorname{div} \operatorname{grad} \phi = 0. \tag{3.22.1}$$

That is, the velocity potential satisfies the Laplace Equation in all dimensions. I like it because various exact solutions are available for the Laplace equation. Two-dimensional flows are even nicer because a potential flow can be described by the velocity potential and the stream function as described in Section 2.9. It is nice especially because the velocity potential and the stream function satisfy the Cauchy-Riemann equations (see Section 2.9.2), and we can generate various exact solutions by the theory of complex variables. Note that the applicability of the theory of complex variables is limited to two-dimensional flows; it doesn't apply to general three-dimensional flows including axisymmetric flows because the stream function does not satisfy the Laplace equation. See Section 4.24 for the governing equation of the stream function.

3.23 Crocco's Equation

Crocco's equation (or Crocco's theorem) is a modified form of the momentum equation. Consider the momentum equation written in the primitive form,

$$\frac{\partial \mathbf{v}}{\partial t} + (\operatorname{grad}\mathbf{v})\mathbf{v} + \frac{1}{\rho}\operatorname{grad} p = 0. \tag{3.23.1}$$

Using the following relations,

$$(\operatorname{grad}\mathbf{v})\mathbf{v} = \operatorname{grad}\left(\frac{\mathbf{v}^2}{2}\right) - \mathbf{v} \times \operatorname{curl}\mathbf{v}, \tag{3.23.2}$$

$$T ds = dh - \frac{1}{\rho}\operatorname{grad} p, \tag{3.23.3}$$

$$H = h + \frac{\mathbf{v}^2}{2}, \tag{3.23.4}$$

the momentum equation can be written as

$$\frac{\partial \mathbf{v}}{\partial t} - \mathbf{v} \times \boldsymbol{\omega} = T \operatorname{grad} s - \operatorname{grad} H, \tag{3.23.5}$$

where $\boldsymbol{\omega} = \operatorname{curl} \mathbf{v}$. This is Crocco's equation. It is a quite interesting equation. For example, we can use it to show that the entropy is constant along a streamline for a steady adiabatic flow. To do so, first, note that for a steady adiabatic flow, the equation for the total enthalpy (3.3.7)

$$\frac{DH}{Dt} = \frac{1}{\rho} \frac{\partial p}{\partial t}, \tag{3.23.6}$$

reduces to

$$\mathbf{v} \cdot \operatorname{grad} H = 0, \tag{3.23.7}$$

which implies, incidentally, that the total enthalpy is preserved along a streamline. Now, take a dot product of the velocity and Crocco's equation to get

$$\mathbf{v} \cdot (\mathbf{v} \times \boldsymbol{\omega}) = \mathbf{v} \cdot \operatorname{grad} H - T\mathbf{v} \cdot \operatorname{grad} s. \tag{3.23.8}$$

Observe that the left hand side vanishes because $\mathbf{v} \perp (\mathbf{v} \times \boldsymbol{\omega})$, and also that the first term on the right vanishes by Equation (3.23.7). Then, we are left with

$$0 = \mathbf{v} \cdot \operatorname{grad} s. \tag{3.23.9}$$

This indicates that the entropy is constant along a streamline. This completes the proof for constant entropy along a streamline in a steady adiabatic flow. On the other hand, if we assume that $\boldsymbol{\omega} = \mathbf{0}$ and H is uniform, then it follows immediately from Equation (3.23.5) that the entropy is constant everywhere, i.e., the flow is homentropic. That is, zero vorticity implies uniform entropy if H is uniform. Basically, Crocco's equation is another form of the Euler equation (i.e., the momentum equation). So, it is perfectly possible to replace the momentum equation by Crocco's equation in the Euler system (then we would have to call it the Crocco system). I like Crocco's equation. It is a quite insightful equation.

3.24   Kelvin's Circulation Theorem

The circulation Γ is defined for a closed curve $\mathcal{C}$ by

$$\Gamma = \oint_{\mathcal{C}} \mathbf{v} \cdot d\mathbf{l}, \tag{3.24.1}$$

where $d\mathbf{l}$ is an element taken counterclockwise along $\mathcal{C}$, and it is related to the vorticity $\boldsymbol{\omega} = \operatorname{curl} \mathbf{v}$ by

$$\Gamma = \oint_{\mathcal{C}} \mathbf{v} \cdot d\mathbf{l} = \iint_{S} \operatorname{curl} \mathbf{v} \cdot \mathbf{n} \, dS = \iint_{S} \boldsymbol{\omega} \cdot \mathbf{n} \, dS, \tag{3.24.2}$$

i.e., the circulation around $\mathcal{C}$ is equal to the in/out-flow of the vorticity across *any* surface S whose boundary is $\mathcal{C}$.

   The rate of change of the circulation of a closed curve $\mathcal{C}$ (moving with the fluid) can be obtained easily from the primitive form of the Euler equations:

$$\frac{D\Gamma}{Dt} = -\oint_{\mathcal{C}} \frac{\operatorname{grad} p}{\rho} \cdot d\mathbf{l}. \tag{3.24.3}$$

If derived from the Navier-Stokes equations, by the way, it will have a contribution from the viscous term on the right hand side:

$$\frac{D\Gamma}{Dt} = \oint_{\mathcal{C}} \frac{1}{\rho} \operatorname{div} \boldsymbol{\tau} \, d\mathbf{l} - \oint_{\mathcal{C}} \frac{\operatorname{grad} p}{\rho} \cdot d\mathbf{l}, \tag{3.24.4}$$

where $\boldsymbol{\tau}$ is the viscous stress tensor (4.2.1). For inviscid incompressible flows, ρ is constant and the right hand side of Equation (3.24.3) vanishes:

$$\oint_{\mathcal{C}} \frac{\operatorname{grad} p}{\rho} \cdot d\mathbf{l} = \frac{1}{\rho} \oint_{\mathcal{C}} \operatorname{grad} p \cdot d\mathbf{l} = \frac{1}{\rho} \oint_{\mathcal{C}} dp = 0. \tag{3.24.5}$$

Therefore, we have from Equation (3.24.3)

$$\frac{D\Gamma}{Dt} = 0, \tag{3.24.6}$$

i.e., the circulation is conserved. This is interesting. Even for compressible flows, where Equation (3.24.3) can be written as

$$\frac{D\Gamma}{Dt} = -\oint_C \frac{\operatorname{grad} p}{\rho} \cdot dl = -\oint_C (\operatorname{grad} h - T \operatorname{grad} s) \cdot dl = \oint_C T \operatorname{grad} s \cdot dl, \tag{3.24.7}$$

if the entropy is constant, then again the circulation is conserved:

$$\frac{D\Gamma}{Dt} = 0. \tag{3.24.8}$$

This shows that the circulation around a closed curve remains constant at all times as the curve moves with the fluid. This is called Kelvin's circulation theorem. I like it because it gives interesting and important explanations about vortical flows. The vorticity defines a vector field, and thus integral curves (i.e., the curves tangent to the vorticity vector) may be defined. These curves are generally called the vortex lines. The surface formed by the vortex lines passing through a closed curve is called a vortex tube (or filaments). The strength of the vortex tube is defined as the integral of the vorticity over a cross-section of a tube, which is equivalent to the line integral of the velocity around the closed curve, i.e., the circulation Γ. Now I begin to see the connection between the vortex tube and Kelvin's circulation theorem. Below are the theorems, known as Helmholz's theorems, obtained as direct consequences of the circulation theorem:

1. *A flow that is initially irrotational remains irrotational if no rotational forces exist.*

2. *The strength of a vortex tube is constant along its length.*

3. *A vortex tube cannot end in the flow, and must extend to the boundary or form a closed loop.*

One of the most interesting behaviors that I like, which is implied by the second theorem, is that the vortex strength increases/decreases as the cross-section of the vortex tube shrinks/expands. This is called the vortex stretching mechanism as the cross-sectional changes are associated with the stretching of the tube.

3.25 Vorticity Equation

Evolution of the vorticity is described by the vorticity equation, which can be derived by taking the curl of the momentum equations. Consider the primitive form of the momentum equations:

$$\frac{\partial \mathbf{v}}{\partial t} + (\operatorname{grad} \mathbf{v})\mathbf{v} + \frac{\operatorname{grad} p}{\rho} = 0, \tag{3.25.1}$$

which can be written by Equation (1.7.7) as

$$\frac{\partial \mathbf{v}}{\partial t} + \operatorname{grad} \frac{\mathbf{v} \cdot \mathbf{v}}{2} - \mathbf{v} \times \operatorname{curl} \mathbf{v} + \frac{\operatorname{grad} p}{\rho} = 0. \tag{3.25.2}$$

Taking the curl, we get

$$\frac{\partial \omega}{\partial t} + \operatorname{curl} \left(\operatorname{grad} \frac{\mathbf{v} \cdot \mathbf{v}}{2} \right) - \operatorname{curl}(\mathbf{v} \times \omega) + \operatorname{curl} \left(\frac{\operatorname{grad} p}{\rho} \right) = 0, \tag{3.25.3}$$

where $\omega = \operatorname{curl} \mathbf{v}$ is the vorticity, which can be written by Equations (1.7.19) and (1.7.20) as

$$\frac{\partial \omega}{\partial t} - \mathbf{v}\operatorname{div}\omega - (\operatorname{grad} \mathbf{v})\omega + (\operatorname{div} \mathbf{v})\omega + (\operatorname{grad} \omega)\mathbf{v} + \frac{1}{\rho}\operatorname{curl}\operatorname{grad} p + \operatorname{grad}\left(\frac{1}{\rho}\right) \times \operatorname{grad} p = 0. \tag{3.25.4}$$

Therefore, we have

$$\frac{D\omega}{Dt} = \frac{\partial \omega}{\partial t} + (\operatorname{grad} \omega)\mathbf{v} = (\operatorname{grad} \mathbf{v})\omega - (\operatorname{div} \mathbf{v})\omega + \frac{\operatorname{grad} \rho \times \operatorname{grad} p}{\rho^2}. \tag{3.25.5}$$

The left hand side is the rate of change of the vorticity along the streamlines. The first terms on the right hand side describes the vortex stretching and twisting (or tilting), and the second term describes stretching due to the compressibility. The third term is called the baroclinic term, describing the change in the vorticity by the misalignment of the gradient of the density and the gradient the pressure. The baloclinic term vanishes for barotropic gases, i.e., if the pressure is a function of the density only, because then the density and pressure gradients become parallel to each other and the cross product vanishes. An example of such flows is the homentropic flow discussed in Section 3.14. Of course, the barocliic term vanishes also for incompressible flows because the density gradient vanishes.

The vorticity equation can be arranged further by using the following relations:

$$\frac{\partial}{\partial t}\left(\frac{\omega}{\rho}\right) = \frac{1}{\rho}\frac{\partial \omega}{\partial t} - \frac{\omega}{\rho^2}\frac{\partial \rho}{\partial t} = \frac{1}{\rho}\frac{\partial \omega}{\partial t} + \left(\frac{\omega}{\rho}\right)\operatorname{div}\mathbf{v} + \frac{\omega}{\rho^2}(\operatorname{grad}\rho)\cdot\mathbf{v},\tag{3.25.6}$$

where the continuity equation (3.3.1) has been substituted, and

$$\begin{aligned}\left[\operatorname{grad}\left(\frac{\omega}{\rho}\right)\right]\mathbf{v} &= \left[\omega\otimes\operatorname{grad}\left(\frac{1}{\rho}\right) + \frac{1}{\rho}(\operatorname{grad}\omega)\right]\mathbf{v}\\ &= -\frac{1}{\rho^2}\left(\omega\otimes\operatorname{grad}\rho\right)\mathbf{v} + \frac{1}{\rho}(\operatorname{grad}\omega)\mathbf{v}\\ &= -\frac{\omega}{\rho^2}(\operatorname{grad}\rho)\cdot\mathbf{v} + \frac{1}{\rho}(\operatorname{grad}\omega)\mathbf{v},\end{aligned}\tag{3.25.7}$$

which yields

$$\frac{D}{Dt}\left(\frac{\omega}{\rho}\right) = \frac{\partial}{\partial t}\left(\frac{\omega}{\rho}\right) + \left[\operatorname{grad}\left(\frac{\omega}{\rho}\right)\right]\mathbf{v} = \frac{1}{\rho}\frac{\partial \omega}{\partial t} + \left(\frac{\omega}{\rho}\right)\operatorname{div}\mathbf{v} + \frac{1}{\rho}(\operatorname{grad}\omega)\mathbf{v}.\tag{3.25.8}$$

Finally, dividing Equation(3.25.5) by the density and substituting Equation (3.25.8), we obtain

$$\frac{D}{Dt}\left(\frac{\omega}{\rho}\right) = (\operatorname{grad}\mathbf{v})\left(\frac{\omega}{\rho}\right) + \frac{\operatorname{grad}\rho\times\operatorname{grad}p}{\rho^3}.\tag{3.25.9}$$

Comparing this form with Equation (3.25.5), we see that the stretching term due to compressibility is gone. In fact, I like this form very much because the compressibility effect is incorporated into the stretching-twisting term and it shows that it is the quantity ω/ρ, not ω, that behaves like the vorticity in the same way as in incompressible flows. That is, ω/ρ satisfies the equation that ω satisfies in incompressible flows.

I like the vorticity equation because it describes very interestingly how the vorticity evolves in a flow and it is one of the basic tools to understand turbulence as well as the basic equation for vortex methods [29].

3.26 Shallow-Water (or Saint Venant) Equations

Like a smooth water surface, when the vertical (z-direction) velocity and the vertical displacement of the surface is sufficiently small (compared with the frequency of water waves; see Figure 3.26.1), the water depth h and the velocity $\mathbf{v} = [u, v]^t$ are governed by the following shallow-water equations (also called the Saint Venant equations):

$$\partial_t h + \operatorname{div}(h\mathbf{v}) = 0,\tag{3.26.1}$$

$$\partial_t(h\mathbf{v}) + \operatorname{div}(h\mathbf{v}\otimes\mathbf{v}) + \operatorname{grad}(gh^2/2) = -gh\operatorname{grad}z_0(x, y),\tag{3.26.2}$$

where $g = 9.8m/s^2$ and $z_0(x, y)$ is a function that defines the shape of the bottom. Note that all quantities are independent of z. It is interesting that the shallow-water equations can be derived from the Euler equations: add a gravitational force term in the z-component, ignore the z-velocity and any variation in z, and then integrate all equations in z from the bottom to the water surface. In doing so, we obtain from the z-momentum equation,

$$p_z = -\rho g \quad\longrightarrow\quad p = p_0 + \rho g(h + z_0 - z).\tag{3.26.3}$$

This is a familiar relationship, called the hydrostatic pressure relation. Conversely, if we write $p = \frac{1}{2}gh^2$, rewrite h as ρ, and assume $z_0 = $ constant, then, the shallow-water equations become

$$\partial_t \rho + \operatorname{div}(\rho\mathbf{v}) = 0,\tag{3.26.4}$$

$$\partial_t(\rho\mathbf{v}) + \operatorname{div}(\rho\mathbf{v}\otimes\mathbf{v}) + \operatorname{grad}p = 0,\tag{3.26.5}$$

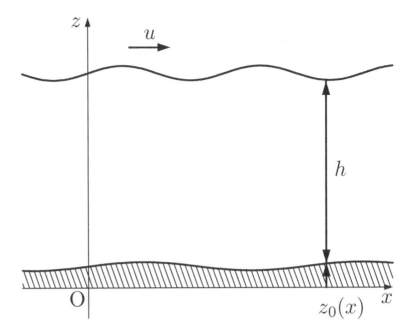

Figure 3.26.1: One-dimensional free surface flow.

which are nothing but the Euler equations without the energy equation. In fact, if we look at the eigenstructure, we find that the quantity $\sqrt{gh}$ corresponds to the speed of sound:

$$\sqrt{gh} = \sqrt{2p/h} = \sqrt{2p/\rho}. \tag{3.26.6}$$

This implies that we have $\gamma = 2$. Therefore, the Euler equations above are in the form of the homentropic/isothermal Euler equations with $\gamma = 2$. Basically, the shallow-water equations are a subset of the Euler equations. This is very interesting. I like it.

Incidentally, it is easy to derive a conservation form for $\phi = gh$, but it requires differentiations and therefore it is not really valid for flows with shocks. See Ref.[81] for details.

3.26.1   1D Shallow-Water Equations

$$h_t + (hu)_x = 0, \tag{3.26.7}$$

$$(hu)_t + \left(hu^2 + \frac{1}{2}gh^2 \right)_x = -gh(z_0)_x, \tag{3.26.8}$$

or in the vector form

$$\mathbf{U}_t + \mathbf{F}_x = \mathbf{S}, \tag{3.26.9}$$

where

$$\mathbf{U} = \begin{bmatrix} h \\ hu \end{bmatrix}, \quad \mathbf{F} = \begin{bmatrix} hu \\ hu^2 + \frac{1}{2}gh^2 \end{bmatrix}, \quad \mathbf{S} = \begin{bmatrix} 0 \\ -gh(z_0)_x \end{bmatrix}. \tag{3.26.10}$$

Jacobian

$$\mathbf{A} = \frac{\partial \mathbf{F}}{\partial \mathbf{U}} = \begin{bmatrix} 0 & 1 \\ gh - u^2 & 2u \end{bmatrix}. \tag{3.26.11}$$

Eigenstructure

$$\mathbf{A} = \mathbf{R}\mathbf{\Lambda}\mathbf{L}, \tag{3.26.12}$$

where

$$\mathbf{\Lambda} = \begin{bmatrix} u - \sqrt{gh} & 0 \\ 0 & u + \sqrt{gh} \end{bmatrix}, \quad \mathbf{R} = \begin{bmatrix} 1 & 1 \\ u - \sqrt{gh} & u + \sqrt{gh} \end{bmatrix}, \tag{3.26.13}$$

$$\mathbf{L} = \begin{bmatrix} \dfrac{u + \sqrt{gh}}{2\sqrt{gh}} & -\dfrac{1}{2\sqrt{gh}} \\ -\dfrac{u - \sqrt{gh}}{2\sqrt{gh}} & \dfrac{1}{2\sqrt{gh}} \end{bmatrix}, \quad \mathbf{L}d\mathbf{U} = \begin{bmatrix} \dfrac{1}{2}dh - \dfrac{h}{2\sqrt{gh}}du \\ \dfrac{1}{2}dh + \dfrac{h}{2\sqrt{gh}}du \end{bmatrix}. \tag{3.26.14}$$

3.26.2   2D Shallow-Water Equations

$$h_t + (hu)_x + (hv)_y = 0, \tag{3.26.15}$$

$$(hu)_t + \left(hu^2 + \frac{1}{2}gh^2 \right)_x + (huv)_y = -gh(z_0)_x, \tag{3.26.16}$$

$$(hv)_t + (huv)_x + \left(hv^2 + \frac{1}{2}gh^2 \right)_y = -gh(z_0)_y, \tag{3.26.17}$$

or in the vector form

$$\mathbf{U}_t + \mathbf{F}_x + \mathbf{G}_y = \mathbf{S}, \tag{3.26.18}$$

where

$$\mathbf{U} = \begin{bmatrix} h \\ hu \\ hv \end{bmatrix}, \quad \mathbf{F} = \begin{bmatrix} hu \\ p + hu^2 \\ huv \end{bmatrix}, \quad \mathbf{G} = \begin{bmatrix} hv \\ huv \\ p + hv^2 \end{bmatrix}, \quad \mathbf{S} = \begin{bmatrix} 0 \\ -gh(z_0)_x \\ -gh(z_0)_y \end{bmatrix}. \tag{3.26.19}$$

Jacobians

$$\mathbf{A} = \frac{\partial \mathbf{F}}{\partial \mathbf{U}} = \begin{bmatrix} 0 & 1 & 0 \\ gh - u^2 & 2u & 0 \\ -uv & v & u \end{bmatrix}, \quad \mathbf{B} = \frac{\partial \mathbf{G}}{\partial \mathbf{U}} = \begin{bmatrix} 0 & 0 & 1 \\ -uv & v & u \\ gh - v^2 & 0 & 2v \end{bmatrix}, \tag{3.26.20}$$

$$\mathbf{A}_n = \mathbf{A}n_x + \mathbf{B}n_y = \begin{bmatrix} 0 & n_x & n_y \\ ghn_x - uq_n & q_n + un_x & un_y \\ ghn_y - vq_n & vn_x & q_n + vn_y \end{bmatrix}, \tag{3.26.21}$$

where

$$q_n = un_x + vn_y. \tag{3.26.22}$$

Eigenstructure

$$\mathbf{A}_n = \mathbf{R}_n \mathbf{\Lambda}_n \mathbf{L}_n, \tag{3.26.23}$$

where

$$\mathbf{\Lambda}_n = \begin{bmatrix} q_n - \sqrt{gh} & 0 & 0 \\ 0 & q_n & 0 \\ 0 & 0 & q_n + \sqrt{gh} \end{bmatrix}, \tag{3.26.24}$$

$$\mathbf{R}_n = \begin{bmatrix} 1 & 0 & 1 \\ u - \sqrt{gh}\, n_x & \ell_x & u + \sqrt{gh}\, n_x \\ v - \sqrt{gh}\, n_y & \ell_y & v + \sqrt{gh}\, n_y \end{bmatrix}, \tag{3.26.25}$$

$$\mathbf{L}_n = \begin{bmatrix} \dfrac{\sqrt{gh} + q_n}{2\sqrt{gh}} & -\dfrac{n_x}{2\sqrt{gh}} & -\dfrac{n_y}{2\sqrt{gh}} \\ -q_\ell & \ell_x & \ell_y \\ \dfrac{\sqrt{gh} - q_n}{2\sqrt{gh}} & \dfrac{n_x}{2\sqrt{gh}} & \dfrac{n_y}{2\sqrt{gh}} \end{bmatrix}, \tag{3.26.26}$$

and

$$q_\ell = u\ell_x + v\ell_y, \tag{3.26.27}$$

$$[\ell_x, \ell_y] = [-n_y, n_x]. \tag{3.26.28}$$

3.26.3 3D Shallow-Water Equations

Unfortunately, I cannot write them because I don't know them. I would definitely like them if the shallow-water equations existed in three dimensions.

Chapter 4

Navier-Stokes Equations

4.1 Navier-Stokes Equations

I like a little bit more realistic model for describing a fluid flow than the Euler equations. So, I will add some missing pieces to the Euler equations: the viscous stress tensor τ and the heat flux vector $\mathbf{q}$. Yes, these are the Navier-Stokes equations.

Conservative form of the Navier-Stokes equations

$$\partial_t \rho + \operatorname{div}(\rho \mathbf{v}) = 0, \tag{4.1.1}$$

$$\partial_t (\rho \mathbf{v}) + \operatorname{div}(\rho \mathbf{v} \otimes \mathbf{v}) = \operatorname{div} \boldsymbol{\sigma}, \tag{4.1.2}$$

$$\partial_t (\rho E) + \operatorname{div}(\rho \mathbf{v} H) = \operatorname{div}(\boldsymbol{\tau} \mathbf{v}) - \operatorname{div} \mathbf{q}, \tag{4.1.3}$$

where $\boldsymbol{\sigma}$ is the total surface stress tensor which consists of the static pressure p and the viscous stress tensor τ,

$$\boldsymbol{\sigma} = -p\mathbf{I} + \boldsymbol{\tau}, \tag{4.1.4}$$

where $\mathbf{I}$ is the identity matrix.

Primitive form of the Navier-Stokes equations

$$\frac{D\rho}{Dt} + \rho \operatorname{div} \mathbf{v} = 0, \tag{4.1.5}$$

$$\rho \frac{D\mathbf{v}}{Dt} + \operatorname{grad} p = \operatorname{div} \boldsymbol{\tau}, \tag{4.1.6}$$

$$\frac{Dp}{Dt} + \gamma p \operatorname{div} \mathbf{v} = (\gamma - 1) [\boldsymbol{\tau} : \operatorname{grad} \mathbf{v} - \operatorname{div} \mathbf{q}]. \tag{4.1.7}$$

The continuity equation (4.1.5) can be derived from the conservative form (4.1.1) simply by expanding the divergence term. The momentum equation (4.1.6) can be derived from the conservative form (4.1.2) by expanding the left hand side and substituting the continuity equation (4.1.5). Note also that

$$\operatorname{div} \boldsymbol{\sigma} = -\operatorname{div}(p\mathbf{I}) + \operatorname{div} \boldsymbol{\tau} = -\operatorname{grad} p + \operatorname{div} \boldsymbol{\tau}. \tag{4.1.8}$$

To derive the pressure equation (4.1.7), first derive the kinetic energy equation by taking the dot product of the velocity vector and the momentum equation (4.1.6),

$$\rho \frac{D}{Dt} \left(\frac{\mathbf{v}^2}{2} \right) + (\operatorname{grad} p) \cdot \mathbf{v} = -\boldsymbol{\tau} : \operatorname{grad} \mathbf{v} + \operatorname{div}(\boldsymbol{\tau} \mathbf{v}), \tag{4.1.9}$$

then subtract the kinetic equation from the energy equation (4.1.3), and finally use the equation of state for a perfect gas, i.e., $p = (\gamma - 1)\rho e$, and the continuity equation to convert the time derivative of the internal energy into that of the pressure.

Which form do you like? I think I like the conservative form better because conservation is a very important property (to ensure a correct shock speed in particular), and it is easier to develop a conservative scheme by directly discretizing the conservative form.

4.2   Viscous Stresses and Heat Fluxes

In Newtonian fluids (e.g., the sea-level air or water), the viscous stress tensor τ depends linearly on the strain and it is given by

$$\tau = \lambda(\text{div }\mathbf{v})\mathbf{I} + \mu \left[\text{grad }\mathbf{v} + (\text{grad }\mathbf{v})^t\right], \tag{4.2.1}$$

where $\text{grad }\mathbf{v} + (\text{grad }\mathbf{v})^t$ is the rate of strain tensor (also called the deformation tensor), μ is the viscosity coefficient, and λ is the second viscosity coefficient. The bulk viscosity μ_B defined by

$$\mu_B = \lambda + \frac{2}{3}\mu, \tag{4.2.2}$$

is usually very small, and therefore we may set $\mu_B = 0$, i.e.,

$$\lambda = -\frac{2}{3}\mu. \tag{4.2.3}$$

This is called Stokes' hypothesis. I like this because it is a reasonable assumption. Consider the average normal stress $\overline{P}$ (also called the mean pressure),

$$\begin{aligned}
\overline{P} &\equiv -\frac{\sigma_{ii}}{3} \\
&= p - \left(\frac{1}{3}\lambda \times 3\,\text{div }\mathbf{v} + \frac{2}{3}\mu\,\text{div }\mathbf{v}\right) \\
&= p - \left(\lambda + \frac{2}{3}\mu\right)\text{div }\mathbf{v} \\
&= p - \mu_B\,\text{div }\mathbf{v},
\end{aligned} \tag{4.2.4}$$

where it has been defined as negative because σ_{ii} is negative when acting onto the fluid element. Basically, this gives, independent of coordinates, an average normal force acting on a fluid element; such a quantity is generally considered as the thermodynamic pressure. Then, it is a state variable. That is, under the assumption of the thermodynamic equilibrium, it should not depend on the rate of the volume change, $\text{div }\mathbf{v}$. So, it would be reasonable to accept Stokes' hypothesis $\mu_B = 0$, so that we have a reasonable relation, $\overline{P} = p$. Of course, this is only a hypothesis, but it works very well for a wide range of conditions. See Refs.[136, 169] for more details. Note that the viscosity coefficient generally depends on the temperature. On the other hand, the heat flux is given by Fourier's law,

$$\mathbf{q} = -\kappa\,\text{grad }T, \tag{4.2.5}$$

where κ is the heat conductivity which is a function of the temperature. The negative sign is put to define $\mathbf{q}$ as positive when the temperature gradient is negative. That is, the heat flows from higher to lower temperature. In fact, this is the basic assumption (or observation) of this empirical law. I think that it is indeed very reasonable.

I like both Newtonian fluids and Fourier's law for the same reason: both the viscous stress and heat flux depend linearly on the derivatives of the primitive variables. This is nice.

4.3   Artificial Bulk Viscosity

As mentioned in the previous section, the bulk viscosity μ_B is typically set to be zero, but there are methods in CFD that bring it back in the governing equations as an artificial viscosity to improve numerical algorithms:

$$\tau = \left(\mu_B^{\text{artificial}} - \frac{2}{3}\mu\right)(\text{div }\mathbf{v})\mathbf{I} + \mu \left[\text{grad }\mathbf{v} + (\text{grad }\mathbf{v})^t\right], \tag{4.3.1}$$

where $\mu_B^{\text{artificial}}(>0)$ is the artificial bulk viscosity. In Ref.[124], the artificial bulk viscosity is introduced to accelerate the convergence of an incompressible Navier-Stokes solver. The solver is based on the pseudo-compressible formulation [21]. This formulation generates pseudo-acoustic-waves (see Section 3.13), and the numerical convergence towards the steady state depends on how quickly the pseudo-acoustic-waves are eliminated via propagation and damping. The artificial bulk viscosity is then introduced to enhance the damping effect. I like the idea very much, especially the fact that the bulk viscosity term has no effect on the steady solution because it is proportional to the divergence of the velocity $\operatorname{div}\mathbf{v}$ that vanishes in the steady state for incompressible flows. Therefore, it only affects the convergence of the numerical method. The idea has been extended to the compressible Euler [90] and Navier-Stokes equations [3], where the artificial bulk viscosity term is formulated by $\operatorname{div}(\rho\mathbf{v})$ so that it vanishes in the steady state for compressible flows. Of course, these methods are designed for steady problems, but time-accurate computations are possible by implicit time-integration schemes where the fast steady solver can be employed to solve the implicit-residual equations over each physical time step.

Another example is the localized artificial diffusivity method [27], where the artificial bulk viscosity is introduced as a means to effectively capture shock waves without deteriorating the resolution of turbulence structure. To effectively accomplish the task, the artificial bulk viscosity is designed to get activated locally at shocks, thus the term "localized". The localized artificial diffusivity method has been applied to compressible shock-turbulent simulations with very high-order numerical schemes. See Ref.[69] and references therein for details.

It is interesting to observe the impact of the artificial bulk viscosity term on the divergence $\delta = \operatorname{div}\mathbf{v}$ and the vorticity $\boldsymbol{\omega} = \operatorname{curl}\mathbf{v}$:

$$\frac{\partial \delta}{\partial t} + \cdots = \operatorname{div}\left[\frac{1}{\rho}\operatorname{grad}\left(\mu_B^{\text{artificial}}\delta\right)\right] + \cdots, \tag{4.3.2}$$

$$\frac{\partial \boldsymbol{\omega}}{\partial t} + \cdots = -\frac{1}{\rho^2}\operatorname{grad}\rho \times \operatorname{grad}\left(\mu_B^{\text{artificial}}\delta\right) + \cdots, \tag{4.3.3}$$

which have been obtained by taking the divergence and the curl of the momentum equations (4.1.6) divided by the density. It is clear that the artificial bulk viscosity plays a role of diffusion for the divergence and is expected to provide damping for faster convergence and for shock capturing. On the other hand, the second equation shows apparently that the artificial bulk viscosity can actually generate the vorticity. In the convergence acceleration methods mentioned above, it actually does not affect the vorticity at all because it vanishes in the steady state. In the localized artificial diffusivity method, it is expected to be small at numerically captured shocks because the cross product of the density gradient and the gradient of the artificial bulk viscosity is expected to be small (they are both dominant locally in the direction normal to a shock front) [70]. It is nice to be small because accurate prediction of vorticity production across a shock can be very important in turbulent simulations [78].

I like these ideas in that they attempt to improve numerical methods by improving the governing equations rather than the discretization. In a way, it makes the construction of numerical schemes easier: given such improved governing equations (with desired features built in), you just have to discretize them accurately.

4.4 Viscosity and Heat Conductivity

The heat conductivity κ can be evaluated through the viscosity coefficient μ by utilizing the definition of the Prandtl number Pr,

$$Pr = \frac{c_p\mu}{\kappa} = \frac{\gamma R\mu}{\kappa(\gamma-1)} \quad \longrightarrow \quad \kappa = \frac{\gamma R\mu}{Pr(\gamma-1)}. \tag{4.4.1}$$

This is very nice because the heat flux vector can then be expressed in any of the following forms:

$$\mathbf{q} = -\kappa\operatorname{grad}T = -\frac{\mu}{Pr(\gamma-1)}\operatorname{grad}\left(a^2\right) = -\frac{\mu\gamma}{Pr}\operatorname{grad}e = -\frac{\mu}{Pr}\operatorname{grad}h, \tag{4.4.2}$$

where the relations, $a^2 = \gamma p/\rho = \gamma R T$, $e = \frac{R}{\gamma-1}T$, and $h = \frac{\gamma R}{\gamma-1}T$ have been used. It is nice to know then that the Prandtl number for the sea-level air is nearly constant for a wide range of temperature; $Pr = 0.72$ is a valid approximation for the temperature between $200K$ and $1000K$. In general, it is known that the Prandtl number is related to γ by

$$Pr = \frac{4\gamma}{9\gamma-5}. \tag{4.4.3}$$

This can be useful for non-ideal gases. See Ref.[159] for more details. Therefore all we need to evaluate the heat flux is the viscosity coefficient μ. The viscosity coefficient can be computed by Sutherland's law,

$$\mu = \mu_0 \frac{T_0 + C}{T + C} \left(\frac{T}{T_0}\right)^{\frac{3}{2}}, \quad C = 110.5\ [K], \quad T_0 = 273.1\ [K], \quad \mu_0 = 1.716 \times 10^{-5}\ [\text{Pa} \cdot s], \tag{4.4.4}$$

for air. This is also a good approximation for a wide range of temperature, between $170K$ and $1900K$. See Ref.[169] for more details.

4.5   Other Energy Equations

We obtain the equation for the internal energy e by subtracting the kinetic energy equation (4.1.9) from the total energy equation (4.1.3):

$$\rho \frac{De}{Dt} = p\,\text{div}\,\mathbf{v} - \text{div}\,\mathbf{q} + \Phi, \tag{4.5.1}$$

where Φ is the dissipation term given by

$$\Phi = \boldsymbol{\tau} : \text{grad}\,\mathbf{v} = \mu \left[\text{grad}\,\mathbf{v} + (\text{grad}\,\mathbf{v})^t\right] : \text{grad}\,\mathbf{v} + \lambda(\text{div}\,\mathbf{v})^2. \tag{4.5.2}$$

It can be shown that the dissipation term can be written as

$$\Phi = \frac{\mu}{2}\left[\text{grad}\,\mathbf{v} + (\text{grad}\,\mathbf{v})^t - \frac{2}{3}(\text{div}\,\mathbf{v})\mathbf{I}\right] : \left[\text{grad}\,\mathbf{v} + (\text{grad}\,\mathbf{v})^t - \frac{2}{3}(\text{div}\,\mathbf{v})\mathbf{I}\right] + \left(\lambda + \frac{2}{3}\mu\right)(\text{div}\,\mathbf{v})^2$$

$$= \frac{\mu}{2}\left[\left(\frac{\partial u_i}{\partial x_j} + \frac{\partial u_j}{\partial x_i}\right) - \frac{2}{3}\delta_{ij}\frac{\partial u_k}{\partial x_k}\right]^2 + \left(\lambda + \frac{2}{3}\mu\right)\left(\frac{\partial u_k}{\partial x_k}\right)^2. \tag{4.5.3}$$

It follows immediately from this that the dissipation term is positive if

$$\mu \ge 0, \quad \lambda + \frac{2}{3}\mu \ge 0. \tag{4.5.4}$$

This is very nice. It shows that Stokes' hypothesis ($\lambda = -\frac{2}{3}\mu$) guarantees the positivity of the dissipation term. If you are interested to derive Equation (4.5.3) from Equation (4.5.2), first you need to understand these identities,

$$\left[\text{grad}\,\mathbf{v} + (\text{grad}\,\mathbf{v})^t\right] : \left[\text{grad}\,\mathbf{v} - (\text{grad}\,\mathbf{v})^t\right] = 0, \tag{4.5.5}$$

$$(\text{div}\,\mathbf{v})\mathbf{I} : \left[\text{grad}\,\mathbf{v} - (\text{grad}\,\mathbf{v})^t\right] = 0. \tag{4.5.6}$$

Then, you write

$$\Phi = \boldsymbol{\tau} : \text{grad}\,\mathbf{v}$$

$$= \boldsymbol{\tau} : \left[\frac{1}{2}\left(\text{grad}\,\mathbf{v} + (\text{grad}\,\mathbf{v})^t\right) + \frac{1}{2}\left(\text{grad}\,\mathbf{v} - (\text{grad}\,\mathbf{v})^t\right)\right], \tag{4.5.7}$$

and expand this with $\boldsymbol{\tau} = \lambda(\text{div}\,\mathbf{v})\mathbf{I} + \mu\left[\text{grad}\,\mathbf{v} + (\text{grad}\,\mathbf{v})^t\right]$ and the identities above to get

$$\Phi = \frac{\mu}{2}\left[\text{grad}\,\mathbf{v} + (\text{grad}\,\mathbf{v})^t\right] : \left[\text{grad}\,\mathbf{v} + (\text{grad}\,\mathbf{v})^t\right] + \frac{\lambda}{2}(\text{div}\,\mathbf{v})\mathbf{I} : \left[\text{grad}\,\mathbf{v} + (\text{grad}\,\mathbf{v})^t\right]$$

$$= \frac{\mu}{2}\left[\text{grad}\,\mathbf{v} + (\text{grad}\,\mathbf{v})^t\right] : \left[\text{grad}\,\mathbf{v} + (\text{grad}\,\mathbf{v})^t\right] + \lambda(\text{div}\,\mathbf{v})^2. \tag{4.5.8}$$

Finally, you somehow arrange this into the final form (4.5.3), e.g., by adding $\left[\frac{2}{3}\lambda(\text{div}\,\mathbf{v})^2 - \frac{2}{3}\lambda(\text{div}\,\mathbf{v})^2\right]$ and refactoring.

Basically, the energy equation in the Navier-Stokes equations can be replaced by any of the followings:

$$\frac{\partial(\rho H)}{\partial t} - \frac{\partial p}{\partial t} + \text{div}(\rho \mathbf{v} H) = \text{div}(\boldsymbol{\tau}\mathbf{v}) - \text{div}\,\mathbf{q}, \tag{4.5.9}$$

$$\rho \frac{DH}{Dt} - \frac{\partial p}{\partial t} = \text{div}(\boldsymbol{\tau}\mathbf{v}) - \text{div}\,\mathbf{q}, \tag{4.5.10}$$

$$\rho \frac{DE}{Dt} + \text{div}(p\mathbf{v}) = \text{div}(\boldsymbol{\tau}\mathbf{v}) - \text{div}\,\mathbf{q}, \tag{4.5.11}$$

$$\rho \frac{De}{Dt} + p\,\text{div}\,\mathbf{v} = \Phi - \text{div}\,\mathbf{q}, \tag{4.5.12}$$

$$\rho \frac{Dh}{Dt} - \frac{Dp}{Dt} = \Phi - \text{div}\,\mathbf{q}, \tag{4.5.13}$$

$$\rho \frac{D}{Dt}\left(\frac{\mathbf{v}^2}{2}\right) + (\text{grad}\,p) \cdot \mathbf{v} = \text{div}(\boldsymbol{\tau}\mathbf{v}) - \Phi. \tag{4.5.14}$$

It is very nice that we have so many options for the energy equation. Naturally, the entropy equation/inequality can also be used:

$$\rho \frac{Ds}{Dt} + \text{div}\left(\frac{\mathbf{q}}{T}\right) = \frac{\Phi}{T} + \frac{\kappa}{T^2}(\text{grad}T) \cdot (\text{grad}T) \geq 0. \tag{4.5.15}$$

4.6 Total Pressure Equation

I find it interesting that the governing equation for the total pressure, $p_t = p + \rho \mathbf{v}^2/2$, can be derived quite easily for steady incompressible flows. Take the dot product of the velocity and the steady version of the momentum equation (4.1.2) to get

$$\mathbf{v} \cdot \text{div}(\rho \mathbf{v} \otimes \mathbf{v}) = -\mathbf{v} \cdot \text{grad}\,p + \mathbf{v} \cdot \text{div}\,\boldsymbol{\tau}, \tag{4.6.1}$$

which becomes by the vector identities (1.7.15) and (1.7.11), and the continuity equation

$$\text{div}\left(\rho\mathbf{v}\frac{\mathbf{v}\cdot\mathbf{v}}{2}\right) + \frac{\mathbf{v}\cdot\mathbf{v}}{2}\,\cancel{\text{div}(\rho\mathbf{v})} = -\text{div}(p\mathbf{v}) + p\,\cancel{\text{div}\,\mathbf{v}} + \mathbf{v} \cdot \text{div}\,\boldsymbol{\tau}. \tag{4.6.2}$$

We thus obtain

$$\text{div}(p_t\mathbf{v}) = \text{div}(\boldsymbol{\tau}\mathbf{v}) - \Phi, \tag{4.6.3}$$

which can be written, by the continuity equation, as

$$(\text{grad}\,p_t)\mathbf{v} = \text{div}(\boldsymbol{\tau}\mathbf{v}) - \Phi. \tag{4.6.4}$$

where Φ is the dissipation term (4.5.2). This is the governing equation for the total pressure p_t of steady incompressible flows. I like it very much because it tells us some interesting physical behaviors of the total pressure in inviscid and viscous flows. For inviscid flows, the right hand side vanishes, and the equation shows that the total pressure is constant along the streamline. For viscous flows, the second term on the right hand side acts as dissipation, and it only decreases the total pressure. On the other hand, the first term on the right hand side can be locally positive and therefore can increase the total pressure although the global effect is zero provided no external forces are applied at the domain boundary. This is an interesting observation, first made in Ref.[55]. In Ref.[55], a local rise in the total pressure is actually demonstrated for a two-dimensional viscous stagnation flow and a three-dimensional viscous flow over a sphere.

4.7 Zero Divergence and Viscous Force on Solid Body

Consider the continuity equation for steady compressible flows:

$$\text{div}(\rho\mathbf{v}) = 0, \tag{4.7.1}$$

which can be written by Equation (1.7.11) as

$$\rho \operatorname{div} \mathbf{v} + \operatorname{grad}\rho \cdot \mathbf{v} = 0. \tag{4.7.2}$$

Now, suppose we apply the no-slip condition on a solid body, i.e., $\mathbf{v} = 0$. Then, we obtain

$$\operatorname{div} \mathbf{v} = 0. \tag{4.7.3}$$

This is interesting. It says that the velocity divergence vanishes on a solid body in steady compressible flows. I like it because it means that compressibility has no impact on the viscous force acting on a solid body:

$$\tau\big|_{\text{body}} = \lambda(\cancel{\operatorname{div}\mathbf{v}})\mathbf{I} + \mu \left[\operatorname{grad} \mathbf{v} + (\operatorname{grad} \mathbf{v})^t \right] = \mu \left[\operatorname{grad} \mathbf{v} + (\operatorname{grad} \mathbf{v})^t \right], \tag{4.7.4}$$

which may be integrated over the body to calculate the viscous force. It is even more interesting to attempt to calculate the viscous force. Consider the viscous force acting on an infinitesimal surface element on the body:

$$\tau\big|_{\text{body}} \, \mathbf{n} \, dS = \mu \left[\operatorname{grad} \mathbf{v} + (\operatorname{grad} \mathbf{v})^t \right] \mathbf{n} \, dS, \tag{4.7.5}$$

where dS is the infinitesimal surface element and $\mathbf{n}$ is the unit outward normal vector. We write it as

$$\tau\big|_{\text{body}} \, \mathbf{n} \, dS = \mu \left[\operatorname{grad} \mathbf{v} - (\operatorname{grad} \mathbf{v})^t \right] \mathbf{n} \, dS + 2\mu(\operatorname{grad} \mathbf{v})^t \, \mathbf{n} \, dS. \tag{4.7.6}$$

Then, we can show that the last term vanishes on the body. First, I can easily show that

$$(\operatorname{grad} \mathbf{v})^t \, \mathbf{n} \, dS = \operatorname{grad} q_n \, dS, \tag{4.7.7}$$

where $q_n = \mathbf{v} \cdot \mathbf{n}$. The gradient of the normal velocity can be decomposed into three orthogonal directions: the normal direction and two tangential directions.

$$\operatorname{grad} q_n = \frac{\partial q_n}{\partial n}\mathbf{n} + \frac{\partial q_n}{\partial \ell}\boldsymbol{\ell} + \frac{\partial q_n}{\partial m}\mathbf{m}, \tag{4.7.8}$$

where the tangential directions are denoted by ℓ and m, and their unit vectors are denoted by $\boldsymbol{\ell}$ and $\mathbf{m}$, respectively. We then immediately find that the tangential derivatives vanish by the no-slip condition (i.e., no suction or injection):

$$(\operatorname{grad} \mathbf{v})^t \, \mathbf{n} \, dS = \frac{\partial q_n}{\partial n} \mathbf{n} \, dS. \tag{4.7.9}$$

Does this vanish also? Yes, it does. Recall that the velocity divergence vanishes on the body, which can be expressed in the orthogonal coordinates taken with respect to the normal direction:

$$\operatorname{div} \mathbf{v} = \frac{\partial q_n}{\partial n} + \frac{\partial q_\ell}{\partial \ell} + \frac{\partial q_m}{\partial m} = 0. \tag{4.7.10}$$

The no-slip condition implies that the tangential velocities are uniformly zero in the tangential directions on the body. Therefore, their derivatives vanish, and we are left with

$$\frac{\partial q_n}{\partial n} = 0, \tag{4.7.11}$$

on the body, and thus we do have

$$(\operatorname{grad} \mathbf{v})^t \, \mathbf{n} \, dS = 0. \tag{4.7.12}$$

We therefore conclude that

$$\tau\big|_{\text{body}} \, \mathbf{n} \, dS = \mu \left[\operatorname{grad} \mathbf{v} - (\operatorname{grad} \mathbf{v})^t \right] \mathbf{n} \, dS. \tag{4.7.13}$$

I like it because it can be written by Equation (1.7.6) as

$$\tau\big|_{\text{body}} \, \mathbf{n} \, dS = \mu \left(\operatorname{curl} \mathbf{v} \times \mathbf{n} \right) dS = \mu \left(\boldsymbol{\omega} \times \mathbf{n} \right) dS. \tag{4.7.14}$$

This is interesting. The viscous force can be calculated by integrating the vorticity over the body. The result is, of course, valid also for incompressible flows [11, 122]. So, if your CFD code produces accurate vorticity on the body, the viscous force can be calculated by integrating it over the body for steady incompressible/compressible flows.

4.8   1D Navier-Stokes Equations

The 1D Navier-Stokes equations in conservative form:

$$\frac{\partial \mathbf{U}}{\partial t} + \frac{\partial \mathbf{F}}{\partial x} = 0, \tag{4.8.1}$$

where

$$\mathbf{U} = \begin{bmatrix} \rho \\ \rho u \\ \rho E \end{bmatrix}, \quad \mathbf{F} = \begin{bmatrix} \rho u \\ \rho u^2 + p - \tau_{xx} \\ \rho u H - \tau_{xx} u + q_x \end{bmatrix}, \tag{4.8.2}$$

$$\tau_{xx} = (2\mu + \lambda)\frac{\partial u}{\partial x}, \tag{4.8.3}$$

$$q_x = -k\frac{\partial T}{\partial x} = -\frac{\mu}{Pr(\gamma - 1)}\frac{\partial a^2}{\partial x} = -\frac{\gamma \mu}{Pr(\gamma - 1)}\frac{\partial}{\partial x}\left(\frac{p}{\rho}\right). \tag{4.8.4}$$

I like that Stokes' hypothesis ($\lambda = -\frac{2}{3}\mu$) simplifies the viscous stress to

$$\tau_{xx} = \frac{4}{3}\mu\frac{\partial u}{\partial x}. \tag{4.8.5}$$

Incidentally, the dissipation term (4.5.2) becomes

$$\Phi = 2\mu\left(\frac{\partial u}{\partial x}\right)^2 + \lambda\left(\frac{\partial u}{\partial x}\right)^2, \tag{4.8.6}$$

which is always positive, of course. I like this dissipation term, but it is not needed in the conservative form (4.8.1).

4.9   2D Navier-Stokes Equations

The 2D Navier-Stokes equations in conservative form:

$$\frac{\partial \mathbf{U}}{\partial t} + \frac{\partial \mathbf{F}}{\partial x} + \frac{\partial \mathbf{G}}{\partial y} = 0, \tag{4.9.1}$$

where

$$\mathbf{U} = \begin{bmatrix} \rho \\ \rho u \\ \rho v \\ \rho E \end{bmatrix}, \tag{4.9.2}$$

$$\mathbf{F} = \begin{bmatrix} \rho u \\ \rho u^2 + p - \tau_{xx} \\ \rho u v - \tau_{yx} \\ \rho u H - \tau_{xx} u - \tau_{xy} v + q_x \end{bmatrix}, \tag{4.9.3}$$

$$\mathbf{G} = \begin{bmatrix} \rho v \\ \rho u v - \tau_{xy} \\ \rho v^2 + p - \tau_{yy} \\ \rho v H - \tau_{yx} u - \tau_{yy} v + q_y \end{bmatrix}, \tag{4.9.4}$$

$$\tau_{xx} = 2\mu\frac{\partial u}{\partial x} + \lambda\left(\frac{\partial u}{\partial x} + \frac{\partial v}{\partial y}\right), \tag{4.9.5}$$

$$\tau_{yy} = 2\mu\frac{\partial v}{\partial y} + \lambda\left(\frac{\partial u}{\partial x} + \frac{\partial v}{\partial y}\right), \tag{4.9.6}$$

$$\tau_{xy} = \tau_{yx} = \mu\left(\frac{\partial u}{\partial y} + \frac{\partial v}{\partial x}\right), \tag{4.9.7}$$

$$q_x = -k\frac{\partial T}{\partial x} = -\frac{\mu}{Pr(\gamma-1)}\frac{\partial a^2}{\partial x} = -\frac{\gamma\mu}{Pr(\gamma-1)}\frac{\partial}{\partial x}\left(\frac{p}{\rho}\right), \tag{4.9.8}$$

$$q_y = -k\frac{\partial T}{\partial y} = -\frac{\mu}{Pr(\gamma-1)}\frac{\partial a^2}{\partial y} = -\frac{\gamma\mu}{Pr(\gamma-1)}\frac{\partial}{\partial y}\left(\frac{p}{\rho}\right). \tag{4.9.9}$$

I like again that Stokes' hypothesis ($\lambda = -\frac{2}{3}\mu$) simplifies the viscous stresses to

$$\tau_{xx} = \frac{2}{3}\mu\left(2\frac{\partial u}{\partial x} - \frac{\partial v}{\partial y}\right), \tag{4.9.10}$$

$$\tau_{yy} = \frac{2}{3}\mu\left(2\frac{\partial v}{\partial y} - \frac{\partial u}{\partial x}\right). \tag{4.9.11}$$

Again, the dissipation term (4.5.2) is always positive:

$$\Phi = \mu\left[2\left(\frac{\partial u}{\partial x}\right)^2 + 2\left(\frac{\partial v}{\partial y}\right)^2 + \left(\frac{\partial v}{\partial x} + \frac{\partial u}{\partial y}\right)^2\right] + \lambda\left(\frac{\partial u}{\partial x} + \frac{\partial v}{\partial y}\right)^2. \tag{4.9.12}$$

I like this dissipation term, but again it has nothing to do with the conservative form (4.9.1).

4.10   3D Navier-Stokes Equations

The 3D Navier-Stokes equations in conservative form:

$$\frac{\partial \mathbf{U}}{\partial t} + \frac{\partial \mathbf{F}}{\partial x} + \frac{\partial \mathbf{G}}{\partial y} + \frac{\partial \mathbf{H}}{\partial z} = 0, \tag{4.10.1}$$

where

$$\mathbf{U} = \begin{bmatrix} \rho \\ \rho u \\ \rho v \\ \rho w \\ \rho E \end{bmatrix}, \tag{4.10.2}$$

$$\mathbf{F} = \begin{bmatrix} \rho u \\ \rho u^2 + p - \tau_{xx} \\ \rho uv - \tau_{yx} \\ \rho uw - \tau_{zx} \\ \rho uH - \tau_{xx}u - \tau_{xy}v - \tau_{xz}w + q_x \end{bmatrix}, \tag{4.10.3}$$

$$\mathbf{G} = \begin{bmatrix} \rho v \\ \rho vu - \tau_{xy} \\ \rho v^2 + p - \tau_{yy} \\ \rho vw - \tau_{zy} \\ \rho vH - \tau_{yx}u - \tau_{yy}v - \tau_{yz}w + q_y \end{bmatrix}, \tag{4.10.4}$$

$$\mathbf{H} = \begin{bmatrix} \rho w \\ \rho w u - \tau_{xz} \\ \rho w v - \tau_{yz} \\ \rho w^2 + p - \tau_{zz} \\ \rho w H - \tau_{zx} u - \tau_{zy} v - \tau_{zz} w + q_z \end{bmatrix}, \tag{4.10.5}$$

$$\tau_{xx} = 2\mu \frac{\partial u}{\partial x} + \lambda \left(\frac{\partial u}{\partial x} + \frac{\partial v}{\partial y} + \frac{\partial w}{\partial z} \right), \tag{4.10.6}$$

$$\tau_{yy} = 2\mu \frac{\partial v}{\partial y} + \lambda \left(\frac{\partial u}{\partial x} + \frac{\partial v}{\partial y} + \frac{\partial w}{\partial z} \right), \tag{4.10.7}$$

$$\tau_{zz} = 2\mu \frac{\partial w}{\partial z} + \lambda \left(\frac{\partial u}{\partial x} + \frac{\partial v}{\partial y} + \frac{\partial w}{\partial z} \right), \tag{4.10.8}$$

$$\tau_{xy} = \tau_{yx} = \mu \left(\frac{\partial u}{\partial y} + \frac{\partial v}{\partial x} \right), \tag{4.10.9}$$

$$\tau_{xz} = \tau_{zx} = \mu \left(\frac{\partial w}{\partial x} + \frac{\partial u}{\partial z} \right), \tag{4.10.10}$$

$$\tau_{yz} = \tau_{zy} = \mu \left(\frac{\partial v}{\partial z} + \frac{\partial w}{\partial y} \right), \tag{4.10.11}$$

$$q_x = -k \frac{\partial T}{\partial x} = -\frac{\mu}{Pr(\gamma-1)} \frac{\partial a^2}{\partial x} = -\frac{\gamma\mu}{Pr(\gamma-1)} \frac{\partial}{\partial x} \left(\frac{p}{\rho} \right), \tag{4.10.12}$$

$$q_y = -k \frac{\partial T}{\partial y} = -\frac{\mu}{Pr(\gamma-1)} \frac{\partial a^2}{\partial y} = -\frac{\gamma\mu}{Pr(\gamma-1)} \frac{\partial}{\partial y} \left(\frac{p}{\rho} \right), \tag{4.10.13}$$

$$q_z = -k \frac{\partial T}{\partial z} = -\frac{\mu}{Pr(\gamma-1)} \frac{\partial a^2}{\partial z} = -\frac{\gamma\mu}{Pr(\gamma-1)} \frac{\partial}{\partial z} \left(\frac{p}{\rho} \right). \tag{4.10.14}$$

Of course, I like Stokes' hypothesis ($\lambda = -\frac{2}{3}\mu$) which simplifies the viscous stresses to

$$\tau_{xx} = \frac{2}{3}\mu \left(2\frac{\partial u}{\partial x} - \frac{\partial v}{\partial y} - \frac{\partial w}{\partial z} \right), \tag{4.10.15}$$

$$\tau_{yy} = \frac{2}{3}\mu \left(2\frac{\partial v}{\partial y} - \frac{\partial u}{\partial x} - \frac{\partial w}{\partial z} \right), \tag{4.10.16}$$

$$\tau_{zz} = \frac{2}{3}\mu \left(2\frac{\partial w}{\partial z} - \frac{\partial u}{\partial x} - \frac{\partial v}{\partial y} \right). \tag{4.10.17}$$

Yes, I like the dissipation term (4.5.2) that is always positive:

$$\Phi = \mu \left[2\left(\frac{\partial u}{\partial x} \right)^2 + 2\left(\frac{\partial v}{\partial y} \right)^2 + 2\left(\frac{\partial w}{\partial z} \right)^2 + \left(\frac{\partial v}{\partial x} + \frac{\partial u}{\partial y} \right)^2 + \left(\frac{\partial w}{\partial y} + \frac{\partial v}{\partial z} \right)^2 + \left(\frac{\partial u}{\partial z} + \frac{\partial w}{\partial x} \right)^2 \right]$$
$$+ \lambda \left(\frac{\partial u}{\partial x} + \frac{\partial v}{\partial y} + \frac{\partial w}{\partial z} \right)^2, \tag{4.10.18}$$

which is again irrelevant to the conservative form (4.10.1).

4.11 Axisymmetric Navier-Stokes Equations

I like the axisymmetric Navier-Stokes equations as much as the axisymmetric Euler equations in Section 3.7. Again, these equations describe a three-dimensional flow but formally two-dimensional equations. Since the inviscid part is already described in Section 3.7, here we focus on the viscous terms.

Consider cylindrical coordinates in Figure 1.5.1, where the velocity is denoted by $\mathbf{v} = [u_r, u_\theta, u_z]^t$ and the heat flux is denoted by $\mathbf{q} = [q_r, q_\theta, q_z]^t$. By taking the z-axis as the axis of symmetry, we assume an axisymmetric flow with $u_\theta = 0$ and $\partial/\partial\theta = 0$. First, the divergence of the heat flux can be written by Equation (1.5.23) as

$$- \operatorname{div} \mathbf{q} = -\frac{1}{r}\frac{\partial(r q_r)}{\partial r} - \frac{1}{r}\frac{\partial q_\theta}{\partial\theta} - \frac{\partial q_z}{\partial z}. \tag{4.11.1}$$

Next, the divergence of the viscous stresses in the momentum equations is given by (see Equation (1.5.45))

$$\operatorname{div}\boldsymbol{\tau} = \operatorname{div}\begin{bmatrix} \tau_{rr} & \tau_{r\theta} & \tau_{rz} \\ \tau_{\theta r} & \tau_{\theta\theta} & \tau_{\theta z} \\ \tau_{zr} & \tau_{z\theta} & \tau_{zz} \end{bmatrix} = \begin{bmatrix} \dfrac{1}{r}\dfrac{\partial(r\tau_{rr})}{\partial r} + \dfrac{1}{r}\dfrac{\partial\tau_{r\theta}}{\partial\theta} + \dfrac{\partial\tau_{rz}}{\partial z} - \dfrac{\tau_{\theta\theta}}{r} \\[2ex] \dfrac{1}{r}\dfrac{\partial(r\tau_{\theta r})}{\partial r} + \dfrac{1}{r}\dfrac{\partial\tau_{\theta\theta}}{\partial\theta} + \dfrac{\partial\tau_{\theta z}}{\partial z} + \dfrac{\tau_{r\theta}}{r} \\[2ex] \dfrac{1}{r}\dfrac{\partial(r\tau_{zr})}{\partial r} + \dfrac{1}{r}\dfrac{\partial\tau_{z\theta}}{\partial\theta} + \dfrac{\partial\tau_{zz}}{\partial z} \end{bmatrix}. \tag{4.11.2}$$

where the viscous stress tensor is given by

$$\boldsymbol{\tau} = \lambda(\operatorname{div}\mathbf{v})\mathbf{I} + \mu\left[\operatorname{grad}\mathbf{v} + (\operatorname{grad}\mathbf{v})^t\right]$$

$$= \lambda\left(\frac{1}{r}\frac{\partial(ru_r)}{\partial r} + \frac{1}{r}\frac{\partial u_\theta}{\partial\theta} + \frac{\partial u_z}{\partial z}\right)\mathbf{I} + \mu\left\{\begin{bmatrix} \dfrac{\partial u_r}{\partial r} & \dfrac{1}{r}\dfrac{\partial u_r}{\partial\theta} - \dfrac{u_\theta}{r} & \dfrac{\partial u_r}{\partial z} \\[2ex] \dfrac{\partial u_\theta}{\partial r} & \dfrac{1}{r}\dfrac{\partial u_\theta}{\partial\theta} + \dfrac{u_r}{r} & \dfrac{\partial u_\theta}{\partial z} \\[2ex] \dfrac{\partial u_z}{\partial r} & \dfrac{1}{r}\dfrac{\partial u_z}{\partial\theta} & \dfrac{\partial u_z}{\partial z} \end{bmatrix} + \begin{bmatrix} \dfrac{\partial u_r}{\partial r} & \dfrac{\partial u_\theta}{\partial r} & \dfrac{\partial u_z}{\partial r} \\[2ex] \dfrac{1}{r}\dfrac{\partial u_r}{\partial\theta} - \dfrac{u_\theta}{r} & \dfrac{1}{r}\dfrac{\partial u_\theta}{\partial\theta} + \dfrac{u_r}{r} & \dfrac{1}{r}\dfrac{\partial u_z}{\partial\theta} \\[2ex] \dfrac{\partial u_r}{\partial z} & \dfrac{\partial u_\theta}{\partial z} & \dfrac{\partial u_z}{\partial z} \end{bmatrix}\right\}$$

$$= \lambda\left(\frac{\partial u_r}{\partial r} + \frac{\partial u_z}{\partial z} + \frac{u_r}{r}\right)\mathbf{I} + \mu\begin{bmatrix} 2\dfrac{\partial u_r}{\partial r} & 0 & \dfrac{\partial u_r}{\partial z} + \dfrac{\partial u_z}{\partial r} \\[2ex] 0 & 0 & 0 \\[2ex] \dfrac{\partial u_z}{\partial r} + \dfrac{\partial u_r}{\partial z} & 0 & 2\dfrac{\partial u_z}{\partial z} \end{bmatrix} + \begin{bmatrix} 0 & 0 & 0 \\[2ex] 0 & 2\mu\dfrac{u_r}{r} & 0 \\[2ex] 0 & 0 & 0 \end{bmatrix}. \tag{4.11.3}$$

Therefore, we have

$$\operatorname{div}\boldsymbol{\tau} = \operatorname{div}\boldsymbol{\tau}' + \mathbf{s}, \tag{4.11.4}$$

where

$$\operatorname{div}\boldsymbol{\tau}' = \operatorname{div}\begin{bmatrix} \tau'_{rr} & \tau'_{r\theta} & \tau'_{rz} \\ \tau'_{\theta r} & \tau'_{\theta\theta} & \tau'_{\theta z} \\ \tau'_{zr} & \tau'_{z\theta} & \tau'_{zz} \end{bmatrix} = \begin{bmatrix} \dfrac{1}{r}\dfrac{\partial(r\tau'_{rr})}{\partial r} + \dfrac{\partial\tau'_{rz}}{\partial z} \\[2ex] 0 \\[2ex] \dfrac{1}{r}\dfrac{\partial(r\tau'_{zr})}{\partial r} + \dfrac{\partial\tau'_{zz}}{\partial z} \end{bmatrix}, \mathbf{s} = \begin{bmatrix} -\dfrac{2\mu u_r}{r^2} - \dfrac{\lambda}{r}\left(\dfrac{\partial u_r}{\partial r} + \dfrac{\partial u_z}{\partial z} + \dfrac{u_r}{r}\right) \\[2ex] 0 \\[2ex] 0 \end{bmatrix}, \tag{4.11.5}$$

$$\boldsymbol{\tau}' = \lambda\left(\frac{\partial u_r}{\partial r} + \frac{\partial u_z}{\partial z} + \frac{u_r}{r}\right)\mathbf{I} + \mu\begin{bmatrix} 2\dfrac{\partial u_r}{\partial r} & 0 & \dfrac{\partial u_r}{\partial z} + \dfrac{\partial u_z}{\partial r} \\[2ex] 0 & 0 & 0 \\[2ex] \dfrac{\partial u_z}{\partial r} + \dfrac{\partial u_r}{\partial z} & 0 & 2\dfrac{\partial u_z}{\partial z} \end{bmatrix}. \tag{4.11.6}$$

It is clear that the θ-component is zero. Hence, we have only two equations: r- and z-components. Finally, the viscous work term in the energy equation can be written as

$$\boldsymbol{\tau}\mathbf{v} = \lambda\left(\frac{\partial u_r}{\partial r} + \frac{\partial u_z}{\partial z} + \frac{u_r}{r}\right)\mathbf{I}\begin{bmatrix} u_r \\ u_\theta \\ u_z \end{bmatrix} + \mu\begin{bmatrix} 2\dfrac{\partial u_r}{\partial r} & 0 & \dfrac{\partial u_r}{\partial z} + \dfrac{\partial u_z}{\partial r} \\ 0 & 0 & 0 \\ \dfrac{\partial u_z}{\partial r} + \dfrac{\partial u_r}{\partial z} & 0 & 2\dfrac{\partial u_z}{\partial z} \end{bmatrix}\begin{bmatrix} u_r \\ u_\theta \\ u_z \end{bmatrix} + \begin{bmatrix} 0 & 0 & 0 \\ 0 & 2\mu\dfrac{u_r}{r} & 0 \\ 0 & 0 & 0 \end{bmatrix}\begin{bmatrix} u_r \\ u_\theta \\ u_z \end{bmatrix}.$$

(4.11.7)

Clearly, the θ-component is zero, and we have $\boldsymbol{\tau}\mathbf{v} = \boldsymbol{\tau'}\mathbf{v}$, and thus

$$\operatorname{div}(\boldsymbol{\tau}\mathbf{v}) = \operatorname{div}(\boldsymbol{\tau'}\mathbf{v}). \tag{4.11.8}$$

We are now ready to write the viscous terms in the two-dimensional form. Switch the z-equation and the r-equation just for convenience, and replace (z,r) by (x,y), (q_z, q_r) by (q_x, q_y), and (u_z, u_r) by (u, v) to write

$$-\operatorname{div}\mathbf{q} = -\frac{1}{y}\left(\frac{\partial(yq_x)}{\partial x} + \frac{\partial(yq_y)}{\partial y}\right), \tag{4.11.9}$$

$$\operatorname{div}\boldsymbol{\tau} = \frac{1}{y}\begin{bmatrix} \dfrac{\partial(y\tau_{xx})}{\partial x} + \dfrac{\partial(y\tau_{xy})}{\partial y} \\ \dfrac{\partial(y\tau_{yx})}{\partial x} + \dfrac{\partial(y\tau_{yy})}{\partial y} \end{bmatrix} + \begin{bmatrix} 0 \\ -\dfrac{2\mu v}{y^2} - \dfrac{\lambda}{y}\left(\dfrac{\partial u}{\partial x} + \dfrac{\partial v}{\partial y} + \dfrac{v}{y}\right) \end{bmatrix}, \tag{4.11.10}$$

$$\boldsymbol{\tau} = \begin{bmatrix} \tau_{xx} & \tau_{xy} \\ \tau_{yx} & \tau_{yy} \end{bmatrix} = \lambda\left(\frac{\partial u}{\partial x} + \frac{\partial v}{\partial y} + \frac{v}{y}\right)\mathbf{I} + \mu\begin{bmatrix} 2\dfrac{\partial u}{\partial x} & \dfrac{\partial u}{\partial y} + \dfrac{\partial v}{\partial x} \\ \dfrac{\partial v}{\partial x} + \dfrac{\partial u}{\partial y} & 2\dfrac{\partial v}{\partial y} \end{bmatrix}, \tag{4.11.11}$$

$$\operatorname{div}(\boldsymbol{\tau}\mathbf{v}) = \frac{1}{y}\begin{bmatrix} \dfrac{\partial(y\tau_{xx}u)}{\partial x} + \dfrac{\partial(y\tau_{xy}v)}{\partial y} \\ \dfrac{\partial(y\tau_{yx}u)}{\partial x} + \dfrac{\partial(y\tau_{yy}v)}{\partial y} \end{bmatrix}, \tag{4.11.12}$$

where $\mathbf{v}$ and $\mathbf{q}$ are now two-dimensional vectors, and $\boldsymbol{\tau}$ is a two-dimensional tensor. It is important to note that the velocity divergence in the viscous stress tensor (4.11.11) has an extra term v/y, and the same for the second component of the source term in the viscous-stress divergence (4.11.10). Finally, the axisymmetric Navier-Stokes equations can be written as

$$(y\mathbf{U})_t + (y\mathbf{F})_x + (y\mathbf{G})_y = (y\mathbf{F}^v)_x + (y\mathbf{G}^v)_y + \mathbf{S}, \tag{4.11.13}$$

where

$$\mathbf{U} = \begin{bmatrix} \rho \\ \rho u \\ \rho v \\ \rho E \end{bmatrix}, \quad \mathbf{F} = \begin{bmatrix} \rho u \\ \rho u^2 + p \\ \rho uv \\ \rho uH \end{bmatrix}, \quad \mathbf{G} = \begin{bmatrix} \rho v \\ \rho uv \\ \rho v^2 + p \\ \rho vH \end{bmatrix}, \tag{4.11.14}$$

$$\mathbf{F}^v = \begin{bmatrix} 0 \\ \tau_{xx} \\ \tau_{yx} \\ \tau_{xx}u + \tau_{xy}v - q_x \end{bmatrix}, \mathbf{G}^v = \begin{bmatrix} 0 \\ \tau_{xy} \\ \tau_{yy} \\ \tau_{yx}u + \tau_{yy}v - q_y \end{bmatrix}, \mathbf{S} = \begin{bmatrix} 0 \\ 0 \\ p - \dfrac{2\mu v}{y} - \lambda\left(\dfrac{\partial u}{\partial x} + \dfrac{\partial v}{\partial y} + \dfrac{v}{y}\right) \\ 0 \end{bmatrix}. \tag{4.11.15}$$

As in the inviscid case, following Ref.[173], we can express the axisymmetric form and the two-dimensional form in a single form with a logical switch, $\overline{\omega} = 0$ for the two-dimensional form and $\overline{\omega} = 1$ for the axisymmetric form:

$$(\overline{\omega}_a \mathbf{U})_t + (\overline{\omega}_a \mathbf{F})_x + (\overline{\omega}_a \mathbf{G})_y = (\overline{\omega}_a \mathbf{F}^v)_x + (\overline{\omega}_a \mathbf{G}^v)_y + \overline{\omega}\,\mathbf{S}, \qquad (4.11.16)$$

where

$$\overline{\omega}_a = (1 - \overline{\omega}) + \overline{\omega}\,y. \qquad (4.11.17)$$

This is really nice. If you have a two-dimensional Navier-Stokes code, you almost have an axisymmetric Navier-Stokes code, which is essentially a three-dimensional CFD code. I really like it very much.

4.12   3D Normal Viscous Flux and Jacobian

Often, we consider the inviscid and viscous fluxes separately. The viscous flux of the Navier-Stokes equations, denoted by $\mathbf{F}^v$, $\mathbf{G}^v$, $\mathbf{H}^v$, are given by

$$\mathbf{F}^v = \begin{bmatrix} 0 \\ -\tau_{xx} \\ -\tau_{yx} \\ -\tau_{zx} \\ -\tau_{vx} + q_x \end{bmatrix}, \quad \mathbf{G}^v = \begin{bmatrix} 0 \\ -\tau_{xy} \\ -\tau_{yy} \\ -\tau_{zy} \\ -\tau_{vy} + q_y \end{bmatrix}, \quad \mathbf{H}^v = \begin{bmatrix} 0 \\ -\tau_{xz} \\ -\tau_{yz} \\ -\tau_{zz} \\ -\tau_{vz} + q_z \end{bmatrix}, \qquad (4.12.1)$$

where

$$\tau_{vx} = \tau_{xx}u + \tau_{xy}v + \tau_{xz}w, \qquad (4.12.2)$$
$$\tau_{vy} = \tau_{yx}u + \tau_{yy}v + \tau_{yz}w, \qquad (4.12.3)$$
$$\tau_{vz} = \tau_{zx}u + \tau_{zy}v + \tau_{zz}w. \qquad (4.12.4)$$

It is so much simpler than the inviscid flux. I like it very much. In CFD, we often deal with the flux projected along a particular direction, e.g., normal to a control volume boundary. Let $\mathbf{n} = [n_x, n_y, n_z]^t$ be a unit vector. Then, the viscous flux along $\mathbf{n}$, denoted by $\mathbf{F}_n$, is given by

$$\overline{\mathbf{F}}_n^v = \overline{\mathbf{F}}^v n_x + \overline{\mathbf{G}}^v n_y + \overline{\mathbf{H}}^v n_z = \begin{bmatrix} 0 \\ -\overline{\tau}_{nx} \\ -\overline{\tau}_{ny} \\ -\overline{\tau}_{nz} \\ -\overline{\tau}_{nv} + \overline{q}_n \end{bmatrix}, \qquad (4.12.5)$$

where

$$\overline{\tau}_{nx} = \overline{\tau}_{xx}n_x + \overline{\tau}_{xy}n_y + \overline{\tau}_{xz}n_z, \qquad (4.12.6)$$
$$\overline{\tau}_{ny} = \overline{\tau}_{yx}n_x + \overline{\tau}_{yy}n_y + \overline{\tau}_{yz}n_z, \qquad (4.12.7)$$
$$\overline{\tau}_{nz} = \overline{\tau}_{zx}n_x + \overline{\tau}_{zy}n_y + \overline{\tau}_{zz}n_z, \qquad (4.12.8)$$
$$\overline{\tau}_{nv} = \overline{\tau}_{nx}\overline{u} + \overline{\tau}_{ny}\overline{v} + \overline{\tau}_{nz}\overline{w}, \qquad (4.12.9)$$
$$\overline{q}_n = \overline{q}_x n_x + \overline{q}_y n_y + \overline{q}_z n_z. \qquad (4.12.10)$$

All quantities above have been assumed to have been evaluated at some average state, which is indicated by the over bar. For example, in finite-volume methods, the normal viscous flux $\mathbf{F}_n^v$ needs to be computed at control-volume boundary faces, and $\mathbf{n}$ denotes the normal vector. I like the normal viscous flux because all we need for the heat flux is the normal gradient:

$$\overline{q}_n = -\overline{\kappa}\,\mathbf{n} \cdot \overline{\mathrm{grad}\,T} = -\frac{\overline{\mu}}{Pr(\gamma - 1)}\,\mathbf{n} \cdot \overline{\mathrm{grad}(a^2)} = -\frac{\gamma\overline{\mu}}{Pr(\gamma - 1)}\,\mathbf{n} \cdot \overline{\mathrm{grad}\left(\frac{p}{\rho}\right)}, \qquad (4.12.11)$$

which we may write simply as

$$\overline{q}_n = -\overline{\kappa}\,\frac{\overline{\partial T}}{\partial n} = -\frac{\overline{\mu}}{Pr(\gamma-1)}\,\frac{\overline{\partial a^2}}{\partial n} = -\frac{\gamma\overline{\mu}}{Pr(\gamma-1)}\,\overline{\frac{\partial}{\partial n}\left(\frac{p}{\rho}\right)} = -\frac{\gamma\overline{\mu}}{Pr(\gamma-1)\overline{\rho}}\left(\frac{\overline{\partial p}}{\partial n} - \frac{\overline{p}}{\overline{\rho}}\frac{\overline{\partial\rho}}{\partial n}\right), \qquad (4.12.12)$$

where $\partial/\partial n$ denotes the derivative along $\mathbf{n}$, i.e., $\partial/\partial n = n_x\,\partial/\partial x + n_y\,\partial/\partial y + n_z\,\partial/\partial z$. So, we need only one derivative, i.e., the one along $\mathbf{n}$. This is nice. I like it. On the other hand, the viscous stresses require the derivatives in the other directions. To see this, recall that the viscous stress tensor is given by

$$\boldsymbol{\tau} = \lambda(\operatorname{div}\mathbf{v})\mathbf{I} + \mu\left[\operatorname{grad}\mathbf{v} + (\operatorname{grad}\mathbf{v})^t\right]. \qquad (4.12.13)$$

Then, the viscous stress tensor projected along $\mathbf{n}$ is given by

$$\overline{\boldsymbol{\tau}}\mathbf{n} = \begin{bmatrix} \overline{\tau}_{nx} \\[4pt] \overline{\tau}_{ny} \\[4pt] \overline{\tau}_{nz} \end{bmatrix} = \overline{\lambda}(\overline{\operatorname{div}\mathbf{v}})\mathbf{n} + \overline{\mu}\left[(\overline{\operatorname{grad}\mathbf{v}})\mathbf{n} + (\overline{\operatorname{grad}\mathbf{v}})^t\mathbf{n}\right]$$

$$= \overline{\lambda}(\overline{\operatorname{div}\mathbf{v}})\mathbf{n} + \overline{\mu}\left[\frac{\overline{\partial\mathbf{v}}}{\partial n} + \overline{\operatorname{grad}u_n}\right]$$

$$= \begin{bmatrix} \overline{\lambda}\left(\dfrac{\overline{\partial u}}{\partial x} + \dfrac{\overline{\partial v}}{\partial y} + \dfrac{\overline{\partial w}}{\partial z}\right)n_x + \overline{\mu}\left(\dfrac{\overline{\partial u}}{\partial n} + \dfrac{\overline{\partial u_n}}{\partial x}\right) \\[10pt] \overline{\lambda}\left(\dfrac{\overline{\partial u}}{\partial x} + \dfrac{\overline{\partial v}}{\partial y} + \dfrac{\overline{\partial w}}{\partial z}\right)n_y + \overline{\mu}\left(\dfrac{\overline{\partial v}}{\partial n} + \dfrac{\overline{\partial u_n}}{\partial y}\right) \\[10pt] \overline{\lambda}\left(\dfrac{\overline{\partial u}}{\partial x} + \dfrac{\overline{\partial v}}{\partial y} + \dfrac{\overline{\partial w}}{\partial z}\right)n_z + \overline{\mu}\left(\dfrac{\overline{\partial w}}{\partial n} + \dfrac{\overline{\partial u_n}}{\partial z}\right) \end{bmatrix}, \qquad (4.12.14)$$

where $\overline{u}_n = \overline{u}\,n_x + \overline{v}\,n_y + \overline{w}\,n_z$. It is clear from the above that it cannot be written in terms of only $\partial/\partial n$, and therefore we need all derivatives of the velocity components.

It is often necessary to differentiate the viscous flux, for example, when implicit schemes are constructed. Consider the normal viscous Jacobian:

$$\frac{\partial\overline{\mathbf{F}}_n^v}{\partial\mathbf{U}} = \frac{\partial\overline{\mathbf{F}}_n^v}{\partial\mathbf{W}}\frac{\partial\mathbf{W}}{\partial\mathbf{U}} = \begin{bmatrix} \mathbf{0}^t \\[6pt] -\dfrac{\partial\overline{\tau}_{nx}}{\partial\mathbf{W}} \\[6pt] -\dfrac{\partial\overline{\tau}_{ny}}{\partial\mathbf{W}} \\[6pt] -\dfrac{\partial\overline{\tau}_{nz}}{\partial\mathbf{W}} \\[6pt] -\dfrac{\partial\overline{\tau}_{nv}}{\partial\mathbf{W}} + \dfrac{\partial\overline{q}_n}{\partial\mathbf{W}} \end{bmatrix} \begin{bmatrix} 1 & 0 & 0 & 0 & 0 \\[6pt] -\dfrac{u}{\rho} & \dfrac{1}{\rho} & 0 & 0 & 0 \\[6pt] -\dfrac{v}{\rho} & 0 & \dfrac{1}{\rho} & 0 & 0 \\[6pt] -\dfrac{w}{\rho} & 0 & 0 & \dfrac{1}{\rho} & 0 \\[6pt] \dfrac{\gamma-1}{2}q^2 & -(\gamma-1)u & -(\gamma-1)v & -(\gamma-1)w & \gamma-1 \end{bmatrix}, \qquad (4.12.15)$$

where $\mathbf{U} = [\rho, \rho u, \rho v, \rho w, \rho E]^t$ and $\mathbf{W} = [\rho, u, v, w, p]^t$. Let us assume Stokes' hypothesis and write the viscous stresses as

$$\overline{\tau}_{nx} = \overline{\mu}\,\overline{s}_{nx}, \quad \overline{\tau}_{ny} = \overline{\mu}\,\overline{s}_{ny}, \quad \overline{\tau}_{nz} = \overline{\mu}\,\overline{s}_{nz}, \quad \overline{\tau}_{nv} = \overline{\mu}\,\overline{s}_{nv}, \qquad (4.12.16)$$

where $\overline{s}_{nv} = \overline{s}_{nx}\overline{u} + \overline{s}_{ny}\overline{v} + \overline{s}_{nz}\overline{w}$, and

$$\overline{s}_{nx} = \overline{s}_{xx}n_x + \overline{s}_{xy}n_y + \overline{s}_{xz}n_z, \quad \overline{s}_{ny} = \overline{s}_{yx}n_x + \overline{s}_{yy}n_y + \overline{s}_{yz}n_z, \quad \overline{s}_{nz} = \overline{s}_{zx}n_x + \overline{s}_{zy}n_y + \overline{s}_{zz}n_z, \qquad (4.12.17)$$

$$\overline{s}_{xx} = \frac{2}{3}\left(2\frac{\overline{\partial u}}{\partial x} - \frac{\overline{\partial v}}{\partial y} - \frac{\overline{\partial w}}{\partial z}\right), \quad \overline{s}_{yy} = \frac{2}{3}\left(2\frac{\overline{\partial v}}{\partial y} - \frac{\overline{\partial u}}{\partial x} - \frac{\overline{\partial w}}{\partial z}\right), \quad \overline{s}_{zz} = \frac{2}{3}\left(2\frac{\overline{\partial w}}{\partial z} - \frac{\overline{\partial u}}{\partial x} - \frac{\overline{\partial v}}{\partial y}\right), \qquad (4.12.18)$$

$$\overline{s}_{xy} = \overline{s}_{yx} = \frac{\overline{\partial u}}{\partial y} + \frac{\overline{\partial v}}{\partial x}, \quad \overline{s}_{xz} = \overline{s}_{zx} = \frac{\overline{\partial w}}{\partial x} + \frac{\overline{\partial u}}{\partial z}, \quad \overline{s}_{yz} = \overline{s}_{zy} = \frac{\overline{\partial v}}{\partial z} + \frac{\overline{\partial w}}{\partial y}. \qquad (4.12.19)$$

The derivative of $\overline{\tau}_{nx}$ can be obtained as follows.

$$
\frac{\partial \overline{\tau}_{nx}}{\partial \mathbf{W}} = \frac{\partial \overline{\mu}}{\partial \mathbf{W}}\overline{s}_{nx} + \overline{\mu}\frac{\partial \overline{s}_{nx}}{\partial \mathbf{W}}
$$

$$
= \left[\frac{\partial \overline{\mu}}{\partial \overline{\rho}}\overline{s}_{nx},\ \overline{\mu}\frac{\partial \overline{s}_{nx}}{\partial u},\ \overline{\mu}\frac{\partial \overline{s}_{nx}}{\partial v},\ \overline{\mu}\frac{\partial \overline{s}_{nx}}{\partial w},\ \frac{\partial \overline{\mu}}{\partial p}\overline{s}_{nx} \right]
$$

$$
= \begin{bmatrix}
\dfrac{\partial \overline{\mu}}{\partial \overline{T}}\dfrac{\partial \overline{T}}{\partial \rho}\overline{s}_{nx} \\[2mm]
\overline{\mu}\dfrac{\partial}{\partial u}\left(\dfrac{1}{3}\dfrac{\overline{\partial u}}{\partial x}n_x + \dfrac{\overline{\partial u}}{\partial n} \right) \\[2mm]
\overline{\mu}\dfrac{\partial}{\partial v}\left(-\dfrac{2}{3}\dfrac{\overline{\partial v}}{\partial y}n_x + \dfrac{\overline{\partial v}}{\partial x}n_y \right) \\[2mm]
\overline{\mu}\dfrac{\partial}{\partial w}\left(-\dfrac{2}{3}\dfrac{\overline{\partial w}}{\partial z}n_x + \dfrac{\overline{\partial w}}{\partial x}n_z \right) \\[2mm]
\dfrac{\partial \overline{\mu}}{\partial \overline{T}}\dfrac{\partial \overline{T}}{\partial p}\overline{s}_{nx}
\end{bmatrix}^{t}. \tag{4.12.20}
$$

Similarly, we have

$$
\frac{\partial \overline{\tau}_{ny}}{\partial \mathbf{W}} = \begin{bmatrix}
\dfrac{\partial \overline{\mu}}{\partial \overline{T}}\dfrac{\partial \overline{T}}{\partial \rho}\overline{s}_{ny} \\[2mm]
\overline{\mu}\dfrac{\partial}{\partial u}\left(-\dfrac{2}{3}\dfrac{\overline{\partial u}}{\partial x}n_y + \dfrac{\overline{\partial u}}{\partial y}n_x \right) \\[2mm]
\overline{\mu}\dfrac{\partial}{\partial u}\left(\dfrac{1}{3}\dfrac{\overline{\partial v}}{\partial y}n_y + \dfrac{\overline{\partial v}}{\partial n} \right) \\[2mm]
\overline{\mu}\dfrac{\partial}{\partial w}\left(-\dfrac{2}{3}\dfrac{\overline{\partial w}}{\partial z}n_y + \dfrac{\overline{\partial w}}{\partial y}n_z \right) \\[2mm]
\dfrac{\partial \overline{\mu}}{\partial \overline{T}}\dfrac{\partial \overline{T}}{\partial p}\overline{s}_{ny}
\end{bmatrix}^{t},\quad
\frac{\partial \overline{\tau}_{nz}}{\partial \mathbf{W}} = \begin{bmatrix}
\dfrac{\partial \overline{\mu}}{\partial \overline{T}}\dfrac{\partial \overline{T}}{\partial \rho}\overline{s}_{nz} \\[2mm]
\overline{\mu}\dfrac{\partial}{\partial u}\left(-\dfrac{2}{3}\dfrac{\overline{\partial u}}{\partial x}n_z + \dfrac{\overline{\partial u}}{\partial z}n_x \right) \\[2mm]
\overline{\mu}\dfrac{\partial}{\partial v}\left(-\dfrac{2}{3}\dfrac{\overline{\partial v}}{\partial y}n_x + \dfrac{\overline{\partial v}}{\partial z}n_y \right) \\[2mm]
\overline{\mu}\dfrac{\partial}{\partial w}\left(\dfrac{1}{3}\dfrac{\overline{\partial w}}{\partial z}n_z + \dfrac{\overline{\partial w}}{\partial n} \right) \\[2mm]
\dfrac{\partial \overline{\mu}}{\partial \overline{T}}\dfrac{\partial \overline{T}}{\partial p}\overline{s}_{nz}
\end{bmatrix}^{t}, \tag{4.12.21}
$$

$$
\frac{\partial \overline{\tau}_{nv}}{\partial \mathbf{W}} = \frac{\partial \overline{\mu}}{\partial \mathbf{W}}\overline{s}_{nv} + \overline{\mu}\frac{\partial \overline{s}_{nv}}{\partial \mathbf{W}}
$$

$$
= \left[\frac{\partial \overline{\mu}}{\partial \overline{\rho}}\overline{s}_{nv},\ \frac{\partial \overline{u}}{\partial u}\overline{\tau}_{nx} + \overline{u}\frac{\partial \overline{\tau}_{nx}}{\partial u},\ \frac{\partial \overline{v}}{\partial v}\overline{\tau}_{ny} + \overline{v}\frac{\partial \overline{\tau}_{ny}}{\partial v},\ \frac{\partial \overline{w}}{\partial w}\overline{\tau}_{nz} + \overline{w}\frac{\partial \overline{\tau}_{nz}}{\partial w},\ \frac{\partial \overline{\mu}}{\partial p}\overline{s}_{nv} \right]
$$

$$
= \begin{bmatrix}
\dfrac{\partial \overline{\mu}}{\partial \overline{T}}\dfrac{\partial \overline{T}}{\partial \rho}\overline{s}_{nv} \\[2mm]
\overline{\mu}\,\overline{s}_{nx}\dfrac{\partial \overline{u}}{\partial u} + \overline{\mu}\left(\overline{u}n_x - \dfrac{2}{3}\overline{u}_n \right)\dfrac{\partial}{\partial u}\dfrac{\overline{\partial u}}{\partial x} + \overline{\mu}\,\overline{u}\dfrac{\partial}{\partial u}\dfrac{\overline{\partial u}}{\partial n} \\[2mm]
\overline{\mu}\,\overline{s}_{ny}\dfrac{\partial \overline{v}}{\partial v} + \overline{\mu}\left(\overline{v}n_y - \dfrac{2}{3}\overline{u}_n \right)\dfrac{\partial}{\partial v}\dfrac{\overline{\partial v}}{\partial y} + \overline{\mu}\,\overline{v}\dfrac{\partial}{\partial v}\dfrac{\overline{\partial v}}{\partial n} \\[2mm]
\overline{\mu}\,\overline{s}_{nz}\dfrac{\partial \overline{w}}{\partial w} + \overline{\mu}\left(\overline{w}n_z - \dfrac{2}{3}\overline{u}_n \right)\dfrac{\partial}{\partial w}\dfrac{\overline{\partial w}}{\partial z} + \overline{\mu}\,\overline{w}\dfrac{\partial}{\partial w}\dfrac{\overline{\partial w}}{\partial n} \\[2mm]
\dfrac{\partial \overline{\mu}}{\partial \overline{T}}\dfrac{\partial \overline{T}}{\partial p}\overline{s}_{nv}
\end{bmatrix}^{t}. \tag{4.12.22}
$$

On the other hand, the heat flux can be differentiated as follows.

$$\frac{\partial \overline{q}_n}{\partial \mathbf{W}} = -\frac{1}{Pr(\gamma-1)} \left[\frac{\partial \overline{\mu}}{\partial \rho} \frac{\overline{\partial T}}{\partial n} + \overline{\mu} \frac{\partial}{\partial \rho} \frac{\overline{\partial T}}{\partial n}, \ 0, \ 0, \ 0, \ \frac{\partial \overline{\mu}}{\partial p} \frac{\overline{\partial T}}{\partial n} + \overline{\mu} \frac{\partial}{\partial p} \frac{\overline{\partial T}}{\partial n} \right]$$

$$= -\frac{1}{Pr(\gamma-1)} \left[\frac{\partial \overline{\mu}}{\partial \overline{T}} \frac{\partial \overline{T}}{\partial \rho} \frac{\overline{\partial T}}{\partial n} + \overline{\mu} \frac{\partial}{\partial \rho} \frac{\overline{\partial T}}{\partial n}, \ 0, \ 0, \ 0, \ \frac{\partial \overline{\mu}}{\partial \overline{T}} \frac{\partial \overline{T}}{\partial p} \frac{\overline{\partial T}}{\partial n} + \overline{\mu} \frac{\partial}{\partial p} \frac{\overline{\partial T}}{\partial n} \right]. \quad (4.12.23)$$

For Sutherland's law (4.4.4), the derivative $\partial \overline{\mu}/\partial T$ is given by

$$\frac{\partial \overline{\mu}}{\partial T} = \frac{\partial}{\partial \overline{T}} \left(\mu_0 \frac{T_0+C}{\overline{T}+C} \left(\frac{\overline{T}}{T_0} \right)^{\frac{3}{2}} \right) \frac{\partial \overline{T}}{\partial T} = \frac{\mu_0 \overline{\mu}}{2(\overline{T}+C/T_0)} \left(1 + 3\frac{C}{T_0} \overline{T} \right) \frac{\partial \overline{T}}{\partial T}. \quad (4.12.24)$$

That is all we need to compute the viscous Jacobian (4.12.15). I like it because it is a general expression; the specific form depends on how the interface values, i.e., the quantities and derivatives with the over bar, are defined. For example, if the temperature $\overline{T}$ is evaluated as the average of two states, T_L and T_R, and the Jacobian is sought with respect to the left state, then

$$\frac{\partial \overline{T}}{\partial T_L} = \frac{\partial}{\partial T_L} \left(\frac{T_L+T_R}{2} \right) = \frac{1}{2}. \quad (4.12.25)$$

Also, if the normal derivative is evaluated by the difference between the two states,

$$\frac{\overline{\partial u}}{\partial n} = \frac{u_R - u_L}{\Delta s}, \quad (4.12.26)$$

where Δs is the distance between the two states, then the derivative with respect to the left state is given by

$$\frac{\partial}{\partial u_L} \frac{\overline{\partial u}}{\partial n} = \frac{-1}{\Delta s}. \quad (4.12.27)$$

It really depends on the discretization method, but the derivatives not in the normal direction such as $\frac{\partial}{\partial u} \frac{\overline{\partial u}}{\partial x}$ can often become a source of difficulty. For example, on unstructured grids, these derivatives (as well as a part of the normal gradient) lead to a non-compact stencil [49] and the Jacobian matrix can become prohibitively large. In such a case, these derivatives are often ignored in the construction of the Jacobian matrix. The corresponding discretization is then not consistent, but it is perhaps OK for the Jacobian matrix as it acts only as a driver in implicit solvers. As long as the residual (i.e., the right hand side of the implicit formulation) is constructed by a consistent scheme, accuracy is obtained at convergence although the convergence may not be optimal [111, 157].

4.13 Expanded Form of Viscous Terms

Sometimes, it is convenient to expand the Navier-Stokes equations in terms of the derivatives of the primitive variables. As we have shown the expanded form already for the Euler equations in Section 3.6.1, we here focus on the viscous part only:

$$\frac{\partial \mathbf{F}^v}{\partial x} + \frac{\partial \mathbf{G}^v}{\partial y} + \frac{\partial \mathbf{H}^v}{\partial z} = \begin{bmatrix} -\dfrac{\partial \tau_{xx}}{\partial x} - \dfrac{\partial \tau_{xy}}{\partial y} - \dfrac{\partial \tau_{xz}}{\partial z} \\[2mm] -\dfrac{\partial \tau_{yx}}{\partial x} - \dfrac{\partial \tau_{yy}}{\partial y} - \dfrac{\partial \tau_{yz}}{\partial z} \\[2mm] -\dfrac{\partial \tau_{zx}}{\partial x} - \dfrac{\partial \tau_{zy}}{\partial y} - \dfrac{\partial \tau_{zz}}{\partial z} \\[2mm] -\dfrac{\partial \tau_{vx}}{\partial x} - \dfrac{\partial \tau_{vy}}{\partial y} - \dfrac{\partial \tau_{vz}}{\partial z} + \dfrac{\partial q_x}{\partial x} + \dfrac{\partial q_y}{\partial y} + \dfrac{\partial q_z}{\partial z} \end{bmatrix}, \quad (4.13.1)$$

where

$$\frac{\partial \tau_{xx}}{\partial x} = 2\frac{\partial \mu}{\partial x}\frac{\partial u}{\partial x} + 2\mu\frac{\partial^2 u}{\partial x^2} + \frac{\partial \lambda}{\partial x}\left(\frac{\partial u}{\partial x} + \frac{\partial v}{\partial y} + \frac{\partial w}{\partial z}\right) + \lambda\frac{\partial}{\partial x}\left(\frac{\partial u}{\partial x} + \frac{\partial v}{\partial y} + \frac{\partial w}{\partial z}\right), \tag{4.13.2}$$

$$\frac{\partial \tau_{yy}}{\partial y} = 2\frac{\partial \mu}{\partial y}\frac{\partial v}{\partial y} + 2\mu\frac{\partial^2 v}{\partial y^2} + \frac{\partial \lambda}{\partial y}\left(\frac{\partial u}{\partial x} + \frac{\partial v}{\partial y} + \frac{\partial w}{\partial z}\right) + \lambda\frac{\partial}{\partial y}\left(\frac{\partial u}{\partial x} + \frac{\partial v}{\partial y} + \frac{\partial w}{\partial z}\right), \tag{4.13.3}$$

$$\frac{\partial \tau_{zz}}{\partial z} = 2\frac{\partial \mu}{\partial z}\frac{\partial w}{\partial z} + 2\mu\frac{\partial^2 w}{\partial z^2} + \frac{\partial \lambda}{\partial z}\left(\frac{\partial u}{\partial x} + \frac{\partial v}{\partial y} + \frac{\partial w}{\partial z}\right) + \lambda\frac{\partial}{\partial z}\left(\frac{\partial u}{\partial x} + \frac{\partial v}{\partial y} + \frac{\partial w}{\partial z}\right), \tag{4.13.4}$$

$$\frac{\partial \tau_{xy}}{\partial x} = \frac{\partial \tau_{yx}}{\partial x} = \frac{\partial \mu}{\partial x}\left(\frac{\partial u}{\partial y} + \frac{\partial v}{\partial x}\right) + \mu\frac{\partial}{\partial x}\left(\frac{\partial u}{\partial y} + \frac{\partial v}{\partial x}\right), \tag{4.13.5}$$

$$\frac{\partial \tau_{xy}}{\partial y} = \frac{\partial \tau_{yx}}{\partial y} = \frac{\partial \mu}{\partial y}\left(\frac{\partial u}{\partial y} + \frac{\partial v}{\partial x}\right) + \mu\frac{\partial}{\partial y}\left(\frac{\partial u}{\partial y} + \frac{\partial v}{\partial x}\right), \tag{4.13.6}$$

$$\frac{\partial \tau_{xz}}{\partial x} = \frac{\partial \tau_{zx}}{\partial x} = \frac{\partial \mu}{\partial x}\left(\frac{\partial w}{\partial x} + \frac{\partial u}{\partial z}\right) + \mu\frac{\partial}{\partial x}\left(\frac{\partial w}{\partial x} + \frac{\partial u}{\partial z}\right), \tag{4.13.7}$$

$$\frac{\partial \tau_{xz}}{\partial z} = \frac{\partial \tau_{zx}}{\partial z} = \frac{\partial \mu}{\partial z}\left(\frac{\partial w}{\partial x} + \frac{\partial u}{\partial z}\right) + \mu\frac{\partial}{\partial z}\left(\frac{\partial w}{\partial x} + \frac{\partial u}{\partial z}\right), \tag{4.13.8}$$

$$\frac{\partial \tau_{yz}}{\partial y} = \frac{\partial \tau_{zy}}{\partial y} = \frac{\partial \mu}{\partial y}\left(\frac{\partial v}{\partial z} + \frac{\partial w}{\partial y}\right) + \mu\frac{\partial}{\partial y}\left(\frac{\partial v}{\partial z} + \frac{\partial w}{\partial y}\right), \tag{4.13.9}$$

$$\frac{\partial \tau_{yz}}{\partial z} = \frac{\partial \tau_{zy}}{\partial z} = \frac{\partial \mu}{\partial z}\left(\frac{\partial v}{\partial z} + \frac{\partial w}{\partial y}\right) + \mu\frac{\partial}{\partial z}\left(\frac{\partial v}{\partial z} + \frac{\partial w}{\partial y}\right), \tag{4.13.10}$$

$$\frac{\partial \tau_{vx}}{\partial x} = u\frac{\partial \tau_{xx}}{\partial x} + v\frac{\partial \tau_{xy}}{\partial x} + w\frac{\partial \tau_{xz}}{\partial x} + \tau_{xx}\frac{\partial u}{\partial x} + \tau_{xy}\frac{\partial v}{\partial x} + \tau_{xz}\frac{\partial w}{\partial x}, \tag{4.13.11}$$

$$\frac{\partial \tau_{vy}}{\partial y} = u\frac{\partial \tau_{yx}}{\partial y} + v\frac{\partial \tau_{yy}}{\partial y} + w\frac{\partial \tau_{yz}}{\partial y} + \tau_{yx}\frac{\partial u}{\partial y} + \tau_{yy}\frac{\partial v}{\partial y} + \tau_{yz}\frac{\partial w}{\partial y}, \tag{4.13.12}$$

$$\frac{\partial \tau_{vz}}{\partial z} = u\frac{\partial \tau_{zx}}{\partial z} + v\frac{\partial \tau_{zy}}{\partial z} + w\frac{\partial \tau_{zz}}{\partial z} + \tau_{zx}\frac{\partial u}{\partial z} + \tau_{zy}\frac{\partial v}{\partial z} + \tau_{zz}\frac{\partial w}{\partial z}, \tag{4.13.13}$$

$$\frac{\partial q_x}{\partial x} = \frac{-\gamma}{Pr(\gamma-1)}\left[\frac{\partial \mu}{\partial x}\left(\frac{1}{\rho}\frac{\partial p}{\partial x} - \frac{p}{\rho^2}\frac{\partial \rho}{\partial x}\right) + \mu\left(-\frac{2}{\rho^2}\frac{\partial \rho}{\partial x}\frac{\partial p}{\partial x} + \frac{1}{\rho}\frac{\partial^2 p}{\partial x^2} + \frac{p}{\rho^3}\frac{\partial \rho}{\partial x}\frac{\partial \rho}{\partial x} - \frac{p}{\rho^2}\frac{\partial^2 \rho}{\partial x^2}\right)\right], \tag{4.13.14}$$

$$\frac{\partial q_y}{\partial y} = \frac{-\gamma}{Pr(\gamma-1)}\left[\frac{\partial \mu}{\partial y}\left(\frac{1}{\rho}\frac{\partial p}{\partial y} - \frac{p}{\rho^2}\frac{\partial \rho}{\partial y}\right) + \mu\left(-\frac{2}{\rho^2}\frac{\partial \rho}{\partial y}\frac{\partial p}{\partial y} + \frac{1}{\rho}\frac{\partial^2 p}{\partial y^2} + \frac{p}{\rho^3}\frac{\partial \rho}{\partial y}\frac{\partial \rho}{\partial y} - \frac{p}{\rho^2}\frac{\partial^2 \rho}{\partial y^2}\right)\right], \tag{4.13.15}$$

$$\frac{\partial q_z}{\partial z} = \frac{-\gamma}{Pr(\gamma-1)}\left[\frac{\partial \mu}{\partial z}\left(\frac{1}{\rho}\frac{\partial p}{\partial z} - \frac{p}{\rho^2}\frac{\partial \rho}{\partial z}\right) + \mu\left(-\frac{2}{\rho^2}\frac{\partial \rho}{\partial z}\frac{\partial p}{\partial z} + \frac{1}{\rho}\frac{\partial^2 p}{\partial z^2} + \frac{p}{\rho^3}\frac{\partial \rho}{\partial z}\frac{\partial \rho}{\partial z} - \frac{p}{\rho^2}\frac{\partial^2 \rho}{\partial z^2}\right)\right]. \tag{4.13.16}$$

The derivatives of the viscosity can be computed for Sutherland's law (4.4.4) by

$$\frac{\partial \mu}{\partial x} = \frac{\partial}{\partial T} \left(\mu_0 \frac{T_0 + C}{T + C} \left(\frac{T}{T_0} \right)^{\frac{3}{2}} \right) \frac{\partial T}{\partial x} = \frac{\mu_0 \mu}{2(T + C/T_0)} \left(1 + 3 \frac{C}{T_0} T \right) \frac{\partial T}{\partial x}, \tag{4.13.17}$$

$$\frac{\partial \mu}{\partial y} = \frac{\partial}{\partial T} \left(\mu_0 \frac{T_0 + C}{T + C} \left(\frac{T}{T_0} \right)^{\frac{3}{2}} \right) \frac{\partial T}{\partial y} = \frac{\mu_0 \mu}{2(T + C/T_0)} \left(1 + 3 \frac{C}{T_0} T \right) \frac{\partial T}{\partial y}, \tag{4.13.18}$$

$$\frac{\partial \mu}{\partial z} = \frac{\partial}{\partial T} \left(\mu_0 \frac{T_0 + C}{T + C} \left(\frac{T}{T_0} \right)^{\frac{3}{2}} \right) \frac{\partial T}{\partial z} = \frac{\mu_0 \mu}{2(T + C/T_0)} \left(1 + 3 \frac{C}{T_0} T \right) \frac{\partial T}{\partial z}. \tag{4.13.19}$$

The derivatives of λ can be easily computed from the above by Stokes' hypothesis, $\lambda = -\frac{2}{3}\mu$. The derivatives of the temperature can be computed from the derivatives of the density and the pressure. Again, I like this expanded form because it can be useful in numerically computing the source term arising from the method of manufactured solutions. Note that the form of the derivatives depends on the nondimensionalization for the viscosity and the temperature, which is the subject of the next section.

4.14 Nondimensionalization

4.14.1 Nondimensionalized Navier-Stokes Equations

It is always nice to nondimensionalize the governing equations. For the Navier-Stokes equations, in addition to the quantities introduced for the Euler equations in Section 3.11:

$$\rho^* = \frac{\rho}{\rho_r}, \quad p^* = \frac{p}{\rho_r V_r^2}, \quad e^* = \frac{e}{V_r^2}, \quad \mathbf{v}^* = \frac{\mathbf{v}}{V_r}, \quad \mathbf{x}^* = \frac{\mathbf{x}}{L}, \quad t^* = \frac{t}{L/V_r}, \tag{4.14.1}$$

we need to introduce nondimensionalized the temperature and the viscosity,

$$T^* = \frac{T}{T_r}, \quad \mu^* = \frac{\mu}{\mu_r}. \tag{4.14.2}$$

Substituting these quantities into the Navier-Stokes equations, (4.1.1), (4.1.2), (4.1.3), we obtain

$$\frac{\partial \rho^*}{\partial t^*} + \mathrm{div}^*(\rho^* \mathbf{v}^*) = 0, \tag{4.14.3}$$

$$\frac{\partial (\rho^* \mathbf{v}^*)}{\partial t^*} + \mathrm{div}^*(\rho^* \mathbf{v}^* \otimes \mathbf{v}^*) + \mathrm{grad}^* p^* = \mathrm{div}^* \boldsymbol{\tau}^*, \tag{4.14.4}$$

$$\frac{\partial (\rho^* E^*)}{\partial t^*} + \mathrm{div}^*(\rho^* \mathbf{v}^* H^*) = \mathrm{div}^* (\boldsymbol{\tau}^* \mathbf{v}^*) - \mathrm{div}^* \mathbf{q}^*, \tag{4.14.5}$$

where $E^* = e^* + \frac{\mathbf{v}^{*2}}{2}$, $H^* = E^* + \frac{p^*}{\rho^*}$. These are exactly the same as in the dimensional version, but the viscous stress tensor and the heat flux vector have some dimensionless parameters,

$$\boldsymbol{\tau}^* = \frac{\mu_r}{\rho_r V_r L} \mu^* \left[-\frac{2}{3}(\mathrm{div}^* \mathbf{v}^*)\mathbf{I} + \mathrm{grad}^* \mathbf{v}^* + (\mathrm{grad}^* \mathbf{v}^*)^t \right], \tag{4.14.6}$$

$$\mathbf{q}^* = -\frac{\gamma R T_r \mu_r}{\rho_r V_r^3 L(\gamma - 1)Pr} \mu^* \mathrm{grad}^* T^*. \tag{4.14.7}$$

Also, note that the equation of state (3.1.1) is now written as

$$p^* = \left(\frac{R T_r}{V_r^2} \right) \rho^* T^*. \tag{4.14.8}$$

Therefore, a specific form of the equation of state as well as the viscous and heat fluxes depend on the choice of the reference values: ρ_r, V_r, T_r, and μ_r. Note that because of the forms of the nondimensional pressure and specific internal energy, the equation of state for the calorically perfect gas holds exactly as in the dimensional form:

$$p^* = (\gamma - 1)\rho^* e^*, \tag{4.14.9}$$

for any choice of the reference states (see Section 3.11), but the one with the temperature, i.e., $p = \rho RT$, depends on the reference states. Note that the governing equations are virtually in the same form as the dimensional form for an appropriately defined scaled viscosity as shown below.

4.14.2   Free Stream Values

We may use the free stream values for the reference quantities, $\rho_r = \rho_\infty$, $V_r = V_\infty$, $T_r = T_\infty$, $\mu_r = \mu_\infty$:

$$\rho^* = \frac{\rho}{\rho_\infty}, \quad p^* = \frac{p}{\rho_\infty V_\infty^2}, \quad e^* = \frac{e}{V_\infty^2}, \quad \mathbf{v}^* = \frac{\mathbf{v}}{V_\infty}, \quad \mathbf{x}^* = \frac{\mathbf{x}}{L}, \quad t^* = \frac{t}{L/V_\infty}, \quad T^* = \frac{T}{T_\infty}, \quad \mu^* = \frac{\mu}{\mu_\infty}. \tag{4.14.10}$$

Then, the equation of state is given by

$$p^* = \frac{\rho^* T^*}{\gamma M_\infty^2}, \tag{4.14.11}$$

and the viscous stress and the heat flux are given by,

$$\boldsymbol{\tau}^* = \hat{\mu}^* \left[-\frac{2}{3}(\text{div}^* \mathbf{v}^*)\mathbf{I} + \text{grad}^* \mathbf{v}^* + (\text{grad}^* \mathbf{v}^*)^t \right], \tag{4.14.12}$$

$$\mathbf{q}^* = -\frac{\hat{\mu}^*}{(\gamma-1)M_\infty^2 Pr} \text{grad}^* T^* = -\frac{\gamma \hat{\mu}^*}{(\gamma-1)Pr} \text{grad}^* \left(\frac{p^*}{\rho^*}\right), \tag{4.14.13}$$

where the scaled viscosity $\hat{\mu}^*$ is defined by

$$\hat{\mu}^* = \frac{\mu^*}{Re_\infty}, \quad Re_\infty = \frac{\rho_\infty V_\infty L}{\mu_\infty}. \tag{4.14.14}$$

The free stream values are set as follows:

$$\rho_\infty^* = 1, \quad p_\infty^* = \frac{1}{\gamma M_\infty^2}, \quad \mathbf{v}_\infty^* = \frac{\mathbf{v}_\infty}{V_\infty} = \mathbf{n}_\infty, \quad e_\infty^* = \frac{1}{\gamma(\gamma-1)M_\infty^2}, \quad T_\infty^* = 1, \quad \mu_\infty^* = 1. \tag{4.14.15}$$

The nondimensionalized viscosity coefficient μ^* is given by

$$\mu^* = \frac{\mu}{\mu_\infty} = \mu_0 \frac{T_0 + C}{T + C} \left(\frac{T}{T_0}\right)^{\frac{3}{2}} \left[\mu_0 \frac{T_0 + C}{T_\infty + C} \left(\frac{T_\infty}{T_0}\right)^{\frac{3}{2}}\right]^{-1} = \frac{1 + C/T_\infty}{T^* + C/T_\infty} (T^*)^{\frac{3}{2}}, \tag{4.14.16}$$

with $C = 110.5 \, [K]$ for air. The derivative of the nondimensionalized viscosity coefficient is given by

$$\frac{\partial \mu^*}{\partial T^*} = \frac{1}{2} \frac{\mu^*}{T^* + C/T_\infty} \left(1 + 3\frac{C/T_\infty}{T^*}\right). \tag{4.14.17}$$

Note that in order to set a free stream inflow condition, we need to specify not only the Mach number M_∞ and the free stream flow direction $\mathbf{n}_\infty$, which are required for the Euler equations (see Section 3.11), but also the free stream temperature T_∞, e.g., $T_\infty = 293.15[K]$ for the sea-level condition. Of course, the free stream Reynolds number Re_∞ must be specified, and also the Prandtl number Pr needs to be given ($Pr = 0.72$ for the sea-level air).

The Reynolds number Re_∞ must be set carefully in relation to the reference length scale L because it is defined based on L as in Equation (4.14.14). It is important to note that if you solve the nondimensionalized system on a computational grid, then Re_∞ is the Reynolds number based on the part having the length of 1 in the grid (i.e., in the nondimensionalized coordinates, $\mathbf{x}^*$). So, for example, if you want to simulate a flow around a cylinder where the Reynolds number based on the diameter of the cylinder is 100, then you generate a grid around the cylinder with the diameter 1 and set $Re_\infty = 100$. On the other hand, if you are given a grid where the diameter of the cylinder is 2 in the grid, then Re_∞ is the Reynolds number based on the radius of the cylinder (which is 1 in the grid). Then, you need to set $Re_\infty = 50$ to simulate a flow with the Reynolds number 100 based on the diameter. Or if the diameter is 0.1 in the grid, then you need to set $Re_\infty = 1000$ to simulate the desired flow.

4.14.3 Speed of Sound

Instead of the free stream velocity V_∞, we may use the free stream speed of sound for the reference velocity, $V_r = a_\infty$:

$$\rho^* = \frac{\rho}{\rho_\infty}, \quad p^* = \frac{p}{\rho_\infty a_\infty^2}, \quad e^* = \frac{e}{a_\infty^2}, \quad \mathbf{v}^* = \frac{\mathbf{v}}{a_\infty}, \quad \mathbf{x}^* = \frac{\mathbf{x}}{L}, \quad t^* = \frac{t}{L/a_\infty}, \tag{4.14.18}$$

$$T^* = \frac{T}{T_\infty}, \quad \mu^* = \frac{\mu}{\mu_\infty}. \tag{4.14.19}$$

Then, the equation of state is given by

$$p^* = \frac{\rho^* T^*}{\gamma}, \tag{4.14.20}$$

and the viscous stress and the heat flux are given by,

$$\boldsymbol{\tau}^* = \hat{\mu}^* \left[-\frac{2}{3}(\text{div}^* \mathbf{v}^*)\mathbf{I} + \text{grad}^* \mathbf{v}^* + (\text{grad}^* \mathbf{v}^*)^t \right], \tag{4.14.21}$$

$$\mathbf{q}^* = -\frac{\hat{\mu}^*}{(\gamma - 1)Pr} \text{grad}^* T^* = -\frac{\gamma \hat{\mu}^*}{(\gamma - 1)Pr} \text{grad}^* \left(\frac{p^*}{\rho^*} \right), \tag{4.14.22}$$

where the scaled viscosity $\hat{\mu}^*$ is defined by

$$\hat{\mu}^* = \frac{M_\infty \mu^*}{Re_\infty}. \tag{4.14.23}$$

The free stream values are set as follows:

$$\rho_\infty^* = 1, \quad p_\infty^* = \frac{1}{\gamma}, \quad \mathbf{v}_\infty^* = \frac{\mathbf{v}_\infty}{a_\infty} = M_\infty \mathbf{n}_\infty, \quad e_\infty^* = \frac{1}{\gamma(\gamma - 1)}, \quad T_\infty^* = 1, \quad \mu_\infty^* = 1. \tag{4.14.24}$$

The nondimensionalized viscosity coefficient μ^* is again given by Equation (4.14.16) as in the previous case. Therefore, at a free stream inflow, again, we need to specify not only the Mach number M_∞ and the free stream flow direction $\mathbf{n}_\infty$, but also the free stream temperature T_∞ (not to mention Re_∞ and Pr). I like this one, again, because the free stream pressure is independent of M_∞. In fact, this one is very widely used for compressible flow computations.

4.14.4 Stagnation Values

Instead of the free stream values, we may use the stagnation values for the reference quantities:

$$\rho^* = \frac{\rho}{\rho_0} = \left[1 + \frac{\gamma - 1}{2}M^2 \right]^{-\frac{1}{\gamma - 1}}, \quad p^* = \frac{p}{\rho_0 a_0^2} = \frac{1}{\gamma} \left[1 + \frac{\gamma - 1}{2}M^2 \right]^{-\frac{\gamma}{\gamma - 1}}, \tag{4.14.25}$$

$$e^* = \frac{e}{a_0^2}, \quad \mathbf{v}^* = \frac{\mathbf{v}}{a_0}, \quad \mathbf{x}^* = \frac{\mathbf{x}}{L}, \quad t^* = \frac{t}{L/a_0}, \tag{4.14.26}$$

$$T^* = \frac{T}{T_0} = \left[1 + \frac{\gamma - 1}{2}M^2 \right]^{-1}, \quad \mu^* = \frac{\mu}{\mu_\infty}. \tag{4.14.27}$$

Then, the equation of state is given by

$$p^* = \frac{\rho^* T^*}{\gamma}, \tag{4.14.28}$$

and the viscous stress and the heat flux are given by,

$$\boldsymbol{\tau}^* = \hat{\mu}^* \left[-\frac{2}{3}(\text{div}^* \mathbf{v}^*)\mathbf{I} + \text{grad}^* \mathbf{v}^* + (\text{grad}^* \mathbf{v}^*)^t \right], \tag{4.14.29}$$

$$\mathbf{q}^* = -\frac{\hat{\mu}^*}{(\gamma - 1)Pr} \text{grad}^* T^* = -\frac{\gamma \hat{\mu}^*}{(\gamma - 1)Pr} \text{grad}^* \left(\frac{p^*}{\rho^*} \right), \tag{4.14.30}$$

where the scaled viscosity $\hat{\mu}^*$ is defined by

$$\hat{\mu}^* = \frac{m_\infty \mu^*}{Re_\infty}, \quad m_\infty = \rho_\infty^* a_\infty^* = \sqrt{\gamma \rho_\infty^* p_\infty^*} = \left[1 + \frac{\gamma - 1}{2} M_\infty^2\right]^{-\frac{\gamma+1}{2(\gamma-1)}}. \tag{4.14.31}$$

In this case, the free stream values are set as follows:

$$\rho_\infty^* = \left[1 + \frac{\gamma - 1}{2} M_\infty^2\right]^{-\frac{1}{\gamma-1}}, \quad p_\infty^* = \frac{1}{\gamma}\left[1 + \frac{\gamma - 1}{2} M_\infty^2\right]^{-\frac{\gamma}{\gamma-1}}, \tag{4.14.32}$$

$$e^* = \frac{T_\infty^*}{\gamma(\gamma - 1)}, \quad \mathbf{v}_\infty^* = \frac{\mathbf{v}_\infty}{a_\infty}\frac{a_\infty}{a_0} = \frac{\mathbf{v}_\infty}{a_\infty}a_\infty^* = M_\infty \mathbf{n}_\infty \sqrt{\gamma \frac{p_\infty^*}{\rho_\infty^*}}, \tag{4.14.33}$$

$$T_\infty^* = \left[1 + \frac{\gamma - 1}{2} M_\infty^2\right]^{-1}, \quad \mu_\infty^* = 1. \tag{4.14.34}$$

The nondimensionalized viscosity coefficient μ^* is again given by Equation (4.14.16) as in the previous case. Therefore, in this case also, at a free stream inflow, we need to specify the Mach number M_∞, the free stream flow direction $\mathbf{n}_\infty$, and the free stream temperature T_∞, in addition to Re_∞ and Pr.

4.15   Reduced Navier-Stokes Equations

I like reduced Navier-Stokes equations such as thin-layer, parabolized, or conical Navier-Stokes equations. They are approximated forms of the full Navier-Stokes equations and hence simpler and cheaper to solve than the full Navier-Stokes equations. If you like them too, see Refs.[51, 155].

4.16   Quasi-Linear Form of the Navier-Stokes Equations

Often for various purposes (e.g., analysis, scheme development, etc.), it is nice to write the Navier-Stokes equations in the quasi-linear form,

$$\mathbf{V}_t + \mathbf{A}^v\mathbf{V}_x + \mathbf{B}^v\mathbf{V}_y + \mathbf{C}^v\mathbf{V}_z = \mathbf{D}^v\mathbf{V}_{xx} + \mathbf{E}^v\mathbf{V}_{yy} + \mathbf{F}^v\mathbf{V}_{zz} + \mathbf{G}^v\mathbf{V}_{xy} + \mathbf{H}^v\mathbf{V}_{yz} + \mathbf{I}^v\mathbf{V}_{zx}, \tag{4.16.1}$$

for some set of variables $\mathbf{V}$, where the superscript v merely indicates that the coefficient matrices are associated with $\mathbf{V}$. To be able to do so, it is necessary to assume that the viscosity μ and the heat conductivity κ are constant. Also, it is necessary to ignore non-linear terms which involve a product of derivatives. It will be, therefore, no longer the full Navier-Stokes system. However, recall that not all terms are necessary to identify the type of partial differential equations as discussed in Section 1.17. In fact, the quasi-linear form of the Navier-Stokes system retains the essential feature of the full Navier-Stokes system; this is the form modeled by the scalar advection-diffusion or the viscous Burgers equation. For example, a symmetric form of the quasi-linear Navier-Stokes system has been successfully used in the development of convergence acceleration techniques [1, 72, 77, 167].

Consider the primitive form of the Navier-Stokes equations,

$$\frac{D\rho}{Dt} + \rho \operatorname{div} \mathbf{v} = 0, \tag{4.16.2}$$

$$\rho \frac{D\mathbf{v}}{Dt} + \operatorname{grad} p = \operatorname{div} \boldsymbol{\tau}, \tag{4.16.3}$$

$$\frac{Dp}{Dt} + \gamma p \operatorname{div} \mathbf{v} = (\gamma - 1)\left[\boldsymbol{\tau} : \operatorname{grad} \mathbf{v} - \operatorname{div} \mathbf{q}\right]. \tag{4.16.4}$$

Obviously, the dissipation term ($\boldsymbol{\tau} \cdot \operatorname{grad} \mathbf{v}$) involves products of derivatives and needs to be ignored. Also, a part of the heat transfer term needs to be ignored as well. To see this, we first express T in terms of p and ρ,

$$T = \frac{1}{R}\frac{p}{\rho} = \frac{\gamma}{(\gamma - 1)c_p}\frac{p}{\rho} = \frac{1}{\kappa}\frac{\mu\gamma}{(\mu c_p/\kappa)(\gamma - 1)}\frac{p}{\rho}, \tag{4.16.5}$$

thus

$$\kappa T = \frac{\mu\gamma}{Pr(\gamma-1)}\frac{p}{\rho}. \tag{4.16.6}$$

Then, the heat transfer term can be written as

$$
\begin{aligned}
\operatorname{div}\mathbf{q} &= -\operatorname{div}\left(\kappa\operatorname{grad}T\right) \\
&= -\frac{\mu\gamma}{Pr(\gamma-1)}\operatorname{div}\operatorname{grad}\left(\frac{p}{\rho}\right) \\
&= -\frac{\mu\gamma}{Pr(\gamma-1)}\operatorname{div}\left(\frac{1}{\rho}\operatorname{grad}p - \frac{p}{\rho^2}\operatorname{grad}\rho\right) \\
&= -\frac{\mu\gamma}{Pr(\gamma-1)}\left[-\frac{1}{\rho^2}\operatorname{grad}\rho\cdot\operatorname{grad}p + \frac{1}{\rho}\operatorname{div}\operatorname{grad}p - \operatorname{grad}\left(\frac{p}{\rho^2}\right)\cdot\operatorname{grad}\rho - \frac{p}{\rho^2}\operatorname{div}\operatorname{grad}\rho\right]. \tag{4.16.7}
\end{aligned}
$$

Hence, the first and the third terms involve products of derivatives and need to be ignored. The pressure equation then becomes

$$\frac{Dp}{Dt} + \gamma p\operatorname{div}\mathbf{v} = \frac{\mu\gamma}{\rho Pr}\left[\operatorname{div}\operatorname{grad}p - \frac{p}{\rho}\operatorname{div}\operatorname{grad}\rho\right]. \tag{4.16.8}$$

In this sense, therefore, the quasi-linear Navier-Stokes equations may be called *linearized* Navier-Stokes equations, or to be precise the Navier-Stokes equations with the *linearized* energy equation. Clearly, the continuity equation is already in the quasi-linear form. The momentum equation is also in the quasi-linear form, which can be clearly seen by expanding the viscous term as follows,

$$
\begin{aligned}
\operatorname{div}\tau &= \operatorname{div}\left[\lambda(\operatorname{div}\mathbf{v})\mathbf{I} + \mu\operatorname{grad}\mathbf{v} + \mu\left(\operatorname{grad}\mathbf{v}\right)^t\right] \\
&= \lambda\operatorname{grad}\left(\operatorname{div}\mathbf{v}\right) + \mu\operatorname{div}\operatorname{grad}\mathbf{v} + \mu\operatorname{div}\left(\operatorname{grad}\mathbf{v}\right)^t \\
&= -\frac{2}{3}\mu\operatorname{grad}\left(\operatorname{div}\mathbf{v}\right) + \mu\operatorname{div}\left(\operatorname{grad}\mathbf{v}\right)^t + \mu\operatorname{div}\operatorname{grad}\mathbf{v} \\
&= -\frac{2}{3}\mu\operatorname{grad}\left(\operatorname{div}\mathbf{v}\right) + \mu\operatorname{grad}\left(\operatorname{div}\mathbf{v}\right) + \mu\operatorname{div}\operatorname{grad}\mathbf{v} \\
&= \frac{1}{3}\mu\operatorname{grad}\left(\operatorname{div}\mathbf{v}\right) + \mu\operatorname{div}\operatorname{grad}\mathbf{v}, \tag{4.16.9}
\end{aligned}
$$

where we have assumed constant viscosity and Stokes' hypothesis. This involves second-derivatives of the velocity components only, and therefore it can be written in the form (4.16.1).

Linearized Navier-Stokes Equations

The linearized Navier-Stokes equations, with $\nu = \mu/\rho$ and $a^2 = \gamma p/\rho$,

$$\frac{D\rho}{Dt} + \rho\operatorname{div}\mathbf{v} = 0, \tag{4.16.10}$$

$$\frac{D\mathbf{v}}{Dt} + \frac{1}{\rho}\operatorname{grad}p = \frac{1}{3}\nu\operatorname{grad}\left(\operatorname{div}\mathbf{v}\right) + \nu\operatorname{div}\operatorname{grad}\mathbf{v}, \tag{4.16.11}$$

$$\frac{Dp}{Dt} + \gamma p\operatorname{div}\mathbf{v} = \frac{\nu\gamma}{Pr}\operatorname{div}\operatorname{grad}p - \frac{\nu a^2}{Pr}\operatorname{div}\operatorname{grad}\rho, \tag{4.16.12}$$

are now ready to be written out in the quasi-linear form (4.16.1) in one, two, and three dimensions.

4.16.1   1D Quasi-Linear Navier-Stokes Equations

Primitive Form

The primitive form in one dimension is given by

$$\rho_t + u\rho_x + \rho u_x \;=\; 0, \tag{4.16.13}$$

$$u_t + uu_x + \frac{1}{\rho}\,p_x \;=\; \frac{4\nu}{3}u_{xx}, \tag{4.16.14}$$

$$p_t + up_x + \gamma p u_x \;=\; \frac{\nu\gamma}{Pr}p_{xx} - \frac{\nu a^2}{Pr}\rho_{xx}. \tag{4.16.15}$$

This can be written in the form:

$$\mathbf{W}_t + \mathbf{A}^w\mathbf{W}_x = \mathbf{D}^w\mathbf{W}_{xx}, \tag{4.16.16}$$

where

$$\mathbf{W} = \begin{bmatrix} \rho \\ u \\ p \end{bmatrix}, \quad \mathbf{A}^w = \begin{bmatrix} u & \rho & 0 \\ 0 & u & 1/\rho \\ 0 & \gamma p & u \end{bmatrix}, \quad \mathbf{D}^w = \begin{bmatrix} 0 & 0 & 0 \\ 0 & 4\nu/3 & 0 \\ -a^2\nu/Pr & 0 & \nu\gamma/Pr \end{bmatrix}. \tag{4.16.17}$$

Symmetric Form

I like the following variables,

$$d\mathbf{V} = \left[\frac{a\,d\rho}{\rho\sqrt{\gamma}},\; du,\; \frac{a\,dT}{T\sqrt{\gamma(\gamma-1)}}\right]^t, \tag{4.16.18}$$

where $a = \sqrt{\gamma\frac{p}{\rho}}$, since the quasi-linear Navier-Stokes system is symmetrized by these variables [1]. Naturally, these variables are called symmetrizing variables, and often called parabolic symmetrizing variables to distinguish them from the symmetrizing variables for the Euler equations (3.5.28) which are in contrast called hyperbolic symmetrizing variables.

The primitive variables $\mathbf{W}$ and the symmetrizing variables $\mathbf{V}$ are transformed into each other as follows.

$$d\mathbf{V} = \frac{\partial\mathbf{V}}{\partial\mathbf{W}}d\mathbf{W}, \quad d\mathbf{W} = \frac{\partial\mathbf{W}}{\partial\mathbf{V}}d\mathbf{V}, \tag{4.16.19}$$

where

$$\frac{\partial\mathbf{V}}{\partial\mathbf{W}} = \begin{bmatrix} \dfrac{a}{\rho\sqrt{\gamma}} & 0 & 0 \\[2ex] 0 & 1 & 0 \\[2ex] -\dfrac{a}{\rho\sqrt{\gamma(\gamma-1)}} & 0 & \dfrac{a}{p\sqrt{\gamma(\gamma-1)}} \end{bmatrix}, \tag{4.16.20}$$

$$\frac{\partial\mathbf{W}}{\partial\mathbf{V}} = \left(\frac{\partial\mathbf{V}}{\partial\mathbf{W}}\right)^{-1} = \begin{bmatrix} \dfrac{\rho\sqrt{\gamma}}{a} & 0 & 0 \\[2ex] 0 & 1 & 0 \\[2ex] \dfrac{\rho a}{\sqrt{\gamma}} & 0 & \dfrac{\rho a\sqrt{\gamma-1}}{\sqrt{\gamma}} \end{bmatrix}. \tag{4.16.21}$$

The primitive form is now transformed, with $\frac{\partial\mathbf{V}}{\partial\mathbf{W}}$ multiplied from the left, into the symmetric form:

$$\mathbf{V}_t + \mathbf{A}^v\mathbf{V}_x = \mathbf{D}^v\mathbf{V}_{xx}, \tag{4.16.22}$$

where

$$\mathbf{A}^v = \frac{\partial \mathbf{V}}{\partial \mathbf{W}} \mathbf{A}^w \frac{\partial \mathbf{W}}{\partial \mathbf{V}} = \begin{bmatrix} u & \dfrac{a}{\sqrt{\gamma}} & 0 \\[2ex] \dfrac{a}{\sqrt{\gamma}} & u & \dfrac{a\sqrt{\gamma-1}}{\sqrt{\gamma}} \\[2ex] 0 & \dfrac{a\sqrt{\gamma-1}}{\sqrt{\gamma}} & u \end{bmatrix}, \quad \mathbf{D}^v = \frac{\partial \mathbf{V}}{\partial \mathbf{W}} \mathbf{D}^w \frac{\partial \mathbf{W}}{\partial \mathbf{V}} = \begin{bmatrix} 0 & 0 & 0 \\[1ex] 0 & \dfrac{4}{3}\nu & 0 \\[1ex] 0 & 0 & \dfrac{\nu\gamma}{Pr} \end{bmatrix}. \quad (4.16.23)$$

It is clear from these matrices that the system is indeed symmetric. Note that in transforming the viscous term we have ignored again non-linear terms involving products of derivatives:

$$\begin{aligned}
\frac{\partial \mathbf{V}}{\partial \mathbf{W}} \mathbf{D}^w \mathbf{W}_{xx} &= \left(\frac{\partial \mathbf{V}}{\partial \mathbf{W}} \mathbf{D}^w \frac{\partial \mathbf{W}}{\partial \mathbf{V}}\right) \frac{\partial \mathbf{V}}{\partial \mathbf{W}} \mathbf{W}_{xx} \\
&= \mathbf{D}^v \left[\frac{\partial}{\partial x}\left(\frac{\partial \mathbf{V}}{\partial \mathbf{W}} \mathbf{W}_x\right) - \frac{\partial}{\partial x}\left(\frac{\partial \mathbf{V}}{\partial \mathbf{W}}\right)\mathbf{W}_x\right] \\
&\approx \mathbf{D}^v \mathbf{V}_{xx}.
\end{aligned} \quad (4.16.24)$$

Or equivalently, it can be thought of as the transformation matrix $\frac{\partial \mathbf{V}}{\partial \mathbf{W}}$ being frozen at a certain constant state (so that its derivative is zero).

4.16.2 2D Quasi-Linear Navier-Stokes Equations

Primitive Form

The primitive form in two dimensions is given by

$$\rho_t + u\rho_x + \rho u_x + v\rho_y + \rho v_y = 0, \quad (4.16.25)$$

$$u_t + uu_x + vu_y + \frac{1}{\rho} p_x = \frac{4\nu}{3} u_{xx} + \frac{\nu}{3} v_{yx} + \nu u_{yy}, \quad (4.16.26)$$

$$v_t + uv_x + vv_y + \frac{1}{\rho} p_y = \nu v_{xx} + \frac{\nu}{3} u_{xy} + \frac{4\nu}{3} v_{yy}, \quad (4.16.27)$$

$$p_t + up_x + vp_y + \gamma pu_x + \gamma pv_y = \frac{\nu\gamma}{Pr}(p_{xx} + p_{yy}) - \frac{\nu a^2}{Pr}(\rho_{xx} + \rho_{yy}). \quad (4.16.28)$$

This can be written in the form:

$$\mathbf{W}_t + \mathbf{A}^w \mathbf{W}_x + \mathbf{B}^w \mathbf{W}_y = \mathbf{D}^w \mathbf{W}_{xx} + \mathbf{G}^w \mathbf{W}_{xy} + \mathbf{E}^w \mathbf{W}_{yy}, \quad (4.16.29)$$

where $\mathbf{W} = [\rho, u, v, p]^t$,

$$\mathbf{A}^w = \begin{bmatrix} u & \rho & 0 & 0 \\ 0 & u & 0 & 1/\rho \\ 0 & 0 & u & 0 \\ 0 & \gamma p & 0 & u \end{bmatrix}, \quad \mathbf{B}^w = \begin{bmatrix} v & 0 & \rho & 0 \\ 0 & v & 0 & 0 \\ 0 & 0 & v & 1/\rho \\ 0 & 0 & \gamma p & v \end{bmatrix}, \quad (4.16.30)$$

$$\mathbf{D}^w = \begin{bmatrix} 0 & 0 & 0 & 0 \\ 0 & 4\nu/3 & 0 & 0 \\ 0 & 0 & \nu & 0 \\ -a^2\nu/Pr & 0 & 0 & \nu\gamma/Pr \end{bmatrix}, \quad \mathbf{G}^w = \begin{bmatrix} 0 & 0 & 0 & 0 \\ 0 & 0 & \nu/3 & 0 \\ 0 & \nu/3 & 0 & 0 \\ 0 & 0 & 0 & 0 \end{bmatrix}, \quad (4.16.31)$$

$$\mathbf{E}^w = \begin{bmatrix} 0 & 0 & 0 & 0 \\ 0 & \nu & 0 & 0 \\ 0 & 0 & 4\nu/3 & 0 \\ -a^2\nu/Pr & 0 & 0 & \nu\gamma/Pr \end{bmatrix}. \quad (4.16.32)$$

Symmetric Form

The symmetrizing variables,

$$dV = \left[\frac{a\,d\rho}{\rho\sqrt{\gamma}},\ du,\ dv,\ \frac{a\,dT}{T\sqrt{\gamma(\gamma-1)}}\right]^t, \tag{4.16.33}$$

and the primitive variables $\mathbf{W} = [\rho, u, v, p]^t$ are transformed to each other by the following matrices:

$$\frac{\partial \mathbf{V}}{\partial \mathbf{W}} = \begin{bmatrix} \dfrac{a}{\rho\sqrt{\gamma}} & 0 & 0 & 0 \\[2mm] 0 & 1 & 0 & 0 \\[2mm] 0 & 0 & 1 & 0 \\[2mm] -\dfrac{a}{\rho\sqrt{\gamma(\gamma-1)}} & 0 & 0 & \dfrac{a}{p\sqrt{\gamma(\gamma-1)}} \end{bmatrix}, \tag{4.16.34}$$

$$\frac{\partial \mathbf{W}}{\partial \mathbf{V}} = \begin{bmatrix} \dfrac{\rho\sqrt{\gamma}}{a} & 0 & 0 & 0 \\[2mm] 0 & 1 & 0 & 0 \\[2mm] 0 & 0 & 1 & 0 \\[2mm] \dfrac{a\rho}{\sqrt{\gamma}} & 0 & 0 & \dfrac{a\rho\sqrt{\gamma-1}}{\sqrt{\gamma}} \end{bmatrix}. \tag{4.16.35}$$

The primitive form is now transformed, with $\frac{\partial \mathbf{V}}{\partial \mathbf{W}}$ multiplied from the left, into the symmetric form:

$$\mathbf{V}_t + \mathbf{A}^v \mathbf{V}_x + \mathbf{B}^v \mathbf{V}_y = \mathbf{D}^v \mathbf{V}_{xx} + \mathbf{G}^v \mathbf{V}_{xy} + \mathbf{E}^v \mathbf{V}_{yy}, \tag{4.16.36}$$

where

$$\mathbf{A}^v = \begin{bmatrix} u & \dfrac{a}{\sqrt{\gamma}} & 0 & 0 \\[2mm] \dfrac{a}{\sqrt{\gamma}} & u & 0 & a\sqrt{\dfrac{\gamma-1}{\gamma}} \\[2mm] 0 & 0 & u & 0 \\[2mm] 0 & a\sqrt{\dfrac{\gamma-1}{\gamma}} & 0 & u \end{bmatrix}, \quad \mathbf{B}^v = \begin{bmatrix} v & 0 & \dfrac{a}{\sqrt{\gamma}} & 0 \\[2mm] 0 & v & 0 & 0 \\[2mm] \dfrac{a}{\sqrt{\gamma}} & 0 & v & a\sqrt{\dfrac{\gamma-1}{\gamma}} \\[2mm] 0 & 0 & a\sqrt{\dfrac{\gamma-1}{\gamma}} & v \end{bmatrix}, \tag{4.16.37}$$

$$\mathbf{D}^v = \begin{bmatrix} 0 & 0 & 0 & 0 \\[2mm] 0 & \dfrac{4}{3}\nu & 0 & 0 \\[2mm] 0 & 0 & \nu & 0 \\[2mm] 0 & 0 & 0 & \dfrac{\nu\gamma}{Pr} \end{bmatrix}, \quad \mathbf{G}^v = \begin{bmatrix} 0 & 0 & 0 & 0 \\[2mm] 0 & 0 & \dfrac{\nu}{3} & 0 \\[2mm] 0 & \dfrac{\nu}{3} & 0 & 0 \\[2mm] 0 & 0 & 0 & 0 \end{bmatrix}, \tag{4.16.38}$$

$$\mathbf{E}^v = \begin{bmatrix} 0 & 0 & 0 & 0 \\[2mm] 0 & \nu & 0 & 0 \\[2mm] 0 & 0 & \dfrac{4}{3}\nu & 0 \\[2mm] 0 & 0 & 0 & \dfrac{\nu\gamma}{Pr} \end{bmatrix}. \tag{4.16.39}$$

4.16.3 3D Quasi-Linear Navier-Stokes Equations

Primitive Form

The primitive form in three dimensions is given by

$$\rho_t + u\rho_x + \rho u_x + v\rho_y + \rho v_y + w\rho_z + \rho w_z = 0, \tag{4.16.40}$$

$$u_t + uu_x + vu_y + wu_z + \frac{1}{\rho}p_x = \frac{4\nu}{3}u_{xx} + \frac{\nu}{3}v_{yx} + \frac{\nu}{3}w_{zx} + \nu u_{yy} + \nu u_{zz}, \tag{4.16.41}$$

$$v_t + uv_x + vv_y + wv_z + \frac{1}{\rho}p_y = \nu v_{xx} + \frac{\nu}{3}u_{xy} + \frac{\nu}{3}w_{zy} + \frac{4\nu}{3}v_{yy} + \nu v_{zz}, \tag{4.16.42}$$

$$w_t + uw_x + vw_y + ww_z + \frac{1}{\rho}p_z = \nu w_{xx} + \frac{\nu}{3}u_{xz} + \frac{\nu}{3}v_{yz} + \nu w_{yy} + \frac{4\nu}{3}w_{zz}, \tag{4.16.43}$$

$$p_t + up_x + vp_y + wp_z + \gamma pu_x + \gamma pv_y + \gamma pw_z = \frac{\nu\gamma}{Pr}(p_{xx} + p_{yy} + p_{zz}) - \frac{\nu a^2}{Pr}(\rho_{xx} + \rho_{yy} + \rho_{zz}). \tag{4.16.44}$$

This can be written in the form:

$$\mathbf{W}_t + \mathbf{A}^w\mathbf{W}_x + \mathbf{B}^w\mathbf{W}_y + \mathbf{C}^w\mathbf{W}_z = \mathbf{D}^w\mathbf{W}_{xx} + \mathbf{E}^w\mathbf{W}_{yy} + \mathbf{F}^w\mathbf{W}_{zz} + \mathbf{G}^w\mathbf{W}_{xy} + \mathbf{H}^w\mathbf{W}_{yz} + \mathbf{I}^w\mathbf{W}_{zx}, \tag{4.16.45}$$

where $\mathbf{W} = [\rho, u, v, w, p]^t$,

$$\mathbf{A}^w = \begin{bmatrix} u & \rho & 0 & 0 & 0 \\ 0 & u & 0 & 0 & 1/\rho \\ 0 & 0 & u & 0 & 0 \\ 0 & 0 & 0 & u & 0 \\ 0 & \gamma p & 0 & 0 & u \end{bmatrix}, \quad \mathbf{B}^w = \begin{bmatrix} v & 0 & \rho & 0 & 0 \\ 0 & v & 0 & 0 & 0 \\ 0 & 0 & v & 0 & 1/\rho \\ 0 & 0 & 0 & v & 0 \\ 0 & 0 & \gamma p & 0 & v \end{bmatrix}, \quad \mathbf{C}^w = \begin{bmatrix} w & 0 & 0 & \rho & 0 \\ 0 & w & 0 & 0 & 0 \\ 0 & 0 & w & 0 & 0 \\ 0 & 0 & 0 & w & 1/\rho \\ 0 & 0 & 0 & \gamma p & w \end{bmatrix}, \tag{4.16.46}$$

$$\mathbf{D}^w = \begin{bmatrix} 0 & 0 & 0 & 0 & 0 \\ 0 & 4\nu/3 & 0 & 0 & 0 \\ 0 & 0 & \nu & 0 & 0 \\ 0 & 0 & 0 & \nu & 0 \\ -a^2\nu/Pr & 0 & 0 & 0 & \nu\gamma/Pr \end{bmatrix}, \quad \mathbf{E}^w = \begin{bmatrix} 0 & 0 & 0 & 0 & 0 \\ 0 & \nu & 0 & 0 & 0 \\ 0 & 0 & 4\nu/3 & 0 & 0 \\ 0 & 0 & 0 & \nu & 0 \\ -a^2\nu/Pr & 0 & 0 & 0 & \nu\gamma/Pr \end{bmatrix}, \tag{4.16.47}$$

$$\mathbf{F}^w = \begin{bmatrix} 0 & 0 & 0 & 0 & 0 \\ 0 & \nu & 0 & 0 & 0 \\ 0 & 0 & \nu & 0 & 0 \\ 0 & 0 & 0 & 4\nu/3 & 0 \\ -a^2\nu/Pr & 0 & 0 & 0 & \nu\gamma/Pr \end{bmatrix}, \quad \mathbf{G}^w = \begin{bmatrix} 0 & 0 & 0 & 0 & 0 \\ 0 & 0 & \nu/3 & 0 & 0 \\ 0 & \nu/3 & 0 & 0 & 0 \\ 0 & 0 & 0 & 0 & 0 \\ 0 & 0 & 0 & 0 & 0 \end{bmatrix}, \tag{4.16.48}$$

$$\mathbf{H}^w = \begin{bmatrix} 0 & 0 & 0 & 0 & 0 \\ 0 & 0 & 0 & 0 & 0 \\ 0 & 0 & 0 & \nu/3 & 0 \\ 0 & 0 & \nu/3 & 0 & 0 \\ 0 & 0 & 0 & 0 & 0 \end{bmatrix}, \quad \mathbf{I}^w = \begin{bmatrix} 0 & 0 & 0 & 0 & 0 \\ 0 & 0 & 0 & \nu/3 & 0 \\ 0 & 0 & 0 & 0 & 0 \\ 0 & \nu/3 & 0 & 0 & 0 \\ 0 & 0 & 0 & 0 & 0 \end{bmatrix}. \tag{4.16.49}$$

Symmetric Form

The symmetrizing variables,

$$d\mathbf{V} = \left[\frac{a\,d\rho}{\rho\sqrt{\gamma}}, \, du, \, dv, \, dw, \, \frac{a\,dT}{T\sqrt{\gamma(\gamma-1)}} \right]^t, \tag{4.16.50}$$

and the primitive variables $\mathbf{W} = [\rho, u, v, w, p]^t$ are transformed to each other by the following matrices:

$$\frac{\partial \mathbf{V}}{\partial \mathbf{W}} = \begin{bmatrix} \dfrac{a}{\rho\sqrt{\gamma}} & 0 & 0 & 0 & 0 \\[2mm] 0 & 1 & 0 & 0 & 0 \\[2mm] 0 & 0 & 1 & 0 & 0 \\[2mm] 0 & 0 & 0 & 1 & 0 \\[2mm] -\dfrac{a}{\rho\sqrt{\gamma(\gamma-1)}} & 0 & 0 & 0 & \dfrac{a}{p\sqrt{\gamma(\gamma-1)}} \end{bmatrix}, \tag{4.16.51}$$

$$\frac{\partial \mathbf{W}}{\partial \mathbf{V}} = \begin{bmatrix} \dfrac{\rho\sqrt{\gamma}}{a} & 0 & 0 & 0 & 0 \\[2mm] 0 & 1 & 0 & 0 & 0 \\[2mm] 0 & 0 & 1 & 0 & 0 \\[2mm] 0 & 0 & 0 & 1 & 0 \\[2mm] \dfrac{a\rho}{\sqrt{\gamma}} & 0 & 0 & 0 & \dfrac{a\rho\sqrt{\gamma-1}}{\sqrt{\gamma}} \end{bmatrix}. \tag{4.16.52}$$

The primitive form is now transformed, with $\frac{\partial \mathbf{V}}{\partial \mathbf{W}}$ multiplied from the left, into the symmetric form:

$$\mathbf{V}_t + \mathbf{A}^v\mathbf{V}_x + \mathbf{B}^v\mathbf{V}_y + \mathbf{C}^v\mathbf{V}_z = \mathbf{D}^v\mathbf{V}_{xx} + \mathbf{E}^v\mathbf{V}_{yy} + \mathbf{F}^v\mathbf{V}_{zz} + \mathbf{G}^v\mathbf{V}_{xy} + \mathbf{H}^v\mathbf{V}_{yz} + \mathbf{I}^v\mathbf{V}_{zx}, \tag{4.16.53}$$

where

$$\mathbf{A}^v = \begin{bmatrix} u & \dfrac{a}{\sqrt{\gamma}} & 0 & 0 & 0 \\[2mm] \dfrac{a}{\sqrt{\gamma}} & u & 0 & 0 & a\sqrt{\dfrac{\gamma-1}{\gamma}} \\[2mm] 0 & 0 & u & 0 & 0 \\[2mm] 0 & 0 & 0 & u & 0 \\[2mm] 0 & a\sqrt{\dfrac{\gamma-1}{\gamma}} & 0 & 0 & u \end{bmatrix}, \quad \mathbf{B}^v = \begin{bmatrix} v & 0 & \dfrac{a}{\sqrt{\gamma}} & 0 & 0 \\[2mm] 0 & v & 0 & 0 & 0 \\[2mm] \dfrac{a}{\sqrt{\gamma}} & 0 & v & 0 & a\sqrt{\dfrac{\gamma-1}{\gamma}} \\[2mm] 0 & 0 & 0 & v & 0 \\[2mm] 0 & 0 & a\sqrt{\dfrac{\gamma-1}{\gamma}} & 0 & v \end{bmatrix}, \tag{4.16.54}$$

$$\mathbf{C}^v = \begin{bmatrix} w & 0 & 0 & \dfrac{a}{\sqrt{\gamma}} & 0 \\[2mm] 0 & w & 0 & 0 & 0 \\[2mm] 0 & 0 & w & 0 & 0 \\[2mm] \dfrac{a}{\sqrt{\gamma}} & 0 & 0 & w & a\sqrt{\dfrac{\gamma-1}{\gamma}} \\[2mm] 0 & 0 & 0 & a\sqrt{\dfrac{\gamma-1}{\gamma}} & w \end{bmatrix}, \quad \mathbf{D}^v = \begin{bmatrix} 0 & 0 & 0 & 0 & 0 \\[2mm] 0 & \dfrac{4}{3}\nu & 0 & 0 & 0 \\[2mm] 0 & 0 & \nu & 0 & 0 \\[2mm] 0 & 0 & 0 & \nu & 0 \\[2mm] 0 & 0 & 0 & 0 & \dfrac{\nu\gamma}{Pr} \end{bmatrix}, \tag{4.16.55}$$

$$\mathbf{E}^v = \begin{bmatrix} 0 & 0 & 0 & 0 & 0 \\[2mm] 0 & \nu & 0 & 0 & 0 \\[2mm] 0 & 0 & \dfrac{4}{3}\nu & 0 & 0 \\[2mm] 0 & 0 & 0 & \nu & 0 \\[2mm] 0 & 0 & 0 & 0 & \dfrac{\nu\gamma}{Pr} \end{bmatrix}, \quad \mathbf{F}^v = \begin{bmatrix} 0 & 0 & 0 & 0 & 0 \\[2mm] 0 & \nu & 0 & 0 & 0 \\[2mm] 0 & 0 & \nu & 0 & 0 \\[2mm] 0 & 0 & 0 & \dfrac{4}{3}\nu & 0 \\[2mm] 0 & 0 & 0 & 0 & \dfrac{\nu\gamma}{Pr} \end{bmatrix}, \tag{4.16.56}$$

$$\mathbf{G}^v = \begin{bmatrix} 0 & 0 & 0 & 0 & 0 \\ 0 & 0 & \dfrac{\nu}{3} & 0 & 0 \\ 0 & \dfrac{\nu}{3} & 0 & 0 & 0 \\ 0 & 0 & 0 & 0 & 0 \\ 0 & 0 & 0 & 0 & 0 \end{bmatrix}, \quad \mathbf{H}^v = \begin{bmatrix} 0 & 0 & 0 & 0 & 0 \\ 0 & 0 & 0 & 0 & 0 \\ 0 & 0 & 0 & \dfrac{\nu}{3} & 0 \\ 0 & 0 & \dfrac{\nu}{3} & 0 & 0 \\ 0 & 0 & 0 & 0 & 0 \end{bmatrix}, \tag{4.16.57}$$

$$\mathbf{I}^v = \begin{bmatrix} 0 & 0 & 0 & 0 & 0 \\ 0 & 0 & 0 & \dfrac{\nu}{3} & 0 \\ 0 & 0 & 0 & 0 & 0 \\ 0 & \dfrac{\nu}{3} & 0 & 0 & 0 \\ 0 & 0 & 0 & 0 & 0 \end{bmatrix}. \tag{4.16.58}$$

4.17 Incompressible Navier-Stokes Equations

As I said before, I like incompressible fluids: constant density or small Mach number. So, of course, I like the incompressible Navier-Stokes equations, which are obtained from the full Navier-Stokes equations by assuming $\rho = $ constant.

Conservative Form of the Incompressible Navier-Stokes Equations

$$\operatorname{div} \rho \mathbf{v} = 0, \tag{4.17.1}$$

$$\partial_t(\rho \mathbf{v}) + \operatorname{div}(\rho \mathbf{v} \otimes \mathbf{v}) + \operatorname{grad} p - \operatorname{div} \boldsymbol{\tau} = 0, \tag{4.17.2}$$

$$\rho c_v \frac{DT}{Dt} - \operatorname{div}(k \operatorname{grad} T) - \Phi = 0. \tag{4.17.3}$$

where

$$\boldsymbol{\tau} = \mu \left[\operatorname{grad} \mathbf{v} + (\operatorname{grad} \mathbf{v})^t \right], \tag{4.17.4}$$

$$\Phi = \boldsymbol{\tau} : \operatorname{grad} \mathbf{v} - \mu \left[\operatorname{grad} \mathbf{v} + (\operatorname{grad} \mathbf{v})^t \right] : \operatorname{grad} \mathbf{v}. \tag{4.17.5}$$

Primitive Form of the Incompressible Navier-Stokes Equations

$$\operatorname{div} \mathbf{v} = 0, \tag{4.17.6}$$

$$\rho \frac{D\mathbf{v}}{Dt} + \operatorname{grad} p - \operatorname{div} \boldsymbol{\tau} = 0, \tag{4.17.7}$$

$$\rho c_v \frac{DT}{Dt} - \operatorname{div}(k \operatorname{grad} T) - \Phi = 0. \tag{4.17.8}$$

If the viscosity coefficient is constant, the viscous term can be written as the Laplacian,

$$\begin{aligned} \operatorname{div}\boldsymbol{\tau} &= \operatorname{div}\left[\mu \left\{ \operatorname{grad} \mathbf{v} + (\operatorname{grad} \mathbf{v})^t \right\} \right] \\ &= \mu \operatorname{div} \operatorname{grad} \mathbf{v} + \mu \operatorname{div}(\operatorname{grad} \mathbf{v})^t \\ &= \mu \operatorname{div} \operatorname{grad} \mathbf{v} + \mu \operatorname{grad}(\operatorname{div} \mathbf{v}) \\ &= \mu \operatorname{div} \operatorname{grad} \mathbf{v}, \end{aligned} \tag{4.17.9}$$

which is nice. Then, assuming also that the heat conductivity is constant, we have

$$\operatorname{div} \mathbf{v} = 0, \tag{4.17.10}$$

$$\frac{D\mathbf{v}}{Dt} + \frac{1}{\rho} \operatorname{grad} p - \nu \operatorname{div} \operatorname{grad} \mathbf{v} = 0, \tag{4.17.11}$$

$$\rho c_v \frac{DT}{Dt} - k \operatorname{div} \operatorname{grad} T - \Phi = 0, \tag{4.17.12}$$

where

$$\nu = \frac{\mu}{\rho}. \tag{4.17.13}$$

I like the fact that the energy equation (the temperature equation) is completely independent from other equations. So, we can solve the continuity and momentum equations first (to determine the velocity field), and then, if necessary, solve the energy equation separately for a given velocity field. Also, I like the idea of pseudo-compressibility. See Section 4.21.

4.18   2D Incompressible Navier-Stokes Equations

I like the 2D incompressible Navier-Stokes equations in any of the following forms.

Conservative Form

$$\frac{\partial(\rho u)}{\partial x} + \frac{\partial(\rho v)}{\partial y} = 0, \tag{4.18.1}$$

$$
\begin{aligned}
\frac{\partial(\rho u)}{\partial t} + \frac{\partial(\rho u^2)}{\partial x} + \frac{\partial(\rho uv)}{\partial y} &= -\frac{\partial p}{\partial x} + \frac{\partial}{\partial x}\left(\mu \frac{\partial u}{\partial x}\right) + \frac{\partial}{\partial y}\left(\mu \frac{\partial u}{\partial y}\right) \\
&\quad + \frac{\partial}{\partial x}\left(\mu \frac{\partial u}{\partial x}\right) + \frac{\partial}{\partial y}\left(\mu \frac{\partial v}{\partial x}\right),
\end{aligned}
\tag{4.18.2}
$$

$$
\begin{aligned}
\frac{\partial(\rho v)}{\partial t} + \frac{\partial(\rho uv)}{\partial x} + \frac{\partial(\rho v^2)}{\partial y} &= -\frac{\partial p}{\partial y} + \frac{\partial}{\partial x}\left(\mu \frac{\partial v}{\partial x}\right) + \frac{\partial}{\partial y}\left(\mu \frac{\partial v}{\partial y}\right) \\
&\quad + \frac{\partial}{\partial x}\left(\mu \frac{\partial u}{\partial y}\right) + \frac{\partial}{\partial y}\left(\mu \frac{\partial v}{\partial y}\right),
\end{aligned}
\tag{4.18.3}
$$

$$\rho c_v \left(\frac{\partial T}{\partial t} + u\frac{\partial T}{\partial x} + v\frac{\partial T}{\partial y}\right) = \frac{\partial}{\partial x}\left(k\frac{\partial T}{\partial x}\right) + \frac{\partial}{\partial y}\left(k\frac{\partial T}{\partial y}\right) + \Phi, \tag{4.18.4}$$

where

$$\Phi = \mu \left[2\left(\frac{\partial u}{\partial x}\right)^2 + 2\left(\frac{\partial v}{\partial y}\right)^2 + \left(\frac{\partial v}{\partial x} + \frac{\partial u}{\partial y}\right)^2\right]. \tag{4.18.5}$$

Primitive Form

$$\frac{\partial u}{\partial x} + \frac{\partial v}{\partial y} = 0, \tag{4.18.6}$$

$$
\begin{aligned}
\rho\frac{\partial u}{\partial t} + \rho u\frac{\partial u}{\partial x} + \rho v\frac{\partial u}{\partial y} &= -\frac{\partial p}{\partial x} + \frac{\partial}{\partial x}\left(\mu \frac{\partial u}{\partial x}\right) + \frac{\partial}{\partial y}\left(\mu \frac{\partial u}{\partial y}\right) \\
&\quad + \frac{\partial}{\partial x}\left(\mu \frac{\partial u}{\partial x}\right) + \frac{\partial}{\partial y}\left(\mu \frac{\partial v}{\partial x}\right),
\end{aligned}
\tag{4.18.7}
$$

$$\rho\frac{\partial v}{\partial t} + \rho u\frac{\partial v}{\partial x} + \rho v\frac{\partial v}{\partial y} = -\frac{\partial p}{\partial y} + \frac{\partial}{\partial x}\left(\mu\frac{\partial v}{\partial x}\right) + \frac{\partial}{\partial y}\left(\mu\frac{\partial v}{\partial y}\right)$$
$$+\frac{\partial}{\partial x}\left(\mu\frac{\partial u}{\partial y}\right) + \frac{\partial}{\partial y}\left(\mu\frac{\partial v}{\partial y}\right), \tag{4.18.8}$$

$$\rho c_v\left(\frac{\partial T}{\partial t} + u\frac{\partial T}{\partial x} + v\frac{\partial T}{\partial y}\right) = \frac{\partial}{\partial x}\left(k\frac{\partial T}{\partial x}\right) + \frac{\partial}{\partial y}\left(k\frac{\partial T}{\partial y}\right) + \Phi. \tag{4.18.9}$$

Case of Constant Viscosity and Heat Conductivity

$$\frac{\partial u}{\partial x} + \frac{\partial v}{\partial y} = 0, \tag{4.18.10}$$

$$\frac{\partial u}{\partial t} + u\frac{\partial u}{\partial x} + v\frac{\partial u}{\partial y} = -\frac{1}{\rho}\frac{\partial p}{\partial x} + \nu\left(\frac{\partial^2 u}{\partial x^2} + \frac{\partial^2 u}{\partial y^2}\right), \tag{4.18.11}$$

$$\frac{\partial v}{\partial t} + u\frac{\partial v}{\partial x} + v\frac{\partial v}{\partial y} = -\frac{1}{\rho}\frac{\partial p}{\partial y} + \nu\left(\frac{\partial^2 v}{\partial x^2} + \frac{\partial^2 v}{\partial y^2}\right), \tag{4.18.12}$$

$$\rho c_v\left(\frac{\partial T}{\partial t} + u\frac{\partial T}{\partial x} + v\frac{\partial T}{\partial y}\right) = k\left(\frac{\partial^2 T}{\partial x^2} + \frac{\partial^2 T}{\partial y^2}\right) + \Phi. \tag{4.18.13}$$

4.19 3D Incompressible Navier-Stokes Equations

Of course, I like the 3D incompressible Navier-Stokes equations also in any of the following forms.

Conservative Form

$$\frac{\partial(\rho u)}{\partial x} + \frac{\partial(\rho v)}{\partial y} + \frac{\partial(\rho w)}{\partial z} = 0, \tag{4.19.1}$$

$$\frac{\partial(\rho u)}{\partial t} + \frac{\partial(\rho u^2)}{\partial x} + \frac{\partial(\rho uv)}{\partial y} + \frac{\partial(\rho uw)}{\partial z} = -\frac{\partial p}{\partial x} + \frac{\partial}{\partial x}\left(\mu\frac{\partial u}{\partial x}\right) + \frac{\partial}{\partial y}\left(\mu\frac{\partial u}{\partial y}\right) + \frac{\partial}{\partial z}\left(\mu\frac{\partial u}{\partial z}\right)$$
$$+\frac{\partial}{\partial x}\left(\mu\frac{\partial u}{\partial x}\right) + \frac{\partial}{\partial y}\left(\mu\frac{\partial v}{\partial x}\right) + \frac{\partial}{\partial z}\left(\mu\frac{\partial w}{\partial x}\right),$$
$$\tag{4.19.2}$$

$$\frac{\partial(\rho v)}{\partial t} + \frac{\partial(\rho uv)}{\partial x} + \frac{\partial(\rho v^2)}{\partial y} + \frac{\partial(\rho vw)}{\partial z} = -\frac{\partial p}{\partial y} + \frac{\partial}{\partial x}\left(\mu\frac{\partial v}{\partial x}\right) + \frac{\partial}{\partial y}\left(\mu\frac{\partial v}{\partial y}\right) + \frac{\partial}{\partial z}\left(\mu\frac{\partial v}{\partial z}\right)$$
$$+\frac{\partial}{\partial x}\left(\mu\frac{\partial u}{\partial y}\right) + \frac{\partial}{\partial y}\left(\mu\frac{\partial v}{\partial y}\right) + \frac{\partial}{\partial z}\left(\mu\frac{\partial w}{\partial y}\right),$$
$$\tag{4.19.3}$$

$$\frac{\partial(\rho w)}{\partial t} + \frac{\partial(\rho uw)}{\partial x} + \frac{\partial(\rho vw)}{\partial y} + \frac{\partial(\rho w^2)}{\partial z} = -\frac{\partial p}{\partial z} + \frac{\partial}{\partial x}\left(\mu\frac{\partial w}{\partial x}\right) + \frac{\partial}{\partial y}\left(\mu\frac{\partial w}{\partial y}\right) + \frac{\partial}{\partial z}\left(\mu\frac{\partial w}{\partial z}\right)$$
$$+\frac{\partial}{\partial x}\left(\mu\frac{\partial u}{\partial z}\right) + \frac{\partial}{\partial y}\left(\mu\frac{\partial v}{\partial z}\right) + \frac{\partial}{\partial z}\left(\mu\frac{\partial w}{\partial z}\right),$$
$$\tag{4.19.4}$$

$$\rho c_v\left(\frac{\partial T}{\partial t} + u\frac{\partial T}{\partial x} + v\frac{\partial T}{\partial y} + w\frac{\partial T}{\partial z}\right) = \frac{\partial}{\partial x}\left(k\frac{\partial T}{\partial x}\right) + \frac{\partial}{\partial y}\left(k\frac{\partial T}{\partial y}\right) + \frac{\partial}{\partial z}\left(k\frac{\partial T}{\partial z}\right) + \Phi, \tag{4.19.5}$$

where

$$
\Phi = \mu \left[2 \left(\frac{\partial u}{\partial x} \right)^2 + 2 \left(\frac{\partial v}{\partial y} \right)^2 + 2 \left(\frac{\partial w}{\partial z} \right)^2 \right.
$$
$$
\left. + \left(\frac{\partial v}{\partial x} + \frac{\partial u}{\partial y} \right)^2 + \left(\frac{\partial w}{\partial y} + \frac{\partial v}{\partial z} \right)^2 + \left(\frac{\partial u}{\partial z} + \frac{\partial w}{\partial x} \right)^2 \right]. \tag{4.19.6}
$$

Primitive Form

$$
\frac{\partial u}{\partial x} + \frac{\partial v}{\partial y} + \frac{\partial w}{\partial z} = 0, \tag{4.19.7}
$$

$$
\rho \frac{\partial u}{\partial t} + \rho u \frac{\partial u}{\partial x} + \rho v \frac{\partial u}{\partial y} + \rho w \frac{\partial u}{\partial z} = -\frac{\partial p}{\partial x} + \frac{\partial}{\partial x}\left(\mu \frac{\partial u}{\partial x}\right) + \frac{\partial}{\partial y}\left(\mu \frac{\partial u}{\partial y}\right) + \frac{\partial}{\partial z}\left(\mu \frac{\partial u}{\partial z}\right)
$$
$$
+ \frac{\partial}{\partial x}\left(\mu \frac{\partial u}{\partial x}\right) + \frac{\partial}{\partial y}\left(\mu \frac{\partial v}{\partial x}\right) + \frac{\partial}{\partial z}\left(\mu \frac{\partial w}{\partial x}\right), \tag{4.19.8}
$$

$$
\rho \frac{\partial v}{\partial t} + \rho u \frac{\partial v}{\partial x} + \rho v \frac{\partial v}{\partial y} + \rho w \frac{\partial v}{\partial z} = -\frac{\partial p}{\partial y} + \frac{\partial}{\partial x}\left(\mu \frac{\partial v}{\partial x}\right) + \frac{\partial}{\partial y}\left(\mu \frac{\partial v}{\partial y}\right) + \frac{\partial}{\partial z}\left(\mu \frac{\partial v}{\partial z}\right)
$$
$$
+ \frac{\partial}{\partial x}\left(\mu \frac{\partial u}{\partial y}\right) + \frac{\partial}{\partial y}\left(\mu \frac{\partial v}{\partial y}\right) + \frac{\partial}{\partial z}\left(\mu \frac{\partial w}{\partial y}\right), \tag{4.19.9}
$$

$$
\rho \frac{\partial w}{\partial t} + \rho u \frac{\partial w}{\partial x} + \rho v \frac{\partial w}{\partial y} + \rho w \frac{\partial w}{\partial z} = -\frac{\partial p}{\partial z} + \frac{\partial}{\partial x}\left(\mu \frac{\partial w}{\partial x}\right) + \frac{\partial}{\partial y}\left(\mu \frac{\partial w}{\partial y}\right) + \frac{\partial}{\partial z}\left(\mu \frac{\partial w}{\partial z}\right)
$$
$$
+ \frac{\partial}{\partial x}\left(\mu \frac{\partial u}{\partial z}\right) + \frac{\partial}{\partial y}\left(\mu \frac{\partial v}{\partial z}\right) + \frac{\partial}{\partial z}\left(\mu \frac{\partial w}{\partial z}\right), \tag{4.19.10}
$$

$$
\rho c_v \left(\frac{\partial T}{\partial t} + u \frac{\partial T}{\partial x} + v \frac{\partial T}{\partial y} + w \frac{\partial T}{\partial z} \right) = \frac{\partial}{\partial x}\left(k \frac{\partial T}{\partial x}\right) + \frac{\partial}{\partial y}\left(k \frac{\partial T}{\partial y}\right) + \frac{\partial}{\partial z}\left(k \frac{\partial T}{\partial z}\right) + \Phi. \tag{4.19.11}
$$

Case of Constant Viscosity and Heat Conductivity

$$
\frac{\partial u}{\partial x} + \frac{\partial v}{\partial y} + \frac{\partial w}{\partial z} = 0, \tag{4.19.12}
$$

$$
\frac{\partial u}{\partial t} + u \frac{\partial u}{\partial x} + v \frac{\partial u}{\partial y} + w \frac{\partial u}{\partial z} = -\frac{1}{\rho}\frac{\partial p}{\partial x} + \nu \left(\frac{\partial^2 u}{\partial x^2} + \frac{\partial^2 u}{\partial y^2} + \frac{\partial^2 u}{\partial z^2} \right), \tag{4.19.13}
$$

$$
\frac{\partial v}{\partial t} + u \frac{\partial v}{\partial x} + v \frac{\partial v}{\partial y} + w \frac{\partial v}{\partial z} = -\frac{1}{\rho}\frac{\partial p}{\partial y} + \nu \left(\frac{\partial^2 v}{\partial x^2} + \frac{\partial^2 v}{\partial y^2} + \frac{\partial^2 v}{\partial z^2} \right), \tag{4.19.14}
$$

$$
\frac{\partial w}{\partial t} + u \frac{\partial w}{\partial x} + v \frac{\partial w}{\partial y} + w \frac{\partial w}{\partial z} = -\frac{1}{\rho}\frac{\partial p}{\partial z} + \nu \left(\frac{\partial^2 w}{\partial x^2} + \frac{\partial^2 w}{\partial y^2} + \frac{\partial^2 w}{\partial z^2} \right), \tag{4.19.15}
$$

$$
\rho c_v \left(\frac{\partial T}{\partial t} + u \frac{\partial T}{\partial x} + v \frac{\partial T}{\partial y} + w \frac{\partial T}{\partial z} \right) = k \left(\frac{\partial^2 T}{\partial x^2} + \frac{\partial^2 T}{\partial y^2} + \frac{\partial^2 T}{\partial z^2} \right) + \Phi. \tag{4.19.16}
$$

4.20 Axisymmetric Incompressible Navier-Stokes Equations

I like the axisymmetric incompressible Navier-Stokes equations because they are much simpler than the compressible version in Section 4.11. I really like the assumptions that ρ and μ are constant, and thus $\nu = \mu/\rho$ is also constant. Then, we write the continuity equation (4.17.1) and the conservative momentum equations (4.17.2) in cylindrical coordinates in Figure 1.5.1, where the velocity is denoted by $\mathbf{v} = (u_r, u_\theta, u_z)$:

$$\frac{1}{r}\frac{\partial(ru_r)}{\partial r} + \frac{\partial u_z}{\partial z} = 0, \tag{4.20.1}$$

$$\frac{\partial u_r}{\partial t} + \frac{1}{r}\frac{\partial(ru_r^2)}{\partial r} + \frac{\partial(u_r u_z)}{\partial z} = -\frac{\partial(p/\rho)}{\partial r} + \nu\left[\frac{1}{r}\frac{\partial}{\partial r}\left(r\frac{\partial u_r}{\partial r}\right) + \frac{\partial^2 u_r}{\partial z^2} - \frac{u_r}{r}\right], \tag{4.20.2}$$

$$\frac{\partial u_z}{\partial t} + \frac{1}{r}\frac{\partial(ru_z u_r)}{\partial r} + \frac{\partial(u_z^2)}{\partial z} = -\frac{\partial(p/\rho)}{\partial z} + \nu\left[\frac{1}{r}\frac{\partial}{\partial r}\left(r\frac{\partial u_z}{\partial r}\right) + \frac{\partial^2 u_z}{\partial z^2}\right], \tag{4.20.3}$$

where the viscous terms are obtained by using the Laplacian form (4.17.9). Note that $u_\theta = 0$ for axisymmetric flows and the momentum equation for u_θ vanishes and therefore it is ignored in the above. Note also that the viscous terms have been obtained by Equation (1.5.61), i.e., the Laplacian of a vector. The momentum equations in the primitive form can be obtained from the primitive form also by directly writing out $(\mathrm{grad}\,\mathbf{v})\mathbf{v}$ or by inserting the continuity equation into the conservative momentum equations:

$$\frac{\partial u_r}{\partial t} + u_r\frac{\partial u_r}{\partial r} + u_z\frac{\partial u_r}{\partial z} = -\frac{\partial(p/\rho)}{\partial r} + \nu\left[\frac{1}{r}\frac{\partial}{\partial r}\left(r\frac{\partial u_r}{\partial r}\right) + \frac{\partial^2 u_r}{\partial z^2} - \frac{u_r}{r}\right], \tag{4.20.4}$$

$$\frac{\partial u_z}{\partial t} + u_r\frac{\partial u_z}{\partial r} + u_z\frac{\partial u_z}{\partial z} = -\frac{\partial(p/\rho)}{\partial z} + \nu\left[\frac{1}{r}\frac{\partial}{\partial r}\left(r\frac{\partial u_z}{\partial r}\right) + \frac{\partial^2 u_z}{\partial z^2}\right]. \tag{4.20.5}$$

Now, as we have done in Section 4.11, we multiply all the equations by r, replace (r, z) by (y, x) and (u_r, u_z) by (v, u), and reverse the order of the momentum equations to obtain the axisymmetric incompressible Navier-Stokes equations in a form similar to the two-dimensional equations:

$$\frac{\partial(yu)}{\partial x} + \frac{\partial(yv)}{\partial y} = 0, \tag{4.20.6}$$

$$\frac{\partial(yu)}{\partial t} + \frac{\partial(yu^2)}{\partial x} + \frac{\partial(yuv)}{\partial y} = -\frac{\partial(yp/\rho)}{\partial x} + \nu\left[\frac{\partial}{\partial x}\left(y\frac{\partial u}{\partial x}\right) + \frac{\partial}{\partial y}\left(y\frac{\partial u}{\partial y}\right)\right], \tag{4.20.7}$$

$$\frac{\partial(yv)}{\partial t} + \frac{\partial(yvu)}{\partial x} + \frac{\partial(yv^2)}{\partial y} = -\frac{\partial(yp/\rho)}{\partial y} + \nu\left[\frac{\partial}{\partial x}\left(y\frac{\partial v}{\partial x}\right) + \frac{\partial}{\partial y}\left(y\frac{\partial v}{\partial y}\right)\right] + \frac{p}{\rho} - \nu\frac{v}{y}. \tag{4.20.8}$$

In the vector form, it can be written as

$$(y\mathbf{U})_t + (y\mathbf{F})_x + (y\mathbf{G})_y = (y\mathbf{F}^v)_x + (y\mathbf{G}^v)_y + \mathbf{S}, \tag{4.20.9}$$

where

$$\mathbf{U} = \begin{bmatrix} 0 \\ u \\ v \end{bmatrix}, \quad \mathbf{F} = \begin{bmatrix} u \\ u^2 + p/\rho \\ uv \end{bmatrix}, \quad \mathbf{G} = \begin{bmatrix} v \\ uv \\ v^2 + p/\rho \end{bmatrix}, \tag{4.20.10}$$

$$\mathbf{F}^v = \nu\begin{bmatrix} 0 \\ \partial u/\partial x \\ \partial v/\partial x \end{bmatrix}, \quad \mathbf{G}^v = \nu\begin{bmatrix} 0 \\ \partial u/\partial y \\ \partial v/\partial y \end{bmatrix}, \quad \mathbf{S} = \begin{bmatrix} 0 \\ 0 \\ p/\rho - \nu\dfrac{v}{y} \end{bmatrix}. \tag{4.20.11}$$

Of course, in the same way as described in Section 4.11, we can express the axisymmetric form and the two-dimensional form in a single form with a logical switch, $\overline{\omega} = 0$ for the two-dimensional form and $\overline{\omega} = 1$ for the axisymmetric form:

$$(\overline{\omega}_a\mathbf{U})_t + (\overline{\omega}_a\mathbf{F})_x + (\overline{\omega}_a\mathbf{G})_y = (\overline{\omega}_a\mathbf{F}^v)_x + (\overline{\omega}_a\mathbf{G}^v)_y + \overline{\omega}\,\mathbf{S}, \tag{4.20.12}$$

where

$$\overline{\omega}_a = (1 - \overline{\omega}) + \overline{\omega}\,y. \tag{4.20.13}$$

This is really nice.

One of the reasons that I like the axisymmetric incompressible Navier-Stokes equations very much is that the stream function ψ for axisymmetric flows can be easily found from the continuity equation (4.20.6):

$$u = \frac{1}{y}\frac{\partial\psi}{\partial y}, \quad v = -\frac{1}{y}\frac{\partial\psi}{\partial x}. \tag{4.20.14}$$

Clearly, then, the continuity equation is automatically satisfied. It's so simple. I like it.

4.21   Pseudo-Compressible Navier-Stokes Equations

Yes, I like the pseudo-compressible Navier-Stokes equations. Introduce the artificial density,

$$\rho^* = p/a^{*2}, \tag{4.21.1}$$

and add its time derivative to the continuity equation,

$$\frac{\partial\rho^*}{\partial t^*} + \operatorname{div}\rho\mathbf{v} = 0, \tag{4.21.2}$$

which can be written also as

$$\frac{\partial P}{\partial t^*} + \operatorname{div}(a^{*2}\mathbf{v}) = 0, \tag{4.21.3}$$

where $P = p/\rho$ which is the kinematic pressure. The incompressible Navier-Stokes equations with this modified continuity equation are the pseudo-compressible Navier-Stokes equations [21, 75, 97, 132]:

$$\frac{\partial P}{\partial t^*} + \operatorname{div}(a^{*2}\mathbf{v}) = 0, \tag{4.21.4}$$

$$\frac{\partial\mathbf{v}}{\partial t^*} + \operatorname{div}(\mathbf{v}\otimes\mathbf{v} + P\mathbf{I}) = \operatorname{div}\tau, \tag{4.21.5}$$

where

$$\tau = \nu\left[\operatorname{grad}\mathbf{v} + (\operatorname{grad}\mathbf{v})^t\right]. \tag{4.21.6}$$

The time derivative in the momentum equation is a real time derivative but not realistic when solved with the pseudo-compressible continuity equation (so t^* instead of t).

In Cartesian coordinates:

$$\mathbf{U}_{t^*} + \mathbf{F}_x + \mathbf{G}_y + \mathbf{H}_z = \mathbf{F}_x^v + \mathbf{G}_y^v + \mathbf{H}_z^v, \tag{4.21.7}$$

$$\mathbf{U} = \begin{bmatrix} P \\ u \\ v \\ w \end{bmatrix}, \quad \mathbf{F} = \begin{bmatrix} a^{*2}u \\ u^2 + P \\ uv \\ uw \end{bmatrix}, \quad \mathbf{G} = \begin{bmatrix} a^{*2}v \\ uv \\ v^2 + P \\ uw \end{bmatrix}, \quad \mathbf{H} = \begin{bmatrix} a^{*2}w \\ uv \\ vw \\ w^2 + P \end{bmatrix}, \tag{4.21.8}$$

$$\mathbf{F}^v = \nu\begin{bmatrix} 0 \\ 2u_x \\ u_y + v_x \\ u_z + w_x \end{bmatrix}, \quad \mathbf{G}^v = \nu\begin{bmatrix} 0 \\ v_x + u_y \\ 2v_y \\ v_z + w_y \end{bmatrix}, \quad \mathbf{H}^v = \nu\begin{bmatrix} 0 \\ w_x + u_z \\ w_y + v_z \\ 2w_z \end{bmatrix}. \tag{4.21.9}$$

If we assume that the viscosity is constant, then the viscous term can be simplified to the Laplacian as shown in Equation (4.17.9),

$$\frac{\partial\mathbf{v}}{\partial t^*} + \operatorname{div}(\mathbf{v}\otimes\mathbf{v} + P\mathbf{I}) = \nu\operatorname{div}\operatorname{grad}\mathbf{v}, \tag{4.21.10}$$

and therefore the viscous fluxes are given by

$$\mathbf{F}^v = \nu\begin{bmatrix} 0 \\ u_x \\ v_x \\ w_x \end{bmatrix}, \quad \mathbf{G}^v = \nu\begin{bmatrix} 0 \\ u_y \\ v_y \\ w_y \end{bmatrix}, \quad \mathbf{H}^v = \nu\begin{bmatrix} 0 \\ u_z \\ v_z \\ w_z \end{bmatrix}. \tag{4.21.11}$$

This is nice. I like it.

4.22 Vorticity Transport Equation and Stream Function

The vorticity transport equation (the curl of the momentum equation (4.17.11)) is given by

$$\omega_t + u\,\omega_x + v\,\omega_y = \nu\left(\omega_{xx} + \omega_{yy}\right), \tag{4.22.1}$$

where $\omega = v_x - u_y$ is the vorticity. To solve this for ω, the velocity components (u, v) must be given or solved simultaneously. A nice way to compute the velocity is to introduce the stream function ψ as

$$u = \psi_y, \quad v = -\psi_x, \tag{4.22.2}$$

so that the continuity equation is automatically satisfied, and that we see from $\omega = v_x - u_y$ that the stream function satisfies the following:

$$\psi_{xx} + \psi_{yy} = -\omega. \tag{4.22.3}$$

That is, we now have two equations (4.22.1) and (4.22.3) for two unknowns (ω, ψ). So, we only need to solve these two equations (4.22.1) and (4.22.3) to get ω and ψ, and determine the velocity field.

Once the velocity is obtained (via ψ), the pressure can be computed as a solution of the following Poisson equation (the divergence of the momentum equations):

$$\operatorname{div}\operatorname{grad} p = 2\rho\left(u_x v_y - u_y v_x\right) = 2\rho\left[\psi_{xx}\psi_{yy} - (\psi_{xy})^2\right]. \tag{4.22.4}$$

These are the fundamental equations for the famous vorticity-stream function approach (see Ref.[155] for details). It is nice that instead of solving three equations simultaneously (continuity and two momentum equations), we solve two equations simultaneously and then solve one equation independently afterwards. I surely like these equations.

4.23 Vorticity Transport Equation and Vector Potential

For a three-dimensional incompressible flow, we write the velocity as follows,

$$\mathbf{v} = \operatorname{grad}\phi + \operatorname{curl}\boldsymbol{\Psi}, \tag{4.23.1}$$

where ϕ is a scalar potential and $\boldsymbol{\Psi}$ is a vector potential. Note that the first term on the right is curl-free while the second term is divergence-free. This is called the Helmholtz decomposition. Substitute Equation (4.23.1) into the continuity equation to find the governing equation of ϕ,

$$\operatorname{div}\operatorname{grad}\phi = 0. \tag{4.23.2}$$

Substitute Equation (4.23.1) into the vorticity, $\boldsymbol{\omega} = \operatorname{curl}\mathbf{v}$, to get the governing equation of $\boldsymbol{\Psi}$,

$$\operatorname{div}\operatorname{grad}\boldsymbol{\Psi} = -\boldsymbol{\omega}. \tag{4.23.3}$$

Also, take the curl of the momentum equations (4.17.11) to get the governing equation of $\boldsymbol{\omega}$,

$$\frac{D\boldsymbol{\omega}}{Dt} = (\operatorname{grad}\mathbf{v})\boldsymbol{\omega} + \nu\operatorname{div}\operatorname{grad}\boldsymbol{\omega}. \tag{4.23.4}$$

Now, we can determine the velocity field by solving the above seven equations, (4.23.2), (4.23.3), and (4.23.4). Once the velocity field is determined, we can compute the pressure by solving the Poisson equation (the divergence of the momentum equations):

$$\operatorname{div}\operatorname{grad} p = -\operatorname{div}\left[\operatorname{div}\left(\rho\mathbf{v}\otimes\mathbf{v}\right)\right], \tag{4.23.5}$$

whose components are

$$\frac{\partial}{\partial x_i}\left(\frac{\partial p}{\partial x_i}\right) = -\frac{\partial}{\partial x_i}\left[\frac{\partial(\rho v_i v_j)}{\partial x_j}\right]. \tag{4.23.6}$$

We may use the continuity equation to write this as

$$\frac{\partial}{\partial x_i}\left(\frac{\partial p}{\partial x_i}\right) = -\rho\,\frac{\partial v_j}{\partial x_i}\frac{\partial v_i}{\partial x_j}. \tag{4.23.7}$$

In Cartesian coordinates, we can write the Poisson equation as

$$\begin{aligned}
p_{xx} + p_{yy} + p_{zz} = & \; -(\rho u^2)_{xx} - (\rho uv)_{xy} - (\rho uw)_{xz} \\
& -(\rho vu)_{yx} - (\rho v^2)_{yy} - (\rho vw)_{yz} \\
& -(\rho wu)_{zx} - (\rho wv)_{zy} - (\rho w^2)_{zz},
\end{aligned} \tag{4.23.8}$$

or we obtain directly from Equation (4.23.7)

$$p_{xx} + p_{yy} + p_{zz} = -\rho\left[(u_x)^2 + (v_y)^2 + (w_z)^2 + 2u_yv_x + 2u_zw_x + 2w_yv_z\right], \tag{4.23.9}$$

which we can rewrite as (by the continuity equation)

$$p_{xx} + p_{yy} + p_{zz} = 2\rho\left[u_xv_y + v_yw_z + w_zu_x - u_yv_x - u_zw_x - w_yv_z\right]. \tag{4.23.10}$$

These equations are fundamental equations for vortex methods [29].

4.24   Stream Functions

I like stream functions. They are useful in generating as well as visualizing flows. In 2D, we have only one stream function, and we get two velocity components from a scalar function. That is nice. However, in 3D, we need two stream functions. That's one more scalar function. But it is still nice compared with the vorticity, which has one component in 2D but three components in 3D.

4.24.1   Three-Dimensional Stream Functions

I like to set $\boldsymbol{\Psi} = \psi\,\mathrm{grad}\,\chi$ and $\phi = 0$ because then I find from Equation (4.23.1) that

$$\begin{aligned}
\mathbf{v} &= \mathrm{curl}\,\boldsymbol{\Psi} \\
&= \mathrm{curl}\,(\psi\,\mathrm{grad}\,\chi) \\
&= \psi\,\mathrm{curl}(\mathrm{grad}\chi) + \mathrm{grad}\psi \times \mathrm{grad}\chi \\
&= \mathrm{grad}\,\psi \times \mathrm{grad}\,\chi.
\end{aligned} \tag{4.24.1}$$

This means that $\mathbf{v}$ is tangent to the two surfaces defined by $\psi =$constant and $\chi =$ constant, and so that these two families of surfaces are composed of streamlines. Therefore, ψ and χ can be thought of as three-dimensional stream functions. Of course, the continuity equation, $\mathrm{div}\,\mathbf{v} = 0$, is automatically satisfied because $\mathrm{div}(\mathrm{curl}\,\boldsymbol{\Psi}) = 0$. This is very interesting. In fact, various well-known two-dimensional stream functions can be derived from the three-dimensional stream functions.

4.24.2   Cartesian Coordinates (2D)

Consider Cartesian coordinates (x, y, z), and denote the velocity components by $\mathbf{v} = [u, v, w]^t$. We can derive a well-known two-dimensional stream function by setting $\psi = \psi(x, y)$ and $\chi = z$, i.e., $\boldsymbol{\Psi} = \psi\,\mathrm{grad}\,\chi = \psi\,\mathbf{e}_z$. Equation (4.24.1) becomes

$$u = \frac{\partial\psi}{\partial y}, \quad v = -\frac{\partial\psi}{\partial x}. \tag{4.24.2}$$

The governing equation of the stream function is given by $\mathrm{curl}\,\mathbf{v} = \boldsymbol{\omega}$, which becomes

$$\frac{\partial^2\psi}{\partial x^2} + \frac{\partial^2\psi}{\partial y^2} = -\omega_z, \tag{4.24.3}$$

where $\boldsymbol{\omega} = (0, 0, \omega_z)$. For irrotational flows, the stream function satisfies the Laplace equation:

$$\frac{\partial^2\psi}{\partial x^2} + \frac{\partial^2\psi}{\partial y^2} = 0. \tag{4.24.4}$$

That is nice.

4.24.3 Polar Coordinates (2D)

Consider cylindrical coordinates (r, θ, z), and denote the velocity components by $\mathbf{v} = [u_r, v_\theta, u_z]^t$. We can derive a well-known two-dimensional stream function in polar coordinates by setting $\psi = \psi(r, \theta)$ and $\chi = z$, i.e., $\mathbf{\Psi} = \psi \operatorname{grad} \chi = \psi \, \mathbf{e}_z$. Equation (4.24.1) becomes

$$u_r = \frac{1}{r}\frac{\partial \psi}{\partial \theta}, \quad u_\theta = -\frac{\partial \psi}{\partial r}. \tag{4.24.5}$$

The governing equation of the stream function is given by $\operatorname{curl} \mathbf{v} = \boldsymbol{\omega}$, which becomes in cylindrical coordinates

$$\frac{1}{r}\frac{\partial}{\partial r}\left(r\frac{\partial \psi}{\partial r}\right) + \frac{1}{r^2}\frac{\partial^2 \psi}{\partial \theta^2} = -\omega_z. \tag{4.24.6}$$

Hence, for irrotational flows, the stream function satisfies, again, the Laplace equation:

$$\frac{1}{r}\frac{\partial}{\partial r}\left(r\frac{\partial \psi}{\partial r}\right) + \frac{1}{r^2}\frac{\partial^2 \psi}{\partial \theta^2} = 0. \tag{4.24.7}$$

4.24.4 Axisymmetric Cylindrical Coordinates

We can derive the stream function for axisymmetric flows (i.e., $u_\theta = 0$, $\partial/\partial\theta = 0$) in cylindrical coordinates by setting $\psi = \psi(r, z)$ and $\chi = \theta$, i.e., $\mathbf{\Psi} = \psi \operatorname{grad} \chi = \psi \, \mathbf{e}_\theta$. Equation (4.24.1) becomes

$$u_r = -\frac{1}{r}\frac{\partial \psi}{\partial z}, \quad u_z = \frac{1}{r}\frac{\partial \psi}{\partial r}, \tag{4.24.8}$$

where the axis of symmetry has been taken along the z-axis (see Figure 1.5.1). Axisymmetric flows do not have variations in θ by definition, and therefore they are essentially two-dimensional. Here, the two-dimensional plane is defined by the (z, r)-plane. The governing equation of the stream function is given by $\operatorname{curl} \mathbf{v} = \boldsymbol{\omega}$, which becomes in cylindrical coordinates for axisymmetric flows

$$\frac{\partial}{\partial z}\left(\frac{1}{r}\frac{\partial \psi}{\partial z}\right) + \frac{\partial}{\partial r}\left(\frac{1}{r}\frac{\partial \psi}{\partial r}\right) = -\omega_\theta, \tag{4.24.9}$$

or equivalently

$$\frac{1}{r}\left(\frac{\partial^2 \psi}{\partial z^2} + \frac{\partial^2 \psi}{\partial r^2} - \frac{1}{r}\frac{\partial \psi}{\partial r}\right) = -\omega_\theta. \tag{4.24.10}$$

Note that the left hand side is not the Laplacian. The Laplacian for axisymmetric flows in cylindrical coordinates is given by

$$
\begin{aligned}
\operatorname{div}\operatorname{grad} &= \operatorname{div}\left(\frac{\partial}{\partial r}, \; \frac{1}{r}\frac{\partial}{\partial\theta}, \; \frac{\partial}{\partial z}\right) \\[2mm]
&= \frac{1}{r}\frac{\partial}{\partial r}\left(r\frac{\partial}{\partial r}\right) + \frac{\partial^2}{\partial z^2} \\[2mm]
&= \frac{\partial^2}{\partial r^2} + \frac{\partial^2}{\partial z^2} + \frac{1}{r}\frac{\partial}{\partial r},
\end{aligned}
\tag{4.24.11}
$$

which does not precisely match the operator on the left hand side of Equation 4.24.10. Therefore, the stream function does not satisfy the Laplace equation in axisymmetric irrotational flows.

4.24.5 Axisymmetric Spherical Coordinates

The stream function for axisymmetric flows can be derived also in spherical coordinates (r, θ, ϕ), where the velocity components are denoted by $\mathbf{v} = [u_r, u_\theta, u_\phi]^t$. Note that ϕ here is the angle, not the scalar potential. By setting $\psi = \psi(r, \phi)$ and $\chi = -\theta$, i.e., $\mathbf{\Psi} = \psi \operatorname{grad} \chi = -\psi \, \mathbf{e}_\theta$, we obtain from Equation (4.24.1)

$$u_r = \frac{1}{r^2\sin\phi}\frac{\partial \psi}{\partial \phi}, \quad u_\phi = -\frac{1}{r\sin\phi}\frac{\partial \psi}{\partial r}, \tag{4.24.12}$$

where again the axis of symmetry has been taken along the z-axis (see Figure 1.5.2). The governing equation of the stream function is given by $\operatorname{curl} \mathbf{v} = \boldsymbol{\omega}$, which becomes in spherical coordinates for axisymmetric flows

$$\frac{1}{r}\left[\frac{\partial}{\partial r}\left(\frac{1}{\sin\phi}\frac{\partial\psi}{\partial r}\right) + \frac{\partial}{\partial\phi}\left(\frac{1}{r^2\sin\phi}\frac{\partial\psi}{\partial\phi}\right)\right] = -\omega_\theta. \tag{4.24.13}$$

Again, the left hand side is not the Laplacian. The stream function does not satisfy the Laplace equation in axisymmetric irrotational flows.

4.24.6   Axisymmetric (x, y) Coordinates

It is straightforward to show that Equation (4.24.8) and Equation (4.24.12) are equivalent, and both can be written as

$$u = \frac{1}{y}\frac{\partial\psi}{\partial y}, \quad v = -\frac{1}{y}\frac{\partial\psi}{\partial x}. \tag{4.24.14}$$

Note that here the x-axis is the horizontal axis of symmetry and the y-axis is the vertical axis. Of course, the above expression is equivalent to the one obtained directly from the axisymmetric continuity equation, i.e., Equation (4.20.14) in Section 4.20. The governing equation of the stream function is

$$\frac{\partial^2\psi}{\partial x^2} + \frac{\partial^2\psi}{\partial y^2} - \frac{1}{y}\frac{\partial\psi}{\partial y} = -y\,\omega_\theta, \tag{4.24.15}$$

which will not reduce to the Laplace equation for irrotational flows as shown in Section 4.24.4. For irrotational flows, we can define the velocity potential, ϕ:

$$u = \frac{\partial\phi}{\partial x}, \quad v = \frac{\partial\phi}{\partial y}. \tag{4.24.16}$$

The system (4.24.14) and (4.24.16) looks like the Cauchy-Riemann system for ϕ and ψ as described in Section 2.9.2, but not really because of the non-constant coefficient $1/y$. Note also that although ψ does not satisfy the Laplace equation, the velocity potential satisfies the Laplace equation. Anyway, I like axisymmetric flows especially expressed in the (x, y)-plane because the equations are very similar to those in two-dimensional flows and it makes it easy to understand the differences between two-dimensional and axisymmetric flows as we have just seen.

4.24.7   Volume Flow and Stream Functions

Of course, the stream functions are associated with the volume flow. Consider a volume flow Q through a section S,

$$Q = \iint_S \mathbf{v} \cdot \mathbf{n}\, dS, \tag{4.24.17}$$

where dS and $\mathbf{n}$ are the infinitesimal area and the unit outward normal vector of the section, respectively. By Stokes' theorem in Section 1.13, it reduces to the line integral over the boundary of S,

$$Q = \oint_{\partial S} \operatorname{curl}\boldsymbol{\Psi} \cdot d\mathbf{l} = \oint_{\partial S} \psi\,\operatorname{grad}\chi \cdot d\mathbf{l}, \tag{4.24.18}$$

where $d\mathbf{l}$ denotes the infinitesimal length vector in the counterclockwise direction. Assume that the section S is bounded by four stream surfaces, $\psi = \psi_1, \psi_2$ and $\chi = \chi_1, \chi_2$:

$$\partial S = \partial S_{\psi_1} + \partial S_{\chi_1} + \partial S_{\psi_2} + \partial S_{\chi_2}, \tag{4.24.19}$$

where ψ_1, ψ_2, χ_1, and χ_2 are constants, ∂S_{ψ_1} denotes the boundary where $\psi = \psi_1$, and similarly for others. Then, on each boundary portion, one of the functions is constant. Noting that $\operatorname{grad}\chi = 0$ on ∂S_{χ_1} and ∂S_{χ_2}, we obtain

$$Q = \psi_1 \int_{\partial S_{\psi_1}} \operatorname{grad}\chi \cdot d\mathbf{l} + \psi_2 \int_{\partial S_{\psi_2}} \operatorname{grad}\chi \cdot d\mathbf{l}. \tag{4.24.20}$$

By definition, $d\chi = \operatorname{grad}\chi \cdot d\mathbf{l}$. We thus finally obtain

$$Q = \psi_1(\chi_1 - \chi_2) + \psi_2(\chi_2 - \chi_1) = (\psi_2 - \psi_1)(\chi_2 - \chi_1). \tag{4.24.21}$$

Therefore, the volume flow through the section S is given by the product of the differences of the stream functions.

I like the stream functions because they are applicable to a wide range of flows: irrotational, rotational, inviscid, and viscous flows. Note also that the stream functions can be defined for compressible flows as $\mathbf{v} = (1/\rho)\text{curl }\mathbf{\Psi}$, where ρ is the density, so that it automatically satisfies the steady continuity equation $\text{div}(\rho\mathbf{v}) = 0$. See, for example, Refs.[18, 67, 120] for more details on three-dimensional stream functions. See Ref.[140] for an application of three-dimensional stream functions to CFD.

4.25 Stokes' Equations

I like Stokes' equations because they are linear. In fact, we obtain Stokes' equations by ignoring the nonlinear convective term from the incompressible Navier-Stokes equations. This is a valid approximation for small Reynolds number flows. These flows are often called Stokes' flows or creeping flows.

Stokes' Equations

Stokes' equations are given by

$$\text{div }\mathbf{v} = 0, \tag{4.25.1}$$

$$\rho\frac{\partial\mathbf{v}}{\partial t} + \text{grad }p - \mu\,\text{div grad }\mathbf{v} = 0, \tag{4.25.2}$$

where ρ and μ are constants. Note that the momentum equation has been obtained by ignoring the convective term $(\text{grad }\mathbf{v})\mathbf{v}$ from Equation (4.17.7). I like the fact that the pressure satisfies the Laplace equation, which can be easily shown by taking the divergence of the momentum equation:

$$\rho\frac{\partial(\text{div }\mathbf{v})}{\partial t} + \text{div grad }p - \mu\,\text{div div grad }\mathbf{v} = 0, \tag{4.25.3}$$

which becomes by Equation (1.7.12)

$$\rho\frac{\partial(\cancel{\text{div }\mathbf{v}})}{\partial t} + \text{div grad }p - \mu\,\text{div }[\text{grad}(\text{div }\mathbf{v}) - \text{curl}(\text{curl }\mathbf{v})] = 0, \tag{4.25.4}$$

$$\text{div grad }p - \mu\,\text{div grad}(\cancel{\text{div }\mathbf{v}}) + \mu\,\cancel{\text{div curl}(\text{curl }\mathbf{v})} = 0, \tag{4.25.5}$$

and thus

$$\text{div grad }p = 0. \tag{4.25.6}$$

This is the Laplace equation for the pressure. That's nice. Recall that the pressure satisfies the Poisson equation (4.22.4) when the convective term is present.

Stokes Equations (pressure and vorticity)

Stokes' equations can be written as a first-order system in terms of the pressure and the vorticity. By Equation (1.7.12), we have

$$\text{div grad }\mathbf{v} = \text{grad}(\text{div }\mathbf{v}) - \text{curl curl }\mathbf{v} = -\text{curl }\boldsymbol{\omega}, \tag{4.25.7}$$

where $\boldsymbol{\omega} = \text{curl }\mathbf{v} = [\omega_x, \omega_y, \omega_z]^t$ is the vorticity. Then, we obtain

$$\text{div }\mathbf{v} = 0, \tag{4.25.8}$$

$$\text{curl }\mathbf{v} = \boldsymbol{\omega}, \tag{4.25.9}$$

$$\rho\frac{\partial\mathbf{v}}{\partial t} + \text{grad }p + \mu\,\text{curl }\boldsymbol{\omega} = 0. \tag{4.25.10}$$

It is nice to be a first-order system because often the discretization is relatively simpler compared with second derivatives. Nevertheless, I also find it interesting that the vorticity obeys the diffusion equation. Take the curl of the momentum equation:

$$\rho\frac{\partial\omega}{\partial t} + \text{curl grad } p + \mu\,\text{curl curl}\,\omega \;=\; 0, \tag{4.25.11}$$

which becomes by Equation (1.7.12)

$$\frac{\partial\omega}{\partial t} = \nu\,\text{div grad}\,\omega, \tag{4.25.12}$$

where $\nu = \mu/\rho$. More interestingly, for steady flows, the vorticity satisfies the Laplace equation:

$$\text{div grad}\,\omega = 0. \tag{4.25.13}$$

4.26   2D Stokes Equations

I like the 2D Stokes equations. Especially I like the steady version because they reduce to two sets of Cauchy-Riemann type equations.

2D Stokes Equations

$$\frac{\partial u}{\partial x} + \frac{\partial v}{\partial y} \;=\; 0, \tag{4.26.1}$$

$$\rho\frac{\partial u}{\partial t} + \frac{\partial p}{\partial x} - \mu\left(\frac{\partial^2 u}{\partial x^2} + \frac{\partial^2 u}{\partial y^2}\right) \;=\; 0, \tag{4.26.2}$$

$$\rho\frac{\partial v}{\partial t} + \frac{\partial p}{\partial y} - \mu\left(\frac{\partial^2 v}{\partial x^2} + \frac{\partial^2 v}{\partial y^2}\right) \;=\; 0. \tag{4.26.3}$$

2D Stokes Equations (pressure and vorticity)

$$\frac{\partial u}{\partial x} + \frac{\partial v}{\partial y} \;=\; 0, \tag{4.26.4}$$

$$\frac{\partial v}{\partial x} - \frac{\partial u}{\partial y} \;=\; \omega_z, \tag{4.26.5}$$

$$\rho\frac{\partial u}{\partial t} + \frac{\partial p}{\partial x} + \mu\frac{\partial\omega_z}{\partial y} \;=\; 0, \tag{4.26.6}$$

$$\rho\frac{\partial v}{\partial t} + \frac{\partial p}{\partial y} - \mu\frac{\partial\omega_z}{\partial x} \;=\; 0. \tag{4.26.7}$$

For steady flows, these are basically two sets of the Cauchy-Riemann equations; they are coupled through the vorticity ω_z. It is nice to be the Cauchy-Riemann system because exact solutions can be found by using the theory of complex variables [54]. I find it interesting that the Stokes equations cannot accurately describe a flow around an object in two dimensions: a far-field free stream condition cannot be satisfied simultaneously with the no-slip condition on the object. This is called Stokes' paradox. To resolve this, we must retain a leading term in a linearized advection term; this is called Oseen's approximation. See Ref.[54] for more details.

4.27   3D Stokes Equations

I like the 3D Stokes equations because Stokes' paradox does not apply and a flow over a body can be properly described. It is also exciting that the vorticity is now a vector having three components (it has only one component in two dimensions).

3D Stokes Equations

$$\frac{\partial u}{\partial x} + \frac{\partial v}{\partial y} + \frac{\partial w}{\partial z} = 0, \tag{4.27.1}$$

$$\rho\frac{\partial u}{\partial t} + \frac{\partial p}{\partial x} - \mu\left(\frac{\partial^2 u}{\partial x^2} + \frac{\partial^2 u}{\partial y^2} + \frac{\partial^2 u}{\partial z^2}\right) = 0, \tag{4.27.2}$$

$$\rho\frac{\partial v}{\partial t} + \frac{\partial p}{\partial y} - \mu\left(\frac{\partial^2 v}{\partial x^2} + \frac{\partial^2 v}{\partial y^2} + \frac{\partial^2 v}{\partial z^2}\right) = 0, \tag{4.27.3}$$

$$\rho\frac{\partial w}{\partial t} + \frac{\partial p}{\partial z} - \mu\left(\frac{\partial^2 w}{\partial x^2} + \frac{\partial^2 w}{\partial y^2} + \frac{\partial^2 w}{\partial z^2}\right) = 0. \tag{4.27.4}$$

3D Stokes Equations (pressure and vorticity)

$$\frac{\partial u}{\partial x} + \frac{\partial v}{\partial y} + \frac{\partial w}{\partial z} = 0, \tag{4.27.5}$$

$$\frac{\partial w}{\partial y} - \frac{\partial v}{\partial z} = \omega_x, \tag{4.27.6}$$

$$\frac{\partial u}{\partial z} - \frac{\partial w}{\partial x} = \omega_y, \tag{4.27.7}$$

$$\frac{\partial v}{\partial x} - \frac{\partial u}{\partial y} = \omega_z, \tag{4.27.8}$$

$$\frac{\partial p}{\partial x} + \mu\left(\frac{\partial \omega_z}{\partial y} - \frac{\partial \omega_y}{\partial z}\right) = 0, \tag{4.27.9}$$

$$\rho\frac{\partial u}{\partial t} + \frac{\partial p}{\partial y} + \mu\left(\frac{\partial \omega_x}{\partial z} - \frac{\partial \omega_z}{\partial x}\right) = 0, \tag{4.27.10}$$

$$\rho\frac{\partial v}{\partial t} + \frac{\partial p}{\partial z} + \mu\left(\frac{\partial \omega_y}{\partial x} - \frac{\partial \omega_x}{\partial y}\right) = 0. \tag{4.27.11}$$

It is really nice to be a first-order system, but now I don't see the Cauchy-Riemann equations for steady flows. The theory of complex variables cannot be applied, and exact solutions need to be found by some other methods [54]. "How about axisymmetric flows?" That's a good point. As we have seen before, axisymmetric equations are almost like two-dimensional equations. It is worth looking at.

4.28   Axisymmetric Stokes Equations

Axisymmetric Stokes Equations

Axisymmetric Stokes equations can be obtained from the axisymmetric incompressible Navier-Stokes equations (4.20.8) by ignoring the convective terms:

$$\frac{\partial(yu)}{\partial x} + \frac{\partial(yv)}{\partial y} = 0, \tag{4.28.1}$$

$$\rho\frac{\partial(yu)}{\partial t} + \frac{\partial(yp)}{\partial x} - \mu\left[\frac{\partial}{\partial x}\left(y\frac{\partial u}{\partial x}\right) + \frac{\partial}{\partial y}\left(y\frac{\partial u}{\partial y}\right)\right] = 0, \tag{4.28.2}$$

$$\rho\frac{\partial(yv)}{\partial t} + \frac{\partial(yp)}{\partial y} - \mu\left[\frac{\partial}{\partial x}\left(y\frac{\partial v}{\partial x}\right) + \frac{\partial}{\partial y}\left(y\frac{\partial v}{\partial y}\right)\right] = p - \mu\frac{v}{y}, \tag{4.28.3}$$

where as before the x-axis has been taken as the axis of symmetry.

Axisymmetric Stokes Equations (pressure and vorticity)

For axisymmetric flows, the first-order system of Equations (4.25.8), (4.25.9), and (4.25.10) becomes

$$\frac{\partial(yu)}{\partial x} + \frac{\partial(yv)}{\partial y} = 0, \tag{4.28.4}$$

$$\frac{\partial v}{\partial x} - \frac{\partial u}{\partial y} = \omega, \tag{4.28.5}$$

$$\rho\frac{\partial(yu)}{\partial t} + \frac{\partial(yp)}{\partial x} + \mu\frac{\partial(y\omega)}{\partial y} = 0, \tag{4.28.6}$$

$$\rho\frac{\partial(yv)}{\partial t} + \frac{\partial(yp)}{\partial y} - \mu\frac{\partial(y\omega)}{\partial x} = p, \tag{4.28.7}$$

where ω is the θ-component of the vorticity, which is the only nonzero component. As expected, the axisymmetric system is very similar to the two-dimensional system, but not similar enough. Unfortunately, the momentum equations do not become purely the Cauchy-Riemann system for steady flows because of the presence of the pressure source term. Nevertheless, the stream function can be defined for the continuity equation as in the axisymmetric incompressible Navier-Stokes equations (see Section 4.20), which can be useful for deriving exact solutions [54].

4.29 Boundary Layer Equations

The boundary layer equations for two-dimensional incompressible flows are given by

$$\frac{\partial u}{\partial x} + \frac{\partial v}{\partial y} = 0, \tag{4.29.1}$$

$$u\frac{\partial u}{\partial x} + v\frac{\partial u}{\partial y} = U\frac{\partial U}{\partial x} + \nu\frac{\partial^2 u}{\partial y^2}, \tag{4.29.2}$$

where $U = U(x)$ is the velocity outside the boundary layer (constant for a flow over a flat plate). This can be solved numerically, for example, as follows [158]. First, we compute v by numerically integrating the continuity equation in y direction with the boundary condition $u(x,0) = v(x,0) = 0$, then solve the momentum equation for given v, and repeat this process until convergence. In this case, the momentum equation behaves like the viscous Burgers' equation (2.15.4). It is nice that a scalar scheme can be used for solving such a nonlinear system. To be even nicer, there exists an exact solution for this system (see Section 7.15.9), so that the accuracy of the numerical solution can be directly measured.

Chapter 5

Turbulence Equations

5.1 Averages and Filters

Express a physical quantity a as the sum of the average $\overline{a}$ and the fluctuation a': for example,

$$\mathbf{v} = \overline{\mathbf{v}} + \mathbf{v}', \quad \rho = \overline{\rho} + \rho', \quad p = \overline{p} + p', \quad h = \overline{h} + h', \quad T = \overline{T} + T', \quad H = \overline{H} + H'. \tag{5.1.1}$$

This is called the Reynolds decomposition. Here we must be careful about the fact that the fluctuation is not necessarily small (compared with the average), and also that usually fluctuations of the viscosity, the heat conductivity, and the specific heats are small and therefore ignored.

In turbulence calculations, typically we average the Navier-Stokes equations and solve for averaged quantities. After averaging, as we will see later, we find extra terms created which involve fluctuations and cannot be evaluated from averaged quantities. So, we derive governing equations for these extra terms, but then we will find similar extra terms again that cannot be evaluated, and this process goes on forever. Therefore, at some point, we must stop and model these terms, i.e., evaluate them somehow in terms of averaged quantities.

The most fundamental in turbulence calculations is the averaging. There are many different ways of averaging. Properties of averages and fluctuations depend on their definitions. Also, terms that need to be modeled, in averaged governing equations, depend on them.

5.1.1 Reynolds Averaging

The Reynolds averaging includes ensemble averaging, time averaging, and space averaging. These are very interesting. For example, they are identical to one another for steady homogeneous turbulence (which is called Ergodic hypothesis). The simulation based on the Reynolds-averaged Navier-Stokes equations is called RANS.

Ensemble Averaging

The ensemble averaging:

$$\overline{f}(\mathbf{x}, t) = \lim_{N \to \infty} \frac{1}{N} \sum_{i=1, N}^{N} f_i(\mathbf{x}, t), \tag{5.1.2}$$

which is the average of measurements, $f_i(\mathbf{x}, t)$ where $i = 1, 2, 3, \cdots N$, obtained by N experiments. I like it, but it is unrealistic to perform this averaging in practice. You can, of course, formally (or theoretically or mathematically, whatever) use this averaging to write down averaged governing equations.

Time Averaging

The time averaging:

$$\overline{f}(\mathbf{x}, t) = \frac{1}{T} \int_{t}^{t+T} f(\mathbf{x}, t) \, dt, \tag{5.1.3}$$

where T must be larger than the fluctuation time scale but smaller than the average time scale. I like this one also, but in real flows, these two time scales are comparable, so such T may not exist. Of course, you can still formally derive averaged governing equations based on this assumption. For a steady flow, by taking $T \to \infty$, we may define

$$\overline{f}(\mathbf{x}) = \lim_{T \to \infty} \frac{1}{T} \int_t^{t+T} f(\mathbf{x}, t)\, dt. \tag{5.1.4}$$

Space Averaging

The space averaging:

$$\overline{f}(t) = \lim_{V \to \infty} \frac{1}{V} \iiint_V f(\mathbf{x}, t)\, dV, \tag{5.1.5}$$

which can be used for homogeneous turbulence in which the mean flow quantities are uniform in space. Of course, many of us are probably not interested in uniform mean flows. However, for any flows, if we look close enough, there could be a limit beyond which the flow looks locally uniform. Then, the homogeneous turbulence does make sense, and the space averaging can be useful. I like it.

Properties of Reynolds Averaging

The Reynolds averaging has the following properties:

$$\overline{\overline{f}} = \overline{f}, \tag{5.1.6}$$

$$\overline{f'} = 0, \tag{5.1.7}$$

$$\overline{\overline{f} g'} = 0, \tag{5.1.8}$$

$$\overline{\overline{f} g} = \overline{f}\, \overline{g}, \tag{5.1.9}$$

$$\overline{f + g} = \overline{f} + \overline{g}. \tag{5.1.10}$$

They all make sense and thus easy to remember. I like them.

5.1.2 Favre Averaging

Favre Averaging

The Favre averaging (denoted by a tilde):

$$\widetilde{f} = \frac{\overline{\rho f}}{\overline{\rho}}, \tag{5.1.11}$$

which may be called the mass-weighted averaging or the mass-averaging. In fact, I like it very much because we can greatly simplify the averaged compressible Navier-Stokes equations using this particular averaging, much simpler than those obtained by using the Reynolds averaging.

Usually, only the velocity and thermodynamic variables are Favre-averaged and the density and the pressure are Reynolds-averaged:

$$\widetilde{\mathbf{v}} = \frac{\overline{\rho \mathbf{v}}}{\overline{\rho}}, \quad \widetilde{h} = \frac{\overline{\rho h}}{\overline{\rho}}, \quad \widetilde{T} = \frac{\overline{\rho T}}{\overline{\rho}}, \quad \widetilde{H} = \frac{\overline{\rho H}}{\overline{\rho}}. \tag{5.1.12}$$

Also note that the fluctuation associated with Favre averaging $\widetilde{f}$ is denoted by f'':

$$\mathbf{v} = \widetilde{\mathbf{v}} + \mathbf{v}'', \quad \rho = \overline{\rho} + \rho', \quad p = \overline{p} + p', \quad h = \widetilde{h} + h'', \quad T = \widetilde{T} + T'', \quad H = \widetilde{H} + H''. \tag{5.1.13}$$

Properties of Favre Averaging

Favre averaging has the following properties:

$$f'' = f - \overline{f} - \frac{\overline{\rho' f'}}{\overline{\rho}}, \tag{5.1.14}$$

$$\overline{f''} = -\frac{\overline{\rho' f'}}{\overline{\rho}}, \tag{5.1.15}$$

$$\overline{f''} \neq 0, \tag{5.1.16}$$

$$\overline{\rho f''} = 0, \tag{5.1.17}$$

$$\overline{\rho f} = \overline{\rho} \widetilde{f}. \tag{5.1.18}$$

These are more complicated than those of the Reynolds Averaging, but I like them. These are very important, anyway.

5.1.3 Spatial Filter

Spatial Filter

The spatial filter is a local space (or a local wave-number space) averaging. A typical spatial filter is the volume-average box filter (also called the top-hat filter):

$$\overline{f}(\mathbf{x}, t) = \frac{1}{\Delta^3} \int_{x-\Delta x/2}^{x+\Delta x/2} \int_{y-\Delta y/2}^{y+\Delta y/2} \int_{z-\Delta z/2}^{z+\Delta z/2} f(\boldsymbol{\xi}, t) \, d\xi d\eta d\zeta, \tag{5.1.19}$$

where

$$\mathbf{x} = (x, y, z)^t = (x_1, x_2, x_3)^t, \quad \boldsymbol{\xi} = (\xi, \eta, \zeta)^t = (\xi_1, \xi_2, \xi_3)^t, \quad \Delta = (\Delta x \Delta y \Delta z)^{1/3}. \tag{5.1.20}$$

We may think of it as the integral over a computational cell: $\overline{f}(\mathbf{x}, t)$ can be thought of as a smooth (low-frequency) component that can be resolved on that grid. This smooth component is called a grid scale (GS) or a resolvable scale, and the rest (high-frequency component that cannot be resolved) is called a sub-grid scale (SGS), i.e.,

$$f = \overline{f} + f', \tag{5.1.21}$$

in which f' is the sub-grid scale. Note that in principle the filter spacing, $(\Delta x, \Delta y, \Delta z)$, does not have to be the grid spacing (as long as larger than the grid spacing). So, the terms like GS and SGS may be confusing. This filtering operation can be thought of as a decomposition of a flow into large eddies(GS) and small eddies(SGS). The smaller the filter spacing is, the more universal the properties of small eddies can be (Kolmogorov's law: a theory by Kolmogorov, which is sometimes called K41 theory because it came out in 1941 [73]. See Refs.[12, 156] for details). If the filter spacing is small enough, then, a universal model can be developed: methods applicable to many different flows can be developed. Yes, a smaller filter spacing implies finer computational grids, and hence such simulations are generally very expensive.

We can generalize the filtering operation by writing it in the form,

$$\overline{f}(\mathbf{x}, t) = \iiint G(\mathbf{x} - \boldsymbol{\xi}) f(\boldsymbol{\xi}, t) \, d\xi d\eta d\zeta, \tag{5.1.22}$$

where the integration is over all space and G is called the filter function. For example, the following filter function corresponds to the volume-average box filter (5.1.19),

$$G(x_i - \xi_i) = \begin{cases} 1/\Delta^3 & |x_i - \xi_i| < \Delta x_i/2, \\ 0 & \text{otherwise.} \end{cases} \tag{5.1.23}$$

There are other types of filters such as the Gaussian filter or the Fourier cut-off filter. See Ref.[171] for details.

The large eddy simulation (LES) refers to a simulation based on filtered governing equations with sufficiently small filter spacing such that the sub-grid scale will be homogeneous. I like the idea, but it requires a very fine computational grid (very expensive). If I use a grid fine enough to resolve all scales, then no turbulence models are needed, and the simulation is called the direct numerical simulations (DNS). I like this one. It's very simple. It is particularly nice that we do not need turbulence modeling. However it is even more expensive than LES: the grid must be extremely or often prohibitively fine especially for high-Reynolds-number flows encountered in many practical applications. There is also a class of methods called detached eddy simulations (DES) [147, 151]. In DES, the near-wall region (too expensive to resolve by LES) is solved by RANS and the rest of the domain (where relatively large eddies dominate) is solved by LES. I like the idea of DES, but I really think that it would be very very nice if DNS would become feasible for practical flows. It is so simple.

Properties of Spatial Filter

Generally, filtering once and twice, we obtain different values:

$$\overline{\overline{f}} \neq \overline{f}, \tag{5.1.24}$$

which is not the case in the Reynolds averaging. Therefore, filtered governing equations and averaged governing equations can be quite different. We must be careful about the following also:

$$\overline{f'} \neq 0, \tag{5.1.25}$$

$$\overline{\overline{f}g'} \neq 0. \tag{5.1.26}$$

Do I like it? Well, sure, of course, I like it.

5.2   Reynolds-Averaged Incompressible Navier-Stokes Equations

Take the Reynolds average of the incompressible Navier-Stokes equations to get the Reynolds-averaged incompressible Navier-Stokes equations ($\rho = $ constant):

$$\text{div}\,(\rho\overline{\mathbf{v}}) = 0, \tag{5.2.1}$$

$$\partial_t(\rho\overline{\mathbf{v}}) + \text{div}(\rho\overline{\mathbf{v}}\otimes\overline{\mathbf{v}}) + \text{grad}\,\overline{p} - \text{div}\,\hat{\boldsymbol{\tau}} = 0, \tag{5.2.2}$$

where

$$\hat{\boldsymbol{\tau}} = \overline{\boldsymbol{\tau}} - \rho\overline{\mathbf{v}'\otimes\mathbf{v}'} = \mu\left[\text{grad}\,\overline{\mathbf{v}} + (\text{grad}\,\overline{\mathbf{v}})^t\right] - \rho\overline{\mathbf{v}'\otimes\mathbf{v}'}. \tag{5.2.3}$$

The extra term,

$$\rho\overline{\mathbf{v}'\otimes\mathbf{v}'}, \tag{5.2.4}$$

has been generated by the averaging operation. This is usually interpreted as a stress and called the Reynolds stress. Without the Reynolds stress, the system would have been closed and solved for the average quantities (just like the laminar equations). It is possible to derive its own governing equations; but it is so much simpler to model this term (i.e., to relate somehow to the average quantities) by the Boussinesq approximation (or the gradient transport hypothesis) [16]:

$$-\rho\overline{\mathbf{v}'\otimes\mathbf{v}'} = \mu_T\left[\text{grad}\,\overline{\mathbf{v}} + (\text{grad}\,\overline{\mathbf{v}})^t\right] - \frac{2}{3}\rho k\mathbf{I}, \tag{5.2.5}$$

where μ_T is called the eddy viscosity or the turbulence viscosity, and

$$k = \frac{1}{2}\overline{\mathbf{v}'\cdot\mathbf{v}'}, \tag{5.2.6}$$

is called the turbulent energy. Basically, we assume that this fluctuation stress is proportional to the gradient of the average quantities (similarly to Newtonian fluids). The second term $-\frac{2}{3}\rho k\mathbf{I}$ is needed in order for the Boussinesq approximation to be valid when traced: the trace of the right hand side must be equal to that of the left hand side $-\rho\overline{v_i'v_i'} = -2\rho k$. But this term is usually ignored if we have no ways to evaluate k, e.g., in zero-equation or one-equation models. To evaluate k, we would probably have to derive and solve the governing equation for k. Typically, two-equation models include such an option (see Section 5.6).

I like ignoring k and employing the Boussinesq approximation because then I can write the viscous stress term as

$$\hat{\boldsymbol{\tau}} = (\mu + \mu_T)\left[\operatorname{grad}\overline{\mathbf{v}} + (\operatorname{grad}\overline{\mathbf{v}})^t\right], \tag{5.2.7}$$

so that practically the form of the Reynolds-averaged incompressible Navier-Stokes equations is the same as that of the original (laminar) incompressible Navier-Stokes equations except that it has one extra viscosity coefficient, μ_T,

$$\mu \longrightarrow \mu + \mu_T. \tag{5.2.8}$$

So, the focus is now on estimating the turbulent viscosity, μ_T. You might want to evaluate it algebraically (algebraic models), by deriving and solving a governing equation for μ_T (one-equation models), or by deriving and solving governing equations for μ_T as well as k (or any relevant two quantities: two-equation models).

5.3   Favre-Averaged Compressible Navier-Stokes Equations

Apply the Favre averaging (instead of the Reynolds averaging) to the compressible Navier-Stokes equations to obtain the following relatively simple form of the averaged Navier-Stokes Equations:

$$\partial_t \overline{\rho} + \operatorname{div}(\overline{\rho}\,\widetilde{\mathbf{v}}) = 0, \tag{5.3.1}$$

$$\partial_t(\overline{\rho}\,\widetilde{\mathbf{v}}) + \operatorname{div}(\overline{\rho}\,\widetilde{\mathbf{v}}\otimes\widetilde{\mathbf{v}}) + \operatorname{grad}\overline{p} = \operatorname{div}\hat{\boldsymbol{\tau}}, \tag{5.3.2}$$

$$\partial_t(\overline{\rho}\widetilde{E}) + \operatorname{div}(\overline{\rho}\,\widetilde{\mathbf{v}}\widetilde{H}) = \operatorname{div}(\hat{\boldsymbol{\tau}}\widetilde{\mathbf{v}}) - \operatorname{div}\hat{\mathbf{q}} + \operatorname{div}\mathbf{K}, \tag{5.3.3}$$

where

$$\overline{p} = \overline{\rho}R\widetilde{T}, \tag{5.3.4}$$

$$\widetilde{E} = \widetilde{e} + \frac{\widetilde{\mathbf{v}}\cdot\widetilde{\mathbf{v}}}{2} + k, \tag{5.3.5}$$

$$\widetilde{H} = \widetilde{h} + \frac{\widetilde{\mathbf{v}}\cdot\widetilde{\mathbf{v}}}{2} + k, \tag{5.3.6}$$

$$k = \frac{\overline{\rho\mathbf{v}''\cdot\mathbf{v}''}}{2\overline{\rho}}, \tag{5.3.7}$$

$$\hat{\boldsymbol{\tau}} = \overline{\boldsymbol{\tau}} - \overline{\rho\mathbf{v}''\otimes\mathbf{v}''} = \mu\left[\operatorname{grad}\widetilde{\mathbf{v}} + (\operatorname{grad}\widetilde{\mathbf{v}})^t\right] - \frac{2}{3}\mu\,(\operatorname{div}\widetilde{\mathbf{v}})\mathbf{I} - \overline{\rho\mathbf{v}''\otimes\mathbf{v}''}, \tag{5.3.8}$$

$$\hat{\mathbf{q}} = -\frac{\mu}{P_r}\operatorname{grad}h - \overline{\rho\mathbf{v}''h''}, \tag{5.3.9}$$

and also

$$\mathbf{K} = \overline{\boldsymbol{\tau}\mathbf{v}''} - \overline{\rho\mathbf{v}''(\mathbf{v}''\cdot\mathbf{v}'')}/2, \tag{5.3.10}$$

but this one is considered to be small and often ignored. Also, as a matter of fact, because $\overline{\mathbf{v}''} \neq 0$, we actually have

$$\overline{\boldsymbol{\tau}} = \mu\left[\operatorname{grad}\widetilde{\mathbf{v}} + (\operatorname{grad}\widetilde{\mathbf{v}})^t\right] - \frac{2}{3}\mu\,(\operatorname{div}\widetilde{\mathbf{v}})\mathbf{I} + \left[\operatorname{grad}\overline{\mathbf{v}''} + (\operatorname{grad}\overline{\mathbf{v}''})^t\right] - \frac{2}{3}\mu\,(\operatorname{div}\overline{\mathbf{v}''})\mathbf{I}, \tag{5.3.11}$$

but the last two terms are considered to be small and often ignored. So, I have done so in Equation (5.3.8). Also, the turbulent energy k in $\widetilde{E}$ and $\widetilde{H}$ are often ignored. Now, with these terms all ignored, we apply the Boussinesq approximation:

$$-\overline{\rho\mathbf{v}''\otimes\mathbf{v}''} = \mu_T\left[\operatorname{grad}\widetilde{\mathbf{v}} + (\operatorname{grad}\widetilde{\mathbf{v}})^t\right] - \frac{2}{3}\mu_T\,(\operatorname{div}\widetilde{\mathbf{v}})\mathbf{I} - \frac{2}{3}\overline{\rho}k\mathbf{I}, \tag{5.3.12}$$

$$-\overline{\rho\mathbf{v}''h''} = -\frac{\mu_T}{Pr_T}\operatorname{grad}\widetilde{h}, \tag{5.3.13}$$

where of course, k will have to be ignored if there are no ways to estimate it; Pr_T is called the turbulent Prandtl number and $Pr_T = 0.9$ is a good approximation for the sea-level air. So, we ignore k and obtain the following Favre-averaged

Navier-Stokes equations:

$$\partial_t \overline{\rho} + \mathrm{div}(\overline{\rho}\,\widetilde{\mathbf{v}}) \;=\; 0, \tag{5.3.14}$$

$$\partial_t(\overline{\rho}\,\widetilde{\mathbf{v}}) + \mathrm{div}(\overline{\rho}\,\widetilde{\mathbf{v}}\otimes\widetilde{\mathbf{v}}) + \mathrm{grad}\,\overline{p} \;=\; \mathrm{div}\,\hat{\boldsymbol{\tau}}, \tag{5.3.15}$$

$$\partial_t(\overline{\rho}\widetilde{E}) + \mathrm{div}(\overline{\rho}\,\widetilde{\mathbf{v}}\widetilde{H}) \;=\; \mathrm{div}\,(\hat{\boldsymbol{\tau}}\widetilde{\mathbf{v}}) - \mathrm{div}\,\hat{\mathbf{q}}, \tag{5.3.16}$$

where

$$\overline{p} \;=\; \overline{\rho}R\widetilde{T}, \tag{5.3.17}$$

$$\widetilde{E} \;=\; \widetilde{e} + \frac{\widetilde{\mathbf{v}}\cdot\widetilde{\mathbf{v}}}{2}, \tag{5.3.18}$$

$$\widetilde{H} \;=\; \widetilde{h} + \frac{\widetilde{\mathbf{v}}\cdot\widetilde{\mathbf{v}}}{2}, \tag{5.3.19}$$

$$\hat{\boldsymbol{\tau}} \;=\; (\mu+\mu_T)\left[\mathrm{grad}\,\widetilde{\mathbf{v}} + (\mathrm{grad}\,\widetilde{\mathbf{v}})^t\right] - \frac{2}{3}(\mu+\mu_T)(\mathrm{div}\,\widetilde{\mathbf{v}})\mathbf{I}, \tag{5.3.20}$$

$$\hat{\mathbf{q}} \;=\; -\left(\frac{\mu}{Pr} + \frac{\mu_T}{Pr_T}\right)\mathrm{grad}\,\widetilde{h}. \tag{5.3.21}$$

Therefore, the differences between these equations and the laminar equations are the turbulent viscosity and the turbulent Prandtl number:

$$\mu \;\longrightarrow\; \mu+\mu_T, \tag{5.3.22}$$

$$\frac{\mu}{Pr} \;\longrightarrow\; \frac{\mu}{Pr} + \frac{\mu_T}{Pr_T}, \tag{5.3.23}$$

But since we may take $Pr_T = 0.9$, it is again a matter of how to estimate μ_T just like in the Reynolds-averaged incompressible Navier-Stokes equations. This is nice and simple. I like it.

5.4   Filtered Incompressible Navier-Stokes Equations

Apply filtering to the incompressible Navier-Stokes equations to get the filtered incompressible Navier-Stokes equations:

$$\mathrm{div}\,(\rho\overline{\mathbf{v}}) \;=\; 0, \tag{5.4.1}$$

$$\partial_t(\rho\overline{\mathbf{v}}) + \mathrm{div}(\rho\overline{\mathbf{v}}\otimes\overline{\mathbf{v}}) + \mathrm{grad}\,\overline{p} - \mathrm{div}\,\hat{\boldsymbol{\tau}} \;=\; 0, \tag{5.4.2}$$

where

$$\hat{\boldsymbol{\tau}}^{SGS} \;=\; -\rho(\overline{\mathbf{v}\otimes\mathbf{v}} - \overline{\mathbf{v}}\otimes\overline{\mathbf{v}}), \tag{5.4.3}$$

is called the SGS stress, which is usually expanded as

$$\hat{\boldsymbol{\tau}}^{SGS} \;=\; (\overline{\overline{\mathbf{v}}\otimes\overline{\mathbf{v}}} - \overline{\mathbf{v}}\otimes\overline{\mathbf{v}}) + (\overline{\mathbf{v}'\otimes\overline{\mathbf{v}} + \overline{\mathbf{v}}\otimes\mathbf{v}'}) + \overline{\mathbf{v}'\otimes\mathbf{v}'}, \tag{5.4.4}$$

and on the right hand side the first term is called the Leonard stress, the second the cross-term stress, the third the SGS Reynolds stress. It may look a little bit more complicated than the Reynolds stress in the Reynolds-averaged incompressible Navier-Stokes equations (5.2.4), but I like it.

5.5   Filtered Compressible Navier-Stokes Equations

Apply mass-averaged filtering to the compressible Navier-Stokes Equations to obtain the simple filtered compressible Navier-Stokes Equations:

$$\partial_t\overline{\rho} + \mathrm{div}(\overline{\rho}\,\widetilde{\mathbf{v}}) \;=\; 0, \tag{5.5.1}$$

$$\partial_t(\overline{\rho}\,\widetilde{\mathbf{v}}) + \mathrm{div}(\overline{\rho}\,\widetilde{\mathbf{v}}\otimes\widetilde{\mathbf{v}}) + \mathrm{grad}\,\overline{p} \;=\; \mathrm{div}\,\hat{\boldsymbol{\tau}}, \tag{5.5.2}$$

$$\partial_t(\overline{\rho}\widetilde{E}) + \mathrm{div}(\overline{\rho}\,\widetilde{\mathbf{v}}\widetilde{H}) \;=\; \mathrm{div}\,(\hat{\boldsymbol{\tau}}\widetilde{\mathbf{v}}) - \mathrm{div}\,\hat{\mathbf{q}} + \mathrm{div}\,\mathbf{K}^{SGS}, \tag{5.5.3}$$

where

$$\overline{p} = \overline{\rho}R\widetilde{T}, \quad \widetilde{E} = \widetilde{e} + \frac{\widetilde{\mathbf{v}} \cdot \widetilde{\mathbf{v}}}{2} + k^{SGS}, \quad \widetilde{H} = \widetilde{h} + \frac{\widetilde{\mathbf{v}} \cdot \widetilde{\mathbf{v}}}{2} + k^{SGS}, \tag{5.5.4}$$

$$\hat{\boldsymbol{\tau}} = \mu\left[\operatorname{grad}\widetilde{\mathbf{v}} + (\operatorname{grad}\widetilde{\mathbf{v}})^t\right] - \frac{2}{3}\mu\,(\operatorname{div}\widetilde{\mathbf{v}})\mathbf{I} - \boldsymbol{\tau}^{SGS}, \quad \hat{\mathbf{q}} = -\frac{\mu}{Pr}\operatorname{grad}\widetilde{h} + \mathbf{q}^{SGS}, \tag{5.5.5}$$

$$k^{SGS} = \frac{1}{2}\overline{\rho}(\widetilde{\mathbf{v}\mathbf{v}} - \widetilde{\mathbf{v}}\widetilde{\mathbf{v}}), \tag{5.5.6}$$

$$\boldsymbol{\tau}^{SGS} = -\rho(\widetilde{\mathbf{v}\otimes\mathbf{v}} - \widetilde{\mathbf{v}}\otimes\widetilde{\mathbf{v}}), \tag{5.5.7}$$

$$\mathbf{q}^{SGS} = -\overline{\rho}(\widetilde{E\mathbf{v}} - \widetilde{E}\widetilde{\mathbf{v}}), \tag{5.5.8}$$

$$\mathbf{K}^{SGS} = \overline{\rho}(\widetilde{\boldsymbol{\tau}\mathbf{v}} - \widetilde{\boldsymbol{\tau}}\widetilde{\mathbf{v}}) - (\widetilde{p\mathbf{v}} - \widetilde{p}\widetilde{\mathbf{v}}). \tag{5.5.9}$$

Usually, if the Mach number is not large, k^{SGS} and $\mathbf{K}^{SGS}$ are ignored and only $\boldsymbol{\tau}^{SGS}$ and $\mathbf{q}^{SGS}$ are modeled. These equations may look more complicated than the Favre-Averaged Navier-Stokes Equations, but essentially the same in that extra terms are introduced in the viscous stress and the heat flux. I like these equations.

5.6 Turbulence Models

In many cases, the Boussinesq type approximation is made, and it leads us to the problem of estimating the eddy viscosity, μ_T. There are various ways to evaluate the eddy viscosity. The simplest would be the algebraic models where the eddy viscosity is computed algebraically in terms of the mean flow quantities. This includes, for examples, Prandtl's mixing-length theory [121], the Cebeci-Smith model [143], the Baldwin-Lomax model [8], and their numerous variants. These methods can give a good estimate for the eddy viscosity in some cases, but its success essentially depends on the flow because the eddy viscosity is not an intrinsic property of the fluid (unlike the molecular viscosity, μ). In other words, the models need to be tuned (e.g., constants somehow adjusted) for each application. Johnson-King model [63] is a little bit more sophisticated model with an ordinary differential equation incorporated (often called 1/2 equation model), but still its applicability is limited because it does not directly estimate μ_T. I like these algebraic models because they are relatively easy to implement, but I must admit that we definitely need more sophisticated methods for successful and general turbulence calculations. Beyond the algebraic model, the simplest method would be to directly solve a governing equation for μ_T. This is good because μ_T will be solved along with the flow equations and thus it now depend on the flow. This includes, for examples, the Baldwin-Barth model [7, 6], the Spalart-Allmaras model [146], and their variants. I like these models. They are still simple. It is possible to couple the turbulence equation with the flow equations, but it is also possible to solve them separately. Personally, I like the latter. In fact, these models work successfully for many different flows, but not so much for separated flows. It looks like that we need more sophisticated models. Then, the simplest thing to do is to solve another equation for yet another turbulence quantity. This leads to two-equation models. This includes, for example, the Menter shear stress transport (SST) model [92], and its variants. These models are widely used; so people must like them. Anyway, I know there are many other turbulence models, e.g., Reynolds stress model, Smagorinsky model for LES, etc., but honestly speaking, I am not really an expert in turbulence modeling (although I like it very much). So, I recommend readers to consult other books such as Ref.[171] for more details.

Chapter 6

Methods for Exact Solutions

I like studying methods for deriving exact solutions. Various exact solutions have been used for accuracy verification in CFD, but it is sometimes difficult to find detailed descriptions on how they were derived. It is nice to know various methods for deriving exact solutions so that I can generate an exact solution myself whenever I need one.

6.1 Exact Solutions with Separation of Variables

I like separation of variables because it makes it possible to obtain exact solutions for some partial differential equations. It is a very powerful technique.

6.1.1 Laplace Equation

Consider the Laplace equation:

$$u_{xx} + u_{yy} = 0. \tag{6.1.1}$$

In separation of variables, we first assume the solution of the form,

$$u = X(x)Y(y), \tag{6.1.2}$$

and substitute this into the Laplace equation to get

$$\frac{1}{X}\frac{d^2X}{dx^2} + \frac{1}{Y}\frac{d^2Y}{dy^2} = 0, \tag{6.1.3}$$

from which we find

$$\frac{1}{X}\frac{d^2X}{dx^2} = -\frac{1}{Y}\frac{d^2Y}{dy^2} = \text{constant}. \tag{6.1.4}$$

So, now the problem is to solve the two ordinary differential equations. If we want the solution to be periodic in x-direction, then set the constant to be negative, say, $-k^2$, and we obtain

$$X(x) = A\sin(kx) + B\cos(kx), \tag{6.1.5}$$

$$Y(y) = Ce^{ky} + De^{-ky}, \tag{6.1.6}$$

where A, B, C, and D are arbitrary constants which are determined by boundary conditions. Therefore, we obtain an exact solution as

$$u(x,y) = X(x)Y(y) = [A\sin(kx) + B\cos(kx)]\left[Ce^{ky} + De^{-ky}\right]. \tag{6.1.7}$$

It is, of course, true that any linear combination of the solution with different k, i.e.,the Fourier series:

$$u(x,y) = \sum_{k=1}^{\infty} X_k(x)Y_k(y) = \sum_{k=1}^{\infty} [A_k\sin(kx) + B_k\cos(kx)]\left[C_ke^{ky} + D_ke^{-ky}\right], \tag{6.1.8}$$

179

is also an exact solution because the equation is linear. This is called the principle of superposition. It may be useful for deriving a solution which will fit a given domain and boundary conditions. But just one term would be enough for the purpose of accuracy verification of numerical schemes.

As an example, we consider deriving a solution from the fundamental solution (6.1.7) in a square domain $[x, y] = [0, 1] \times [0, 1]$ with the following boundary conditions:

$$u(x, 0) = u(0, y) = u(1, y) = 0, \tag{6.1.9}$$
$$u(x, 1) = \sin(\pi x). \tag{6.1.10}$$

These boundary conditions require that we set

$$B = 0, \quad C = -D, \quad AC = \frac{1}{\sinh(\pi)}, \tag{6.1.11}$$

in Equation (6.1.7), and thus the exact solution is given by

$$u(x, y) = \frac{\sin(\pi x) \sinh(\pi y)}{\sinh(\pi)}. \tag{6.1.12}$$

Now, consider another set of boundary conditions,

$$u(0, y) = u(x, 0) = u(x, 1) = 0, \tag{6.1.13}$$
$$u(1, y) = \sin(\pi y), \tag{6.1.14}$$

to obtain the exact solution,

$$u(x, y) = \frac{\sin(\pi y) \sinh(\pi x)}{\sinh(\pi)}, \tag{6.1.15}$$

and add this to the previous one to generate a new exact solution,

$$u(x, y) = \frac{\sinh(\pi x) \sin(\pi y) + \sinh(\pi y) \sin(\pi x)}{\sinh(\pi)}, \tag{6.1.16}$$

which is the exact solution to the following boundary condition:

$$u(0, y) = u(x, 0) = 0, \tag{6.1.17}$$
$$u(x, 1) = \sin(\pi x), \tag{6.1.18}$$
$$u(1, y) = \sin(\pi y). \tag{6.1.19}$$

This superposition is possible because the Laplace equation is linear. It is very nice. I like linear partial differential equations.

6.1.2   Diffusion Equation

I like the two-dimensional diffusion equation also,

$$u_t = \nu \left(u_{xx} + u_{yy} \right). \tag{6.1.20}$$

This can also be easily solved by separation of variables. First, we assume

$$u(x, y, t) = T(t) X(x) Y(y), \tag{6.1.21}$$

substitute it into the diffusion equation to find

$$\frac{1}{T(t)} \frac{dT}{dt} = \frac{\nu}{X(x)} \frac{d^2 X}{dx^2} + \frac{\nu}{Y(y)} \frac{d^2 Y}{dy^2}, \tag{6.1.22}$$

and so by setting

$$\frac{1}{T(t)} \frac{dT}{dt} = -\nu \pi^2 (a^2 + b^2), \quad \frac{1}{X(x)} \frac{d^2 X}{dx^2} = -(a\pi)^2, \quad \frac{1}{Y(y)} \frac{d^2 Y}{dy^2} = -(b\pi)^2, \tag{6.1.23}$$

where a and b are arbitrary constants, we find

$$T(t) = Ce^{-\nu\pi^2(a^2+b^2)t}, \tag{6.1.24}$$

$$X(x) = A_1 \sin(a\pi x) + A_2 \cos(a\pi x), \tag{6.1.25}$$

$$Y(y) = B_1 \sin(b\pi y) + B_2 \cos(b\pi y), \tag{6.1.26}$$

where C, A_1, A_2, B_1, and B_2 are arbitrary constants. Therefore, we obtain an exact solution as

$$u(x,y,t) = T(t)X(x)Y(y) = Ce^{-\nu\pi^2(a^2+b^2)t}\left[A_1 \sin(a\pi x) + A_2 \cos(a\pi x)\right]\left[B_1 \sin(b\pi y) + B_2 \cos(b\pi y)\right], \quad (6.1.27)$$

where C, A_1, A_2, B_1, and B_2 are arbitrary constants which can be determined by specifying boundary conditions. Of course, any linear combination of the solution with different a and b can also be an exact solution. Constants may be determined to produce a desired solution for given domain and boundary conditions. But I like a simple one, even simpler than the solution above. For example, I would set $A_1 = 1$, $A_2 = 0$, $B_1 = 1$, and $B_2 = 0$:

$$u(x,y,t) = T(t)X(x)Y(y) = Ce^{-\nu\pi^2(a^2+b^2)t} \sin(a\pi x)\sin(b\pi y). \tag{6.1.28}$$

This solution in a square domain with the Dirichlet boundary condition is enough for verifying the accuracy of a diffusion scheme.

6.1.3 Advection-Diffusion Equation

Consider the two-dimensional advection-diffusion equation,

$$u_t + au_x + bu_y = \nu(u_{xx} + u_{yy}), \tag{6.1.29}$$

where a, b, and ν are constants, and $a^2 + b^2 \neq 0$. This can be transformed into the following form,

$$\frac{u_t}{a^2+b^2} + u_\xi = \nu(u_{\xi\xi} + u_{\eta\eta}), \tag{6.1.30}$$

where $\xi = ax + by$ and $\eta = bx - ay$ (see Section 1.15 for the transformation). Basically, the coordinate system has been chosen along and normal to the advection direction. So, there is only one advection term present, which is a derivative along the advection direction. It is always nice to have less terms. I like it. To solve this, I first assume

$$u(\xi,\eta,t) = T(t)X(\xi)Y(\eta), \tag{6.1.31}$$

substitute it into the transformed advection-diffusion equation (6.1.30) to find

$$\frac{1}{(a^2+b^2)T(t)}\frac{dT}{dt} = -\frac{1}{X(\xi)}\frac{dX}{d\xi} + \frac{\nu}{X(\xi)}\frac{d^2X}{d\xi^2} + \frac{\nu}{Y(\eta)}\frac{d^2Y}{d\eta^2}, \tag{6.1.32}$$

and so by setting,

$$\frac{1}{(a^2+b^2)T(t)}\frac{dT}{dt} = -\nu\pi^2(B^2 - A^2), \tag{6.1.33}$$

$$-\frac{1}{\nu X(\xi)}\frac{dX}{d\xi} + \frac{1}{X(\xi)}\frac{d^2X}{d\xi^2} = (A\pi)^2, \quad \frac{1}{Y(\eta)}\frac{d^2Y}{d\eta^2} = -(B\pi)^2, \tag{6.1.34}$$

where A and B are arbitrary constants (the constants on the right hand sides have been chosen to make the solution periodic in η), we find

$$T(t) = C_0 e^{-\nu\pi^2(a^2+b^2)(B^2-A^2)t}, \tag{6.1.35}$$

$$X(\xi) = C_1 \exp(\lambda_1\xi) + C_2 \exp(\lambda_2\xi), \quad Y(\eta) = C_3 \sin(B\pi\eta) + C_4 \cos(B\pi\eta), \tag{6.1.36}$$

where C_0, C_1, C_2, C_3, C_4 are arbitrary constants, and

$$\lambda_1 = \frac{1 + \sqrt{1 + 4A^2\pi^2\nu^2}}{2\nu}, \quad \lambda_2 = \frac{1 - \sqrt{1 + 4A^2\pi^2\nu^2}}{2\nu}. \tag{6.1.37}$$

Therefore, we obtain an exact solution as

$$u = T(t)X(\xi)Y(\eta) = C_0 e^{-\nu\pi^2(a^2+b^2)(B^2-A^2)t} \left[C_1 \exp(\lambda_1\xi) + C_2 \exp(\lambda_2\xi)\right]\left[C_3 \sin(B\pi\eta) + C_4 \cos(B\pi\eta)\right]. \quad (6.1.38)$$

Note that it is easy to obtain a steady solution by taking $A = B$ (see Section 7.3). This is very nice. But this solution looks complicated to me. I cannot even think about taking a linear combination of this solution (which is also exact, though). I like a simple one such as

$$u = C_0 \cos(B\pi\eta) \exp\left(\frac{1 - \sqrt{1 + 4A^2\pi^2\nu^2}}{2\nu}\xi\right), \quad (6.1.39)$$

which has been obtained by taking $C_1 = 0$, $C_2 = 1$, $C_3 = 0$, $C_4 = 1$. I can now imagine what this solution looks like. Can you?

6.1.4   Advection-Diffusion-Reaction Equation

I like the one-dimensional advection-diffusion-reaction equation,

$$u_t + a u_x = \nu\, u_{xx} - ku, \quad (6.1.40)$$

where a, ν, and k are positive constants. I like it because I can easily derive an exact solution. Assume

$$u(x,t) = T(t)X(x), \quad (6.1.41)$$

and substitute it into the advection-diffusion-reaction equation to find

$$\frac{1}{T(t)}\frac{dT}{dt} = -\frac{a}{X(x)}\frac{dX}{dx} + \frac{\nu}{X(x)}\frac{d^2X}{dx^2} - k. \quad (6.1.42)$$

I like solutions that decay in time, and so I set

$$\frac{1}{T(t)}\frac{dT}{dt} = -A^2 \qquad -\frac{a}{X(x)}\frac{dX}{dx} + \frac{\nu}{X(x)}\frac{d^2X}{dx^2} - k = A^2, \quad (6.1.43)$$

where A is an arbitrary constant. Then, I obtain

$$T(t) = C_0 e^{-A^2 t}, \quad (6.1.44)$$

$$X(x) = C_1 \exp(\lambda_1 x) + C_2 \exp(\lambda_2 x), \quad (6.1.45)$$

where C_0, C_1, C_2 are arbitrary constants, and

$$\lambda_1 = \frac{a + \sqrt{a^2 + 4\nu(k + A^2)}}{2\nu}, \quad \lambda_2 = \frac{a - \sqrt{a^2 + 4\nu(k + A^2)}}{2\nu}. \quad (6.1.46)$$

Therefore, we obtain

$$u = T(t)X(x) = C_0 e^{-A^2 t}\left[C_1 \exp\left(\frac{a + \sqrt{a^2 + 4\nu(k + A^2)}}{2\nu}x\right) + C_2 \exp\left(\frac{a - \sqrt{a^2 + 4\nu(k + A^2)}}{2\nu}x\right)\right]. \quad (6.1.47)$$

I would set $A^2 = k$ so that the solution decays at the reaction rate k:

$$u = C_0 e^{-kt}\left[C_1 \exp\left(\frac{a + \sqrt{a^2 + 8\nu k}}{2\nu}x\right) + C_2 \exp\left(\frac{a - \sqrt{a^2 + 8\nu k}}{2\nu}x\right)\right]. \quad (6.1.48)$$

I like this exact solution because a steady solution is easily obtained by taking $A = 0$ in Equation (6.1.47):

$$u = C_1 \exp\left(\frac{a + \sqrt{a^2 + 4\nu k}}{2\nu}x\right) + C_2 \exp\left(\frac{a - \sqrt{a^2 + 4\nu k}}{2\nu}x\right). \quad (6.1.49)$$

6.1.5 Oscillating Solutions with Complex Variables: Diffusion

Notice that the exact solutions in the previous subsections will decay as time goes on and eventually vanish (because of the exponential term). It would be nice also to have solutions which keep changing and never vanish. Then, I like the idea of using the complex variables in separation of variables because it makes it possible to obtain such solutions [54]. Consider the one-dimensional diffusion equation,

$$u_t = \nu u_{yy}, \quad \text{in } y \in (0, d), \tag{6.1.50}$$

with the time-dependent boundary conditions,

$$u(0) = 0, \tag{6.1.51}$$
$$u(d) = U\cos(\omega t), \tag{6.1.52}$$

where U and ω are arbitrary constants. Note that it is very hard to apply the solution obtained in Section 6.1.2 to this problem because it is not compatible with the boundary condition at $y = d$. Now, an interesting strategy for deriving an exact solution to this problem is to first write the boundary condition as

$$u(d) = U\exp(i\omega t), \tag{6.1.53}$$

where $i = \sqrt{-1}$, so that its real part describes the original condition; then, derive a solution by separation of variables, which possibly involves complex variables; and finally extract the real part of the solution, which is the exact solution for the original boundary condition. So, assume $u(y, t) = Y(y)T(t)$ and substitute into the diffusion equation to get

$$\frac{1}{T(t)}\frac{dT}{dt} = \frac{\nu}{Y(y)}\frac{d^2Y}{dy^2} = \text{constant.} \tag{6.1.54}$$

The boundary condition suggests that the constant should be $i\omega$, which leads to

$$T(t) = C\exp(i\omega t), \tag{6.1.55}$$
$$Y(y) = A\exp(\lambda y) + B\exp(-\lambda y), \tag{6.1.56}$$

where C, A, B are arbitrary constants and

$$\lambda = \sqrt{\frac{i\omega}{\nu}} = (1+i)k, \quad k = \sqrt{\frac{\omega}{2\nu}}, \tag{6.1.57}$$

(note that $\sqrt{i} = [\exp(i\pi/2)]^{1/2} = \exp(i\pi/4) = (1+i)/\sqrt{2}$). The constants are determined by the boundary conditions:

$$B = -A, \quad AC = \frac{U}{\exp(\lambda d) - \exp(-\lambda d)} = \frac{U}{2\sinh(\lambda d)}. \tag{6.1.58}$$

We thus obtain the complex solution,

$$u(y, t) = T(t)Y(y) = \frac{\sinh(\lambda y)}{\sinh(\lambda d)}U\exp(i\omega t). \tag{6.1.59}$$

Finally, we take the real part of this to get the solution to the original problem,

$$u(y, t) = Q\left[\sinh(ky)\cos(ky)\sinh(kd)\cos(kd) + \cosh(ky)\sin(ky)\cosh(kd)\sin(kd)\right]\cos(\omega t)$$
$$+ Q\left[\sinh(ky)\cos(ky)\cosh(kd)\sin(kd) - \cosh(ky)\sin(ky)\sinh(kd)\cos(kd)\right]\sin(\omega t), \tag{6.1.60}$$

where

$$Q = \frac{U}{\cosh^2(kd) - \cos^2(kd)}. \tag{6.1.61}$$

It looks a little complicated, but I like it. It is really nice that this solution does not decay in time and oscillates forever. It can be a good test case for studying a long time behavior of a diffusion scheme.

6.1.6   Oscillating Solutions with Complex Variables: Advection-Diffusion

It is possible to derive an exact oscillating solution to the one-dimensional advection-diffusion equation,

$$u_t + au_y = \nu u_{yy}, \quad \text{in } y \in (0, d), \tag{6.1.62}$$

with the time-dependent boundary conditions,

$$u(0) = 0, \tag{6.1.63}$$

$$u(d) = U\cos(\omega t), \tag{6.1.64}$$

where U and ω are arbitrary constants. The derivation follows exactly the same procedure as in the previous section. So, assume $u(y, t) = Y(y)T(t)$ and substitute into the advection-diffusion equation to get

$$\frac{1}{T(t)}\frac{dT}{dt} = -\frac{a}{Y(y)}\frac{d^2}{dy} + \frac{\nu}{Y(y)}\frac{d^2Y}{dy^2} = i\omega, \tag{6.1.65}$$

which gives

$$T(t) = C\exp(i\omega t), \tag{6.1.66}$$

$$Y(y) = A\exp(\lambda_1 y) + B\exp(\lambda_2 y), \tag{6.1.67}$$

where A, B, and C are arbitrary constants, and

$$\lambda_1 = \frac{a + \sqrt{a^2 + 4\nu\omega i}}{2\nu}, \quad \lambda_2 = \frac{a - \sqrt{a^2 + 4\nu\omega i}}{2\nu}. \tag{6.1.68}$$

Therefore, by applying the boundary conditions, we obtain the following complex solution:

$$u(y, t) = T(t)Y(y) = \frac{\exp(\lambda_1 y) - \exp(\lambda_2 y)}{\exp(\lambda_1 d) - \exp(\lambda_2 d)}U\exp(i\omega t), \tag{6.1.69}$$

whose real part is the exact solution to the original problem. I haven't been able to analytically extract the real part of the complex solution, but I like it because I can easily compute it numerically and it is useful for testing time-accurate advection-diffusion schemes.

6.2   Exact Solutions with Complex Variables

I like the Cauchy-Riemann system for the velocity potential ϕ and the stream function ψ,

$$\psi_x + \phi_y = 0, \tag{6.2.1}$$

$$\phi_x - \psi_y = 0, \tag{6.2.2}$$

because this implies that ϕ and ψ are the real and imaginary parts of an analytic function F of a complex variable $Z = x + iy$:

$$F(Z) = \phi(z, y) + i\psi(z, y), \tag{6.2.3}$$

where $i = \sqrt{-1}$. F is called the complex potential. The velocity is obtained by the derivative of the complex potential:

$$W(Z) = F' = dF/dZ = u - iv, \tag{6.2.4}$$

which is called the complex velocity. Note that *any* analytic function of Z represents a flow. In other words, we can generate an exact solution for a steady, incompressible, inviscid, irrotational two-dimensional flow, just by picking up an analytic function of Z [68]. This is very nice. In particular, I like the following basic examples:

(a) Flow around a corner of angle α, where α is measured with respect to the x-axis,

$$F = CZ^{\pi/\alpha}, \tag{6.2.5}$$

$$W = u - iv = \frac{\pi C}{\alpha}Z^{\pi/\alpha - 1}. \tag{6.2.6}$$

If $\alpha = \pi$, this is a uniform stream with the velocity C which is a constant. If $\alpha = \pi/2$, it represents a flow against a horizontal wall (stagnation flow).

(b) Uniform Stream at an angle α:

$$F = U_\infty Z e^{-i\alpha}, \tag{6.2.7}$$

$$W = U_\infty e^{-i\alpha}, \tag{6.2.8}$$

$$u = U_\infty \cos\alpha, \tag{6.2.9}$$

$$v = U_\infty \sin\alpha. \tag{6.2.10}$$

(c) Source/Sink with a strength σ located at $Z = Z_0 = x_0 + iy_0$:

$$F = \frac{\sigma}{2\pi} \ln(Z - Z_0), \tag{6.2.11}$$

$$W = \frac{\sigma}{2\pi(Z - Z_0)}, \tag{6.2.12}$$

$$u = \frac{\sigma}{2\pi} \frac{x - x_0}{(x - x_0)^2 + (y - y_0)^2}, \tag{6.2.13}$$

$$v = \frac{\sigma}{2\pi} \frac{y - y_0}{(x - x_0)^2 + (y - y_0)^2}. \tag{6.2.14}$$

(d) Doublet with a strength m at an angle α located at $Z = Z_0 = x_0 + iy_0$:

$$F = -\frac{me^{i\alpha}}{2\pi(Z - Z_0)}, \tag{6.2.15}$$

$$W = \frac{me^{i\alpha}}{2\pi(Z - Z_0)^2}, \tag{6.2.16}$$

$$u = \frac{m}{2\pi} \frac{\left[(x - x_0)^2 - (y - y_0)^2\right]\cos\alpha + 2(x - x_0)(y - y_0)\sin\alpha}{\left[(x - x_0)^2 + (y - y_0)^2\right]^2}, \tag{6.2.17}$$

$$v = \frac{m}{2\pi} \frac{2(x - x_0)(y - y_0)\cos\alpha - \left[(x - x_0)^2 - (y - y_0)^2\right]\sin\alpha}{\left[(x - x_0)^2 + (y - y_0)^2\right]^2}. \tag{6.2.18}$$

(e) Clockwise Vortex with a positive circulation Γ located at $Z = Z_0 = x_0 + iy_0$:

$$F = \frac{i\Gamma}{2\pi} \ln(Z - Z_0), \tag{6.2.19}$$

$$W = \frac{i\Gamma}{2\pi(Z - Z_0)}, \tag{6.2.20}$$

$$u = \frac{\Gamma}{2\pi} \frac{y - y_0}{(x - x_0)^2 + (y - y_0)^2}, \tag{6.2.21}$$

$$v = -\frac{\Gamma}{2\pi} \frac{x - x_0}{(x - x_0)^2 + (y - y_0)^2}. \tag{6.2.22}$$

(f) Superposition:

Above solutions can be superposed to create a new flow: e.g., a uniform flow around a cylinder (of unit radius) at the origin is obtained by the sum of the uniform stream and the doublet.

$$F = U_\infty Z + \frac{U_\infty}{Z}, \tag{6.2.23}$$

$$W = U_\infty - \frac{U_\infty}{Z^2}, \tag{6.2.24}$$

$$u = U_\infty - \frac{U_\infty \cos 2\theta}{r^2}, \tag{6.2.25}$$

$$v = \frac{U_\infty \sin 2\theta}{r^2}. \tag{6.2.26}$$

Here, the doublet strength m has been set $m = 2\pi U_\infty$, so that the cylinder itself becomes a streamline. Also, note that $\alpha = \pi$ has been taken, i.e. the doublet pointing against the uniform stream.

More examples are given in Section 7.11.

6.3   Superposition for Nonlinear Equations

Generally, the principle of superposition (a sum of solutions is also a solution) is valid only for linear equations. But it is sometimes possible to 'superpose' a solution onto another solution for nonlinear equations such as the Euler equations. Consider superposing a perturbation velocity field $\mathbf{v}'$ to a mean flow with a constant velocity $\mathbf{v}_\infty$:

$$\mathbf{v}(\mathbf{x}, t) = \mathbf{v}_\infty + \mathbf{v}'(\mathbf{x} - \mathbf{x}_0, t), \tag{6.3.1}$$

where $\mathbf{x}_0$ indicates some reference center location of the perturbation field (which may be the origin). More specifically, here we consider a type of solution in which the perturbation is convected at the mean flow velocity, i.e.,

$$\mathbf{v}(\mathbf{x}, t) = \mathbf{v}_\infty + \mathbf{v}'(\mathbf{x} - \mathbf{x}_0 - \mathbf{v}_\infty t). \tag{6.3.2}$$

If such a solution exists, it will be very interesting. The perturbation flow field will not affect the mean flow at all; it will be simply convected in the uniform stream. Of course, the density, the pressure, and the temperature will also be convected together at the mean flow velocity:

$$\rho(\mathbf{x}, t) = \rho(\mathbf{x} - \mathbf{x}_0 - \mathbf{v}_\infty t), \tag{6.3.3}$$

$$p(\mathbf{x}, t) = p(\mathbf{x} - \mathbf{x}_0 - \mathbf{v}_\infty t), \tag{6.3.4}$$

$$T(\mathbf{x}, t) = T_\infty + T'(\mathbf{x} - \mathbf{x}_0 - \mathbf{v}_\infty t), \tag{6.3.5}$$

where only the temperature has been decomposed into the mean T_∞ and the perturbation T' for the sake of convenience; others can be obtained from the temperature. In particular, we consider only a flow with constant entropy (homentropic flow), i.e., the following relations are valid,

$$\frac{p}{\rho^\gamma} = \frac{p_\infty}{\rho_\infty{}^\gamma}, \quad \frac{\rho}{\rho_\infty} = \left(\frac{T}{T_\infty}\right)^{\frac{1}{\gamma-1}}, \quad \frac{p}{p_\infty} = \left(\frac{T}{T_\infty}\right)^{\frac{\gamma}{\gamma-1}}, \tag{6.3.6}$$

which can be used to calculate p and ρ from T. Then, we seek such a solution for the Euler equations:

$$\frac{\partial \rho}{\partial t} + \operatorname{div}(\rho \mathbf{v}) = 0, \tag{6.3.7}$$

$$\frac{\partial \mathbf{v}}{\partial t} + (\operatorname{grad} \mathbf{v})\mathbf{v} + \frac{1}{\rho}\operatorname{grad} p = 0. \tag{6.3.8}$$

Note that the energy equation is not needed for homentropic flows. Now, substituting the velocity (6.3.2), the density (6.3.3), the pressure (6.3.4), and the temperature (6.3.5) into the Euler equations, we obtain

$$\operatorname{div}(\rho \mathbf{v}') = 0, \tag{6.3.9}$$

$$(\operatorname{grad} \mathbf{v}')\mathbf{v}' + \frac{\gamma R}{\gamma - 1}\operatorname{grad} T' = 0, \tag{6.3.10}$$

where the isentropic relation $\frac{dp}{\rho} = \frac{\gamma R}{\gamma - 1} dT$ has been used to replace the pressure gradient by the temperature gradient. This is interesting. This means that our desired *unsteady* solution exists if the perturbations satisfy the above *steady* Euler equations. A good example is a vortex-type perturbation in two-dimensions:

$$u_r' = 0, \tag{6.3.11}$$

$$u_\theta' = f(r), \tag{6.3.12}$$

where (u_r', u_θ') is the perturbation velocity in polar coordinates and $f(r)$ is a function to be determined later. Likewise, the density, the pressure, and the temperature depend only on r. Then, the continuity equation and the θ-momentum equation are identically satisfied, and we are left with the r-momentum equation,

$$\frac{(u_\theta')^2}{r} = \frac{\gamma R}{\gamma - 1} \frac{dT'}{dr}. \tag{6.3.13}$$

Integrating this to the far-field $(r \to \infty)$, where the perturbation is assumed to vanish, we get

$$-T' = \frac{\gamma - 1}{\gamma R} \int_r^\infty \frac{f(r)^2}{r} dr. \tag{6.3.14}$$

Now, the form of $f(r)$ needs to be chosen so that this can be fully integrated. A good choice for $f(r)$ is

$$f(r) = \frac{K}{2\pi} r e^{\alpha(1-r^2)/2}, \tag{6.3.15}$$

which describes a vortex perturbation with the strength K and the rate of decay α (it quickly decays to the far-field for a large value of α). Integrating Equation (6.3.14) with this $f(r)$, we obtain

$$T' = -\frac{K^2(\gamma - 1)}{8\alpha\pi^2\gamma R} e^{\alpha(1-r^2)}. \tag{6.3.16}$$

Using this result, we finally obtain the exact solution of a vortex, initially centered at (x_0, y_0), convected in a uniform flow:

$$\frac{u(x,y,t)}{a_\infty} = \frac{u_\infty}{a_\infty} + \frac{u'(\overline{x},\overline{y})}{a_\infty} = \frac{u_\infty}{a_\infty} - \frac{K}{2\pi a_\infty}\overline{y}e^{\alpha(1-\overline{r}^2)/2}, \tag{6.3.17}$$

$$\frac{v(x,y,t)}{a_\infty} = \frac{v_\infty}{a_\infty} + \frac{v'(\overline{x},\overline{y})}{a_\infty} = \frac{v_\infty}{a_\infty} + \frac{K}{2\pi a_\infty}\overline{x}e^{\alpha(1-\overline{r}^2)/2}, \tag{6.3.18}$$

$$\frac{T(x,y,t)}{T_\infty} = \frac{T_\infty}{T_\infty} + \frac{T'(\overline{x},\overline{y})}{T_\infty} = 1 - \frac{K^2(\gamma - 1)}{8\alpha\pi^2 a_\infty^2}e^{\alpha(1-\overline{r}^2)}, \tag{6.3.19}$$

$$\frac{\rho(x,y,t)}{\rho_\infty} = \left(\frac{T(x,y,t)}{T_\infty}\right)^{\frac{1}{\gamma-1}}, \tag{6.3.20}$$

$$\frac{p(x,y,t)}{p_\infty} = \left(\frac{T(x,y,t)}{T_\infty}\right)^{\frac{\gamma}{\gamma-1}}, \tag{6.3.21}$$

where $a_\infty = \sqrt{\gamma p_\infty / \rho_\infty}$,

$$\overline{x} = x - x_0 - u_\infty t, \tag{6.3.22}$$

$$\overline{y} = y - y_0 - v_\infty t, \tag{6.3.23}$$

$$\overline{r} = \sqrt{\overline{x}^2 + \overline{y}^2}. \tag{6.3.24}$$

I like this solution very much. It is really interesting that it is possible to superpose a vortex on a uniform flow and that such a simple flow is an exact solution to the Euler equations. Moreover, other exact solutions could be obtained in a similar way, i.e., by choosing some other steady solution as a perturbation.

6.4   Exact Solutions with Transformations

6.4.1   Cole-Hopf Transformations

One-Dimensional Viscous Burgers' Equation

Consider the one-dimensional viscous Burgers' equation,

$$u_t + uu_x = \nu u_{xx}. \tag{6.4.1}$$

It is well known that the transformation

$$u = -2\nu \frac{\partial(\ln \phi)}{\partial x} = -2\nu \frac{\phi_x}{\phi}, \tag{6.4.2}$$

called the Cole-Hopf transformation, relates $u(x,t)$ and $\phi(x,t) > 0$ in such a way that if ϕ is a positive solution of the linear diffusion equation,

$$\phi_t = \nu \phi_{xx}, \tag{6.4.3}$$

then, u is a solution of the Burgers' equation (6.4.1). This means that we can generate various exact solutions of the viscous Burgers' equation simply by substituting exact solutions of the linear diffusion equation into Equation (6.4.2). This is nice. I like it. See Ref.[13] for various exact solutions, and also see Section 7.8.1 for a few examples.

Two-Dimensional Viscous Burgers' Equation

Consider the two-dimensional viscous Burgers' equations,

$$u_t + uu_x + vu_y - \nu(u_{xx} + u_{yy}) = 0, \tag{6.4.4}$$
$$v_t + uv_x + vv_y - \nu(v_{xx} + v_{yy}) = 0. \tag{6.4.5}$$

Again, we have a nice transformation [37],

$$u = -2\nu \frac{\partial(\ln \phi)}{\partial x} = -2\nu \frac{\phi_x}{\phi}, \quad v = -2\nu \frac{\partial(\ln \phi)}{\partial y} = -2\nu \frac{\phi_y}{\phi}, \tag{6.4.6}$$

which relate $u(x,t)$, $v(x,t)$ and $\phi(x,t) > 0$ in such a way that if ϕ is a positive solution of the linear diffusion equation,

$$\phi_t = \nu(\phi_{xx} + \phi_{yy}), \tag{6.4.7}$$

then, u and v are solutions of the Burgers' equations (6.4.4) and (6.4.5). See Ref.[37] for examples, or Section 7.8.3 for an example.

Vector Form of the Viscous Burgers' Equation (3D)

Consider the vector form of the viscous Burgers' equation,

$$\mathbf{U}_t + (\operatorname{grad} \mathbf{U})\mathbf{U} = \nu \operatorname{div} \operatorname{grad} \mathbf{U}, \tag{6.4.8}$$

where $\mathbf{U} = [u, v, w]^t$. It is nice to know that there is in fact a vector version of the Cole-Hopf transformation. That is, the transformation

$$\mathbf{U} = -2\nu \operatorname{grad}(\ln \phi) = -2\nu \frac{\operatorname{grad} \phi}{\phi}, \tag{6.4.9}$$

relates the solution vector $\mathbf{U}$ and a scalar function $\phi(x, y, z, t) > 0$ in such a way that if ϕ is a positive solution of the linear diffusion equation,

$$\phi_t = \nu \operatorname{div} \operatorname{grad} \phi, \tag{6.4.10}$$

then, $\mathbf{U}$ is a solution of the vector form of the viscous Burgers' equation (6.4.8). This is very nice. I like it very much.

Viscous Burgers' Equation with Source Terms

Consider the one-dimensional viscous Burgers' equation with various source terms,

$$u_t + uu_x = \nu u_{xx} + \epsilon u^2 - ku + d, \tag{6.4.11}$$

where ν, ϵ, k, and d are constants. It is nice to be able to find exact solutions to this equation. I like the following derivation. The equation is transformed by

$$u = p\frac{\partial(\ln \phi)}{\partial x} = p\frac{\phi_x}{\phi}, \tag{6.4.12}$$

into

$$[(\phi_t - \nu\phi_{xx})_x + k\phi_x - d\phi/p]\,\phi^2 + [\phi_t - (p+3\nu)\phi_{xx} + \epsilon p\phi_x]\,\phi\,\phi_x - [p+2\nu](\phi_x)^3 = 0, \tag{6.4.13}$$

which is true for any ϕ if

$$(\phi_t - \nu\phi_{xx})_x + k\phi_x - d\phi/p = 0, \tag{6.4.14}$$
$$\phi_t - (p+3\nu)\phi_{xx} + \epsilon p\phi_x = 0, \tag{6.4.15}$$
$$p + 2\nu = 0, \tag{6.4.16}$$

i.e., $p = -2\nu$, and

$$(\phi_t - \nu\phi_{xx})_x + k\phi_x + \frac{d}{2\nu}\phi = 0, \tag{6.4.17}$$
$$\phi_t - \nu\phi_{xx} - 2\epsilon\nu\phi_x = 0, \tag{6.4.18}$$

thus resulting

$$2\epsilon\nu\phi_{xx} + k\phi_x + \frac{d}{2\nu}\phi = 0. \tag{6.4.19}$$

Insert a trial solution $\phi = C(t)\,e^{\lambda x}$, where $C(t)$ is an arbitrary function of time, into this to find the characteristic equation,

$$2\epsilon\nu\lambda^2 + k\lambda + \frac{d}{2\nu}\phi = 0. \tag{6.4.20}$$

Then, this can be solved easily, i.e.,

$$\lambda = \frac{-k \pm \sqrt{k^2 - 4\epsilon d}}{4\epsilon\nu}. \tag{6.4.21}$$

We now consider three cases: $k^2 - 4\epsilon d > 0$, $k^2 - 4\epsilon d < 0$, and $k^2 - 4\epsilon d = 0$.

(a) **Two distinct real roots:** $k^2 - 4\epsilon d > 0$

Denote the two distinct real roots by λ_1 and λ_2, i.e.,

$$\lambda_1 = \frac{-k - \sqrt{k^2 - 4\epsilon d}}{4\epsilon\nu}, \quad \lambda_2 = \frac{-k + \sqrt{k^2 - 4\epsilon d}}{4\epsilon\nu}, \tag{6.4.22}$$

then one solution ϕ_1 is given by

$$\phi_1 = C_1(t)\,e^{\lambda_1 x}, \tag{6.4.23}$$

where $C_1(t)$ depends only on time, and it can be determined by inserting the solution into Equation (6.4.15) and solving the resulting ordinary differential equation:

$$C_1(t) = K_1\,e^{\nu\lambda_1(\lambda_1 + 2\epsilon)t}, \tag{6.4.24}$$

where K_1 is an arbitrary constant. Therefore, the solution is given by

$$\phi_1 = K_1\exp[\lambda_1 x + \nu\lambda_1(\lambda_1 + 2\epsilon)t]. \tag{6.4.25}$$

Now, we obtain a similar solution for the other root λ_2, and superpose the two to get the general solution,

$$\phi = K_1 \exp\left[\lambda_1 x + \nu\lambda_1(\lambda_1 + 2\epsilon)t\right] + K_2 \exp\left[\lambda_2 x + \nu\lambda_2(\lambda_2 + 2\epsilon)t\right], \tag{6.4.26}$$

where K_2 is another arbitrary constant. Finally, we obtain an exact solution to the Burgers equation (6.4.11) from the solution (6.4.12),

$$u = -2\nu \frac{\lambda_1 K_1 \exp\left[\lambda_1 x + \nu\lambda_1(\lambda_1 + 2\epsilon)t\right] + \lambda_2 K_2 \exp\left[\lambda_2 x + \nu\lambda_2(\lambda_2 + 2\epsilon)t\right]}{K_1 \exp\left[\lambda_1 x + \nu\lambda_1(\lambda_1 + 2\epsilon)t\right] + K_2 \exp\left[\lambda_2 x + \nu\lambda_2(\lambda_2 + 2\epsilon)t\right]}. \tag{6.4.27}$$

(b) **Complex conjugate roots:** $k^2 - 4\epsilon d < 0$

Express the complex roots as

$$\lambda_{1,2} = \lambda_r \pm i\lambda_i, \tag{6.4.28}$$

where

$$i = \sqrt{-1}, \quad \lambda_r = \frac{-k}{4\epsilon\nu}, \quad \lambda_i = \frac{\sqrt{4\epsilon d - k^2}}{4\epsilon\nu}, \tag{6.4.29}$$

then insert these roots directly into Equation (6.4.26), split it into the real and imaginary parts, and form a linear combination of them to get the general solution,

$$\phi = e^{\lambda_r x + \nu(\lambda_r^2 - \lambda_i^2 + 2\epsilon\lambda_r)t}\left[K_1' \cos\theta + K_2' \sin\theta\right], \tag{6.4.30}$$

where K_1' and K_2' are arbitrary constants, and

$$\theta = \lambda_i x + 2\nu(\lambda_r\lambda_i + \epsilon\lambda_i)t. \tag{6.4.31}$$

Therefore an exact solution is given by

$$u = -2\nu\frac{\phi_x}{\phi} = -2\nu\frac{(\lambda_r K_1' + \lambda_i K_2')\cos\theta + (\lambda_r K_2' - \lambda_i K_1')\sin\theta}{K_1' \cos\theta + K_2' \sin\theta}. \tag{6.4.32}$$

(c) **Multiple roots:** $k^2 - 4\epsilon d = 0$

Denote the single root by λ_s,

$$\lambda_s = \frac{-k}{4\epsilon\nu}, \tag{6.4.33}$$

and find the general solution of Equation (6.4.19),

$$\phi = e^{\lambda_s x}\left[C_1(t) + x\,C_2(t)\right]. \tag{6.4.34}$$

Insert this into Equation (6.4.15) and solve the resulting system of ordinary differential equations for $C_1(t)$ and $C_2(t)$ straightforwardly to get

$$C_1(t) = 2K_1''\nu(\lambda_s + \epsilon)t\,e^{\nu\lambda_s(\lambda_s + 2\epsilon)t}, \tag{6.4.35}$$

$$C_2(t) = K_2''\,e^{\nu\lambda_s(\lambda_s + 2\epsilon)t}, \tag{6.4.36}$$

where K_1'' and K_2'' are arbitrary constants, and thus obtain

$$\phi = e^{\lambda_1 x + \nu\lambda_s(\lambda_s + 2\epsilon)t}\left[K_2''x + 2K_1''\nu(\lambda_s + \epsilon)t\right]. \tag{6.4.37}$$

Finally, we obtain an exact solution as follows,

$$u = -2\nu\frac{\phi_x}{\phi} = -2\nu\frac{K_2''(1 + \lambda_s x) + 2K_1''\nu\lambda_s(\lambda_s + \epsilon)t}{K_2''x + 2K_1''\nu(\lambda_s + \epsilon)t}. \tag{6.4.38}$$

Finally, I remark that the method described above is originally studied and exact solutions of a more general equation is discussed in Ref.[9].

6.4.2 Hodograph Transformation

Consider the compressible continuity equation in two dimensions,

$$\frac{\partial(\rho u)}{\partial x} + \frac{\partial(\rho v)}{\partial y} = 0. \tag{6.4.39}$$

This is automatically satisfied by the stream function ψ defined by

$$u = \frac{\rho_0}{\rho}\frac{\partial \psi}{\partial y} = \frac{\rho_0}{\rho}\psi_y, \tag{6.4.40}$$

$$v = -\frac{\rho_0}{\rho}\frac{\partial \psi}{\partial x} = -\frac{\rho_0}{\rho}\psi_x, \tag{6.4.41}$$

where the subscript 0 denotes the stagnation state. This stream function is governed by the nonlinear equation,

$$\left[1 - \left(\frac{\rho_0}{\rho}\right)^2 \frac{\psi_y^2}{c^2}\right]\psi_{xx} + \left[1 - \left(\frac{\rho_0}{\rho}\right)^2 \frac{\psi_x^2}{c^2}\right]\psi_{yy} + 2\left(\frac{\rho_0}{\rho}\right)^2 \frac{\psi_x\psi_y}{c^2}\psi_{xy} = 0, \tag{6.4.42}$$

where c is the speed of sound given by

$$c^2 = c_0^2 - \frac{\gamma - 1}{2}\left(\frac{\rho_0}{\rho}\right)^2 (\psi_x^2 + \psi_y^2), \tag{6.4.43}$$

and

$$\frac{\rho_0}{\rho} = \left[1 + \frac{\gamma - 1}{2}\left(\frac{\rho_0}{\rho}\right)^2 \frac{\psi_x^2 + \psi_y^2}{c^2}\right]^{\frac{1}{\gamma-1}}. \tag{6.4.44}$$

These are basically derived from the irrotationality condition $v_x - u_y = 0$ and the momentum equations (see Ref.[138]). Obviously, it is not very easy to solve this nonlinear equation; it looks even more complicated than the nonlinear potential equation (3.19.14). However, there is a nice way to get around it. That is, we exchange the independent variables (x, y) and the dependent variables (V, θ), where V is the flow speed and θ is the flow angle, and transform the equation (6.4.42) into

$$V^2\frac{\partial^2 \psi}{\partial V^2} + V\left(1 + \frac{V^2}{c^2}\right)\frac{\partial \psi}{\partial V} + \left(1 - \frac{V^2}{c^2}\right)\frac{\partial^2 \psi}{\partial \theta^2} = 0, \tag{6.4.45}$$

(see Ref.[138] for derivation). This is a *linear* differential equation in ψ and therefore can be solved relatively easily. In fact, to find analytical solutions, we don't really need to solve this in a formal way. Pick any function and see if it satisfies the equation. If it does, then it is a solution. Moreover, any linear combination of such solutions (i.e., superposition) is also a solution because the equation is linear. This is very nice. Especially, I like the solution called Ringleb's flow described in Section 7.13.6. See Ref.[138] for other examples.

6.5 Exact Solutions for Linear Systems of Conservation Laws

Consider the linear hyperbolic system of one-dimensional conservation laws of the form,

$$\frac{\partial \mathbf{U}}{\partial t} + \mathbf{A}\frac{\partial \mathbf{U}}{\partial x} = 0, \tag{6.5.1}$$

where $\mathbf{U} = \mathbf{U}(x, t)$ and $\mathbf{A}$ is a constant coefficient matrix having real eigenvalues and linearly independent eigenvectors. If we are given an initial solution as

$$\mathbf{U}(x, 0) = \mathbf{U}_0(x), \tag{6.5.2}$$

then, we can find the exact solution as follows. First, we multiply Equation (6.5.1) by the left-eigenvector matrix $\mathbf{L}$ (of $\mathbf{A}$) from the left,

$$\mathbf{L}\frac{\partial \mathbf{U}}{\partial t} + \mathbf{LARL}\frac{\partial \mathbf{U}}{\partial x} = 0, \tag{6.5.3}$$

where $\mathbf{R}$ is the right-eigenvector matrix (the inverse of $\mathbf{L}$). Because the matrix $\mathbf{L}$ is constant, we can write

$$\frac{\partial(\mathbf{LU})}{\partial t} + \mathbf{LAR}\frac{\partial(\mathbf{LU})}{\partial x} = 0. \tag{6.5.4}$$

Defining the characteristic variables (or the Riemann invariants) by

$$\mathbf{W} = \mathbf{LU}, \tag{6.5.5}$$

and the diagonal matrix $\mathbf{\Lambda}$ by $\mathbf{\Lambda} = \mathbf{LAR}$, we arrive at the diagonalized form of the system (6.5.1).

$$\frac{\partial \mathbf{W}}{\partial t} + \mathbf{\Lambda}\frac{\partial \mathbf{W}}{\partial x} = 0. \tag{6.5.6}$$

We point out that each component is now a linear scalar advection equation

$$\frac{\partial w_k}{\partial t} + \lambda_k \frac{\partial w_k}{\partial x} = 0, \tag{6.5.7}$$

where w_k denotes the k-th component of $\mathbf{W}$ and λ_k is the k-th eigenvalue. Therefore, the exact solution of the k-th component is given simply by

$$w_k(x,t) = w_k(x - \lambda_k t, 0). \tag{6.5.8}$$

Then, by definition (6.5.5), we obtain

$$\mathbf{U}(x,t) = \mathbf{RW}(x,t) = \sum_k w_k(x,t)\mathbf{r}_k = \sum_k w_k(x - \lambda_k t, 0)\mathbf{r}_k, \tag{6.5.9}$$

or equivalently, (by Equation (6.5.5) again),

$$\mathbf{U}(x,t) = \sum_k (\ell_k \mathbf{U}_0(x - \lambda_k t))\mathbf{r}_k, \tag{6.5.10}$$

where ℓ_k is the k-th row of $\mathbf{L}$ (the k-th left-eigenvector) and $\mathbf{r}_k$ is the k-th column of $\mathbf{R}$ (the k-th right-eigenvector). This is the exact solution for the initial solution (6.5.2).

Now we consider a simple wave. A simple wave is defined as a solution whose variation is confined in a one-dimensional subspace spanned by a single eigenvector. For example, if the initial solution is a j-th simple wave, we project it onto the j-th eigenspace to write

$$\begin{aligned}
\mathbf{U}_0(x) &= \sum_k (\ell_k \mathbf{U}_0(x))\mathbf{r}_k \\
&= (\ell_j \mathbf{U}_0(x))\mathbf{r}_j + \sum_{k:k\neq j} c_k \mathbf{r}_k, \tag{6.5.11}
\end{aligned}$$

where c_k are constant. Then, the exact solution (6.5.10) can be written as

$$\begin{aligned}
\mathbf{U}(x,t) &= \sum_k (\ell_k \mathbf{U}_0(x - \lambda_j t))\mathbf{r}_k \\
&= (\ell_j \mathbf{U}_0(x - \lambda_j t))\mathbf{r}_j + \sum_{k:k\neq j} c_k \mathbf{r}_k \\
&= \mathbf{U}_0(x - \lambda_j t). \tag{6.5.12}
\end{aligned}$$

Therefore, the initial solution is preserved perfectly for a simple wave. This is natural because only one advection speed is relevant in a simple wave. Finally, we give an example: set the following initial solution,

$$\mathbf{U}_0(x) = \overline{\mathbf{U}} + \sigma \sin(x)\,\mathbf{r}_j, \tag{6.5.13}$$

where $\overline{\mathbf{U}}$ is a constant state and σ is an amplitude of the sine wave, then the exact solution is given by

$$\mathbf{U}(x,t) = \mathbf{U}_0(x - \lambda_j t) = \overline{\mathbf{U}} + \sigma \sin(x - \lambda_j t)\,\mathbf{r}_j. \tag{6.5.14}$$

It is indeed simple and also general enough to be applicable to arbitrary linear hyperbolic systems of the form Equation (6.5.1). Particularly, I like the fact that the result (and all the above discussion) is valid even for steady systems, i.e., Equation (6.5.1) with (x,t) replaced by (x,y) or (y,x). This is very nice.

6.6 Simple Wave Solutions for Nonlinear Systems of Conservation Laws

Consider the nonlinear hyperbolic system of one-dimensional conservation laws of the form,

$$\frac{\partial \mathbf{U}}{\partial t} + \frac{\partial \mathbf{F}(\mathbf{U})}{\partial x} = 0, \tag{6.6.1}$$

or

$$\frac{\partial \mathbf{U}}{\partial t} + \mathbf{A}(\mathbf{U})\frac{\partial \mathbf{U}}{\partial x} = 0, \tag{6.6.2}$$

where $\mathbf{U} = \mathbf{U}(x,t)$ and $\mathbf{A}(\mathbf{U})$ is a coefficient matrix which is diagonalizable but no longer constant. Unlike the linear case, for a given initial solution such as

$$\mathbf{U}(x,0) = \mathbf{U}_0(x), \tag{6.6.3}$$

we cannot find the exact solution generally in a closed form. To see this, as in the linear case, we multiply Equation (6.6.2) by the left-eigenvector matrix $\mathbf{L}$ from the left,

$$\mathbf{L}\frac{\partial \mathbf{U}}{\partial t} + \mathbf{LARL}\frac{\partial \mathbf{U}}{\partial x} = 0. \tag{6.6.4}$$

Here is a difficulty. We can no longer define the characteristic variables as in Equation (6.5.5) because $\mathbf{L}$ is not constant any more. Instead, I define the characteristic variables by

$$\partial \mathbf{W} = \mathbf{L}\partial \mathbf{U}, \tag{6.6.5}$$

and arrive at the diagonalized form of the system.

$$\frac{\partial \mathbf{W}}{\partial t} + \mathbf{\Lambda}\frac{\partial \mathbf{W}}{\partial x} = 0. \tag{6.6.6}$$

Each component is now a nonlinear scalar advection equation with the exact solution,

$$w_k(x,t) = w_k(x - \lambda_k t, 0), \tag{6.6.7}$$

which is valid up to the time when a shock forms. Note that this solution is given implicitly in general because λ_k depends on the solution. Then, we use Equation (6.6.5) to get the solution in the conservative variables, (6.6.5)

$$\partial \mathbf{U}(x,t) = \mathbf{R}\partial \mathbf{W}(x,t) = \sum_k \partial w_k(x,t)\mathbf{r}_k = \sum_k \partial w_k(x - \lambda_k t, 0)\mathbf{r}_k, \tag{6.6.8}$$

and so

$$\partial \mathbf{U}(x,t) = \sum_k (\ell_k \partial \mathbf{U}_0(x - \lambda_k t))\mathbf{r}_k. \tag{6.6.9}$$

I like it, but it is too bad that we cannot get $\mathbf{U}(x,t)$ unless this is integrable. We cannot project even the initial solution onto the space of an eigenvector exactly:

$$\mathbf{U}_0(x) \neq \sum_k (\ell_k \mathbf{U}_0(x))\mathbf{r}_k. \tag{6.6.10}$$

Then, we consider a simple wave solution because it gives us a hope to get exact solutions. Let $h(x,t)$ be any smooth function that satisfies the j-th component of Equation (6.6.6),

$$\frac{\partial h}{\partial t} + \lambda_j(\mathbf{U})\frac{\partial h}{\partial x} = 0, \tag{6.6.11}$$

so that $h(x,t)$ is the j-th characteristic variable $w_j(x,t) = h(x - \lambda_j(\mathbf{U})\,t) = h(\xi)$, and assume that other characteristic variables are constant. Then, we have from Equation (6.6.5)

$$\partial \mathbf{U}(x,t) = \mathbf{r}_j(\mathbf{U})\,\partial h(\xi), \tag{6.6.12}$$

or because $\mathbf{U}(x,t) = \mathbf{U}(\xi)$, we may write

$$\frac{d\mathbf{U}(\xi)}{d\xi} = \mathbf{r}_j(\mathbf{U})\frac{dh(\xi)}{d\xi}. \tag{6.6.13}$$

If this is integrable, we can find an exact solution $\mathbf{U}$ for the j-th simple wave. This is possible in some cases. Note that the variable $\mathbf{U}$ can be taken to be any variable (e.g., conservative or primitive) as long as the eigenvector $\mathbf{r}_j$ is defined consistently. Some examples are given in the next section.

6.7   Some Exact Simple Wave Solutions

6.7.1   Case 1: $\mathbf{r}_j = $ constant

I like eigenvectors that are independent of the solution because the ordinary differential equation (6.6.13) is then trivial. We readily obtain a simple wave solution as

$$\mathbf{U}(\xi) = h(\xi)\,\mathbf{r}_j + \text{constant.} \tag{6.7.1}$$

The entropy wave in the Euler equations is a good example: the eigenvector, based on the primitive variables $\mathbf{V} = [\rho, u, p]^t$, is given by

$$\mathbf{r}_j = \begin{bmatrix} 1 \\ 0 \\ 0 \end{bmatrix}, \tag{6.7.2}$$

with eigenvalue $\lambda_j = u$. So, it follows immediately from Equation (6.7.1) that

$$\rho \;=\; h(x - u\,t) + \text{constant,} \tag{6.7.3}$$
$$u \;=\; \text{constant,} \tag{6.7.4}$$
$$p \;=\; \text{constant.} \tag{6.7.5}$$

This shows that the wave does not deform because the speed u is constant (the entropy wave is linearly degenerate). Any function (even a discontinuous one) can be chosen for $h(x)$. It will be simply convected at the velocity u.

6.7.2   Case 2: $\mathbf{r}_j \propto \mathbf{U}$

Consider the case where the eigenvector is proportional to the solution vector,

$$\mathbf{r}_j = \overline{\mathbf{M}}\,\mathbf{U}, \tag{6.7.6}$$

where $\overline{\mathbf{M}}$ is a constant matrix. In this case, we have from Equation (6.6.13),

$$\frac{d\mathbf{U}(\xi)}{d\xi} = \overline{\mathbf{M}}\,\mathbf{U}\frac{dh(\xi)}{d\xi} = \overline{\mathbf{M}}\,\mathbf{U}h^{'}(\xi). \tag{6.7.7}$$

This is a linear system of ordinary differential equations and therefore can be solved analytically by a standard technique. This is nice. A good example is the Alfvén wave of the ideal magnetohydrodynamic equations. For the primitive variables $\mathbf{U} = [\rho, v_x, v_y, v_z, B_x, B_y, B_z, p]^t$, where (v_x, v_y, v_z) is the velocity vector and (B_x, B_y, B_z) is the magnetic field, with the coefficient matrix,

$$\mathbf{A} = \begin{bmatrix} v_x & \rho & 0 & 0 & 0 & 0 & 0 \\ 0 & v_x & 0 & 0 & \dfrac{B_y}{4\pi\rho} & \dfrac{B_z}{4\pi\rho} & \dfrac{1}{\rho} \\ 0 & 0 & v_x & 0 & -\dfrac{B_x}{4\pi\rho} & 0 & 0 \\ 0 & 0 & 0 & v_x & 0 & -\dfrac{B_x}{4\pi\rho} & 0 \\ 0 & B_y & -B_x & 0 & v_x & 0 & 0 \\ 0 & B_z & 0 & -B_x & 0 & v_x & 0 \\ 0 & \rho a^2 & 0 & 0 & 0 & 0 & v_x \end{bmatrix}, \tag{6.7.8}$$

where a is the speed of sound ($a^2 = \gamma p / \rho$), the eigenvector for the Alfvén waves is given by

$$\mathbf{r}_j^{\pm} = \begin{bmatrix} 0 \\ 0 \\ \pm B_z \\ \mp B_y \\ -\sqrt{4\pi\rho}\,B_z \\ \sqrt{4\pi\rho}\,B_y \\ 0 \end{bmatrix}, \tag{6.7.9}$$

with the eigenvalues

$$\lambda_j^{\pm} = v_x \pm \frac{B_x}{\sqrt{4\pi\rho}}. \tag{6.7.10}$$

Note that zeroes in Equation (6.7.9) correspond to the density, the velocity component v_x, and the pressure. This means that these variables are constant through the Alfvén waves (see Section 1.16.2). Therefore, the eigenvalues are constant; this wave is linearly degenerate. Now, because the density is constant, the eigenvector is linear in the primitive variable,

$$\mathbf{r}_j = \overline{\mathbf{M}}\,\mathbf{U}, \tag{6.7.11}$$

where $\mathbf{U}$ is taken to be the primitive variable and

$$\overline{\mathbf{M}} = \begin{bmatrix} 0 & 0 & 0 & 0 & 0 & 0 & 0 \\ 0 & 0 & 0 & 0 & 0 & 0 & 0 \\ 0 & 0 & 0 & 0 & 0 & 0 & \pm 1 \\ 0 & 0 & 0 & 0 & 0 & \mp 1 & 0 \\ 0 & 0 & 0 & 0 & 0 & 0 & 0 \\ 0 & 0 & 0 & 0 & 0 & 0 & -\sqrt{4\pi\rho} \\ 0 & 0 & 0 & 0 & 0 & \sqrt{4\pi\rho} & 0 \end{bmatrix}. \tag{6.7.12}$$

Since only four variables, v_y, v_z, p, B_y, B_z, are relevant, we ignore other components in the following. Now, we derive an exact solution by solving Equation (6.6.13):

$$\frac{d}{d\xi} \begin{bmatrix} v_y \\ v_z \\ B_y \\ B_z \end{bmatrix} = \begin{bmatrix} 0 & 0 & 0 & \pm 1 \\ 0 & 0 & \mp 1 & 0 \\ 0 & 0 & 0 & -\sqrt{4\pi\rho} \\ 0 & 0 & \sqrt{4\pi\rho} & 0 \end{bmatrix} \begin{bmatrix} v_y \\ v_z \\ B_y \\ B_z \end{bmatrix} h'(\xi). \tag{6.7.13}$$

The standard diagonalization yields the decoupled system,

$$\frac{d(v_y \pm B_y/\sqrt{4\pi\rho})}{d\xi} = 0, \tag{6.7.14}$$

$$\frac{d(v_z \pm B_z/\sqrt{4\pi\rho})}{d\xi} = 0, \tag{6.7.15}$$

$$\frac{d(B_z + i\,B_y)}{d\xi} = -i\,a\,(B_z + i\,B_y)\,h'(\xi), \tag{6.7.16}$$

$$\frac{d(B_z - i\,B_y)}{d\xi} = i\,a\,(B_z - i\,B_y)\,h'(\xi), \tag{6.7.17}$$

whose general solution is

$$v_y(\xi) = \mp B_y(\xi)/\sqrt{4\pi\rho} + c_{vy}, \tag{6.7.18}$$

$$v_z(\xi) = \mp B_z(\xi)/\sqrt{4\pi\rho} + c_{vz}, \tag{6.7.19}$$

$$B_y(\xi) = K_R \cos(\sqrt{4\pi\rho}\,h(\xi)) + K_I \sin(\sqrt{4\pi\rho}\,h(\xi)), \tag{6.7.20}$$

$$B_z(\xi) = K_I \cos(\sqrt{4\pi\rho}\,h(\xi)) - K_R \sin(\sqrt{4\pi\rho}\,h(\xi)), \tag{6.7.21}$$

where c_{vy}, c_{vz}, K_R and K_I are arbitrary constants. This is the exact simple wave solution for the Alfvén wave in the most general form. Note that we can use only one solution, i.e., we can take only either positive or negative sign in the formula; otherwise it will not be a simple wave. Here is an example: take $h(\xi) = \xi/\sqrt{4\pi\rho}$ and $K_I = 0$, and obtain the simple wave associated with λ_j^+,

$$v_y(x - \lambda_j^+ t) = -K_R' \cos(x - \lambda_j^+ t) + c_{vy}, \tag{6.7.22}$$

$$v_z(x - \lambda_j^+ t) = K_R' \sin(x - \lambda_j^+ t) + c_{vz}, \tag{6.7.23}$$

$$B_y(x - \lambda_j^+ t) = \sqrt{4\pi\rho}\, K_R' \cos(x - \lambda_j^+ t), \tag{6.7.24}$$

$$B_z(x - \lambda_j^+ t) = -\sqrt{4\pi\rho}\, K_R' \sin(x - \lambda_j^+ t), \tag{6.7.25}$$

where c_{vy}, c_{vz}, and K_R' are arbitrary constants, and $K_R = \sqrt{4\pi\rho}\, K_R'$. I like this simple Alfvén wave solution. It is linearly degenerate, and therefore the wave must be preserved at all times. This can be a good test case for code verification.

6.7.3   Case 3: Other cases

The acoustic wave in the Euler equations is a good example. In terms of the primitive variables, the acoustic eigenvectors are given by

$$\mathbf{r}_j^\pm = \begin{bmatrix} \pm\rho/a \\ 1 \\ \pm\rho a \end{bmatrix}, \tag{6.7.26}$$

where a is the speed of sound. The associated eigenvalues are

$$\lambda_j^\pm = u \pm a. \tag{6.7.27}$$

It must be kept in mind that there are two acoustic waves, $+$ and $-$, but we can take only either $+$ or $-$ for this solution to be a simple wave. Now, we insert these into Equation (6.6.13) to get

$$\frac{d}{d\xi} \begin{bmatrix} \rho \\ u \\ p \end{bmatrix} = \begin{bmatrix} \pm\rho/a \\ 1 \\ \pm\rho a \end{bmatrix} \frac{dh(\xi)}{d\xi}. \tag{6.7.28}$$

This is nonlinear, except for the velocity which can be integrated easily,

$$\int_{u_\infty}^{u} du = \int_{h_\infty}^{h} dh, \tag{6.7.29}$$

to give

$$u(\xi) = h(\xi) + u_\infty, \tag{6.7.30}$$

where we have set $h_\infty = 0$ so that u_∞ is a mean flow value and any non-zero h can be thought of as a perturbation to the mean flow. Now, the density component can also be integrated if the flow is adiabatic, i.e. if

$$p/\rho^\gamma = p_\infty/\rho_\infty^\gamma. \tag{6.7.31}$$

We take advantage of this adiabatic relation to write the speed of sound in terms of the density as

$$\frac{a}{a_\infty} = \left(\frac{\rho}{\rho_\infty}\right)^{\frac{\gamma-1}{2}}, \tag{6.7.32}$$

which is substituted into the first component of Equation (6.7.28) to yield

$$\frac{d\rho}{d\xi} = \pm\left(\frac{\rho}{\rho_\infty}\right)^{\frac{3-\gamma}{2}} \frac{\rho_\infty}{a_\infty} \frac{dh(\xi)}{d\xi}. \tag{6.7.33}$$

This can be integrated easily,

$$\int_{1}^{\rho/\rho_\infty} \left(\frac{\rho}{\rho_\infty}\right)^{\frac{\gamma-3}{2}} d\left(\frac{\rho}{\rho_\infty}\right) = \pm\frac{1}{a_\infty} \int_{h_\infty=0}^{h} dh, \tag{6.7.34}$$

resulting

$$\frac{\rho(\xi)}{\rho_\infty} = \left[1 \pm \frac{\gamma - 1}{2}\frac{h(\xi)}{a_\infty}\right]^{\frac{2}{\gamma-1}}, \tag{6.7.35}$$

or by Equation (6.7.30)

$$\frac{\rho(\xi)}{\rho_\infty} = \left[1 \pm \frac{\gamma - 1}{2}\left(\frac{u(\xi)}{a_\infty} - M_\infty\right)\right]^{\frac{2}{\gamma-1}}, \tag{6.7.36}$$

where $M_\infty = u_\infty/a_\infty$. The pressure follows from the adiabatic relation (6.7.31),

$$\frac{p(\xi)}{p_\infty} = \left(\frac{\rho(\xi)}{\rho_\infty}\right)^\gamma = \left[1 \pm \frac{\gamma - 1}{2}\left(\frac{u(\xi)}{a_\infty} - M_\infty\right)\right]^{\frac{2\gamma}{\gamma-1}}. \tag{6.7.37}$$

This completes the derivation of the exact simple wave solution for the acoustic wave. Now, we summarize the exact acoustic simple wave solution:

$$\frac{u(x - \lambda_j^\pm t)}{a_\infty} = M_\infty + \frac{h(x - \lambda_j^\pm t)}{a_\infty}, \tag{6.7.38}$$

$$\frac{\rho(x - \lambda_j^\pm t)}{\rho_\infty} = \left[1 \pm \frac{\gamma - 1}{2}\left(\frac{u(x - \lambda_j^\pm t)}{a_\infty} - M_\infty\right)\right]^{\frac{2}{\gamma-1}}, \tag{6.7.39}$$

$$\frac{p(x - \lambda_j^\pm t)}{p_\infty} = \left[1 \pm \frac{\gamma - 1}{2}\left(\frac{u(x - \lambda_j^\pm t)}{a_\infty} - M_\infty\right)\right]^{\frac{2\gamma}{\gamma-1}}, \tag{6.7.40}$$

where $\lambda_j^\pm = u \pm a$ and h is an arbitrary function which must be chosen to keep the density and the pressure positive. This is a very nice solution, but it is given implicitly, i.e., $\lambda_j^\pm$ depends on the solution itself, and also it is valid only when there are no shocks.

We point out here that the simple wave solution satisfies Burgers' equation as shown in Section 3.4.3. For example, the simple wave solution corresponding to $\lambda_j^+ = u + a$ satisfies Equation (3.4.36) which is repeated here,

$$V_t + V V_x = 0, \tag{6.7.41}$$

where $V = u + a$. I like this because $V = u + a$ can be written in terms of u only, by using the fact that $u - \frac{2}{\gamma-1}a = u_\infty - \frac{2}{\gamma-1}a_\infty = \text{constant}$:

$$V = \frac{\gamma + 1}{2}u + \left(a_\infty - \frac{\gamma - 1}{2}u_\infty\right), \tag{6.7.42}$$

so that, by solving this for u, a solution V of Burgers' equation (6.7.41) can be translated into the simple wave solution of the Euler equation by

$$\frac{u(\xi)}{a_\infty} = \frac{2}{(\gamma + 1)}\frac{V(\xi)}{a_\infty} - \frac{2}{\gamma + 1}\left(1 - \frac{\gamma - 1}{2}M_\infty\right), \tag{6.7.43}$$

where $\xi = x - Vt$, and the density and the pressure follow from Equations (6.7.39) and (6.7.40). Again, V must be chosen such that the density and the pressure stay positive. I like this solution because basically I can add any constant to u (and still V satisfies Burgers' equation), and so we may set up a solution as

$$\frac{u(\xi)}{a_\infty} = M_\infty + \frac{2}{(\gamma + 1)}\frac{V(\xi)}{a_\infty}, \tag{6.7.44}$$

i.e., we use the solution V of Burgers' equation as a perturbation to the mean flow. Compare this with Equation (6.7.38) and see how you can determine $h(\xi)$ by a solution $V(\xi)$ of Burgers' equation. See Section 7.13.1 for some examples.

The bottom-line is that the exact solution can be obtained if Equation (6.6.5) is integrable so that the Riemann invariants can be defined explicitly. For example, the shallow-water equations and the isothermal Euler equations have this property. I like such equations.

6.8   Manufactured Solutions

Pick any function, substitute it into a governing equation you wish to solve, and define whatever left as a source term. This way, you can cook up an exact solution to any equations (by introducing a source term). It is so simple. I like it very much. Many people call such solutions manufactured solutions [127]. Here are some simple examples.

(a) 1D Linear Advection:

Substitute $u = \sin x$ into the advection equation (plus a possible source term f), $au_x = f$, to get

$$a \cos x = f, \tag{6.8.1}$$

which suggests to define $f = a \cos x$. Therefore, we can say that the advection equation,

$$au_x = f \quad \text{with} \quad f = a \cos x, \tag{6.8.2}$$

has the following exact solution,

$$u = \sin x. \tag{6.8.3}$$

(b) Poisson Equation:

Substitute $u = \sin x + \cos y$ into the Poisson equation, $u_{xx} + u_{yy} = f$, to get

$$-\sin x - \cos y = f, \tag{6.8.4}$$

which suggests to define $f = -\sin x - \cos y$. Therefore, the Poisson equation,

$$u_{xx} + u_{yy} = f \quad \text{with} \quad f = -\sin x - \cos y, \tag{6.8.5}$$

has the following exact solution,

$$u = \sin x + \cos y. \tag{6.8.6}$$

I like cooking up exact solutions this way, but it always introduces a source term, i.e., an additional complication to the original equation. Honestly speaking, I don't really like source terms because a scheme could lose its formal accuracy unless the source term discretization is carefully designed [47, 82, 174].

(c) 2D Linear Advection:

In this example, we manufacture a solution by choosing appropriate coefficients rather than introducing a source term. Substitute $u = x^2 + y^2$ into the two-dimensional advection equation, $au_x + bu_y = 0$, to get

$$2xa + 2yb = 0, \tag{6.8.7}$$

which can be true if we take $a = y$ and $b = -x$. Hence, the advection equation,

$$yu_x - xu_y = 0, \tag{6.8.8}$$

has the following exact solution,

$$u = x^2 + y^2. \tag{6.8.9}$$

After all, I like the method of manufactured solutions very much because it is very convenient and useful especially for complicated equations, such as the Navier-Stokes equations [83, 133] and turbulence models [34, 135]. Also note that exact solutions for compressible flow equations can be manufactured by substituting exact incompressible flow solutions into compressible flow equations and generating appropriate source terms. This will be useful for manufacturing a physically realistic exact solution, which does look like a flow. It is very important to have a physically meaningful manufactured solution because a solution will not be useful if it gives a negative viscosity, for example.

6.9 Fine Grid Solutions

Some people use a numerical solution on a very fine grid as an exact solution: for example, compute the error of a numerical solution against the fine grid solution, and estimate the order of accuracy of a scheme. I like the idea, but this is wrong because the fine grid solution is not an exact solution. Such an error convergence study shows only how quickly the solution converge to this particular fine grid solution, not to the exact solution. Note that the error or the order of accuracy would not make sense at all if the scheme turned out to be inconsistent. You would then only know how close your solution is to a wrong solution or how quickly your solution converges to the wrong solution. Anyway, always remember that a fine grid solution is not an exact solution.

Chapter 7

Exact Solutions

I like exact solutions because I can directly measure the errors of numerical solutions. In doing so, in many cases, the domain can be in any shape, and the boundary condition can be specified directly by using the exact solutions (the Dirichlet condition). Also, a regular domain may be used with a periodic boundary condition so that any boundary issues can be avoided. In either way, exact solutions are very convenient and useful for code verification. Note that the verification is to check if a numerical method is correctly implemented, i.e., to see if it is solving the target equation with a design accuracy. Even if a code has been verified, it still needs to be validated, for example, with experimental data, i.e., to see if it is solving the right equation. Exact solutions are useful generally for verification, not for validation. See Ref.[127] for more details on the difference between verification and validation.

7.1 Linear Advection

7.1.1 1D Linear Advection

I like the linear advection equation,

$$u_t + au_x = 0, \tag{7.1.1}$$

where a is a constant, because it simply means that any initial solution is merely convected at the speed a, preserving its initial profile. The exact solution in the general form is therefore given by

$$u = f(x - at), \tag{7.1.2}$$

for the initial solution $u(x,0) = f(x)$. Note that we can avoid, in many cases, any boundary issues by using a periodic boundary condition: $u(1) = u(0)$ in the domain $x \in [0,1]$. I like this periodic boundary condition very much. It is very interesting that it can be interpreted as transforming the plane domain into a cylinder surface by rolling up the plane with respect to the vertical axis and connecting the left and right boundaries. Here are some examples.

(a) Smooth solution (Figure 7.1.1):

$$u = \sin[2\pi(x - at)], \tag{7.1.3}$$

for the initial solution $u(x,0) = \sin(2\pi x)$.

(b) Steep solution (Figure 7.1.2):

$$u = \exp\left[-\frac{(x - x_0 - at)^2}{d}\right], \tag{7.1.4}$$

for the initial solution $u(x,0) = \exp\left[-\frac{(x-x_0)^2}{d}\right]$ where d is a constant, e.g., $d = 0.01$.

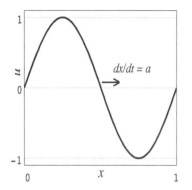

Figure 7.1.1: Smooth solution.

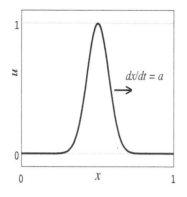

Figure 7.1.2: Steep solution.

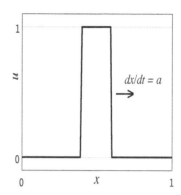

Figure 7.1.3: Discontinuous Solution.

(c) Discontinuous solution (Figure 7.1.3):

$$u = \begin{cases} 1 & \text{if } x_L + at < x < x_R + at, \\ 0 & \text{elsewhere,} \end{cases} \tag{7.1.5}$$

where x_L and x_R are arbitrary constants that specify the width of the non-zero part of the solution. Remember that discontinuous solutions do not satisfy the differential form of the advection equation but do satisfy the integral form (2.1.8).

7.1.2   2D Linear Advection

Of course, I like the linear advection equation in two dimensions also,

$$u_t + au_x + bu_y = 0, \tag{7.1.6}$$

where (a, b) is a constant advection vector. Any initial solution is simply convected at the velocity (a, b), preserving its initial profile. The exact solution in the general form is given by

$$u(x, y, t) = f(x - at, y - bt), \tag{7.1.7}$$

for the initial solution of the form, $u(x, y, 0) = f(x, y)$. Here are some examples.

(a) One-dimensional wave solution: $(a, b) = (a, 0)$,

$$u(x, y, t) = U_\infty + A \sin[2\pi(x - at)], \tag{7.1.8}$$

where U_∞ and A are arbitrary constants.

(b) Two-dimensional plane wave solution:

$$u(x, y, t) = U_\infty + A \sin(x + y - ct), \tag{7.1.9}$$

where U_∞ and A are arbitrary constants, and $c = a + b$.

(c) Discontinuous solution:

In the square domain, $(x, y) \in [0, 1] \times [0, 1]$, with $a > 0$ and $b > 0$,

$$u = \begin{cases} u_L & \text{if } ay - bx > 0, \\ u_R & \text{elsewhere,} \end{cases} \tag{7.1.10}$$

where u_L and u_R are arbitrary constants, with the inflow condition, $u(0, y) = u_L$ and $u(x, 0) = u_R$. See Figure 7.1.4.

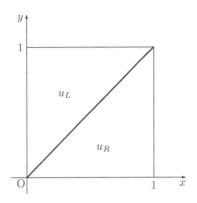

Figure 7.1.4: Discontinuous Solution. Two constant states u_L and u_R are separated along the characteristic passing through the origin with the slope 1.

(d) Steady solution:

I like the characteristic coordinates,

$$\xi = ax + by, \tag{7.1.11}$$
$$\eta = bx - ay, \tag{7.1.12}$$

which give

$$\partial_x = \xi_x \partial_\xi + \eta_x \partial_\eta = a\partial_\xi + b\partial_\eta, \tag{7.1.13}$$
$$\partial_y = \xi_y \partial_\xi + \eta_y \partial_\eta = b\partial_\xi - a\partial_\eta, \tag{7.1.14}$$

and

$$a\partial_x + b\partial_y = (a^2 + b^2)\partial_\xi, \tag{7.1.15}$$

because in these coordinates, the advection equation (7.1.6) is written as

$$u_t + (a^2 + b^2)u_\xi = 0, \tag{7.1.16}$$

so that the steady equation is simply

$$u_\xi = 0. \tag{7.1.17}$$

Hence, any function of η is an exact solution to the two-dimensional linear advection equation: e.g.,

$$u = \eta^3 + \eta^2 + \eta + 1, \quad u = \sin(\pi\eta), \quad u = \exp(\eta). \tag{7.1.18}$$

This is very nice, isn't it? I like it so much. Note that boundary values must be specified on inflow boundaries, according to the flow direction (a, b).

7.1.3 2D Circular Advection

The circular advection equation,

$$u_t + au_x + bu_y = 0, \tag{7.1.19}$$

where $(a, b) = (\omega y, -\omega x)$, describes a clockwise circular advection around the origin at the angular speed ω rad/sec. I like it because it is a very simple and good test case to see how long a numerical scheme can accurately preserve an initial profile. Here are some examples.

(a) Rotating Smooth Cone:

In the domain $[-1, 1] \times [-1, 1]$, start with the initial solution,

$$u(x, y, 0) = \begin{cases} \exp^{-r^2/d} & \text{for } r \leq 0.25, \\ 0 & \text{otherwise,} \end{cases} \tag{7.1.20}$$

where $r = \sqrt{(x + 0.25)^2 + y^2}$ and $(a, b) = (2\pi y, -2\pi x)$. The cone returns to its initial position at $t = 1$.

(b) Rotating Cylinder:

In the same setting as above, we may set instead

$$u(x, y, 0) = \begin{cases} 1 & \text{for } r \leq 0.25, \\ 0 & \text{otherwise,} \end{cases} \tag{7.1.21}$$

to rotate a cylinder which is discontinuous and more difficult compute accurately than a cone.

(c) Steady solutions:

Here is a steady test case where the initial solution is set up as

$$u(x, 0, 0) = \begin{cases} f(x) & \text{for } -0.65 \leq x \leq -0.35, \\ 0 & \text{elsewhere,} \end{cases} \tag{7.1.22}$$

in the domain $[-1, 1] \times [0, 1]$. The initial profile $f(x)$ is then convected in a clockwise circular direction; the exact steady solution is given, in the polar coordinates (r, θ), by

$$u(r, \theta) = \begin{cases} f(r) & \text{for } -0.65 \leq r \leq -0.35, \\ 0 & \text{elsewhere.} \end{cases} \tag{7.1.23}$$

At a steady state, for example, we may compare the solution at the outflow boundary where $x > 0$ and $y = 0$ with the initial profile to see how well its initial shape has been preserved. Here are examples of $f(x)$:

(c-1) Smooth solution:

$$f(x) = \exp^{-(x+0.5)^2/0.01}. \tag{7.1.24}$$

(c-2) Discontinuous solution:

$$f(x) = \begin{cases} 1 & \text{for } -0.65 \leq x \leq -0.35, \\ 0 & \text{elsewhere.} \end{cases} \tag{7.1.25}$$

I like the smooth one better because the numerical error should converge to zero as I refine the mesh whereas it will not converge generally if the solution is discontinuous. But, of course, if I am interested to study the non-oscillatory behavior of numerical solutions, I would use the discontinuous one for which numerical oscillations are more likely to occur.

7.1.4   3D Circular (Helicoidal) Advection

I like the 3D circular advection equation,

$$u_t + au_x + bu_y + cu_z = 0, \tag{7.1.26}$$

where

$$(a, b, c) = (\omega z, V, -\omega x), \tag{7.1.27}$$

and $\omega > 0$ and $V > 0$. This describes a flow rotating clockwise around (at the angular speed ω rad/sec) and simultaneously transported (at the speed V) along the y-axis. So, any initial solution imposed on the (x, y)-plane at $z = 0$ is rotated around and transported along the y-axis. For example, in the domain defined by

$$(x, y, z) \in [-1, 0] \times [0, 1] \times [0, 1], \tag{7.1.28}$$

if we specify the initial solution,

$$u(x, y, 0) = \exp\left[-50\left\{(x + 0.5)^2 + (y - 0.5)^2\right\}\right], \tag{7.1.29}$$

in the (x, y)-plane at $z = 0$, then the exact steady solution in the (y, z)-plane at $x = 0$ is given by

$$u(0, y, z) = \exp\left[-50\left\{(y - 0.5 - V)^2 + (z - 0.5)^2\right\}\right]. \tag{7.1.30}$$

It is a very simple and useful solution for evaluating the accuracy of a three-dimensional advection scheme.

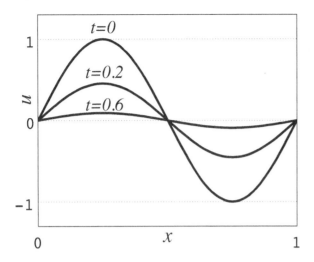

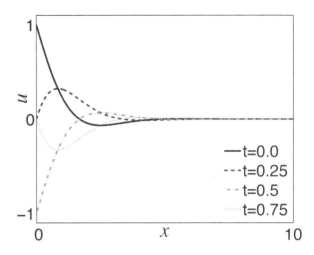

Figure 7.1.5: Unsteady solution of the 1D diffusion equation. $n = 2$.

Figure 7.1.6: Unsteady solution that oscillates forever. $U_0 = 1$ and $\omega = 2\pi$.

7.2 Diffusion Equation

I like the diffusion equation because it is easy to obtain exact solutions by separation of variables (see Section 6.1).

7.2.1 1D Diffusion Equation

The one-dimensional diffusion equation,

$$u_t = \nu u_{xx}, \quad \text{in } (0, 1), \tag{7.2.1}$$

has the following exact solutions.

(a) Time dependent solution I (Figure 7.1.5):

$$u = e^{-\nu n^2 \pi^2 t} \sin(n\pi x), \tag{7.2.2}$$

for the initial solution $u(x, 0) = \sin(n\pi x)$ with the boundary conditions $u(0, t) = u(1, t) = 0$. where n is an integer. I like this simple solution, but if you are looking for something more involved, you can generate it yourself by superposition (see Section 6.1).

(b) Time dependent solution II:

$$u(x, t) = \sqrt{\frac{A}{t + B}} \, e^{-\frac{(x-1)^2}{4\nu(t+B)}}, \tag{7.2.3}$$

where A, B and x_0 are arbitrary constants. for the initial solution and the boundary conditions:

$$u(x, 0) = \sqrt{\frac{A}{B}} \, e^{-\frac{(x-1)^2}{4\nu B}}, \quad u(0, t) = \sqrt{\frac{A}{t + B}} \, e^{-\frac{1}{4\nu(t+B)}}, \quad u(1, t) = \sqrt{\frac{A}{t + B}}. \tag{7.2.4}$$

Note that the boundary condition changes in time at both ends. I like it. This type of solution is used for accuracy verification of high-order schemes [119].

(c) Time dependent solution that will never decay to zero:

As derived in Section 6.1.5, there exists a time-dependent solution that oscillates forever:

$$u(x, t) = Q \left[\sinh(kx) \cos(kx) \sinh(k) \cos(k) + \cosh(kx) \sin(kx) \cosh(k) \sin(k) \right] \cos(\omega t)$$
$$+ Q \left[\sinh(kx) \cos(kx) \cosh(k) \sin(k) - \cosh(kx) \sin(kx) \sinh(k) \cos(k) \right] \sin(\omega t), \tag{7.2.5}$$

where k and ω are arbitrary positive constants, and

$$Q = \frac{U}{\cosh^2(k) - \cos^2(k)}. \tag{7.2.6}$$

It satisfies the diffusion equation for the following boundary conditions:

$$
\begin{align}
u(0) &= 0, \tag{7.2.7}\\
u(1) &= U\cos(\omega t). \tag{7.2.8}
\end{align}
$$

I like this one since it can be a good test case for studying a long time behavior of a diffusion scheme.

(d) Another time dependent solution that will never decay to zero:

There exists another time-dependent solution that oscillates forever:

$$u(x,t) = U\exp\left(-\sqrt{\frac{\omega}{2\nu}}x\right)\cos\left[\omega\left(t - \sqrt{\frac{1}{2\nu\omega}}x\right)\right], \tag{7.2.9}$$

where U and ω are arbitrary constants. This exact solution can be easily derived by a trial solution of the form:

$$u = Ue^{i(\omega t + \xi x)}, \tag{7.2.10}$$

where ξ is determined so as to satisfy the diffusion equation[1]. That is, you substitute it into the diffusion equation and solve the resulting equation for ξ. I like this solution, especially, the way it is derived. You pick a function and make it satisfy the target equation. That's a wonderful way of deriving exact solutions. It's like you are creating a solution rather than solving the equation. Of course, the solution itself is also very interesting in that it represents a traveling wave entering the domain at $x = 0$ and quickly damped towards $x \to \infty$. See Figure 7.1.6. The difference between the previous solution (7.2.5) and this solution (7.2.9) lies in the boundary condition: the former has $u = 0$ precisely at $x = 0$ whereas the latter has $u = 0$ theoretically at $x \to \infty$. At the other end, both solutions have the same boundary condition: $u = U\cos(\omega t)$.

(e) Steady solution (a simple linear function):

$$u(x) = (1-x)u_L + xu_R, \tag{7.2.11}$$

with the boundary conditions $u(0) = u_L$ and $u(1) = u_R$. This is too simple. Any diffusion scheme should be able to compute this exactly.

7.2.2 2D Diffusion Equation

The two-dimensional diffusion equation,

$$u_t = \nu(u_{xx} + u_{yy}), \tag{7.2.12}$$

has exact solutions of the form,

$$u(x,y,t) = e^{-\nu\pi^2(n^2+m^2)t}\sin(n\pi x)\sin(m\pi y), \tag{7.2.13}$$

where n and m are integers so that $u = 0$ on the boundary. It is straightforward to set up the problem in a square domain. But sometimes it is too simple. Then, I would consider the following highly anisotropic problem,

$$u_t = \nu(u_{xx} + u_{yy}), \quad (x,y) \in (0,1) \times (0, 0.005), \tag{7.2.14}$$

with the initial solution,

$$u(x,y,0) = 5\sin(\pi x)\sin(4000\pi y), \tag{7.2.15}$$

and the boundary condition, $u = 0$ on the boundary. The exact solution is given by

$$u(x,y,t) = 5\exp(-16000001\pi^2\nu t)\sin(\pi x)\sin(4000\pi y). \tag{7.2.16}$$

---

[1] This derivation is due to Professor P. L. Roe.

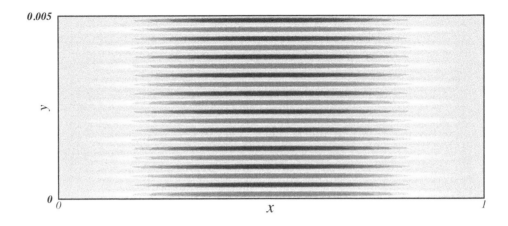

Figure 7.2.1: Anisotropic solution of the unsteady diffusion equation ($t = 0$).

The solution at $t = 0$ is shown in Figure 7.2.1. This is a very good test case for diffusion schemes because it involves high-aspect-ratio and possibly highly-skewed grids as typically encountered in practical CFD computations for viscous problems. Many diffusion schemes work very well for isotropic problems, but not necessarily for this problem. In Refs.[101, 108], this problem is used to illustrate numerical difficulties such as oscillations and instability with highly stretched and skewed grids.

Various other exact solutions can be obtained by superposition as mentioned earlier in Section 6.1. I like the two-dimensional diffusion equation especially because it is nothing but the Laplace equation at a steady state and therefore it has many many exact steady solutions (see Section 7.9).

7.2.3 3D Diffusion Equation

The three-dimensional diffusion equation,

$$u_t = \nu(u_{xx} + u_{yy} + u_{zz}), \quad \text{in } (0,1) \times (0,1) \times (0,1), \tag{7.2.17}$$

has exact solutions such as

$$u(x,y,z,t) = e^{-\nu\pi^2(n^2 + m^2 + l^2)t} \sin(n\pi x) \sin(m\pi y) \sin(l\pi z), \tag{7.2.18}$$

where n, m, and l are integers so that $u = 0$ on the boundary. Other exact solutions can be obtained by superposition as mentioned earlier in Section 6.1. Of course, I like this equation because it is again nothing but the Laplace equation at a steady state and it has many many many exact solutions (see Section 7.9). Also, I like highly anisotropic problems, which can be easily constructed from the above solution as in the two-dimensional case.

7.3 Advection-Diffusion Equation

7.3.1 1D Advection-Diffusion Equation

The one-dimensional advection-diffusion equation,

$$u_t + au_x = \nu u_{xx}, \tag{7.3.1}$$

where a and $\nu(> 0)$ are constants, has the following exact solutions.

(a) Time-dependent solution:

$$u(x,t) = \exp(-k^2\nu t) \sin k(x - at) \quad \text{in } [0,1], \tag{7.3.2}$$

where k is an arbitrary constant, for the initial condition $u(x) = \sin(kx)$ with a periodic boundary condition.

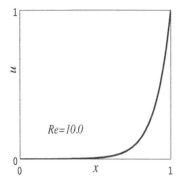

Figure 7.3.1: Boundary Layer.

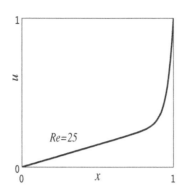

Figure 7.3.2: Friedrichs' model. $c = 0.3$.

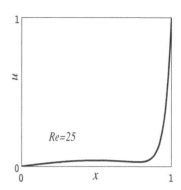

Figure 7.3.3: Manufactured solution. $C = 1/R_e$.

(b) Time-dependent solution that will never decay to zero:

As derived in Section 6.1.6, there exists an exact solution that oscillates forever:

$$u(x, t) = \mathcal{R} \left[\frac{\exp(\lambda_1 x) - \exp(\lambda_2 x)}{\exp(\lambda_1) - \exp(\lambda_2)} U \exp(i\omega t) \right], \tag{7.3.3}$$

where $i = \sqrt{-1}$, $\mathcal{R}$ indicates the real part of the complex solution (the expression in the square brackets), U and ω are arbitrary constants, and

$$\lambda_1 = \frac{a + \sqrt{a^2 + 4\nu\omega i}}{2\nu}, \quad \lambda_2 = \frac{a - \sqrt{a^2 + 4\nu\omega i}}{2\nu}. \tag{7.3.4}$$

The boundary conditions are

$$u(0) = 0, \tag{7.3.5}$$
$$u(1) = U \cos(\omega t). \tag{7.3.6}$$

I like this one because it can be useful for studying a long-time behavior of time-accurate advection-diffusion schemes. I would like it better if it were possible to extract the real part of the solution analytically.

(c) Steady boundary-layer type solution (Figure 7.3.1):

$$u(x) = \frac{1 - e^{xRe}}{1 - e^{Re}} \quad \text{in } [0, 1], \tag{7.3.7}$$

where $Re = a/\nu$, with the boundary conditions $u(0) = 0$ and $u(1) = 1$.

(d) Steady boundary-layer type solution (Figure 7.3.2):

The Friedrichs' model is given by

$$\nu u_{xx} + u_x - c = 0 \quad \text{in } (0, 1), \tag{7.3.8}$$

where c is an arbitrary constant (modeling the pressure gradient). It has the exact solution,

$$u(x) = cx + (1 - c)\frac{1 - e^{xRe}}{1 - e^{Re}}, \tag{7.3.9}$$

where $Re = 1/\nu$, with the boundary conditions $u(0) = 0$ and $u(1) = 1$. Particularly I like this one because it has an interesting geometrical interpretation. The solution curve in the three-dimensional space with coordinates $(u, x, p = u_x)$ is an intersection of two surfaces defined by

$$\nu p + u - cx = \text{constant}, \tag{7.3.10}$$
$$x + \nu \ln(p - c) = \text{constant}, \tag{7.3.11}$$

where the constants are determined by the boundary condition. See Ref.[107] for details. The solution shown in Figure 7.3.2 is a projection of this curve onto (x, u)-plane.

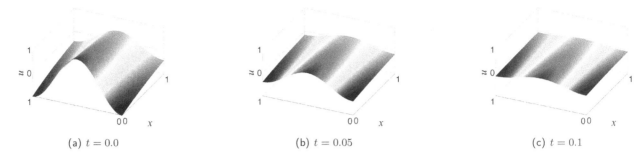

(a) $t = 0.0$ (b) $t = 0.05$ (c) $t = 0.1$

Figure 7.3.4: Unsteady solution (7.3.16): $C = -0.01, \nu = 0.5, a = 1, b = 0, A = 0.5, B = 2.0$.

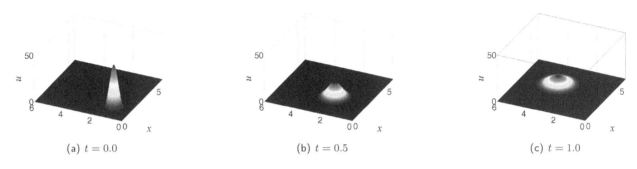

(a) $t = 0.0$ (b) $t = 0.5$ (c) $t = 1.0$

Figure 7.3.5: Unsteady solution (7.3.17): $C = 50.0, \nu = 0.15, a = 2.5, b = 2.5, c_1 = 1.0, c_2 = 1.0$.

(e) Manufactured solution (Figure 7.3.3):

$$u(x) = \frac{1 - e^{xRe}}{1 - e^{Re}} + C\sin(\pi x) \quad \text{in } [0, 1], \tag{7.3.12}$$

where C is an arbitrary constant, is the exact solution of the following steady equation,

$$au_x = \nu u_{xx} + C\pi \left[a\cos(\pi x) + \pi\nu\sin(\pi x) \right]. \tag{7.3.13}$$

I like this solution because it is non-trivial in the diffusion limit ($Re \rightarrow 0$).

7.3.2   2D Advection-Diffusion Equation

The two-dimensional advection-diffusion equation,

$$u_t + au_x + bu_y = \nu(u_{xx} + u_{yy}), \tag{7.3.14}$$

where a, b, and $\nu(> 0)$ are constants, has the following exact solutions.

(a) Unsteady solution 1 (Figure 7.3.4):

As derived in Section 6.1.3, we have the following exact unsteady solution,

$$u(x, y, t) = Ce^{-\nu\pi^2(a^2+b^2)(B^2-A^2)t}\cos(B\pi\eta)\exp\left(\frac{1 - \sqrt{1 + 4A^2\pi^2\nu^2}}{2\nu}\xi \right), \tag{7.3.15}$$

where $\xi = ax + by$, $\eta = bx - ay$, and A, B, and C are arbitrary constants. If we rationalize the exponent, we obtain

$$u(x, y, t) = Ce^{-\nu\pi^2(a^2+b^2)(B^2-A^2)t}\cos(B\pi\eta)\exp\left(\frac{-2A^2\pi^2\nu}{1 + \sqrt{1 + 4A^2\pi^2\nu^2}}\xi \right), \tag{7.3.16}$$

which can be used safely in the diffusion limit, i.e., as $\nu \rightarrow 0$. So, I like this one better.

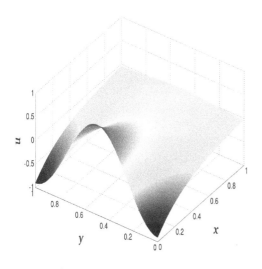

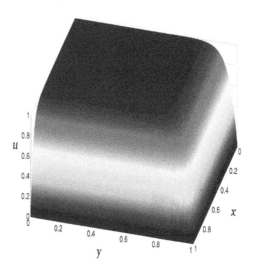

Figure 7.3.6: Exponential steady solution (7.3.21): $C = -0.009, \nu = 0.1, a = 1.0, b = 0.0, A = 2.0$.

Figure 7.3.7: Boundary-layer type solution (7.3.22): $\nu = 0.05, a = 1.0, b = 1.0$.

(b) Unsteady solution 2 (Figure 7.3.5):

Here is another exact unsteady solution to the advection-diffusion equation (7.3.14):

$$u(x,y,t) = \frac{C}{4t + c_1} \exp\left[-\frac{(x - at - c_2)^2}{\nu(4t + c_1)} - \frac{(y - bt - c_2)^2}{\nu(4t + c_1)} \right], \tag{7.3.17}$$

where C, c_1, and c_2 are arbitrary constants. I like the fact that this will reduce to an exact solution of the two-dimensional diffusion equation if we set $a = b = 0$.

(c) Polynomial solutions:

Quadratic: $$u(x,y) = u_0 + C_1\eta + C_2\left(2\nu\xi + \eta^2\right), \tag{7.3.18}$$

Cubic: $$u(x,y) = u_0 + C_3\eta + C_4\left(\nu\xi + \frac{\eta^2}{2}\right) + C_5\left(\nu\xi\eta + \frac{\eta^3}{6}\right), \tag{7.3.19}$$

Quartic: $$u(x,y) = u_0 + C_6\eta + C_7\left(\nu\xi + \frac{\eta^2}{2}\right) + C_8\left(\nu\xi\eta + \frac{\eta^3}{6}\right)$$
$$+ C_9\left(\nu^3\xi + \frac{\nu^2\xi^2}{2} + \frac{\nu\xi\eta^2}{2} + \frac{\eta^4}{24}\right), \tag{7.3.20}$$

where u_0, C_1, C_2, C_3, C_4, C_5, C_6, C_7, C_8, C_9 are all arbitrary constants. I like these polynomial solutions because I can directly check, for example, whether a numerical scheme can preserve a quadratic/cubic/quartic exact solution. Such a polynomial-preserving property can be a useful measure of the accuracy of a scheme, especially on unstructured grids.

(d) Exponential solution (Figure 7.3.6):

Simply by setting $A = B$ in the solution (7.3.16), we obtain the following steady exact solution:

$$u(x,y) = C\cos(A\pi\eta)\exp\left(\frac{-2A^2\pi^2\nu}{1 + \sqrt{1 + 4A^2\pi^2\nu^2}}\xi \right), \tag{7.3.21}$$

where C and A are arbitrary constants. I like this solution because it is very smooth and thus can be easily used to verify the accuracy of numerical schemes.

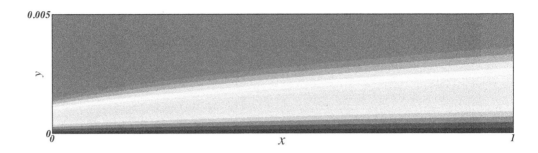

Figure 7.3.8: Boundary-layer type manufactured solution (7.3.24): $Re = 10^4$, $x_0 = 0.2$.

(e) Boundary-layer type solution I (Figure 7.3.7):

$$u(x,y) = \frac{\left[1 - \exp\left((x-1)\frac{a}{\nu}\right)\right]\left[1 - \exp\left((y-1)\frac{b}{\nu}\right)\right]}{\left[1 - \exp\left(-\frac{a}{\nu}\right)\right]\left[1 - \exp\left(-\frac{b}{\nu}\right)\right]} \quad \text{in } [0,1] \times [0,1]. \tag{7.3.22}$$

This solution can be useful when we test a scheme for resolving boundary layers [155]. How would you avoid spurious oscillations? Would you use a stretched mesh or an upwind scheme?

(f) Boundary-layer type solution II:

It is possible to construct an exact solution by superposing two one-dimensional exact solutions: e.g.,

$$u(x,y) = C_x \frac{1 - e^{ax/\nu}}{1 - e^{a/\nu}} + C_y \frac{1 - e^{by/\nu}}{1 - e^{b/\nu}} \quad \text{in } [0,1] \times [0,1], \tag{7.3.23}$$

where C_x and C_y are arbitrary constants. Note that the first term satisfies $au_x = \nu u_{xx}$ and the second term satisfies $bu_y = \nu u_{yy}$: the sum of them satisfies the two-dimensional advection-diffusion equation, $au_x + bu_y = \nu(u_{xx} + u_{yy})$. I like this type of solution because it has no cross-derivatives such as u_{xy} or u_{xxy}.

(g) Boundary-layer type solution III (manufactured solution):

An interesting manufactured solution is used in Ref.[153], which is essentially the same as the following:

$$u(x,y) = 1 - \exp(-\eta), \tag{7.3.24}$$

where

$$\eta = y\sqrt{\frac{Re}{x + x_0}}, \quad Re = \frac{\sqrt{a^2 + b^2}}{\nu}. \tag{7.3.25}$$

The reference location x_0 can be chosen to avoid the singularity, e.g., $x_0 = 0.2$. It resembles the x-component of the velocity for the laminar flat-plate boundary layer flow in Section 7.15.9. It is exact for the steady advection-diffusion equation with a source term, $f(\eta)$:

$$au_x + bu_y = \nu(u_{xx} + u_{yy}) + f(\eta), \tag{7.3.26}$$

where

$$f(\eta) = \exp(-\eta)\left[a\eta_x + b\eta_y - \nu(\eta_{xx} + \eta_{yy} - (\eta_x^2 + \eta_y^2))\right]. \tag{7.3.27}$$

Look at Figure 7.3.8. It is just like a boundary layer in a very thin domain. It can be a good test case for advection-diffusion schemes. I like it very much. Of course, the same solution can be made exact for the compressible Navier-Stokes equations with a suitable source term.

7.4   Advection-Reaction Equation(3D)

The advection-reaction equation,

$$u_t + au_x + bu_y + cu_z = -ku, \tag{7.4.1}$$

has the exact solution of the form,

$$u(x, y, z, t) = e^{-kt} f(x - at, y - bt, z - ct), \tag{7.4.2}$$

where $u(x, y, z, 0) = f(x, y, z)$; and its two-dimensional version,

$$u_t + au_x + bu_y = -ku, \tag{7.4.3}$$

has the exact solution of the form,

$$u(x, y, t) = e^{-kt} g(x - at, y - bt), \tag{7.4.4}$$

where $u(x, y, 0) = g(x, y)$; and its one-dimensional version,

$$u_t + au_x = -ku, \tag{7.4.5}$$

has the exact solution of the form,

$$u(x, t) = e^{-kt} h(x - at), \tag{7.4.6}$$

where $u(x, 0) = h(x)$. So, as soon as you choose the initial solution, you immediately have the exact solution. I like it. Basically, these exact solutions can be derived very easily. All are derived from the fact that

$$\frac{du}{dt} = -ku, \tag{7.4.7}$$

along the characteristic $(dx/dt, dy/dt, dz/dt) = (a, b, c)$. Or equivalently, we may apply the change of variables,

$$u = e^{-\frac{kx}{a}} v, \qquad \text{in one dimension}, \tag{7.4.8}$$

$$u = e^{-\frac{k}{2}\left(\frac{x}{a} + \frac{y}{b}\right)} v, \qquad \text{in two dimensions}, \tag{7.4.9}$$

$$u = e^{-\frac{k}{3}\left(\frac{x}{a} + \frac{y}{b} + \frac{z}{c}\right)} v, \qquad \text{in three dimensions}, \tag{7.4.10}$$

to transform the advection-reaction equation into a pure advection equation of v, which can be solved easily. Note that $t = x/a = y/b = z/c$ along the characteristic. Either way, it is very simple to derive the exact solution.

7.5   Advection-Diffusion-Reaction Equation

(a) Steady Solution:

Consider the steady advection-diffusion-reaction equation:

$$au_x = \nu u_{xx} - ku + s(x), \tag{7.5.1}$$

where a, $\nu(> 0)$, and $k(> 0)$ are constants, and $s(x)$ is an arbitrary function of x. I like this equation because the exact solution can be easily derived. Substitute $u(x) = A\exp(\lambda_1 x) + B\exp(\lambda_2 x) + s(x)/k$, A and B are constants, into Equation (7.5.1), and find that the solution is exact if

$$\nu\lambda_{1,2}^2 - a\lambda_{1,2} - k = 0, \tag{7.5.2}$$

which yields

$$\lambda_{1,2} = \frac{a \pm \sqrt{a^2 + 4\nu k}}{2\nu}.$$ (7.5.3)

Therefore, the exact solution is given by

$$u(x) = A \exp\left(\frac{a + \sqrt{a^2 + 4\nu k}}{2\nu}x\right) + B \exp\left(\frac{a - \sqrt{a^2 + 4\nu k}}{2\nu}x\right) + \frac{s(x)}{k},$$ (7.5.4)

where A and B can be determined from the boundary conditions, e.g., by solving $u(x = 0) = 0$ and $u(x = 1) = 1$ for A and B. Interestingly, this exact solution is used to construct a finite-difference scheme in Ref.[141].

(b) Unsteady Solution I:

As derived in Section 6.1.4, the following solution,

$$u = C_0 e^{-kt}\left[C_1 \exp\left(\frac{a + \sqrt{a^2 + 8\nu k}}{2\nu}x\right) + C_2 \exp\left(\frac{a - \sqrt{a^2 + 8\nu k}}{2\nu}x\right)\right],$$ (7.5.5)

where C_0, C_1, C_2 are arbitrary constants, is the exact solution to the unsteady advection-diffusion-reaction equation,

$$u_t + au_x = \nu u_{xx} - ku.$$ (7.5.6)

(c) Unsteady Solution II (manufactured solution):

The exact solution to the unsteady version of the advection-diffusion-reaction equation can be constructed by the method of manufactured solutions. For example, the solution [141]

$$u(x, t) = x^2 e^{-t},$$ (7.5.7)

satisfies the advection-diffusion-reaction equation,

$$u_t + au_x = \nu u_{xx} - ku + s(x, t),$$ (7.5.8)

where

$$s(x, t) = e^{-t}\left[(k - 1)x^2 + 2ax - 2\nu\right].$$ (7.5.9)

Of course, various other exact solutions can be constructed by the method of manufactured solutions.

7.6 Spherical Advection Equation(3D)

Consider the spherical advection equation:

$$u_t + u_r = -\frac{u}{r},$$ (7.6.1)

where r is the distance from the origin. This equation was used in Ref.[71] as a model for aeroacoustics for studying the long-time behavior of advection schemes in a large domain $r \in [5, 400]$. I like this equation because it can be written as

$$(ru)_t + (ru)_r = 0,$$ (7.6.2)

so that the exact solution is simply $ru(r, t) = f(r - t)$, i.e.,

$$u(r, t) = \frac{1}{r} f(r - t),$$ (7.6.3)

where f is determined by the initial condition, say $u(r, 0) = \frac{1}{r}\sin(2\pi r)$, meaning $f(r) = \sin(2\pi r)$. This is very nice.

7.7 Burgers Equation

7.7.1   1D Burgers Equation

For the one-dimensional Burgers equation,

$$\partial_t u + \partial_x\left(u^2/2\right) = 0, \tag{7.7.1}$$

or

$$\partial_t u + u\,\partial_x u = 0, \tag{7.7.2}$$

basically a shock is developed if the initial solution is monotonically decreasing (characteristics converging) while a rarefaction is developed if monotonically increasing (characteristics diverging). What is nice about this equation is that u is constant along the characteristics $dx/dt = u$, which in turn means that the characteristics are straight lines. So, it is very easy to find the exact solution (simply trace back along a characteristic line) until characteristics collide with each other to form a shock. In particular, I like the following exact solutions.

(a) Smooth solution turning into a shock (Figure 7.7.1):

A smooth initial solution such as

$$u = \frac{1}{2\pi t_s}\sin(2\pi x), \quad x \in [0,1], \tag{7.7.3}$$

with the periodic boundary condition $u(0,t) = u(1,t) = 0$, leads to the following exact solution,

$$u = \frac{1}{2\pi t_s}\sin[2\pi(x - ut)], \tag{7.7.4}$$

where t_s is a free parameter that specifies the time when a shock is formed, i.e., $\partial_x u \to \infty$ at $x = 0.5$. Note that it is given implicitly and therefore must be solved numerically at each point and time (x,t): for example, for a given location x_i at time t_n, iterate on u_i^k by a simple fixed-point iteration,

$$u_i^{k+1} = \frac{1}{2\pi t_s}\sin[2\pi(x_i - u_i^k t_n)], \tag{7.7.5}$$

until $|u_i^{k+1} - u_i^k| < 10^{-15}$. I like this solution because it can be used continuously to measure the numerical accuracy up to the shock formation as well as to test the shock-capturing capability after the shock formation.

(b) Smooth rarefaction (Figure 7.7.2):

A smooth initial solution such as

$$u = A\tanh(k(x - 0.5)), \tag{7.7.6}$$

leads to the following exact solution,

$$u = A\tanh(k(x - 0.5 - ut)), \tag{7.7.7}$$

where A and k are arbitrary constants. Of course, it is given implicitly again, but can be computed numerically as in the previous one.

(c) Compression turning to a shock (Figure 7.7.3):

A linear initial condition such as

$$u(x,0) = \begin{cases} \dfrac{3}{2} & \text{if } x \le 0, \\[2mm] \dfrac{3}{2} - 2x & \text{if } 0 < x < 1, \\[2mm] -\dfrac{1}{2} & \text{if } x \ge 1, \end{cases} \tag{7.7.8}$$

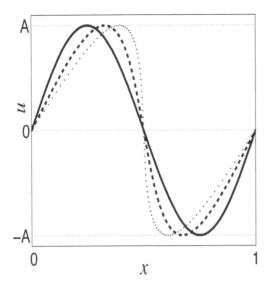
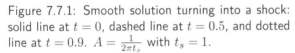

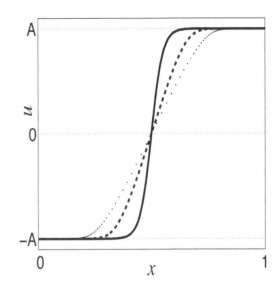

Figure 7.7.1: Smooth solution turning into a shock: solid line at $t = 0$, dashed line at $t = 0.5$, and dotted line at $t = 0.9$. $A = \frac{1}{2\pi t_s}$ with $t_s = 1$.

Figure 7.7.2: Smooth solution spreading out: solid line at $t = 0$, dashed line at $t = 0.5$, and dotted line at $t = 0.9$. $A = 0.2$ and $k = 20$.

leads to the following exact solution: the solution before a shock is formed,

$$u(x, t < 1/2) = \begin{cases} \dfrac{3}{2} & \text{if } x \leq \frac{3t}{2}, \\[2mm] \dfrac{3 - 4x}{2 - 4t} & \text{if } \frac{3t}{2} < x < 1 - \frac{t}{2}, \\[2mm] -\dfrac{1}{2} & \text{if } x \geq 1 - \frac{t}{2}, \end{cases} \tag{7.7.9}$$

and the solution after the shock is formed at $t = 1/2$,

$$u(x, t \geq 1/2) = \begin{cases} \dfrac{3}{2} & \text{if } x \leq \frac{1+t}{2}, \\[2mm] -\dfrac{1}{2} & \text{if } x \geq \frac{1+t}{2}. \end{cases} \tag{7.7.10}$$

It is quite simple. I like it.

(d) Accelerating shock (Figure 7.7.4):

Consider a linear initial condition with a shock,

$$u(x, 0) = \begin{cases} u_L & \text{if } x \leq 0, \\[2mm] u_R(x - 1) & \text{if } x > 0, \end{cases} \tag{7.7.11}$$

where u_L and u_R are arbitrary constants. Assuming $u_L > 0$, we would like a solution where the shock travels to the right keeping the left side of the shock uniform $u = u_L$. On the right side of the shock, the solution is continuous, and therefore the exact solution is obtained by tracing back the characteristics:

$$u(x, t) = u(x - ut, 0) = u_R((x - ut) - 1), \tag{7.7.12}$$

which can be solved for u to yield

$$u(x, t) = \frac{u_R(x - 1)}{1 + u_R t}. \tag{7.7.13}$$

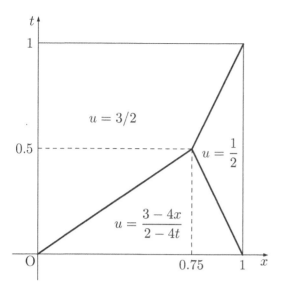

Figure 7.7.3: Compression wave turning into a straight shock at $(x, t) = (0.5, 0.75)$.

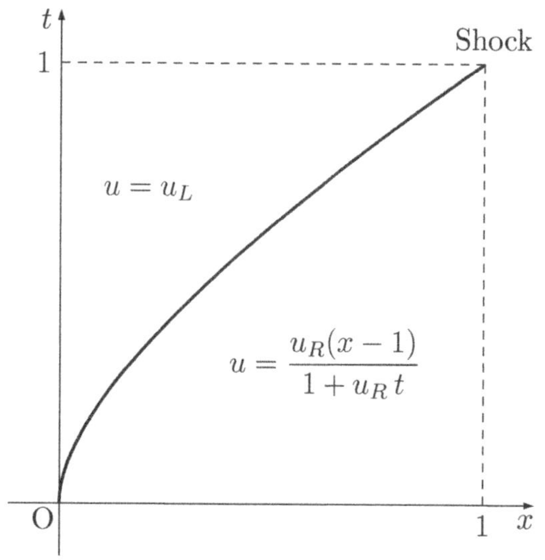

Figure 7.7.4: Accelerating shock wave solution with the parameters: $u_R = 2u_L$ and $u_L = \frac{2}{3}(1 + \sqrt{3})$.

Therefore, the exact solution to the problem can be written as

$$u(x, t) = \begin{cases} u_L & \text{if } x \leq x_s(t), \\ \dfrac{u_R(x - 1)}{1 + u_R t} & \text{if } x \geq x_s(t), \end{cases} \qquad (7.7.14)$$

where $x_s(t)$ is the location of the shock. See Figure 7.7.4; observe that the shock is accelerating. The shock path can be obtained by solving the ordinary differential equation for the shock speed (2.13.3),

$$V_s(t) = \frac{dx_s(t)}{dt} = \frac{1}{2} \left[u_L + \frac{u_R(x_s - 1)}{1 + u_R t} \right], \qquad (7.7.15)$$

with the initial condition $x_s(0) = 0$. The solution is

$$x_s(t) = \left(\frac{u_L}{u_R} + 1 \right) \left(1 - \sqrt{1 + u_R t} \right) + u_L t. \qquad (7.7.16)$$

In order for the shock to travel to the right, we must choose the parameters such that

$$u_R \leq 2u_L. \qquad (7.7.17)$$

And I personally like the following values:

$$u_L = \frac{2}{3}(1 + \sqrt{3}), \quad u_R = 2u_L, \qquad (7.7.18)$$

because it will lead to $x_s(1) = 1$, i.e., the shock will reach $x = 1$ at $t = 1$ as in Figure 7.7.4. In this case, the shock path can be expressed by

$$t = \frac{x + \sqrt{3x}}{1 + \sqrt{3}}. \qquad (7.7.19)$$

I like this solution very much. This is a very good test case to check if a numerical scheme can compute a shock wave with the correct traveling speed. It is well-known that conservation is very important to ensure a correct shock speed; the shock computed by a non-conservative scheme may not reach $x = 1$ at $t = 1$.

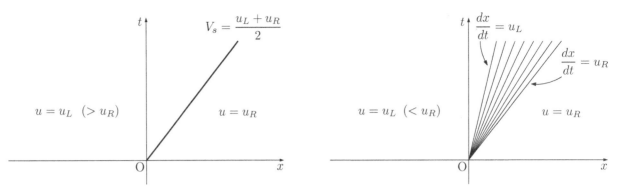

Figure 7.7.5: Isolated Shock for the 1D Burgers equation. Figure 7.7.6: Rarefaction for the 1D Burgers equation.

7.7.2   1D Burgers Equation: Riemann Problems

The Riemann problem consists of a hyperbolic equation and piecewise constant initial data with a single discontinuity. I like it especially because I can cook up exact solutions easily. Consider the Riemann problem for Burgers' equation,

$$\partial_t u + \partial_x \left(u^2 / 2 \right) = 0, \tag{7.7.20}$$

$$u(x,0) = \begin{cases} u_L & \text{if } x < 0, \\ u_R & \text{if } x > 0. \end{cases} \tag{7.7.21}$$

This has the following exact solutions.

(a) Shock (Figure 7.7.5):

Choose u_L and u_R such that

$$u_L > u_R, \tag{7.7.22}$$

then a shock is created and it will travel at the speed,

$$V_s = \frac{u_L + u_R}{2}, \tag{7.7.23}$$

and so, the exact solution is given by

$$u(x,t) = \begin{cases} u_L & \text{if } x/t < V_s, \\ u_R & \text{if } x/t > V_s. \end{cases} \tag{7.7.24}$$

(b) Rarefaction (Figure 7.7.6):

Choose u_L and u_R such that

$$u_L < u_R, \tag{7.7.25}$$

then a rarefaction wave is created, and the exact solution is given by

$$u(x,t) = \begin{cases} u_L & \text{if } x/t < u_L, \\ \dfrac{x/t - u_R}{u_L - u_R} u_L + \dfrac{x/t - u_L}{u_R - u_L} u_R & \text{if } u_L < x/t < u_R, \\ u_R & \text{if } x/t > u_R. \end{cases} \tag{7.7.26}$$

(c) Sonic Rarefaction (Figure 7.7.7):

Choose u_L and u_R such that

$$u_L < 0, \quad u_R > 0, \quad \text{and} \quad u_L = -u_R, \tag{7.7.27}$$

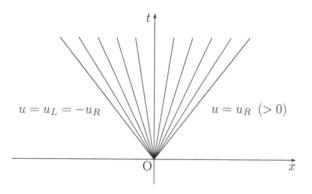

Figure 7.7.7: Sonic Rarefaction for the 1D Burgers equation.

then a rarefaction wave is created at the center, expanding to the left and the right at the same speed, which is called a sonic rarefaction. The exact solution is given by

$$
u(x,t) = \begin{cases} u_L & \text{if } x/t < u_L, \\ x/t & \text{if } u_L < x/t < u_R, \\ u_R & \text{if } x/t > u_R. \end{cases} \tag{7.7.28}
$$

Some numerical schemes fail to compute this solution, preserving the initial discontinuity which must break up immediately.

7.7.3   2D Burgers Equation

The two-dimensional Burgers equation of the form,

$$
uu_x + u_y = 0, \tag{7.7.29}
$$

is basically a one-dimensional equation with y considered as time. So, any time-dependent one-dimensional solutions can be exact solutions of this equation if t is replaced by y. Namely, any exact solutions for $u_t + uu_x = 0$ in (x, t)-space, such as those in Sections 7.7.1 and 7.7.2, are exact solutions for $u_y + uu_x = 0$ in (x, y)-space. For example, those in Figures 7.7.3, 7.7.4, 7.7.5, 7.7.6, and 7.7.7 can be directly interpreted, with t replaced by y, as the exact solutions of the two-dimensional equation (7.7.29). This is very convenient. I love it. Actually, the curved shock solution in Figure 7.7.4 is my favorite, and I have used it as a test case for mesh adaptation [114].

7.8   Viscous Burgers Equation

7.8.1   1D Viscous Burgers Equation

The one-dimensional viscous Burgers equation,

$$
u_t + uu_x = \nu u_{xx}, \tag{7.8.1}
$$

has the following exact solutions.

(a) Unsteady/Steady solution (Figure 7.8.1):

$$
u = \frac{1 + (2a - 1) \exp((1 - a)\xi/\nu)}{1 + \exp((1 - a)\xi/\nu)}, \tag{7.8.2}
$$

$$
\xi = x - at - x_0, \tag{7.8.3}
$$

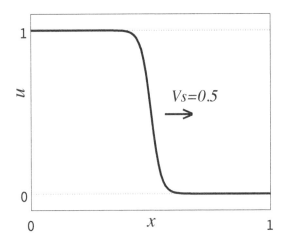

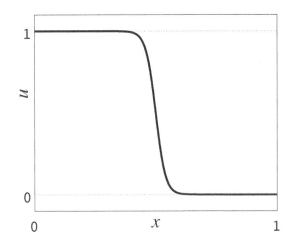

Figure 7.8.1: Unsteady solution of the 1D viscous Burgers' equation. $u_L = 1$, $u_R = 0$, $\nu = 0.01$.

Figure 7.8.2: Steady solution of the 1D viscous Burgers' equation. $x_0 = 0.5$, $b = 1.0$, $c = 0.5$, $\nu = 0.01$.

x_0, a, and ν are constants, e.g., $x_0 = 0.1$, $a = 0.5$, $\nu = 0.001$ (see Ref.[64]). This solution is nice because it can be expressed also in the following form,

$$u = u_R + \frac{1}{2}(u_L - u_R)\left[1 - \tanh\left(\frac{(u_L - u_R)(x - V_s t)}{4\nu}\right)\right], \tag{7.8.4}$$

where

$$V_s = \frac{u_L + u_R}{2}, \quad u_L = \lim_{x \to -\infty} u, \quad u_R = \lim_{x \to \infty} u, \quad u_L > u_R. \tag{7.8.5}$$

Personally, I like this form better. It is very simple and intuitive. Moreover, if we take $u_R = -u_L$, then the shock speed V_s vanishes and thus we obtain a steady solution. This is also very nice.

(b) Steady solution (Figure 7.8.2):

A slightly modified version of the viscous Burgers equation [155],

$$u_t + (bu - c)u_x = \nu u_{xx}, \tag{7.8.6}$$

has the exact steady solution,

$$u = \frac{c}{b}\left[1 - \tanh\left(\frac{c(x - x_0)}{2\nu}\right)\right], \tag{7.8.7}$$

where $b(\neq 0)$, $c(\neq 0)$, and x_0 are arbitrary constants.

I like this version also [154],

$$u_t + uu_x = \frac{\nu}{2}(uu_x)_x, \quad x \in (0,1), \tag{7.8.8}$$

because it has the following very smooth steady exact solution:

$$u(x) = e^{\nu x}. \tag{7.8.9}$$

7.8.2   1D Viscous Burgers Equation with Source Terms

As shown in Section 6.4.1, the one-dimensional Burgers equation with source terms,

$$u_t + uu_x = \nu u_{xx} + \epsilon u^2 - ku + d, \tag{7.8.10}$$

where $\nu(> 0)$, ϵ, k, and d are arbitrary constants, has the following analytical solutions, depending on the value of $D = k^2 - 4\epsilon d$.

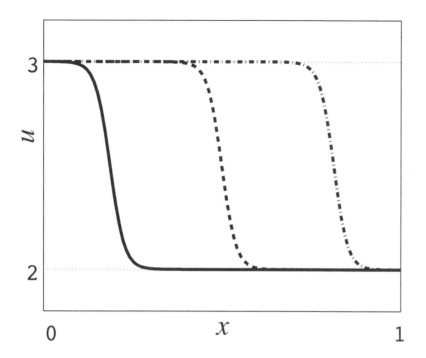

Figure 7.8.3: Exact solution of the 1D viscous Burgers equation with source terms: Solid t=0.0, Dash t=0.125, Dash-Dot t=0.25.

(a) $D > 0$

$$u = -2\nu \frac{\lambda_1 K_1 \exp\left[\lambda_1 x + \nu \lambda_1 (\lambda_1 + 2\epsilon)t\right] + \lambda_2 K_2 \exp\left[\lambda_2 x + \nu \lambda_2 (\lambda_2 + 2\epsilon)t\right]}{K_1 \exp\left[\lambda_1 x + \nu \lambda_1 (\lambda_1 + 2\epsilon)t\right] + K_2 \exp\left[\lambda_2 x + \nu \lambda_2 (\lambda_2 + 2\epsilon)t\right]},\tag{7.8.11}$$

where K_1 and K_2 are arbitrary constants, and

$$\lambda_1 = \frac{-k - \sqrt{D}}{4\epsilon\nu}, \quad \lambda_2 = \frac{-k + \sqrt{D}}{4\epsilon\nu}.\tag{7.8.12}$$

(b) $D < 0$

$$u = -2\nu \frac{(\lambda_r K_1' + \lambda_i K_2')\cos\theta + (\lambda_r K_2' - \lambda_i K_1')\sin\theta}{K_1' \cos\theta + K_2' \sin\theta},\tag{7.8.13}$$

where K_1' and K_2' are arbitrary constants, and

$$\theta = \lambda_i x + 2\nu(\lambda_r \lambda_i + \epsilon\lambda_i)t, \quad \lambda_r = \frac{-k}{4\epsilon\nu}, \quad \lambda_i = \frac{\sqrt{-D}}{4\epsilon\nu}.\tag{7.8.14}$$

(c) $D = 0$

$$u = -2\nu \frac{K_2''(1 + \lambda_s x) + 2K_1'' \nu \lambda_s(\lambda_s + \epsilon)t}{K_2'' x + 2K_1'' \nu(\lambda_s + \epsilon)t},\tag{7.8.15}$$

where K_1'' and K_2'' are arbitrary constants, and

$$\lambda_s = \frac{-k}{4\epsilon\nu}.\tag{7.8.16}$$

In particular, I like the case (a) because the following set of parameters generates a nice and smooth traveling wave solution shown in Figure 7.8.3:

$$\nu = 0.01, \quad K_1 = 100.0, \quad K_2 = 0.01, \quad \epsilon = 0.1, \quad k = 0.5, \quad d = 0.6.\tag{7.8.17}$$

I really like this case because in other cases I have not been able to discover a set of parameters which produces a solution with no singularities.

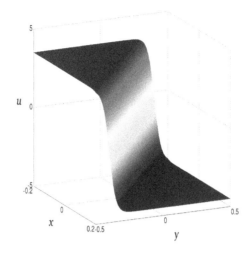

Figure 7.8.4: Steady solution for the 2D viscous Burgers equation, (7.8.24) with $\nu = 0.05$.

7.8.3   2D Viscous Burgers Equations

The two-dimensional viscous Burgers equations,

$$u_t + uu_x + vu_y - \nu(u_{xx} + u_{yy}) = 0, \tag{7.8.18}$$
$$v_t + uv_x + vv_y - \nu(v_{xx} + v_{yy}) = 0, \tag{7.8.19}$$

have an exact steady solution in the form [37],

$$u = \frac{-2\nu\left[a_2 + a_4 y + \lambda a_5\left(e^{\lambda(x-x_0)} - e^{-\lambda(x-x_0)}\right)\cos(\lambda y)\right]}{a_1 + a_2 x + a_3 y + a_4 xy + a_5\left(e^{\lambda(x-x_0)} + e^{-\lambda(x-x_0)}\right)\cos(\lambda y)}, \tag{7.8.20}$$

$$v = \frac{-2\nu\left[a_3 + a_4 x - \lambda a_5\left(e^{\lambda(x-x_0)} - e^{-\lambda(x-x_0)}\right)\sin(\lambda y)\right]}{a_1 + a_2 x + a_3 y + a_4 xy + a_5\left(e^{\lambda(x-x_0)} + e^{-\lambda(x-x_0)}\right)\cos(\lambda y)}, \tag{7.8.21}$$

where a_1, a_2, a_3, a_4, a_5, x_0 are arbitrary constants. In Ref.[37], the following set of parameters,

$$a_1 = a_2 = 1.3 \times 10^{13}, \quad a_3 = a_4 = 0, \quad a_5 = 1, \quad \lambda = 25, \quad x_0 = 1, \quad \nu = 0.04, \tag{7.8.22}$$

are used to generate a smooth boundary-layer type solutions. Well, in fact, my favorite is the following two-dimensional viscous Burgers equation,

$$u_t + uu_x + u_y - \nu u_{xx} = 0. \tag{7.8.23}$$

This is nice because it is a scalar equation and also it has a simple and interesting exact steady solution:

$$u(x, y, t = \infty) = -4\tanh\left(\frac{y + 2x}{\nu}\right) - \frac{1}{2}. \tag{7.8.24}$$

As shown in Figure 7.8.4, this solution has a very simple boundary-layer type feature. A similar solution was used in Ref.[23] to verify the accuracy of high-order schemes.

7.9   Laplace Equation

I like the Laplace equation very much because it has a variety of solutions in two and three dimensions although it has only a trivial linear solution in one dimension (7.2.11).

7.9.1   2D Laplace Equation

I like the two-dimensional Laplace equation:

$$u_{xx} + u_{yy} = 0. \tag{7.9.1}$$

It has various smooth exact solutions:

$$\text{(a)} \quad u = xy, \tag{7.9.2}$$

$$\text{(b)} \quad u = x^2 - y^2, \tag{7.9.3}$$

$$\text{(c)} \quad u = \frac{2y}{(1+x)^2 + y^2}, \tag{7.9.4}$$

$$\text{(d)} \quad u = \exp(kx)\sin(ky), \tag{7.9.5}$$

$$\text{(e)} \quad u = \frac{\sinh(\pi x)\sin(\pi y) + \sinh(\pi y)\sin(\pi x)}{\sinh(\pi)}, \tag{7.9.6}$$

where k is an arbitrary constant. We may take the domain to be $[0,1] \times [0,1]$ for all, and specify the boundary condition directly by using the exact solutions (the Dirichlet condition).

7.9.2   3D Laplace Equation

The three-dimensional Laplace equation,

$$u_{xx} + u_{yy} + u_{zz} = 0, \tag{7.9.7}$$

or in spherical coordinates,

$$\frac{1}{r^2}\frac{\partial}{\partial r}\left(r^2 \frac{\partial \alpha}{\partial r}\right) + \frac{1}{r^2 \sin^2\phi}\frac{\partial^2 \alpha}{\partial \theta^2} + \frac{1}{r^2 \sin\phi}\frac{\partial}{\partial \phi}\left(\sin\phi \frac{\partial \alpha}{\partial \phi}\right) = 0, \tag{7.9.8}$$

has the following exact solutions.

$$\text{(a)} \quad u = \frac{k}{r}, \tag{7.9.9}$$

$$\text{(b)} \quad u = \exp(\sqrt{2}kx)\sin(ky)\cos(kz), \tag{7.9.10}$$

where $r = \sqrt{x^2 + y^2 + z^2}$ and k is an arbitrary constant. In addition, it is possible to generate more solutions by superposition: e.g.,

$$\text{(c)} \quad u = A_1 \exp(\sqrt{2}k_1 x)\sin(k_1 y)\cos(k_1 z) + A_2 \exp(\sqrt{2}k_2 z)\sin(k_2 x)\cos(k_2 y)$$
$$+ A_3 \exp(\sqrt{2}k_3 y)\sin(k_3 z)\cos(k_3 x), \tag{7.9.11}$$

where A_i and k_i with $i = 1, 2, 3$ are arbitrary constants. This is nice. I like it.

Note also that any two dimensional solutions in the previous subsection and their linear combinations are exact solutions to the three-dimensional Laplace equation. For example, these solutions,

$$\text{(d)} \quad u = A_1 xy + A_2 yz + A_3 zx, \tag{7.9.12}$$

$$\text{(e)} \quad u = A_1(x^2 - y^2) + A_2(y^2 - z^2) + A_3(z^2 - x^2), \tag{7.9.13}$$

$$\text{(f)} \quad u = A_1 \exp(k_1 x)\sin(k_1 y) + A_2 \exp(k_2 y)\sin(k_2 z) + A_3 \exp(k_3 z)\sin(k_3 x), \tag{7.9.14}$$

are all exact solutions to the three-dimensional Laplace equation. There are many many many possible exact solutions for the Laplace equation.

7.10 Poisson Equations

I like the Poisson equation because exact solutions can be easily manufactured (see Section 6.8). Basically, you can easily cook up an exact solution simply by picking up a function $u(x, y, z)$, substituting it into the left hand side of the Poisson equation (i.e., the Laplacian part), and defining whatever left as the source term $f(x, y, z)$. Just like the Laplace equation, the Poisson equation is often used as a model equation for developing numerical schemes for the viscous terms.

7.10.1 2D Poisson Equations

The two-dimensional Poisson equation,

$$u_{xx} + u_{yy} = f(x, y), \tag{7.10.1}$$

has the following manufactured solutions:

(a) $u = e^{xy}$ with $f = (x^2 + y^2)e^{xy}$, $\tag{7.10.2}$

(b) $u = \exp(kx)\sin(ky) + \dfrac{A}{4}(x^2 + y^2)$ with $f = A$, $\tag{7.10.3}$

(c) $u = \sinh x$ with $f = \sinh x$, $\tag{7.10.4}$

(d) $u = \exp\left[\dfrac{x^2 + y^2}{d}\right]$ with $f = \dfrac{4(x^2 + y^2 + d)}{d^2}\exp\left[\dfrac{x^2 + y^2}{d}\right]$, $\tag{7.10.5}$

where k, A and d are arbitrary constants. Of course, as in two dimensions, we may take the domain to be $[0, 1] \times [0, 1]$ for all, and specify the boundary conditions directly by using the exact solutions. Also, we can superpose some of these solutions to generate a new exact solution such as

(e) $u = Ae^{xy} + B\sinh x$ with $f = A(x^2 + y^2)e^{xy} + B\sinh x$, $\tag{7.10.6}$

where A and B are arbitrary constants.

Smooth isotropic solutions are nice, but if numerical schemes are to be applied eventually to practical high-Reynolds-number viscous simulations involving boundary layers, they must be tested for anisotropic solutions. I like the following anisotropic solution:

(f) $u = \sin(\pi x)\sin(2000\pi y)$ with $f = -(1 + 2000^2)\pi^2 \sin(\pi x)\sin(2000\pi y)$, $\tag{7.10.7}$

where the domain is taken to be $(x, y) \in (0, 1) \times (0, 0.005)$, so that $u = 0$ on the boundary. This solution is very similar to the anisotropic solution to the diffusion equation (7.2.16).

7.10.2 3D Poisson Equations

The three-dimensional Poisson equation,

$$u_{xx} + u_{yy} + u_{zz} = f(x, y, z), \tag{7.10.8}$$

has the following manufactured solutions:

(a) $u = e^{\sqrt{k}xyz}$ with $f = k(x^2 + y^2 + z^2)e^{xyz}$, $\tag{7.10.9}$

(b) $u = \dfrac{A}{6}(x^2 + y^2 + z^2)$ with $f = A$, $\tag{7.10.10}$

(c) $u = C_x \sinh x + C_y \sinh y + C_z \sinh z$ with $f = C_x \sinh x + C_y \sinh y + C_z \sinh z$, $\tag{7.10.11}$

(d) $u = \exp\left[\dfrac{r^2}{d}\right]$ with $f = \dfrac{2(2r^2 + 3d)}{d^2}\exp\left[\dfrac{r^2}{d}\right]$, $\tag{7.10.12}$

where k, A, C_x, C_y, C_z, and d are arbitrary constants, and $r^2 = x^2 + y^2 + z^2$.

7.11   Incompressible-Euler/Cauchy-Riemann/Laplace Equations

Two-dimensional incompressible irrotational flows are governed by the Cauchy-Riemann equations, in terms of the stream-function ψ and the velocity potential ϕ,

$$\psi_x + \phi_y = 0, \tag{7.11.1}$$
$$\phi_x - \psi_y = 0, \tag{7.11.2}$$

or in terms of the velocity components u and v,

$$u_x + v_y = 0, \tag{7.11.3}$$
$$v_x - u_y = 0, \tag{7.11.4}$$

or the Laplace equations:

$$\phi_{xx} + \phi_{yy} = 0, \tag{7.11.5}$$
$$\psi_{xx} + \psi_{yy} = 0, \tag{7.11.6}$$

or

$$u_{xx} + u_{yy} = 0, \tag{7.11.7}$$
$$v_{xx} + v_{yy} = 0. \tag{7.11.8}$$

Note that exact solutions to these equations are exact for the Euler equations in the incompressible limit, and therefore can be used to test an Euler code. As mentioned in Section 6.2, I find it very interesting that a flow is obtained simply by picking up an analytic function F of a complex variable $Z = x + iy = re^{i\theta}$. The velocity potential and the stream function are immediately obtained as the real and imaginary parts, respectively: $F(Z) = \phi(z, y) + i\psi(z, y)$. The velocity components (u, v) are obtained as the real and imaginary parts of the complex velocity, respectively: $W(Z) = F' = dF/dZ = u - iv$. Some basic flows are given in Section 6.2. More practical flows such as a flow over an airfoil can also be obtained by a mapping technique (the conformal mapping) as shown in this section. We also consider a rotational flow for which the velocity potential does not exist but the stream function exists.

7.11.1   Stagnation Flow

Stagnation flow is obtained by the following complex potential:

$$F = -\frac{Z^2}{2}. \tag{7.11.9}$$

The velocity potential and the stream function are given by

$$\phi = \frac{1}{2}(y^2 - x^2), \quad \psi = -xy. \tag{7.11.10}$$

The complex velocity is given by

$$W = \frac{dF}{dZ} = -Z, \tag{7.11.11}$$

which thus yields by $W = u - iv$

$$u = -x, \quad v = y. \tag{7.11.12}$$

Streamlines are plotted in Figure 7.11.1 where the flow goes from the right to the left towards the wall represented by the y-axis. I like the stagnation flow because the velocity components are linear function of space. So, a second-order Euler code should be able to produce the exact solution. It sounds like an easy flow to compute. However, for practical problems, the stagnation region is often a major source of convergence difficulty due to the vanishing velocity. See Refs.[32, 162] for details.

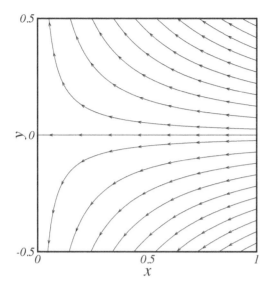

Figure 7.11.1: Inviscid Stagnation Flow in 2D.

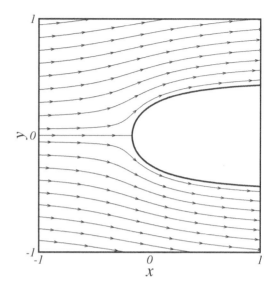

Figure 7.11.2: Flow over a Rankine Half-Body in 2D.

7.11.2 Flow over a Rankine Half-Body

Superposing a source and a free stream, we can generate a flow over a half-body:

$$F(Z) = V_\infty Z + \frac{\sigma}{2\pi} \ln Z, \tag{7.11.13}$$

where σ is the source strength. The velocity potential and the stream function are given by

$$\phi = V_\infty r \cos \theta + \frac{\sigma}{2\pi} \ln r = V_\infty x + \frac{\sigma}{4\pi} \ln(x^2 + y^2), \tag{7.11.14}$$

$$\psi = V_\infty r \sin \theta + \frac{\sigma}{2\pi} \theta = V_\infty y + \frac{\sigma}{2\pi} \tan^{-1}(y/x). \tag{7.11.15}$$

The complex velocity is given by

$$W = \frac{dF}{dZ} = V_\infty + \frac{\sigma}{2\pi Z} = \left(V_\infty + \frac{\sigma}{2\pi} \frac{\cos \theta}{r} \right) - i \frac{\sigma}{2\pi} \frac{\sin \theta}{r}, \tag{7.11.16}$$

which thus yields by $W = u - iv$

$$u = V_\infty + \frac{\sigma}{2\pi} \frac{\cos \theta}{r} = V_\infty + \frac{\sigma}{2\pi} \frac{x}{x^2 + y^2}, \tag{7.11.17}$$

$$v = -\frac{\sigma}{2\pi} \frac{\sin \theta}{r} = \frac{\sigma}{2\pi} \frac{y}{x^2 + y^2}. \tag{7.11.18}$$

Streamlines are plotted in Figure 7.11.2. Clearly, we see a flow over a half-body, which is called the Rankine half-body. The equation of the half-body is given by $\psi = \sigma/2$:

$$y = \frac{\sigma}{2\pi V_\infty} (\pi - \theta), \tag{7.11.19}$$

which can be used to discretize the boundary of the half-body. The above expression represents the thickness of the half-body. It shows that the thickness converges to

$$y_{\max} = \frac{\sigma}{2V_\infty}, \tag{7.11.20}$$

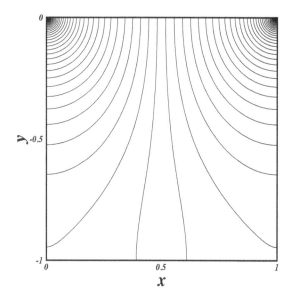

Figure 7.11.3: Cavity Flow: Velocity potential.

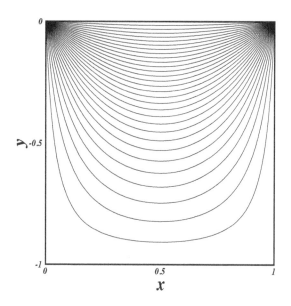

Figure 7.11.4: Cavity Flow: Streamlines.

as $x \rightarrow \infty$ (i.e., $\theta \rightarrow 0$). The location of the stagnation point can be easily found by solving $u = 0$ at $y = 0$:

$$x_{\text{stagnation}} = -\frac{\sigma}{2\pi V_\infty}. \tag{7.11.21}$$

Since we also have $v = 0$ at $y = 0$, it is a stagnation point.

A flow over a Rankine half-body can be useful for testing a CFD code and checking the performance for a flow with a stagnation point. You may want to take only the upper half of the domain as the flow is symmetric anyway. I like the upper half. It may be called a Rankine quater-body of which I've never heard. Also, it is possible to create a closed body by placing a sink behind the source. The resulting flow is known as a Rankine oval. See your favorite fluid dynamics textbook for details.

7.11.3 Cavity Flow

In a square domain $[0,1] \times [-1,0]$, a flow that comes out at the upper left corner and sinks in the upper right corner is given by the following complex potential[2],

$$F(Z) = \phi + i\psi = m \ln\left(\frac{g(Z)}{g(\frac{a}{2})}\right), \tag{7.11.22}$$

$$g(Z) = \sinh\left(\frac{\pi Z}{2a}\right) \frac{\Pi_{k=\text{even}}\left(1 - \dfrac{\sinh^2 \frac{\pi Z}{2}}{\sinh^2 \frac{k\pi}{2}}\right)}{\Pi_{k=\text{odd}}\left(1 - \dfrac{\sinh^2 \frac{\pi Z}{2}}{\sinh^2 \frac{k\pi}{2}}\right)}, \tag{7.11.23}$$

where $Z = x + iy$, $k = 1, 2, 3, \cdots$, and m and a are arbitrary constants. This solution converges very fast: say, $k = 20$ is sufficient in practice. For example, we take

$$m = -\frac{2}{\pi}, \tag{7.11.24}$$

$$a = 1, \tag{7.11.25}$$

---

[2]This solution was derived by Professor P. L. Roe at the University of Michigan.

so that $\psi(x,y)$ and $\phi(x,y)$ take the following boundary values,

$$\psi(x,0) = 0, \tag{7.11.26}$$

$$\psi(x,-1) = \psi(0,y) = \psi(1,y) = 1, \tag{7.11.27}$$

$$\phi(0.5,0) = 0, \tag{7.11.28}$$

i.e., the boundary becomes a wall (see Figures 7.11.3 and 7.11.4: the solutions with $k = 22$.).

The solution is singular at the upper corners: obviously, the stream function is multi-valued and the potential goes positive/negative infinity. But I like this solution. I can easily avoid the singularity, if I really wish, by restricting the domain with the upper boundary taken somewhere a bit below $y = 0$.

7.11.4 Flow over a Circular Cylinder

A flow over a circular cylinder is obtained by superposing a uniform stream, a doublet, and possibly a vortex, in the form of the complex potential,

$$F(Z) = \phi + i\psi = V_\infty \left[Ze^{-i\alpha} + \frac{R^2 e^{i\alpha}}{Z} \right] + \frac{i\Gamma}{2\pi} \ln Z, \tag{7.11.29}$$

i.e., with $Z = re^{i\theta}$,

$$\phi = V_\infty \cos(\theta - \alpha) \left[r + \frac{R^2}{r} \right] - \frac{\Gamma}{2\pi}\theta, \tag{7.11.30}$$

$$\psi = V_\infty \sin(\theta - \alpha) \left[r - \frac{R^2}{r} \right] + \frac{\Gamma}{2\pi} \ln r, \tag{7.11.31}$$

which give

$$u_r = \frac{\partial \phi}{\partial r} = \frac{1}{r}\frac{\partial \psi}{\partial \theta} = V_\infty \cos(\theta - \alpha) \left[1 - \frac{R^2}{r^2} \right], \tag{7.11.32}$$

$$u_\theta = \frac{1}{r}\frac{\partial \phi}{\partial \theta} = -\frac{\partial \psi}{\partial r} = -V_\infty \sin(\theta - \alpha) \left[1 + \frac{R^2}{r^2} \right] - \frac{\Gamma}{2\pi r}. \tag{7.11.33}$$

Alternatively, either from the complex velocity $W(Z)$,

$$W(Z) = \frac{dF}{dZ} = u - iv = V_\infty \left[e^{-i\alpha} - \frac{R^2 e^{i\alpha}}{Z^2} \right] + \frac{i\Gamma}{2\pi Z}, \tag{7.11.34}$$

or by the transformation (1.8.4), we obtain the Cartesian velocity components,

$$u = V_\infty \cos\alpha - \frac{(x^2 - y^2)\cos\alpha + 2xy\sin\alpha}{r^4} V_\infty R^2 + \frac{y\Gamma}{2\pi r^2}, \tag{7.11.35}$$

$$v = V_\infty \sin\alpha + \frac{(x^2 - y^2)\sin\alpha - 2xy\cos\alpha}{r^4} V_\infty R^2 - \frac{x\Gamma}{2\pi r^2}. \tag{7.11.36}$$

This describes a flow past a circular cylinder of radius R with the free stream V_∞ coming from the left at an angle of attack α. With non-zero Γ, it can be a flow over a rotating cylinder whose rotating speed is given by $\frac{\Gamma}{2\pi R}$. I like it very much because I think that it is one of the most beautiful flows. The pressure can be obtained directly from the Bernoulli's equation (3.18.4), or in the form of the pressure coefficient C_p,

$$C_p = \frac{p - p_\infty}{\frac{1}{2}\rho V_\infty^2} = 1 - \frac{u_r^2 + u_\theta^2}{V_\infty^2} = 1 - \frac{u^2 + v^2}{V_\infty^2}. \tag{7.11.37}$$

The flow is symmetric with respect to both x-axis and y-axis if Γ is zero. If it is not zero, the symmetry breaks down (see Figures 7.11.5 and 7.11.6). The stagnation points (zero tangential velocity at $r = R$) are given by

$$\sin\theta_{stag} = -\frac{\Gamma}{4\pi R V_\infty}, \tag{7.11.38}$$

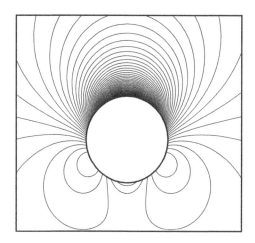

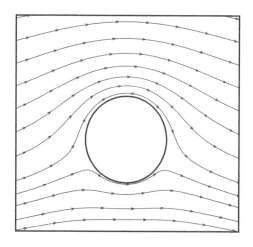

Figure 7.11.5: Pressure coefficient C_p of the cylinder flow: $V_\infty = 1$, $\alpha = 0$, $\Gamma = \pi$.

Figure 7.11.6: Streamlines of the cylinder flow: $V_\infty = 1$, $\alpha = 0$, $\Gamma = \pi$.

where θ_{stag} gives the location of a stagnation point. Note that the stagnation points do not exist on the cylinder surface if

$$|\Gamma| > 4\pi R V_\infty. \tag{7.11.39}$$

Note also that a force $\mathbf{F} = (F_x, F_y)$ acting on the cylinder can be obtained by the Blasius formula,

$$F_x - iF_y = \frac{i\rho}{2} \oint_{Cylinder} [W(Z)]^2 \, dZ, \tag{7.11.40}$$

giving

$$F_x - iF_y = \rho V_\infty \Gamma e^{i(\pi/2 + \alpha)}. \tag{7.11.41}$$

Therefore, the force acts perpendicular to the free stream, and so the drag D and the lift L are given by

$$L = \rho V_\infty \Gamma, \tag{7.11.42}$$
$$D = 0. \tag{7.11.43}$$

This result is called the Kutta-Joukowsky theorem. Of course, I like it. It is very nice that the drag is zero. This is actually called d'Alembert's paradox since the drag will be non-zero in reality. But an Euler code (being without any viscous effects) should be able to produce this paradoxical result. Can your Euler code predict a zero-drag potential flow?

Finally, I find it interesting that the line integral over a closed curve $\mathcal{C}$ of the complex velocity gives the volume flow Q across $\mathcal{C}$ and the circulation Γ around $\mathcal{C}$:

$$\Gamma + iQ = \oint_{\mathcal{C}} W \, dz, \tag{7.11.44}$$

which is very easy to prove. So, I'll leave it to you. Have fun.

7.11.5 Flow over an Elliptic Cylinder

Analytic function g that maps a point $\zeta = \xi + i\eta$ into a point $Z = x + iy$ (see Figure 7.11.7):

$$Z = g(\zeta), \tag{7.11.45}$$

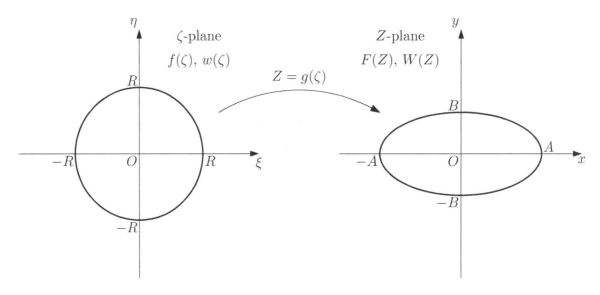

Figure 7.11.7: Conformal mapping from a circle to an ellipse. $A = R + \ell^2/R^2$ and $B = R - \ell^2/R^2$.

is called a conformal mapping. Consider a flow in Z-plane defined by the complex potential $F(Z)$. This flow can be mapped onto ζ-plane easily by $Z = g(\zeta)$. The corresponding flow in ζ-plane is then described by the complex potential $f(\zeta)$ given by

$$f(\zeta) = F(g(\zeta)), \tag{7.11.46}$$

and the complex velocity $w(\zeta)$ is given by

$$w(\zeta) = \frac{df(\zeta)}{d\zeta} = \frac{dF}{dZ}\frac{dZ}{d\zeta} = W(g(\zeta))\,g'(\zeta). \tag{7.11.47}$$

Now, reverse the viewpoint. As a matter of fact, the mapping is useful in obtaining a complicated flow in Z-plane from a simple flow in ζ-plane: $f(\zeta) \to F(Z)$ and $w(\zeta) \to W(Z)$. For example, the Joukowsky transformation,

$$Z = \zeta + \frac{\ell^2}{\zeta}, \tag{7.11.48}$$

maps a circle in ζ-plane onto an ellipse in Z-plane with the foci at $Z = \pm 2\ell$, and a circle with radius ℓ onto a line segment of length 4ℓ. I like the Joukowsky transformation, but I like its inverse better because it is more useful. Suppose we wish to compute a velocity at a point Z_p in Z-plane, but we know only $f(\zeta)$, i.e., the complex velocity in ζ-plane (the simpler flow). Then, first, we must map a point Z_p back to the corresponding point ζ_p in ζ-plane,

$$\zeta_p = g^{-1}(Z_p), \tag{7.11.49}$$

and then compute the desired velocity by

$$W(Z_p) = W(g(\zeta_p)) = w(\zeta_p)/g'(\zeta_p) = f'(\zeta_p)/g'(\zeta_p). \tag{7.11.50}$$

This way, we can compute the velocity at a point Z_p in Z-plane. The key is to have the inverse transformation g^{-1} at hand. It is generally difficult to obtain the inverse transformation, but it is often possible to do so. For the Joukowsky transformation (7.11.48), since it can be written in the form,

$$\frac{Z - 2\ell}{Z + 2\ell} = \left(\frac{\zeta - \ell}{\zeta + \ell}\right)^2, \tag{7.11.51}$$

an inverse is found to be

$$\zeta = \frac{(Z + 2\ell)^{\frac{1}{2}} + (Z - 2\ell)^{\frac{1}{2}}}{(Z + 2\ell)^{\frac{1}{2}} - (Z - 2\ell)^{\frac{1}{2}}}\ell. \tag{7.11.52}$$

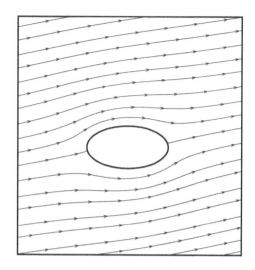

Figure 7.11.8: Pressure coefficient C_p of the flow over an ellipse: $V_\infty = 1$, $R = 3/4$, $\alpha = 10°$, $\Gamma = 0$, $\ell = \sqrt{3}/4$.

Figure 7.11.9: Streamlines of the flow over an ellipse: $V_\infty = 1$, $R = 3/4$, $\alpha = 10°$, $\Gamma = 0$, $\ell = \sqrt{3}/4$.

With this, it is now possible to obtain the velocity in a flow over an elliptic cylinder (or a flat plate) in Z-plane from a flow over a circular cylinder in ζ-plane, by setting up the complex potential,

$$f(\zeta) = V_\infty \left[\zeta e^{-i\alpha} + \frac{R^2 e^{i\alpha}}{\zeta} \right] + \frac{i\Gamma}{2\pi} \ln \zeta, \tag{7.11.53}$$

where $R > \ell$ (or $R = \ell$ for a flat plate), leading to the velocity

$$W(Z) = u - iv = f'(\zeta)/g'(\zeta)$$

$$= \frac{V_\infty \left[e^{-i\alpha} - \dfrac{R^2 e^{i\alpha}}{\zeta^2} \right] + \dfrac{i\Gamma}{2\pi\zeta}}{1 - \dfrac{\ell^2}{\zeta^2}}. \tag{7.11.54}$$

Note that we would have to use L'Hopital's rule to evaluate it at $\zeta = \ell$.

The surface of the elliptic cylinder is defined by

$$x = A \cos \theta, \tag{7.11.55}$$
$$y = B \sin \theta, \tag{7.11.56}$$

where

$$A = R + \frac{\ell^2}{R}, \tag{7.11.57}$$

$$B = R - \frac{\ell^2}{R}, \tag{7.11.58}$$

which can be inverted to get

$$R = \frac{A + B}{2}, \tag{7.11.59}$$

$$\ell = \frac{A^2 - B^2}{4}. \tag{7.11.60}$$

So, now we can do the following.

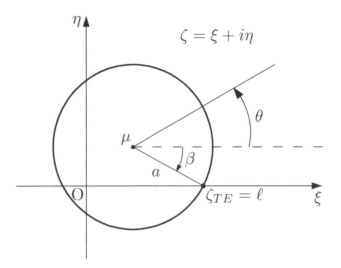

Figure 7.11.10: Circle that is mapped onto an airfoil.

Construction of an Ellipse:

1. Choose the major axis A and the minor axis B of the ellipse.

2. Generate an ellipse by Equations (7.11.55) and (7.11.56) in Z-plane.

Computation of Exact Solution:

1. Determine R and ℓ by

$$R = \frac{A+B}{2}, \quad \ell = \frac{A^2 - B^2}{4}. \tag{7.11.61}$$

2. Choose V_∞, α, and Γ.

3. Pick any point $Z_p = x_p + iy_p$ (on or outside the ellipse), and get the corresponding ζ_p by Equation (7.11.52).

4. Compute the exact velocity at the point Z_p by Equation (7.11.54), using ζ_p, and the pressure by

$$C_p = \frac{p - p_\infty}{\frac{1}{2}\rho V_\infty^2} = 1 - \frac{u^2 + v^2}{V_\infty^2}. \tag{7.11.62}$$

An example is shown in Figures 7.11.8 and 7.11.9.

7.11.6 Flow over an Airfoil

To generate a flow over an airfoil, first set up a flow over a shifted circular cylinder in ζ-plane as follows. Consider a circle of radius a, centered at

$$\zeta = \mu, \tag{7.11.63}$$

and passing through the point,

$$\zeta = \ell, \tag{7.11.64}$$

where ℓ is a real value (nearly a half chord of the resulting airfoil) while μ is a complex value (see Figure 7.11.10). This circle is represented by the formula,

$$\zeta = \mu + ae^{i\theta}, \tag{7.11.65}$$

where

$$0 \le \theta \le 2\pi, \tag{7.11.66}$$

$$\mu = \ell(-\epsilon + i\kappa), \tag{7.11.67}$$

$$a = \ell\sqrt{(1+\epsilon)^2 + \kappa^2}, \tag{7.11.68}$$

and ϵ and κ are positive constants that are related to the thickness and the camber respectively of the resulting airfoil. In fact, the point $\zeta_{TE} = \ell$ corresponds to the trailing edge of the resulting airfoil and if we denote the angle that indicates this point by β, i.e.,

$$\zeta_{TE} = \ell = \mu + ae^{-i\beta}, \tag{7.11.69}$$

then we find

$$\beta = \sin^{-1}\left(\frac{\ell\kappa}{a}\right) = \sin^{-1}\left(\frac{\kappa}{\sqrt{(1+\epsilon)^2 + \kappa^2}}\right). \tag{7.11.70}$$

Note that the point here is to shift the circle off from the origin. Recall that a circle centered at the origin ζ-plane is Joukowsky-transformed onto an ellipse in Z-plane. Now imagine that a circle off from the origin will be Joukowsky-transformed onto something different from an ellipse, which turns out to be an airfoil-like shape in Z-plane.

The complex potential that describes a flow over the circle defined above is given by

$$\frac{f(\zeta)}{V_\infty} = \zeta e^{-i\alpha} + i\,2a\sin(\alpha + \beta)\ln(\zeta - \mu) + \frac{a^2 e^{i\alpha}}{\zeta - \mu}, \tag{7.11.71}$$

which leads to the complex velocity,

$$\frac{w(\zeta)}{V_\infty} = e^{-i\alpha} + i\frac{2a\sin(\alpha + \beta)}{\zeta - \mu} - \frac{a^2 e^{i\alpha}}{(\zeta - \mu)^2}. \tag{7.11.72}$$

Note that the circulation Γ has been determined such that the velocity be finite at the trailing edge (the Kutta condition). In fact, the velocity at the trailing edge is given by $\frac{w(\zeta_{TE})}{dg/d\zeta}$, but $dg/d\zeta$ vanishes for the Joukowsky transformation. Therefore, in order for the velocity to be finite, it is required that $w(\zeta_{TE}) = 0$. This gives $\Gamma = 4\pi aV_\infty \sin(\alpha + \beta)$. For other types of transformations, even if $dg/d\zeta$ does not vanish, we must have $w(\zeta_{TE}) = 0$ (i.e., the trailing edge must be a stagnation point) because otherwise a flow is not uniquely determined and there are a plenty of experimental evidences that support this.

Now, we are ready to transform the circle (and a flow around it) onto something like an airfoil (and a flow around it). Specifically, we create a flow around an airfoil from a flow around the circle by a mapping $Z = g(\zeta)$, and compute a velocity at any point in the airfoil flow by transforming the point back to the ζ-plane by $\zeta = g^{-1}(Z)$ and then evaluate the velocity by

$$\frac{W(Z)}{V_\infty} = \frac{u - iv}{V_\infty} = \frac{w(\zeta)/g'(\zeta)}{V_\infty}. \tag{7.11.73}$$

For this purpose, I give two examples for the mapping $Z = g(\zeta)$: the Joukowsky transformation and the Kármán-Trefftz transformation. Airfoils generated by these transformations are called the Joukowsky and Kármán-Trefftz airfoils respectively.

Joukowsky Airfoils

The Joukowsky transformation:

$$Z = g_1(\zeta) = \zeta + \frac{\ell^2}{\zeta}. \tag{7.11.74}$$

An inverse is given by

$$\zeta = g_1^{-1}(\zeta) = \frac{(Z + 2\ell)^{\frac{1}{2}} + (Z - 2\ell)^{\frac{1}{2}}}{(Z + 2\ell)^{\frac{1}{2}} - (Z - 2\ell)^{\frac{1}{2}}}\ell. \tag{7.11.75}$$

The derivative of the transformation,

$$g_1'(\zeta) = 1 - \frac{\ell^2}{\zeta^2}, \tag{7.11.76}$$

is needed to compute the velocity by Equation (7.11.73).

I like the Joukowsky airfoils because they look very smooth, especially near the trailing edge. In fact, the trailing edge is cusped, i.e., zero thickness, and so the velocity there is not zero but finite (not a stagnation point). This can be nice because the flow is then smooth everywhere.

Note that if κ is not zero (cambered airfoils), there may be a region (below the airfoil but above the x axis) where the inverse (7.11.75) does not work. I would suggest the use of the Kármán-Trefftz transformation to avoid this trouble.

Kármán-Trefftz Airfoil

The Kármán-Trefftz transformation:

$$\frac{Z - n\ell}{Z + n\ell} = \left(\frac{\zeta - \ell}{\zeta + \ell}\right)^n, \quad n = 2 - \frac{\tau}{\pi}, \tag{7.11.77}$$

where τ is a free parameter to specify the trailing edge angle. This transformation can be written as

$$Z = g_2(\zeta) = \frac{(\zeta + \ell)^n + (\zeta - \ell)^n}{(\zeta + \ell)^n - (\zeta - \ell)^n} n\ell. \tag{7.11.78}$$

An inverse transformation is then given by

$$\zeta = g_2^{-1} = \frac{(Z + n\ell)^{\frac{1}{n}} + (Z - n\ell)^{\frac{1}{n}}}{(Z + n\ell)^{\frac{1}{n}} - (Z - n\ell)^{\frac{1}{n}}} \ell. \tag{7.11.79}$$

Note that the inverse transformation does not work if κ is too large for the same reason as in the inverse Joukowsky transformation (increasing the thickness parameter ϵ may help in keeping a valid inverse transformation with a relatively large κ). The derivative of the transformation is given by

$$g_2'(\zeta) = \frac{4n^2\ell^2 \left[(\zeta + \ell)(\zeta - \ell)\right]^{n-1}}{\left[(\zeta + \ell)^n - (\zeta - \ell)^n\right]^2}, \tag{7.11.80}$$

which is needed to compute the velocity by Equation (7.11.73).

I like Kármán-Trefftz airfoils because these airfoils have a trailing edge with a finite angle. This is more realistic and practical than the cusped trailing edge. Note also that the Joukowsky transformation is in fact a special case of the Kármán-Trefftz transformation: the Kármán-Trefftz transformation with $\tau = 0$ is the Joukowsky transformation. Anyway, so, I focus on the Kármán-Trefftz transformation in the rest of the section.

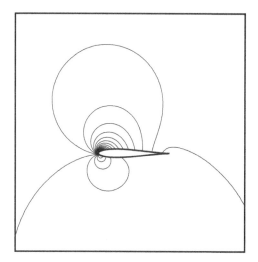

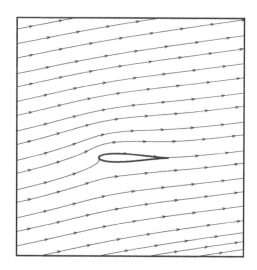

Figure 7.11.11: Pressure coefficient C_p of the flow over a Kármán-Trefftz airfoil: $\ell = 0.5$, $\alpha = 10°$, $\epsilon = 0.1$, $\kappa = 0$, $\tau = 0$.

Figure 7.11.12: Streamlines of the flow over a Kármán-Trefftz airfoil: $\ell = 0.5$, $\alpha = 10°$, $\epsilon = 0.1$, $\kappa = 0$, $\tau = 0$.

Construction of a Kármán-Trefftz Airfoil:

1. Specify the parameters:

$$\ell : \text{ Chord length} \approx 2\ell, \tag{7.11.81}$$
$$\alpha : \text{ Angle of attack (radian)}, \tag{7.11.82}$$
$$\epsilon : \text{ Thickness}, \tag{7.11.83}$$
$$\kappa : \text{ Camber}, \tag{7.11.84}$$
$$\tau : \text{ Trailing edge angle (radian)}. \tag{7.11.85}$$

2. Compute the following quantities:

$$a = \ell\sqrt{(1+\epsilon)^2 + \kappa^2} \qquad : \text{ Radius of the circle,} \tag{7.11.86}$$
$$\beta = \sin^{-1}\left(\frac{\kappa}{a}\right) \qquad : \text{ Trailing edge location,} \tag{7.11.87}$$
$$(\xi_c, \eta_c) = (-\ell\epsilon, \ell\kappa) \qquad : \text{ Center of the circle.} \tag{7.11.88}$$

This completely defines the circle by Equation (7.11.65).

3. Substitute points on the circle, $\zeta_i = (\xi_i, \eta_i)$, $i = 1, 2, 3, \ldots, N$, represented by

$$\xi_i = a\cos(\theta_i - \beta) + \xi_c, \tag{7.11.89}$$
$$\eta_i = a\sin(\theta_i - \beta) + \eta_c, \tag{7.11.90}$$

with θ_i in $[0, 2\pi]$ into the transformation (7.11.78) to obtain the points Z that form an airfoil (NB: $\theta = 0$ corresponds to the trailing edge).

4. (Optional) Compute the velocity at the airfoil points by substituting the points generated in the previous step into Equation (7.11.73). Note that the inverse transformation g_2^{-1} is not needed here, so that the velocity on the surface of the airfoil can be safely computed for any values of camber, thickness, and trailing edge angle.

Computation of Exact Solution (the whole flow field):

1. Pick a point $Z_p = x_p + iy_p$ (on and anywhere outside the airfoil!) where you want to compute the velocity.

2. Transform the point Z_p by Equation (7.11.79) onto the corresponding point ζ_p in ζ-plane.

3. Compute the velocity by substituting ζ_p into Equation (7.11.73).

An example, which was generated by the sample program available at http://www.cfdbooks.com, is given in Figures 7.11.11 and 7.11.12; it is in fact a Joukowsky airfoil since $\tau = 0$. A flow over a cylinder can be generated by taking $\tau = 180°$. Note that the flow has a stagnation point is at the rear of the cylinder because the circulation in in Equation (7.11.72) has been determined by the Kutta condition. It is interesting to compute the lift force, which is given by

$$L = \rho V_\infty \Gamma = 4\rho V_\infty^2 \pi a \sin(\alpha + \beta). \tag{7.11.91}$$

Can your CFD code predict the lift to the design accuracy?

7.11.7 Flow over Airfoils

It is possible to compute an exact solution for a flow around multiple airfoils by the conformal mapping technique. But the solution is usually available only on the surface of the airfoils (not the whole flow field). See Ref.[48, 56, 152, 172] for more details. In particular, the four-element configuration computed in Ref.[152] is widely used. The geometry of the four airfoils and the pressure coefficient over the airfoils are tabulated (with a limited precision) in Ref.[152].

7.11.8 Fraenkel's Flow: Constant Vorticity Flow over a Cylinder

Fraenkel's flow [38] is an exact solution of the incompressible Euler equations with a non-zero constant vorticity, i.e., linearly varying free stream. It was used to study the accuracy of an Euler code in Refs.[113, 123]. Although it is basically a flow over a cylinder (of radius 1 located at the origin), it contains recirculation zones in the front and back of the cylinder (see Figure 7.11.13). These recirculation zones can be quite difficult to capture numerically if a scheme is very dissipative. Can your code capture the recirculation zones (especially those in the rear)?

Here is the stream function ψ, which is presented with complex variables in Ref.[38], written in terms of real variables:

$$\psi = \frac{1}{2}\omega y^2 + \frac{\omega}{2\pi}\left[\left(1 - \frac{1}{r^2}\right)y + \frac{xy(r^4 - 1)}{2r^4}\ln\left(\frac{r^2 - 2x + 1}{r^2 + 2x + 1}\right)\right.$$
$$\left. + \frac{1}{2}\left\{\left(1 + \frac{1}{r^4}\right)(x^2 - y^2) - 2\right\}\arctan\left(\frac{2y}{r^2 - 1}\right)\right] + U_\infty\left(1 - \frac{1}{r^2}\right)y, \tag{7.11.92}$$

where $r = \sqrt{x^2 + y^2}$, ω is the constant vorticity, and U_∞ is the free stream velocity of the irrotational part of the flow. The velocity (u, v) is obtained simply by differentiating the stream function,

$$u = \frac{\partial \psi}{\partial y}, \quad v = -\frac{\partial \psi}{\partial x}, \tag{7.11.93}$$

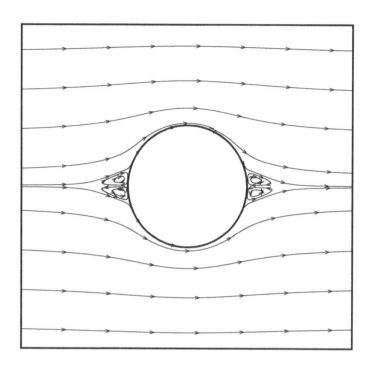

Figure 7.11.13: Streamlines of Fraenkel's flow.

but the result is very complicated:

$$u = \omega y + \frac{\omega}{2\pi}\left[\frac{2y^2}{r^4} + \left(1 - \frac{1}{r^2}\right) + \frac{x}{2}\left(1 - \frac{1}{r^4} + \frac{4y^2}{r^6}\right)\ln\left(\frac{r^2 - 2x + 1}{r^2 + 2x + 1}\right)\right.$$

$$+\frac{4x^2y^2\left(1 - \frac{1}{r^4}\right)}{(r^2 - 2x + 1)(r^2 + 2x + 1)} - \left\{\frac{2y(x^2 - y^2)}{r^6} + \left(1 + \frac{1}{r^4}\right)y\right\}\arctan\left(\frac{2y}{r^2 - 1}\right)$$

$$\left.+\frac{\left\{\left(1 + \frac{1}{r^4}\right)(x^2 - y^2) - 2\right\}(x^2 - y^2 - 1)}{(r^2 - 1)^2 + 4y^2}\right] + U_\infty\left(1 - \frac{x^2 - y^2}{r^4}\right), \tag{7.11.94}$$

$$v = -\frac{\omega}{2\pi}\left[\frac{2xy}{r^4} + \frac{y}{2}\left(1 - \frac{1}{r^4} + \frac{4x^2}{r^6}\right)\ln\left(\frac{r^2 - 2x + 1}{r^2 + 2x + 1}\right)\right.$$

$$+\frac{2xy(x^2 - y^2 - 1)\left(1 - \frac{1}{r^4}\right)}{(r^2 - 2x + 1)(r^2 + 2x + 1)} - \left\{\frac{2x(x^2 - y^2)}{r^6} - \left(1 + \frac{1}{r^4}\right)x\right\}\arctan\left(\frac{2y}{r^2 - 1}\right)$$

$$\left.-\frac{2xy\left\{\left(1 + \frac{1}{r^4}\right)(x^2 - y^2) - 2\right\}}{(r^2 - 1)^2 + 4y^2}\right] - U_\infty\frac{2xy}{r^4}. \tag{7.11.95}$$

These functions were used to generate the streamline plot in Figure 7.11.13. Note that the above exact solution is valid only on the upper half of the domain, i.e., $y \geq 0$, and to make it valid on the lower half, I have to change the sign of y, i.e., $y \to -y$. Also, Note also that the above exact solution is singular at $(x, y) = (-1, 0)$ and $(1, 0)$ where it should be set to be a stagnation point $u = v = 0$. By the way, it is very interesting that such a complicated flow (as in Figure 7.11.13) can be represented by an analytical formula.

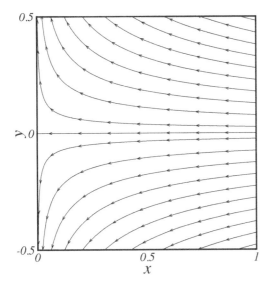

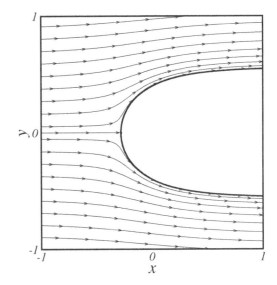

Figure 7.12.1: Inviscid Stagnation Flow in 3D.   Figure 7.12.2: Flow over a Rankine Half-Body in 3D.

7.12 3D Incompressible Euler Equations (Axisymmetric Flows)

I like 3D flows because they are much more realistic than 1D and 2D flows. 3D solutions can be used to verify practical CFD codes, which solve 3D problems. Although the theory of complex variables cannot be applied, various exact solutions can be derived for 3D irrotational flows because the velocity potential satisfies the Laplace equation. Especially, I like axisymmetric flows because these flows can be described by a single stream function just like 2D flows. Below, axisymmetric flows are described in (x, y)-plane with the x-axis as the axis of symmetry. Note that these are 3D solutions and therefore should not be used for 2D CFD codes that solve 2D equations.

7.12.1 Stagnation Flow

A three-dimensional stagnation flow is an axisymmetric flow. It is described by the following velocity potential:

$$\phi = \frac{1}{4}(y^2 - 2x^2).$$
(7.12.1)

It is easy to verify that the above velocity potential satisfies the Laplace equation for axisymmetric flows:

$$\frac{\partial^2 \phi}{\partial x^2} + \frac{\partial^2 \phi}{\partial y^2} + \frac{1}{y}\frac{\partial \phi}{\partial y} = 0.$$
(7.12.2)

The velocity components (u, v) are given by

$$u = \frac{\partial \phi}{\partial x} = -x, \quad v = \frac{\partial \phi}{\partial y} = \frac{1}{2}y,$$
(7.12.3)

which differ from the two-dimensional case in Section 7.11.1 by the factor $1/2$ in v. As in the two-dimensional case, the velocity components are linear function of space. I like that. A second-order scheme should still give the exact result in 3D. The stream function can be derived by integrating Equation (4.24.14):

$$\psi = -\frac{1}{2}xy^2,$$
(7.12.4)

which is different from the two-dimensional version (7.11.10). Streamlines are plotted in Figure 7.12.1, which should be compared with Figure 7.11.1.

7.12.2 Flow over a Rankine Half-Body

A flow over a Rankine half-body can be obtained in 3D. As in 2D, we superpose a source and a free stream:

$$\phi = V_\infty x - \frac{\sigma}{4\pi r}, \tag{7.12.5}$$

where $r = \sqrt{x^2 + y^2}$ and the second term represents the three-dimensional source with the strength σ. It is straightforward to verify that the above potential satisfies the Laplace equation (7.12.2). The velocity components (u, v) are given by

$$u = \frac{\partial \phi}{\partial x} = V_\infty + \frac{\sigma x}{4\pi r^3}, \quad v = \frac{\partial \phi}{\partial y} = \frac{\sigma y}{4\pi r^3}. \tag{7.12.6}$$

The stream function can be derived as

$$\psi = \frac{1}{2} V_\infty y^2 + \frac{\sigma}{4\pi} \left(1 - \frac{x}{r} \right). \tag{7.12.7}$$

Streamlines are plotted in Figure 7.12.2. This is the Rankine half-body in three dimensions. Compare the streamlines and the half-body with those in two dimensions, Figure 7.11.2. They are similar but different. In 3D, the equation of the half-body is given by $\psi = \sigma/(2\pi)$:

$$y = \pm \sqrt{\frac{2\sigma}{\pi V_\infty} (\cos \theta + 1)}, \tag{7.12.8}$$

where $\theta = \tan^{-1}(y/x)$, which can be used to discretize the boundary of the half-body by varying θ. It follows that the thickness converges to

$$y_{\text{max}} = 2 \sqrt{\frac{\sigma}{V_\infty}}, \tag{7.12.9}$$

as $x \to \infty$ (i.e., $\theta \to 0$), which is slightly thicker than the two-dimensional body. The location of the stagnation point can be easily found by solving $u = 0$ at $y = 0$:

$$x_{\text{stagnation}} = -2 \sqrt{\frac{\sigma}{\pi V_\infty}}. \tag{7.12.10}$$

Of course, we have $v = 0$ at $y = 0$, and thus it is indeed a stagnation point.

7.12.3 Flow over a Sphere

A flow over a sphere at zero angle of attack is a valid axisymmetric flow. The following velocity potential, which consists of a uniform flow and a doublet, is an exact solution that represents a flow over a sphere of radius R,

$$\phi = V_\infty x \left[1 + \frac{1}{2} \left(\frac{R}{r} \right)^3 \right]. \tag{7.12.11}$$

It is quite straightforward to verify that the above potential satisfies the Laplace equation (7.12.2). The velocity components are given by

$$u = \frac{\partial \phi}{\partial x} = V_\infty \left[1 + \frac{R^3}{2r^3} \left(1 - \frac{3x^2}{r^2} \right) \right], \tag{7.12.12}$$

$$v = \frac{\partial \phi}{\partial y} = -\frac{3xy V_\infty R^3}{2r^5}. \tag{7.12.13}$$

The stream function can be derived in the form:

$$\psi = \frac{1}{2} V_\infty y^2 \left(1 - \frac{R^3}{r^3} \right). \tag{7.12.14}$$

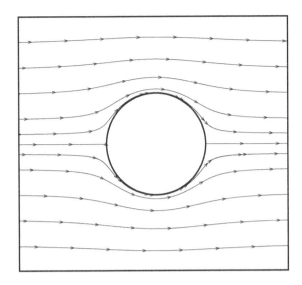

Figure 7.12.3: Flow over a Sphere (3D).

Streamlines are plotted in Figure 7.12.3. Note that the maximum speed on the sphere is

$$|u|_{x=0, y=R} = \frac{3}{2}V_\infty,$$

(7.12.15)

which is smaller than that for a flow over a cylinder obtained from Equation (7.11.35),

$$|u| = 2V_\infty.$$

(7.12.16)

On the sphere $(x^2 + y^2 = R^2)$, we have

$$u^2 + v^2 = \frac{9V_\infty^2}{4}\frac{y^2}{R^2}.$$

(7.12.17)

Therefore, the pressure coefficient over the sphere is given by

$$C_p = \frac{p - p_\infty}{\frac{1}{2}\rho V_\infty^2} = 1 - \left(\frac{u^2 + v^2}{V_\infty^2}\right)\Bigg|_{r=R} = 1 - \frac{9}{4}\frac{y^2}{R^2}.$$

(7.12.18)

I like the flow over a sphere. It is very simple and smooth. But it is not really simple to compute numerically. First, it is not trivial to generate a computational grid around a sphere (unlike generating a grid around a two-dimensional cylinder): structured grids need some effort to avoid possible geometric singularities. Second, it is often hard to obtain a clean symmetric solution with respect to the (y, z)-plane (i.e., fore-aft symmetry) by an Euler code, particularly with unstructured grids. How accurately can your Euler code compute this flow?

7.13   Euler Equations

I like the compressible Euler Equations. Basically, all the incompressible flows described in the previous section are the exact solutions to the compressible Euler Equations with constant density. Here, we consider exact solutions for the compressible Euler Equations with variable density.

7.13.1   One-Dimensional Simple Acoustic Waves

Consider the simple wave solution derived in Section 6.7.3:

$$\frac{u(x,t)}{a_\infty} = M_\infty + \frac{2}{(\gamma+1)}\frac{V(\xi)}{a_\infty}, \tag{7.13.1}$$

$$\frac{\rho(x,t)}{\rho_\infty} = \left[1 + \frac{\gamma-1}{2}\left(\frac{u(\xi)}{a_\infty} - M_\infty\right)\right]^{\frac{2}{\gamma-1}}, \tag{7.13.2}$$

$$\frac{p(x,t)}{p_\infty} = \left[1 + \frac{\gamma-1}{2}\left(\frac{u(\xi)}{a_\infty} - M_\infty\right)\right]^{\frac{2\gamma}{\gamma-1}}, \tag{7.13.3}$$

where $\xi = x - Vt$, $V = u + a$, and V is a solution of Burgers' equation. This is valid for smooth flows and the exact solutions of Burgers' equation given in Section 7.7.1 can be directly employed for V.

(a) Smooth solution turning into a shock:

Set an initial solution with

$$V(x) = \frac{1}{2\pi t_s}\sin(2\pi x), \tag{7.13.4}$$

then it gives the exact solution with

$$V(\xi) = \frac{1}{2\pi t_s}\sin(2\pi(x - Vt)), \tag{7.13.5}$$

where t_s is a free parameter that specifies the time when a shock is formed. This solution is given implicitly and therefore must be computed numerically (see Section 7.7.1).

(b) Smooth rarefaction:

Set an initial solution with

$$V(x) = A\tanh(k(x - 0.5)), \tag{7.13.6}$$

then it gives the exact solution with

$$V(\xi) = A\tanh(k(x - 0.5 - Vt)), \tag{7.13.7}$$

where A and k are arbitrary constants. Again, this is given implicitly and therefore must be computed numerically.

These solutions are very similar to those in Figures 7.7.1 and 7.7.2. These types of solutions were actually used in Ref.[87] for verification of higher-order schemes. It is nice to have smooth exact solutions for nonlinear systems such as the Euler equations. In particular, it is very nice that we can generate various exact solutions through the solutions of Burgers' equation. I really like it.

7.13.2   Entropy Waves

The entropy wave is an exact simple wave solution of the compressible Euler equations. The one-dimensional entropy wave is given by

$$\begin{aligned}
\rho &= \rho_\infty + A\sin\left[\pi(x - U_\infty t)\right], &\tag{7.13.8}\\
u &= U_\infty, &\tag{7.13.9}\\
p &= P_\infty, &\tag{7.13.10}
\end{aligned}$$

where ρ_∞, U_∞ and P_∞ are constants. The two-dimensional version is given by

$$\begin{aligned}
\rho &= \rho_\infty + A\sin\left[\pi(x + y - Q_\infty t)\right], &\tag{7.13.11}\\
u &= U_\infty, &\tag{7.13.12}\\
v &= V_\infty, &\tag{7.13.13}\\
p &= P_\infty, &\tag{7.13.14}
\end{aligned}$$

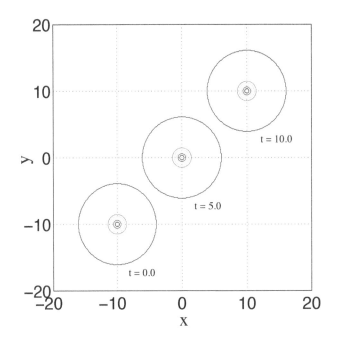

Figure 7.13.1: Density contours for the two-dimensional isentropic vortex convection. $\alpha = 1$, $K = 5.0$, $\rho_\infty = 1$, $p_\infty = 1/\gamma$, $u_\infty = v_\infty = 2.0$, $(x_0, y_0) = (-10, -10)$.

where V_∞ is a constant and $Q_\infty = U_\infty + V_\infty$. The three-dimensional version is given by

$$\rho = \rho_\infty + A \sin\left[\pi(x + y + z - Q_\infty t)\right], \tag{7.13.15}$$

$$u = U_\infty, \tag{7.13.16}$$

$$v = V_\infty, \tag{7.13.17}$$

$$w = W_\infty, \tag{7.13.18}$$

$$p = P_\infty, \tag{7.13.19}$$

where W_∞ is a constant and $Q_\infty = U_\infty + V_\infty + W_\infty$. With a periodic boundary condition, the entropy wave should travel through the domain, preserving its initial profile. Certainly, I like the entropy wave. It is very useful in verifying the accuracy of a time-dependent Euler code.

7.13.3   Two-Dimensional Unsteady Isentropic Vortex Convection

As derived in Section 6.3, the following is an exact solution to the full unsteady two-dimensional compressible Euler equations:

$$\frac{u(x, y, t)}{a_\infty} = \frac{u_\infty}{a_\infty} - \frac{K}{2\pi a_\infty} \bar{y} e^{\alpha(1 - \bar{r}^2)/2}, \tag{7.13.20}$$

$$\frac{v(x, y, t)}{a_\infty} = \frac{v_\infty}{a_\infty} + \frac{K}{2\pi a_\infty} \bar{x} e^{\alpha(1 - \bar{r}^2)/2}, \tag{7.13.21}$$

$$\frac{T(x, y, t)}{T_\infty} = 1 - \frac{K^2(\gamma - 1)}{8\alpha \pi^2 a_\infty^2} e^{\alpha(1 - \bar{r}^2)}, \tag{7.13.22}$$

$$\frac{\rho(x, y, t)}{\rho_\infty} = \left(\frac{T(x, y, t)}{T_\infty}\right)^{\frac{1}{\gamma - 1}}, \tag{7.13.23}$$

$$\frac{p(x, y, t)}{p_\infty} = \left(\frac{T(x, y, t)}{T_\infty}\right)^{\frac{\gamma}{\gamma - 1}}, \tag{7.13.24}$$

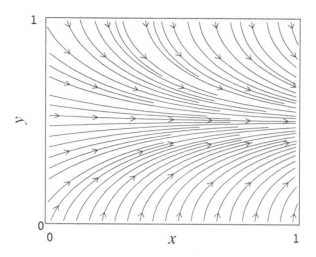

Figure 7.13.2: Subsonic solution. Streamlines calculated by the total velocity $(U_\infty + u, v)$ with $U_\infty = 1.0$, $k = \pi/\sqrt{1 - M_\infty^2}$ and $M_\infty = 0.5$.

where

$$\overline{x} = x - x_0 - u_\infty t, \tag{7.13.25}$$
$$\overline{y} = y - y_0 - v_\infty t, \tag{7.13.26}$$
$$\overline{r} = \sqrt{\overline{x}^2 + \overline{y}^2}. \tag{7.13.27}$$

This exact solution describes a vortex simply convected isentropically with the free stream velocity (u_∞, v_∞). It is often employed to measure the accuracy of an Euler code [28, 142, 177]. For example, I like the following set of parameters: $\alpha = 1$, $\rho_\infty = 1$, $p_\infty = 1/\gamma$, $u_\infty = v_\infty = 2$, $(x_0, y_0) = (-10, -10)$, $K = 5$, because then the vortex starts at $(x, y) = (-10, -10)$, moves to the origin at $t = 5$, and then will finally reach at $(x, y) = (10, 10)$ precisely at $t = 10$ as shown in Figure 7.13.1. You may measure the accuracy of your Euler code at the final location.

7.13.4   Linearized Potential Equations

The linearized potential equation (see Section 3.20),

$$(1 - M_\infty^2) \phi'_{xx} + \phi'_{yy} = 0, \tag{7.13.28}$$

or the equivalent system for the perturbation velocity components (see Section 3.21),

$$(1 - M_\infty^2) u_x + v_y = 0, \tag{7.13.29}$$
$$v_x - u_y = 0, \tag{7.13.30}$$

has various exact solutions.

Subsonic solution

As discussed in Section 3.21, the Prandtl-Glauert transformation

$$\xi = x, \quad \eta = y \sqrt{1 - M_\infty^2}, \quad \Phi' = \sqrt{1 - M_\infty^2} \, \phi', \quad U = \sqrt{1 - M_\infty^2} \, u, \quad V = v, \tag{7.13.31}$$

turns the governing equations into the Laplace equation for the potential,

$$\Phi'_{\xi\xi} + \Phi'_{\eta\eta} = 0, \tag{7.13.32}$$

and the Cauchy-Riemann system for the velocity components,

$$U_\xi + V_\eta = 0, \tag{7.13.33}$$
$$V_\xi - U_\eta = 0, \tag{7.13.34}$$

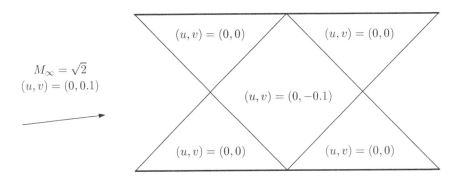

Figure 7.13.3: Supersonic solution. Inflow perturbation velocity components are $u = 0$ and $v = 0.1$, with the mean-flow Mach number $M_\infty = \sqrt{2}$. Thin lines are characteristics dividing the constant states.

where we have assumed here that $M_\infty < 1$. This means that any incompressible flow solution $\Phi'(\xi, \eta)$, i.e., the solution of the Laplace equation (7.13.32), can be transformed into a compressible flow solution, i.e., the solution of Equation (7.13.28), as follows,

$$\phi'(x, y) = \frac{\Phi'(x, y\sqrt{1 - M_\infty^2})}{\sqrt{1 - M_\infty^2}}. \tag{7.13.35}$$

Alternatively, we may start from a set of solutions for the Cauchy-Riemann system, $U(\xi, \eta)$ and $V(\xi, \eta)$, and then obtain the velocity components in the equivalent compressible flow as follows,

$$u(x, y) = \frac{U(x, y\sqrt{1 - M_\infty^2})}{\sqrt{1 - M_\infty^2}}, \tag{7.13.36}$$

$$v(x, y) = V(x, y\sqrt{1 - M_\infty^2}). \tag{7.13.37}$$

For example, the exact solution (7.9.5) from Section 7.9.1,

$$\Phi'(\xi, \eta) = \exp(k\xi)\sin(k\eta), \tag{7.13.38}$$

where k is an arbitrary constant, can be transformed into the corresponding compressible solution as follows,

$$\phi'(x, y) = \frac{\exp(kx)\sin\left(ky\sqrt{1 - M_\infty^2}\right)}{\sqrt{1 - M_\infty^2}}. \tag{7.13.39}$$

To see what flow this represents, we derive the perturbation velocity components,

$$u(x, y) = \phi'_x(x, y) = \frac{k\exp(kx)\sin\left(ky\sqrt{1 - M_\infty^2}\right)}{\sqrt{1 - M_\infty^2}}, \tag{7.13.40}$$

$$v(x, y) = \phi'_y(x, y) = k\exp(kx)\cos\left(ky\sqrt{1 - M_\infty^2}\right). \tag{7.13.41}$$

The streamlines calculated from these are plotted in Figure 7.13.2. It does look like a flow. This is simple and smooth enough for accuracy verification, possibly with the Dirichlet condition. I like this very much. It is really nice that any solution of Laplace's equation can be transformed into a subsonic flow solution.

Supersonic solution

As shown in Section 3.21, for supersonic flows ($M_\infty > 1$), the governing equations become hyperbolic. The method of characteristics can then be used to generate various exact solutions: applying the following characteristic relations,

$$u\sqrt{M_\infty^2 - 1} - v = \text{constant}, \quad \text{along} \quad dy/dx = 1/\sqrt{M_\infty^2 - 1}, \tag{7.13.42}$$

$$u\sqrt{M_\infty^2 - 1} + v = \text{constant}, \quad \text{along} \quad dy/dx = -1/\sqrt{M_\infty^2 - 1}, \tag{7.13.43}$$

with appropriate boundary conditions, we can construct a flow field, starting from a supersonic free stream of the Mach number $M_\infty > 1$ (see Ref.[155] for more details if needed). An example is a supersonic flow through a dust: a mean-flow coming from the left at Mach number $M_\infty = \sqrt{2}$ with the perturbation velocity $(u, v) = (0, 0.1)$, through a parallel duct. The small incidence introduced by the non-zero v causes compression and expansion waves created at the leading edges and reflected through the duct, recovering the inflow condition at the exit. This is sketched in Figure 7.13.3. Solution at each region can be easily found by drawing characteristic lines from a known upstream state and applying the appropriate characteristic relation above. I like this solution because it can be a very simple test case (with a simple rectangular domain) for solving linear hyperbolic systems involving discontinuities. In fact, this solution was used to demonstrate the capability of a particular mesh adaptation technique in Ref.[130].

7.13.5   Nozzle Flows

Basic Equations

Nozzle flows are governed by the equations in Section 3.16 which for a steady adiabatic flow reduce to

$$\rho u A = \text{constant}, \tag{7.13.44}$$

$$\frac{u^2}{2} + \frac{a^2}{\gamma - 1} = \text{constant}, \tag{7.13.45}$$

$$p/\rho^\gamma = \text{constant}. \tag{7.13.46}$$

We then obtain from these (see Ref.[84] for derivation) the fundamental relations for generating exact solutions:

$$\left(\frac{A}{A^*}\right)^2 = \frac{1}{M}\left[\frac{2}{\gamma + 1}\left(1 + \frac{\gamma - 1}{2}M^2\right)\right]^{(\gamma+1)/(\gamma-1)}, \tag{7.13.47}$$

$$\frac{\rho}{\rho_0} = \left(1 + \frac{\gamma - 1}{2}M^2\right)^{\frac{-1}{\gamma-1}}, \tag{7.13.48}$$

$$\frac{p}{p_0} = \left(1 + \frac{\gamma - 1}{2}M^2\right)^{\frac{-\gamma}{\gamma-1}}, \tag{7.13.49}$$

$$\frac{T}{T_0} = \left(1 + \frac{\gamma - 1}{2}M^2\right)^{-1}, \tag{7.13.50}$$

where $\gamma = 1.4$ for air, A^* is the critical area corresponding to sonic speed $u = a$, $M = u/a$, and ρ_0, p_0, and T_0 are the stagnation density, pressure, and temperature. It follows from Equation (7.13.47) that there exist two different Mach numbers for a given area-ratio $\frac{A}{A^*}$: subsonic and supersonic solutions (see Ref.[84]).

Differentiate the continuity equation (7.13.44),

$$\frac{1}{\rho}\frac{d\rho}{dx} + \frac{1}{u}\frac{du}{dx} + \frac{1}{A}\frac{dA}{dx} = 0, \tag{7.13.51}$$

and substitute the momentum equation,

$$\rho u \frac{du}{dx} = -\frac{dp}{dx} = a^2 \frac{d\rho}{dx}, \tag{7.13.52}$$

to obtain

$$\frac{1}{u}\frac{du}{dx} = -\frac{1}{1 - M^2}\frac{1}{A}\frac{dA}{dx}. \tag{7.13.53}$$

This shows that a subsonic flow is accelerated $(du > 0)$ through a converging nozzle $(dA < 0)$; and a supersonic flow accelerated $(du > 0)$ through a diverging nozzle $(dA > 0)$. This is very interesting. I like it.

Now, suppose that a nozzle shape $A(x)$ is given in $x \in [0, 1]$, having the minimum (or throat) area at $x = x_{throat}$. Then, there exist three types of solutions: subsonic-supersonic, fully subsonic, and subsonic-supersonic with a shock.

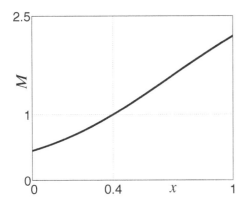

Figure 7.13.4: Mach number: the unique subsonic-supersonic flow.

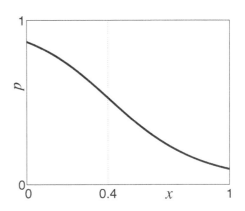

Figure 7.13.5: Pressure: the unique subsonic-supersonic flow.

Subsonic-Supersonic Flow

This is a unique solution for a given nozzle shape: a subsonic inflow achieves $M = 1$ precisely at the throat $x = x_{throat}$ so that $A^* = A(x_{throat})$, and then smoothly becomes supersonic and accelerates towards the exit. It should be noted that the inflow Mach number is not zero but a small value. Theoretically, we see from Equation (7.13.47) that $M \to 0$ when $A \to 0$. That is, a reservoir exists at $x \to -\infty$. In a way, we are looking at only a small part of a big nozzle coming out from a reservoir. Or it may be interpreted as a flow through the nozzle in $[0, 1]$ caused by a pressure difference applied to the inlet and outlet of the nozzle with a finite length.

   Computation of Exact Solution:
   We can numerically compute this unique flow solution at a set of N points $\{x_i\} = \{x_1 = 0, x_2, x_3, \cdots, x_N = 1\}$ as follows:

   1. Set $A^* = A(x_{throat})$.

   2. Compute $M(x_i)$ by solving Equation (7.13.47) for given $A(x_i)$ and A^* for all i.
      (subsonic for $x_i \in [0, x_{throat}]$; supersonic for $x_i \in [x_{throat}, 1]$.)

   3. Compute $\rho(x_i)$, $p(x_i)$, and $T(x_i)$ by Equations (7.13.48), (7.13.49), and (7.13.50).

Note that Equation (7.13.47) needs to be solved for the Mach number numerically. For this purpose, we may employ the bisection method for the following equation,

$$f(M) = \left(\frac{A}{A^*}\right)^2 - \frac{1}{M}\left[\frac{2}{\gamma+1}\left(1 + \frac{\gamma-1}{2}M^2\right)\right]^{(\gamma+1)/(\gamma-1)} = 0, \qquad (7.13.54)$$

with adaptive initial bounds: $[0, 1]$ to find a subsonic solution or $[1, 100]$ to find a supersonic solution. Of course, other methods can be used, but I like the bisection method because it is very simple. In fact, I actually used it to obtain the solution shown in Figures 7.13.4 and 7.13.4.

Fully Subsonic Flow

In this case, there exist an infinite number of solutions for a given nozzle shape; the exit pressure p_e must be specified to make the solution unique. Observe also that in this case a flow can never reach $M = 1$ ($A^* \neq A(x)$ anywhere in the nozzle). Therefore, A^* is just a parameter determined by first computing M_e from (for a specified p_e)

$$\frac{p_e}{p_0} = \left(1 + \frac{\gamma-1}{2}M_e^2\right)^{\frac{-\gamma}{\gamma-1}}, \qquad (7.13.55)$$

and then solving

$$\left(\frac{A_e}{A^*}\right)^2 - \frac{1}{M_e}\left[\frac{2}{\gamma+1}\left(1 + \frac{\gamma-1}{2}M_e^2\right)\right]^{(\gamma+1)/(\gamma-1)} = 0, \qquad (7.13.56)$$

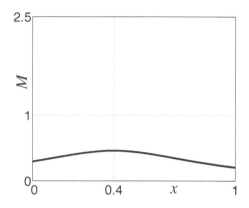

Figure 7.13.6: Mach number: a fully subsonic flow.

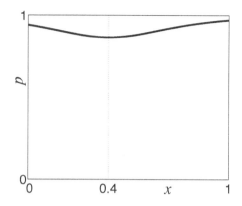

Figure 7.13.7: Pressure: a fully subsonic flow.

for A^*. Here, obviously, $A_e = A(x_N) = A(1.0)$, $M_e = M(x_N) = M(1.0)$, and $p_e = p(x_N) = p(1.0)$.

Note that the exit pressure p_e must satisfy

$$p_e^{crit} \leq p_e < p_0, \qquad (7.13.57)$$

where p_e^{crit} is the pressure corresponding to the ideal subsonic solution, i.e., the one with $A^* = A(x_{throat})$ of Equation (7.13.56), at the exit This p_e^{crit} can be computed by first solving Equation (7.13.56) with $A^* = A(x_{throat})$ to obtain M_e^{crit}, and then substituting this Mach number M_e^{crit} into Equation (7.13.49).

Computation of Exact Solution:

A fully subsonic flow solution can be computed numerically as follows:

1. Compute M_e^{crit} by solving Equation (7.13.56) and $A^* = A(x_{throat})$.
   (Assume that the flow is subsonic.)

2. Compute p_e^{crit} (7.13.48).

3. Specify the exit pressure p_e under the condition (7.13.57).

4. Compute M_e from p_e by Equation (7.13.49).

5. Re-compute A^* by solving Equation (7.13.47) with $A = A_e$ and $M = M_e$.

6. Compute $M(x_i)$ by solving Equation (7.13.47) for given $A(x_i)$ and A^* for all i.
   (Note that the flow is subsonic everywhere.)

7. Compute $\rho(x_i)$, $p(x_i)$, and $T(x_i)$ by Equations (7.13.48), (7.13.49), and (7.13.50).

An example is given in Figures 7.13.6 and 7.13.7.

Shocked Flow

In shocked flows also, the exit pressure p_e must be specified to make the solution unique. This time, the exit pressure, p_e, must satisfy

$$p_{ps} \leq p_{exit} < p_e^{crit}, \qquad (7.13.58)$$

where p_{ps} is a post-shock pressure given by

$$p_{ps} = \left[1 + \frac{\gamma - 1}{2} M_{ps}^2 \right]^{\frac{-\gamma}{\gamma - 1}}, \qquad (7.13.59)$$

$$M_{ps} = \sqrt{\frac{1 + \dfrac{\gamma - 1}{2} \left(M_e^{crit} \right)^2}{\gamma \left(M_e^{crit} \right)^2 - \dfrac{\gamma - 1}{2}}}. \qquad (7.13.60)$$

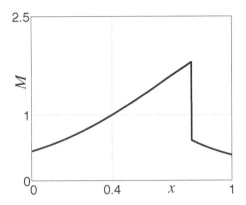

Figure 7.13.8: Mach number: a shocked flow.

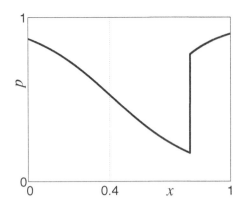

Figure 7.13.9: Pressure: a shocked flow.

This is a pressure obtained based on the assumption that there exists a shock exactly at the exit in the ideal subsonic-supersonic flow, i.e., the flow with $A^* = A(x_{throat})$. The condition (7.13.58) ensures the presence of a shock somewhere between the throat ($x = x_{throat}$) and the exit.

Computation of Exact Solution:
A shocked flow solution can be computed numerically as follows:

1. Specify the exit pressure p_e under the condition (7.13.58).

2. Compute M_e from p_e by Equation (7.13.49).

3. Compute and store the ideal solution $M_{ideal}(x_i)$ by solving Equation (7.13.47) with $A^* = A(x_{throat})$ for all i.
   (subsonic for $x_i \in [0, x_{throat}]$; supersonic for $x_i \in [x_{throat}, 1.0]$.)

4. Re-compute A^* by solving Equation (7.13.47) with $A = A_e$ and $M = M_e$.

5. From the exit (towards the throat), start computing the subsonic solution $M_{sub}(x_i)$ by solving Equation (7.13.47) for each $A(x_i)$.

6. In doing so, check if the condition,

$$M_{post\text{-}shock}(x_{i-1}) < M_{sub}(x_i) < M_{post\text{-}shock}(x_i), \qquad (7.13.61)$$

where $M_{post\text{-}shock}(x_i)$ is a Mach number behind an assumed shock located at x_i in the ideal subsonic-supersonic flow given by

$$M_{post\text{-}shock}(x_i) = \sqrt{\frac{1 + \dfrac{\gamma - 1}{2}\left[M_{ideal}(x_i)\right]^2}{\gamma \left[M_{ideal}(x_i)\right]^2 - \dfrac{\gamma - 1}{2}}}, \qquad (7.13.62)$$

is satisfied. If it is, there is a shock between x_{i-1} and x_i, and you define $i = i_{shock}$ and stop calculating $M(x_i)$.

7. The solution is then given by, for $i = 1, 2, 3, \cdots, N$,

$$M(x_i) = \begin{cases} M_{ideal}(x_i) & \text{if } i < i_{shock}, \\ M_{sub}(x_i) & \text{otherwise}, \end{cases} \qquad (7.13.63)$$

and $\rho(x_i)$, $p(x_i)$, and $T(x_i)$ are computed by Equations (7.13.48), (7.13.49), and (7.13.50).

An example is given in Figures 7.13.8 and 7.13.9.

	Entrance	Exit
M	0.4517	2.1972
p	0.8694	0.0939

Table 7.13.1: Data for the subsonic-supersonic flow in Figures 7.13.4 and 7.13.4.

	Entrance	Exit
M	0.2972	0.2091
p	0.9405	0.9700

Table 7.13.2: Data for the fully subsonic flow in Figures 7.13.6 and 7.13.7. $p_e = 0.97$ was specified.

	Entrance	Exit
M	0.4517	0.3909
p	0.8694	0.9000

Table 7.13.3: Data for the shocked flow in Figures 7.13.8 and 7.13.9. $p_e = 0.9$ was specified. The shock was found to be at $x = 0.8$.

Figure 7.13.10: Upper half of a quadratic nozzle (7.13.64) with $A_e = 0.4$ and $A_t = 0.2$.

Example

Consider a quadratic shape nozzle,

$$A(x) = \frac{25}{9}(A_e - A_t)\left(x - \frac{2}{5}\right)^2 + A_t, \tag{7.13.64}$$

where A_e and A_t are the exit area and the throat area respectively (Figure 7.13.10). Note that $x_{throat} = 2/5$. Applying the algorithms described above, I obtained the results in Figures 7.13.6 to 7.13.9, and the data in Tables 7.13.1 to 7.13.3.

These exact solutions can be used to verify a numerical code for the quasi-1D Euler equations in Section 3.16. But it is also possible to use these exact solutions as references in two-dimensional calculations. Generate a two-dimensional grid based on the area function (only the upper half of the nozzle will suffice because of the symmetry), and perform a two-dimensional CFD calculation on it for one of the exact solutions described above. Then, we may compare the numerical solution with the exact solution at the centerline of the nozzle. In this case, the exact solution is no longer strictly exact. But it should be close for a smooth nozzle with $A_e \approx A_t$.

I like the nozzle flow very much because it is useful also in magnetohydrodynamic flows: the solution can be used without any modification for aligned flows (the magnetic field parallel to the stream line) [42], and also with minor modifications for transverse flows (the magnetic field perpendicular to the stream line) [85].

7.13.6   Ringleb's Flow

Ringleb's flow is an exact solution to the Euler equations for $\gamma = 1.4$ obtained by Ringleb in 1940 [126]. It is often used to measure the accuracy of an Euler code (see Ref.[26] and references therein). This flow is obtained as a solution (a stream function ψ) of the hodograph equation (6.4.45), so that the physical location (x, y) is given in terms of the

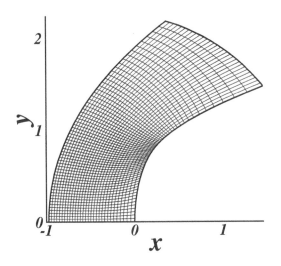

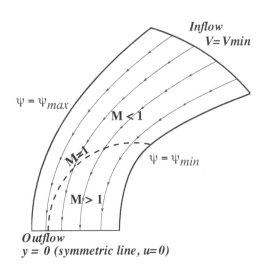

Figure 7.13.11: A computational grid for Ringleb's flow, generated with a uniform spacing in θ rather than V, with parameters $\psi_{min} = 0.69$, $\psi_{max} = 1.2$, $V_{min} = 0.5$.

Figure 7.13.12: Ringleb's flow: a flow around the edge of a flat plate or a flow in a channel, being smoothly accelerated from subsonic to supersonic. The supersonic region can be avoided by taking sufficiently large values of ψ.

physical variables (V, θ), i.e., the flow speed and angle (see Ref.[139] for details):

$$\psi = \frac{\sin \theta}{V}, \tag{7.13.65}$$

$$x(\psi, V) = \frac{1}{\rho}\left[\frac{1}{2V^2} - \psi^2\right] + \frac{L}{2}, \tag{7.13.66}$$

$$y(\psi, V) = \pm\frac{\psi}{\rho V}\sqrt{1 - V^2\psi^2}, \tag{7.13.67}$$

where ρ and L are functions of V:

$$\rho = b^5, \tag{7.13.68}$$

$$L = \frac{1}{b} + \frac{1}{3b^3} + \frac{1}{5b^5} - \frac{1}{2}\ln\left(\frac{1+b}{1-b}\right), \tag{7.13.69}$$

$$b = \sqrt{1 - 0.2V^2}, \tag{7.13.70}$$

and the pressure p is given by $p = b^7$. The isotachs (lines of constant speed) are circles given by

$$\left(x - \frac{L}{2}\right)^2 + y^2 = \left(\frac{1}{2\rho V^2}\right)^2. \tag{7.13.71}$$

Note that the variables have been nondimensionalized by the stagnation values; and so they should be understood actually as

$$\frac{x}{l}, \quad \frac{y}{l}, \quad \frac{\psi}{c_0 l}, \quad \frac{V}{c_0}, \quad \frac{\rho}{\rho_0}, \quad \frac{p}{p_0}, \tag{7.13.72}$$

where l is a reference length, ρ_0, p_0, and c_0 are the stagnation density, pressure, and the speed of sound respectively.

Now, choose some values for ψ_{min}, ψ_{max}, V_{min}, and allow V to increase up to a limit of the domain, e.g., $y = 0$, thus Equation (7.13.67) giving $V_{max} = 1/\psi$. Then, we will have a domain like those in Figures 7.13.11 and 7.13.12; we can generate its boundary points very easily.

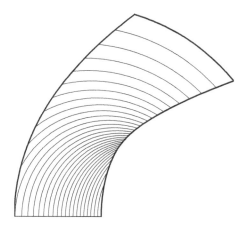

Figure 7.13.13: Flow speed, density, and pressure have the same contours .

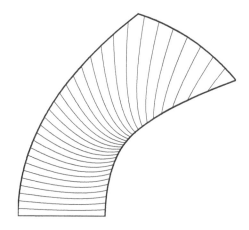

Figure 7.13.14: Flow angle θ.

Generation of Boundary Points:

The following generates N_R points on the right boundary, N_T points on the top boundary, N_L points on the left boundary, N_B points on the bottom boundary (with corner points being shared, i.e., the total number of points on the whole boundary is $N_L + N_R + N_B + N_T - 4$).

1. Right Boundary:

   Generate (x_i, y_i) by Equations (7.13.66) and (7.13.67) with $\psi = \psi_{min}$ and $V_i \in [V_{min}, 1/\psi_{min}]$, $i = 1, 2, 3, \cdots, N_R$.

2. Top Boundary:

   Generate (x_i, y_i) by Equations (7.13.66) and (7.13.67) with $V = V_{min}$ and $\psi_i \in [\psi_{min}, \psi_{max}]$, $i = 1, 2, 3, \cdots, N_T$.

3. Left Boundary:

   Generate (x_i, y_i) by Equations (7.13.66) and (7.13.67) with $\psi = \psi_{max}$ and $V_i \in [V_{min}, 1/\psi_{max}]$, $i = 1, 2, 3, \cdots, N_L$.

4. Bottom Boundary:

   Generate (x_i, y_i) by Equations (7.13.66) and (7.13.67) with $V = 1/\psi_i$ and $\psi_i \in [\psi_{min}, \psi_{max}]$, $i = 1, 2, 3, \cdots, N_B$. (y_i should be zero for all i here.)

Note that we might want to reverse the node ordering in the step 2, so that boundary points are generated and ordered counterclockwise over the entire boundary. Also, note that a uniform distribution of points in $[V_{min}, 1/\psi_{min}]$ does not lead to the uniform distribution in the physical plane (points will be clustered near the midpoint). First, generate a uniform grid in $\theta \in [\sin^{-1}(\psi_{min}V_{min}), \pi/2]$, and then convert it to V_i by $V_i = \frac{\sin\theta_i}{\psi_{min}}$. This will generate a smoother grid in the physical plane such as the one shown in Figure 7.13.11.

Once we generate boundary points, we can generate a computational grid over the domain by connecting them in some way (structured or unstructured). Or you can directly generate a structured grid by applying the above boundary-point generation algorithm to the interior points with varying stream function values over the interval $[\psi_{min}, \psi_{max}]$.

Whatever kind of grid you generate, you would want to run an Euler code on it with the appropriate boundary conditions: uniform inflow at the top boundary; wall boundary conditions on the left and right boundaries; outflow condition on the bottom boundary. Then, of course, you would certainly want to compute the error by comparing the numerical solution with the exact solution.

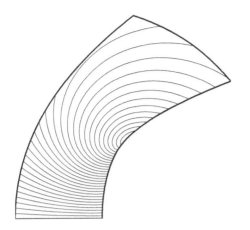

Figure 7.13.15: Velocity component, $u = -V \cos\theta$.

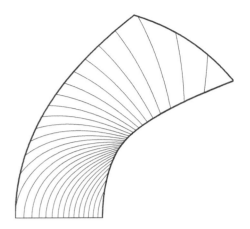

Figure 7.13.16: Velocity component, $v = -V \sin\theta$.

Computation of Exact Solution:

The exact solution at a given location (x_p, y_p) can be computed as follows.

1. Numerically solve Equation (7.13.71) for V with the bounds $[V_{min}, 1/\psi_{min}]$.

2. Compute b by Equation (7.13.70), and then ρ, p, and L.

3. Compute ψ by Equation (7.13.66) in the form,

$$\psi = \sqrt{\frac{1}{2V^2} - \rho\left(x_p - \frac{L}{2}\right)}. \tag{7.13.73}$$

4. Compute θ by Equation (7.13.65) in the form,

$$\theta = \sin^{-1}(\psi V), \quad \theta \in [-\pi/2, \pi/2]. \tag{7.13.74}$$

5. Compute the velocity $(u, v) = -V(\cos\theta, \sin\theta)$ where the negative sign is put to make the upper boundary the inflow).

6. Compute the actual Mach number (if you wish) by

$$M = \sqrt{1.4}\frac{V}{\sqrt{1.4\,p/\rho}} = \frac{V}{\sqrt{p/\rho}}, \tag{7.13.75}$$

where the factor $\sqrt{1.4}$ is needed because V, p, and ρ are in fact V/c_0, p/p_0, and ρ/ρ_0.

For solving Equation (7.13.71), we may employ the fixed-point iteration: write the equation as

$$V^{k+1} = \sqrt{\frac{1}{2\rho^k\sqrt{(x - L^k/2)^2 + y^2}}}, \tag{7.13.76}$$

where ρ^k and L^k are evaluated with the previous value V^k, then iterate on V^k starting with $V^1 = (V_{min} + 1/\psi_{min})/2$ until $|V^{k+1} - V^k| < 10^{-15}$, for example. I like this method because it is very simple.

I like Ringleb's flow very much because in particular the flow accelerates from subsonic to supersonic very smoothly without shocks. That is, it can be very useful to verify the accuracy of an Euler code simultaneously for both subsonic and supersonic flows. A sample program, which was used to generate the grid in Figure 7.13.11 and the solution in Figures 7.13.13, 7.13.14, 7.13.15, and 7.13.16, is available at http://www.cfdbooks.com.

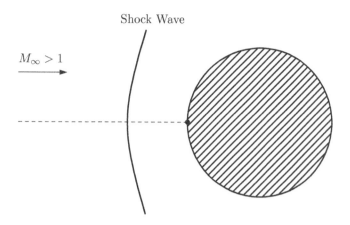

Figure 7.13.17: Supersonic flow over a cylinder: the shock is locally normal to the streamline along the centerline indicated by the dashed line.

7.13.7 Supersonic Flow over a Cylinder (Stagnation Pressure)

The formula that relates the upstream Mach number M_∞ and stagnation pressure $(p_0)_\infty$ to the stagnation pressure p_0 behind the shock is given by

$$\frac{p_0}{(p_0)_\infty} = \left[1 + \frac{2\gamma}{\gamma+1}(M_\infty^2 - 1)\right]^{-1/(\gamma-1)} \left[\frac{(\gamma+1)M_\infty^2}{(\gamma-1)M_\infty^2 + 2}\right]^{\gamma/(\gamma-1)}, \tag{7.13.77}$$

(see Ref.[84] for details). This will give the exact pressure at the stagnation point of a cylinder behind a bow shock created by a supersonic flow (the bow shock is a normal shock at the line of symmetry). This is useful, for example, when I want to see how accurately my Euler code can compute a supersonic flow over a cylinder (so that there will be a bow shock in front of it) by comparing a computed pressure at the front stagnation point of the cylinder with the exact value. For this purpose, I like to use the nondimensionalization by the stagnation values in Section 3.11 because it allows me to directly compare the computed pressure value with the exact value.

7.13.8 Normal/Oblique Shock Waves

Propagating Normal Shock Wave

I like propagating normal shock waves, especially the one running to the right at the speed V_s into the gas at rest (this state is denoted by R),

$$\mathbf{W}_R = [\rho_R, u_R, p_R]^t = [1, 0, 1/\gamma]^t. \tag{7.13.78}$$

It is nice that once I specify the shock speed $V_s(> 0)$, then all upstream quantities, $\mathbf{W}_L = [\rho_L, u_L, p_L]^t$, are automatically determined (by the Rankine-Hugoniot relation) as follows:

$$p_L = \left[1 + \frac{2\gamma}{\gamma+1}\left(V_s^2 - 1\right)\right]\frac{1}{\gamma}, \tag{7.13.79}$$

$$\rho_L = \frac{1 + \dfrac{\gamma+1}{\gamma-1}p_L}{\dfrac{\gamma+1}{\gamma-1} + p_L}, \tag{7.13.80}$$

$$u_L = \frac{\gamma p_L - 1}{\gamma\sqrt{1 + \frac{\gamma+1}{2\gamma}(\gamma p_L - 1)}}. \tag{7.13.81}$$

This can be used in the same form also in two and three dimensions with $v = w = 0$. It is very useful. Note that it is very important to be able to compute a moving shock because the predicted shock speed could be wrong if a method is not conservative.

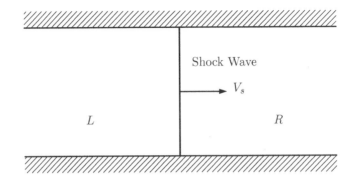

Figure 7.13.18: Normal shock wave: propagating $V_s > 0$, or stationary $V_s = 0$.

Steady Normal Shock Wave

I like steady normal shock waves. In fact, it is very easy to cook up one: specify the upstream state, $\mathbf{W}_L$,

$$\mathbf{W}_L = [\rho_L, u_L, p_L]^t = \left[1, 1, \frac{1}{\gamma M_L^2}\right]^t, \tag{7.13.82}$$

where $M_L > 1$, then the downstream state, $\mathbf{W}_R = [\rho_R, u_R, p_R]^t$ is automatically determined in terms of M_L,

$$\rho_R = \frac{(\gamma + 1)M_L^2}{(\gamma - 1)M_L^2 + 2}, \tag{7.13.83}$$

$$u_R = \frac{1}{\rho_R} = \frac{(\gamma - 1)M_L^2 + 2}{(\gamma + 1)M_L^2}, \tag{7.13.84}$$

$$p_R = \left[1 + \frac{2\gamma}{\gamma + 1}(M_L^2 - 1)\right]\frac{1}{\gamma M_L^2}. \tag{7.13.85}$$

Besides, if we express the two states in terms of the conservative variables, $\mathbf{U}_L$ and $\mathbf{U}_R$, then they look rather simple:

$$\mathbf{U}_L = [\rho_L, \rho_L u_L, \rho_L E_L]^t = \left[1, 1, \frac{1}{\gamma(\gamma - 1)M_L^2} + \frac{1}{2}\right]^t, \tag{7.13.86}$$

$$\mathbf{U}_R = [\rho_R, \rho_R u_R, \rho_R E_R]^t = \left[\rho_R, 1, \frac{p_R}{\gamma - 1} + \frac{u_R^2}{2}\right]^t, \tag{7.13.87}$$

where ρ_R, p_R, and u_R are given by Equations (7.13.83), (7.13.84), and (7.13.85). Of course, this steady shock solution can be used in the same form also in two and three dimensions with $v = w = 0$.

Oblique Shock Waves

We can easily create an oblique shock solution in a square domain by setting a free stream on the left at an angle $\alpha(< 0)$ coming towards the bottom wall, thus creating a oblique shock at the bottom-left corner. Specify the upstream angle α and a desired oblique shock angle θ_s (see Figure 7.13.19), and then we can determine the corresponding upstream Mach number by

$$M_\infty = \sqrt{\frac{2}{\sin(\theta_s + \alpha)\cos(\theta_s + \alpha)\left[(\gamma + 1)\tan(\theta_s) - (\gamma - 1)\tan(\theta_s + \alpha)\right]}}, \tag{7.13.88}$$

where it is safe to use the values that satisfy $\theta_s + \alpha < 60°$ (this is a rough estimate not a precise condition) so that only the weak solution, i.e., the one with the supersonic downstream, is allowed. If we specify this Mach number and

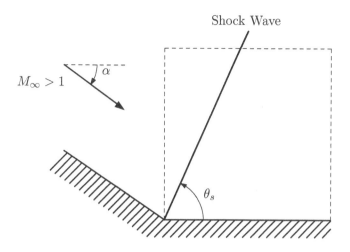

Figure 7.13.19: Oblique shock wave.

$\rho_\infty = 1$ and $p_\infty = 1/\gamma$, for example, then we can determine the downstream state M_2, ρ_2, and p_2 by

$$M_2 = \frac{1}{\sin\theta_s}\sqrt{\frac{1 + \dfrac{\gamma - 1}{2}M_\infty^2 \sin^2(\theta_s + \alpha)}{\gamma M_\infty^2 \sin^2(\theta_s + \alpha) - \dfrac{\gamma - 1}{2}}},\qquad(7.13.89)$$

$$\rho_2 = \frac{(\gamma + 1)M_\infty^2 \sin^2(\theta_s + \alpha)}{(\gamma - 1)M_\infty^2 \sin^2(\theta_s + \alpha) + 2},\qquad(7.13.90)$$

$$p_2 = 1 + \frac{2\gamma}{\gamma + 1}(M_\infty^2 \sin^2(\theta_s + \alpha) - 1).\qquad(7.13.91)$$

(see Ref.[84] for details). It is very nice that the domain is a square; it will be very easy to generate a grid and also to implement a numerical scheme.

7.14   Riemann Problems for the Euler Equations

The Riemann problem is an initial value problem with a single discontinuity that initially separates two constant states, the left and right states. In particular, the Riemann problem with zero velocity on both states describes a shock tube problem in gas dynamics: interactions of gases at two different states suddenly made to interact in a tube, e.g., by a diaphragm raptured at $t = 0$. I like Riemann problems for two reasons. First, it can be solved exactly even for nonlinear equations such as the Euler equations. I find it very interesting that the problem can be solved exactly for highly nonlinear equations with discontinuous initial data, which sounds very difficult in general. Second, it can be thought of as a building block of finite-volume methods. For example, two adjacent control volumes (computational cells) define a Riemann problem at the control-volume boundary, and the solution to the Riemann problem gives the flux at the control-volume boundary that is required to compute the residual in finite-volume methods.

7.14.1   Basic Equations

The 1D Euler equations have three characteristic fields:

$$\lambda_1 = u - c, \quad \lambda_2 = u, \quad \lambda_3 = u + c.\qquad(7.14.1)$$

λ_1-field and λ_3-field are genuinely nonlinear and therefore have nonlinear waves (shocks/rarefactions), but λ_2-field is linearly degenerate and so has a contact wave only. See Section 1.23.

In the following equations, whenever they have double signs, the negative sign is associated with waves in λ_1-field and the positive sign is associated with waves in λ_3-field. Also note that the subscript 0 indicates an undisturbed state.

Shocks

The Rankine-Hugoniot relation across a shock is given by

$$p - p_0 \;=\; \pm \rho_0 c_0 \sqrt{1 + \frac{\gamma + 1}{2\gamma}\left(\frac{p}{p_0} - 1\right)}\,(u - u_0), \tag{7.14.2}$$

$$\frac{\rho}{\rho_0} \;=\; \frac{1 + \dfrac{\gamma + 1}{\gamma - 1}\dfrac{p}{p_0}}{\dfrac{\gamma + 1}{\gamma - 1} + \dfrac{p}{p_0}}. \tag{7.14.3}$$

These are obtained directly from the Rankine-Hugoniot relation in Section 1.22 (See Ref.[159] for derivation). Equation (7.14.2) defines a curve in (p, u)-space, and it is called the Hugoniot curve. The shock speed V_s is given by

$$V_s \;=\; u_0 \pm c_0 \sqrt{1 + \frac{\gamma + 1}{2\gamma}\left(\frac{p}{p_0} - 1\right)}. \tag{7.14.4}$$

Note that Equation (7.14.2) defines a curve that passes through a point (p_0, u_0) and only the branch where $p > p_0$ is physically admissible. This is typically called the entropy condition: the entropy cannot decrease across the shock. You will actually see that $p > p_0$ for $s > s_0$ by deriving the entropy jump, $s - s_0 = c_v[\ln(p/\rho^\gamma) - \ln(p_0/\rho_0^\gamma)]$ from the above equations. The other branch, where $p < p_0$, is often called the non-physical branch of the Hugoniot curve. Such a change is caused by a rarefaction wave.

Rarefactions

Integral relation across a rarefaction is given by

$$u \pm \frac{2}{\gamma - 1}c \;=\; u_0 \pm \frac{2}{\gamma - 1}c_0, \tag{7.14.5}$$

$$p - p_0 \;=\; \pm \rho_0 c_0 \frac{\gamma - 1}{2\gamma}\frac{1 - (p/p_0)}{1 - (p/p_0)^{\frac{\gamma - 1}{2\gamma}}}(u - u_0). \tag{7.14.6}$$

These equations are obtained from the isentropic relations and the Riemann invariant $u + \frac{2c}{\gamma - 1} = $ constant (See Ref.[159] for derivation). Equation (7.14.6) defines a curve in (p, u)-space, and it is called the Poisson curve. Of course, only the branch where $p < p_0$ is physically admissible. Then, a curve describing all admissible jumps is obtained by connecting the Poisson and Hugoniot curves at (p_0, u_0), which will be the main tool for solving a Riemann problem. Across the rarefaction, the solution does not jump but varies smoothly: u and c vary linearly, and p and ρ are given by

$$p \;=\; p_0 \left(\frac{c}{c_0}\right)^{\frac{2\gamma}{\gamma - 1}}, \tag{7.14.7}$$

$$\rho \;=\; \rho_0 \left(\frac{p}{p_0}\right)^{\frac{1}{\gamma}}. \tag{7.14.8}$$

7.14.2   Construction of a Riemann Problem

I like the Riemann problem because it is simple. In fact, it is very easy to construct one. For example, if I want to create a flow where a shock travels to the right, a contact wave follows to the right, and a rarefaction wave travels to the left (see Figure 7.14.1), I can create it as follows.

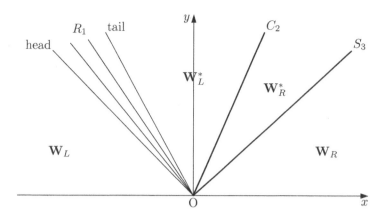

Figure 7.14.1: The Riemann problem I want to create: λ_3-shock (S_3), the contact (C_2), and λ_1-rarefaction (R_1).

Construction of a Riemann Problem:

1. Specify the right state $\mathbf{W}_R = [\rho_R, u_R, p_R]^t$.

2. Specify $p^*(> p_R)$, and determine u^* by the shock relation (7.14.2), i.e.,

$$u^* = u_R + \frac{p^* - p_R}{\rho_R c_R \sqrt{1 + \frac{\gamma+1}{2\gamma}\left(\frac{p^*}{p_R} - 1\right)}}, \tag{7.14.9}$$

and determine ρ_R^* by Equation (7.14.3), i.e.,

$$\rho_R^* = \rho_R \frac{1 + \dfrac{\gamma+1}{\gamma-1}\dfrac{p^*}{p_R}}{\dfrac{\gamma+1}{\gamma-1} + \dfrac{p^*}{p_R}}. \tag{7.14.10}$$

3. Specify ρ_L^*. Across the contact, the velocity and the pressure do not change: $u_L^* = u_R^* = u^*$ and $p_L^* = p_R^* = p^*$.

4. Specify $p_L(> p^*)$, and determine ρ_L by Equation (7.14.8), i.e.,

$$\rho_L = \rho_L^* \left(\frac{p_L}{p^*}\right)^{\frac{1}{\gamma}}, \tag{7.14.11}$$

and the velocity by Equation (7.14.6), i.e.,

$$u_L = u^* + \frac{2\gamma}{(\gamma-1)\rho_L c_L} \frac{1 - (p^*/p_L)^{\frac{\gamma-1}{2\gamma}}}{1 - (p^*/p_L)}(p^* - p_L). \tag{7.14.12}$$

This completes the construction of a Riemann problem.

For the Riemann problem created above, the shock speed is given by

$$V_{S_3} = u_R + c_R \sqrt{1 + \frac{\gamma+1}{2\gamma}\left(\frac{p^*}{p_R} - 1\right)}, \tag{7.14.13}$$

the contact speed by

$$V_{C_2} = u^*, \tag{7.14.14}$$

and the tail and head speeds of the rarefaction by

$$V_{R_1}^{\text{tail}} = u_L^* - c_L^*, \quad V_{R_1}^{\text{head}} = u_L - c_L. \tag{7.14.15}$$

These wave speeds may be useful when we choose the quantities like p^* so that we create a problem with a desired shock speed, for example. By the way, in order to compute the solution inside the rarefaction wave, first compute the velocity (which varies linearly), and then compute all others:

$$u(\xi) = \frac{\xi - \xi_L^*}{\xi_L - \xi_L^*} u_L + \frac{\xi - \xi_L}{\xi_L^* - \xi_L} u^*, \quad c(\xi) = u(\xi) - \xi, \tag{7.14.16}$$

$$p(\xi) = p_L \left(\frac{c(\xi)}{c_L} \right)^{\frac{2\gamma}{\gamma - 1}}, \quad \rho(\xi) = \rho_L \left(\frac{p(\xi)}{p_L} \right)^{\frac{1}{\gamma}}, \tag{7.14.17}$$

where $\xi_L = u_L - c_L$ and $\xi_L^* = u^* - c_L^*$. Note that you can construct any Riemann problem you want. For example, if I stop after determine $\mathbf{W}_R^*$ and set $\mathbf{W}_L = \mathbf{W}_R^*$, then I end up with a Riemann problem with a single shock; or if I set $\rho_L^* = \rho_R^*$, then there will be no contact waves.

7.14.3 Exact Riemann Solver

I like solving a given Riemann problem also. Sometimes the solution doesn't even exist, and that's fun. Given a set of initial data, $\mathbf{W}_L = [\rho_L, u_L, p_L]^t$ and $\mathbf{W}_R = [\rho_R, u_R, p_R]^t$, I will try to find a solution as follows.

Exact Riemann Solver:

1. Find the pressure p^* in the middle by solving the equation,

$$p^* = \frac{m_R p_L + m_L p_R + m_L m_R (u_L - u_R)}{m_L + m_R}, \tag{7.14.18}$$

where for $Q = L$ or R, and

$$m_Q = \begin{cases} \rho_Q c_Q \sqrt{1 + \frac{\gamma + 1}{2\gamma} \left(\frac{p^*}{p_Q} - 1 \right)} & \text{if } \frac{p^*}{p_Q} > 1 : \text{shock}, \\[4ex] \rho_Q c_Q \frac{\gamma - 1}{2\gamma} \frac{1 - \dfrac{p^*}{p_Q}}{1 - \left(\dfrac{p^*}{p_Q} \right)^{\frac{\gamma - 1}{2\gamma}}} & \text{if } \frac{p^*}{p_Q} < 1 + \epsilon : \text{rarefaction}. \end{cases} \tag{7.14.19}$$

Note that ϵ, say $\epsilon = 10^{-15}$ for double precision, is necessary to avoid the zero division. This must be solved iteratively as the right hand side depends on p^*.

2. Compute u^* by the equation,

$$u^* = \frac{m_L u_L + m_R u_R + (p_L - p_R)}{m_L + m_R}. \tag{7.14.20}$$

3. Compute ρ_L^* and ρ_R^* by

$$\rho_Q^* = \begin{cases} \rho_Q \dfrac{1 + \dfrac{\gamma + 1}{\gamma - 1} \dfrac{p^*}{p_Q}}{\dfrac{\gamma + 1}{\gamma - 1} + \dfrac{p^*}{p_Q}} & \text{if } \frac{p^*}{p_Q} > 1 : \text{shock}, \\[4ex] \rho_Q \left(\dfrac{p^*}{p_Q} \right)^{\frac{1}{\gamma}} & \text{if } \frac{p^*}{p_Q} < 1 : \text{rarefaction}, \end{cases} \tag{7.14.21}$$

where $Q = L$ or R. Now, the middle states have been completely determined, and this completes the solution of a Riemann problem.

Some remarks are in order. First, if a rarefaction wave is detected, the solution inside the wave can be computed by Equations (7.14.16) and (7.14.17). Second, it may happen that

$$u_L + \frac{2}{\gamma - 1}c_L < u_R + \frac{2}{\gamma - 1}c_R, \tag{7.14.22}$$

then the pressure p^* will be negative and thus a solution does not exist. We need to check this condition and make sure that a solution exists, before solving a Riemann problem. Third, the intermediate pressure equation (7.14.18) is obtained by solving the following system for p^*,

$$
\begin{aligned}
p^* - p_L &= -m_L(u^* - u_L), \quad \lambda_1\text{-shock/rarefaction}, & (7.14.23) \\
p^* - p_R &= m_R(u^* - u_R), \quad \lambda_3\text{-shock/rarefaction}. & (7.14.24)
\end{aligned}
$$

These are the shock/rarefaction relations across the right-running and left-running waves respectively. Solving this for u^* gives the equation to determine u^* for a given p^*, i.e., Equation (7.14.20). Fourth, to solve Equation (7.14.18), any iterative method, such as the fixed-point iteration, Newton's method, or the bisection method can be employed. In doing so, it is good to have an upper bound and a lower bound for the solution p^*:

$$\text{Upper bound:} \qquad p^* = \left[\frac{\frac{\gamma-1}{2}(u_L - u_R) + (c_L + c_R)}{c_L\,(p_L)^{\frac{1-\gamma}{2\gamma}} + c_R\,(p_R)^{\frac{1-\gamma}{2\gamma}}} \right]^{\frac{2\gamma}{\gamma-1}}, \tag{7.14.25}$$

$$\text{Lower bound:} \qquad p^* = \frac{m_R p_L + m_L p_R + m_L m_R(u_L - u_R)}{m_L + m_R}, \tag{7.14.26}$$

where $m_L = \rho_L c_L$ and $m_R = \rho_R c_R$. The upper bound comes from a solution for two rarefaction waves (the intersection of the two Poisson curves); the lower bound comes from a linearization, i.e., the solution of Equations (7.14.23) and (7.14.24) with $m_L = \rho_L c_L$ and $m_R = \rho_R c_R$. Fifth, the condition (7.14.22) indicates that the intersection between the two Poisson curves (i.e., the upper bound) goes down to the negative part of the p-axis in the (p, u)-space.

Here are some examples where I assume a domain $[-5, 5]$ for all problems and denote the final time by t_f.

(a) Sod's Problem 1 (Figures 7.14.2):

I like this problem. This is a famous shock tube problem by Sod [144].

$$
\begin{aligned}
\mathbf{W}_L &= [\rho_L, u_L, p_L]^t = [1.0,\, 0,\, 1.0]^t, & (7.14.27) \\
\mathbf{W}_R &= [\rho_R, u_R, p_R]^t = [0.125,\, 0,\, 0.1]^t, & (7.14.28) \\
t_f &= 1.7. & (7.14.29)
\end{aligned}
$$

(b) Sod's Problem 2 (Figures 7.14.3):

I like this one also. This is another famous shock tube problem by Sod [144].

$$
\begin{aligned}
\mathbf{W}_L &= [\rho_L, u_L, p_L]^t = [1.0,\, 0,\, 1.0]^t, & (7.14.30) \\
\mathbf{W}_R &= [\rho_R, u_R, p_R]^t = [0.01,\, 0,\, 0.01]^t, & (7.14.31) \\
t_f &= 1.4. & (7.14.32)
\end{aligned}
$$

I like this one particularly because it contains a sonic point in the rarefaction wave and some schemes create a small gap there.

(c) Sonic Rarefaction (Figures 7.14.4):

This is a sonic rarefaction and some schemes cannot break the initial discontinuity at all.

$$
\begin{aligned}
\mathbf{W}_L &= [\rho_L, u_L, p_L]^t = [3.857,\, 0.92,\, 10.333]^t, & (7.14.33) \\
\mathbf{W}_R &= [\rho_R, u_R, p_R]^t = [1.000,\, 3.55,\, 1.000]^t, & (7.14.34) \\
t_f &= 0.7. & (7.14.35)
\end{aligned}
$$

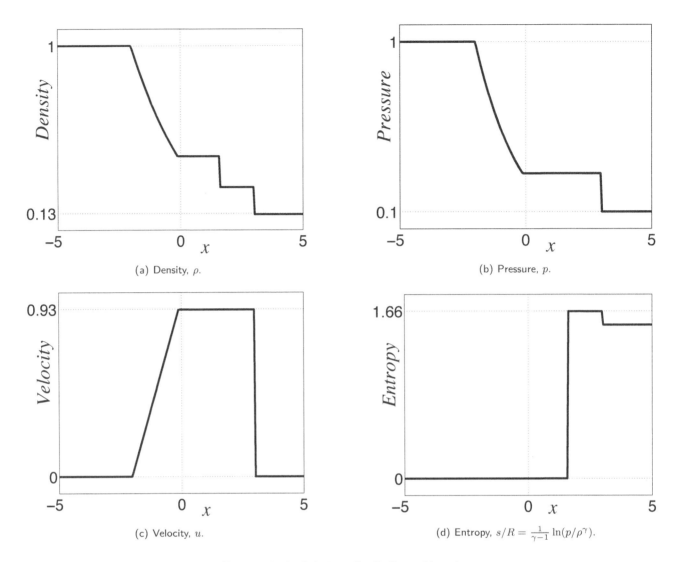

(a) Density, ρ.

(b) Pressure, p.

(c) Velocity, u.

(d) Entropy, $s/R = \frac{1}{\gamma-1} \ln(p/\rho^\gamma)$.

Figure 7.14.2: Solutions for Sod's problem 1.

(d) Slowly Moving Shock (Figures 7.14.5):

For this one, many schemes produce oscillations behind the shock [4].

$$\mathbf{W}_L = [\rho_L, u_L, p_L]^t = [3.86, -0.81, 10.33]^t, \tag{7.14.36}$$

$$\mathbf{W}_R = [\rho_R, u_R, p_R]^t = [1.00, -3.44, 1.00]^t, \tag{7.14.37}$$

$$t_f = 1.7. \tag{7.14.38}$$

(e) Left-Moving Waves in a Right-Moving Supersonic Flow:

I like this one,

$$\mathbf{W}_L = [\rho_L, u_L, p_L]^t = [1.4, 1.1, 1]^t, \tag{7.14.39}$$

$$\mathbf{W}_R = [\rho_R, u_R, p_R]^t = [1.4, 110, 100]^t, \tag{7.14.40}$$

thus $M_L = M_R = 1.1$. This is an interesting case where both states are supersonic to the right but some waves will propagate to the left. This is indeed confusing because we tend to assume that all waves propagate to the right if both states are supersonic to the right. That is true for scalar conservation laws but not for systems. See it for yourself using your Euler code, or see Ref.[86] for details.

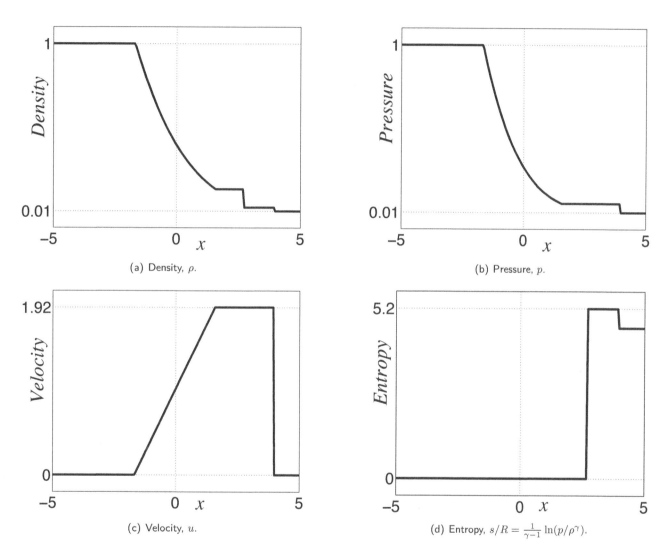

(a) Density, ρ.

(b) Pressure, p.

(c) Velocity, u.

(d) Entropy, $s/R = \frac{1}{\gamma-1}\ln(p/\rho^\gamma)$.

Figure 7.14.3: Solutions for Sod's problem 2.

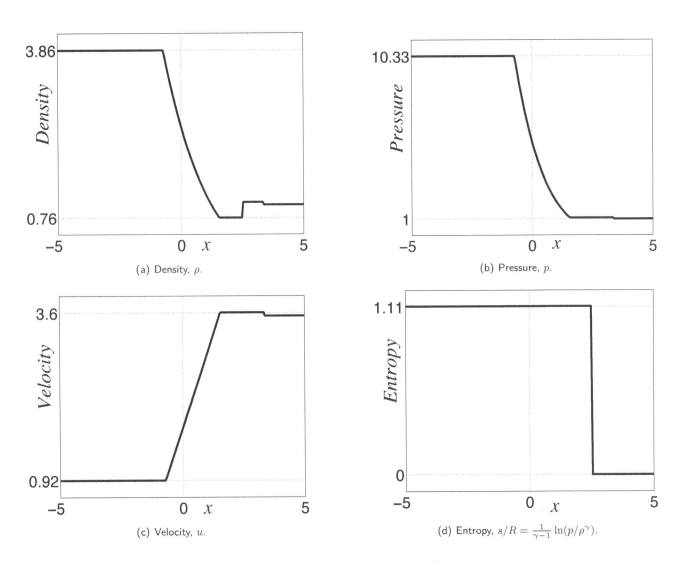

(a) Density, ρ.

(b) Pressure, p.

(c) Velocity, u.

(d) Entropy, $s/R = \frac{1}{\gamma-1}\ln(p/\rho^\gamma)$.

Figure 7.14.4: Solutions for a sonic rarefaction.

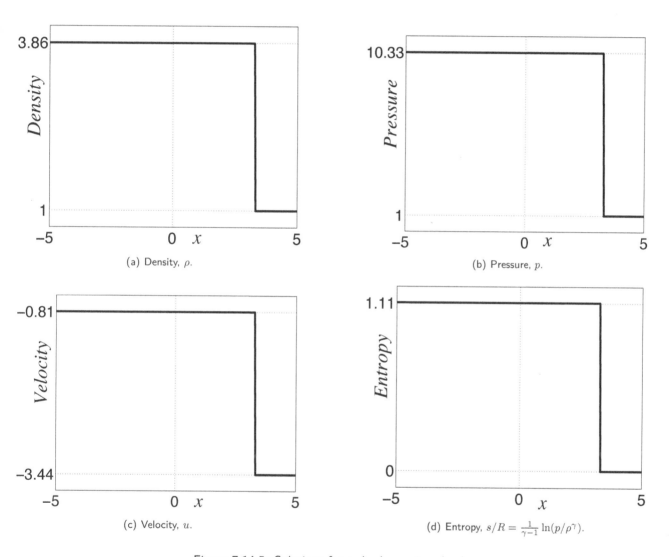

(a) Density, ρ.

(b) Pressure, p.

(c) Velocity, u.

(d) Entropy, $s/R = \frac{1}{\gamma-1}\ln(p/\rho^{\gamma})$.

Figure 7.14.5: Solutions for a slowly moving shock.

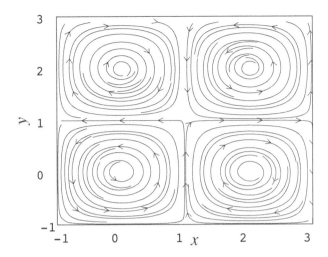

Figure 7.15.1: Unsteady solution for the 2D incompressible Navier-Stokes equations. The velocity field at $t = 0.0$ for $k = 0.5\pi$ and $\nu = 0.01$. The velocity will be damped out as time goes on.

7.15 Incompressible Navier-Stokes Equations

I like incompressible Navier-Stokes equations because it does have exact solutions. Although often limited to simple geometries, these exact solutions can be useful for CFD code verification. Of course, I like Stokes' equations also because there exist a variety of exact solutions as mentioned in Section 4.25. Especially in two dimensions, the theory of complex variables can be applied to generate various exact solutions. There are also various exact solutions available for the axisymmetric Stokes flows as we will see in this section. Note that I like those exact solutions of the Stokes equations where the convective terms vanish because then the solutions are also exact for the Navier-Stokes equations.

7.15.1 Smooth 2D Time-Dependent Solution

The two-dimensional incompressible Navier-Stokes equations,

$$u_x + v_y = 0, \tag{7.15.1}$$

$$u_t + uu_x + vu_y = -p_x + \nu(u_{xx} + u_{yy}), \tag{7.15.2}$$

$$v_t + uv_x + vv_y = -p_y + \nu(v_{xx} + v_{yy}), \tag{7.15.3}$$

have the exact solution given by

$$u(x, y, t) = -\cos(kx)\sin(ky)\,e^{-2\nu k^2 t}, \tag{7.15.4}$$

$$v(x, y, t) = \sin(kx)\cos(ky)\,e^{-2\nu k^2 t}, \tag{7.15.5}$$

$$p(x, y, t) = -\frac{1}{4}[\cos(2kx) + \cos(2ky)]\,e^{-4\nu k^2 t}, \tag{7.15.6}$$

where k is an arbitrary constant [22, 33]. This solution can also be expressed in terms of the vorticity $\omega = v_x - u_y$ and the stream function ψ,

$$\omega(x, y, t) = 2k\,\cos(kx)\cos(ky)\,e^{-2\nu k^2 t}, \tag{7.15.7}$$

$$\psi(x, y, t) = \frac{1}{k}\cos(kx)\cos(ky)\,e^{-2\nu k^2 t}, \tag{7.15.8}$$

which is exact for the equation set (4.22.1) and (4.22.3), i.e.,

$$\omega_t + u\,\omega_x + v\,\omega_y = \nu\,(\omega_{xx} + \omega_{yy}), \tag{7.15.9}$$

$$\psi_{xx} + \psi_{yy} = -\omega, \tag{7.15.10}$$

where $u = \psi_y$ and $v = -\psi_x$. I like this solution very much because it is a rather simple solution for such a complicated nonlinear system. See Figure 7.15.1 for the velocity field generated by this solution. It is very interesting, isn't it?

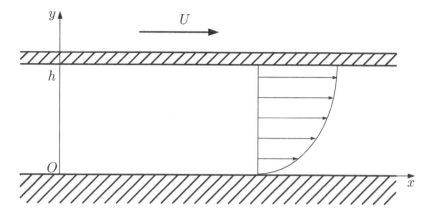

Figure 7.15.2: Couette or Couette-Poiseuille flow: $dp/dx = 0$ or $dp/dx \neq 0$.

7.15.2   Smooth 3D Time-Dependent Solution

The three-dimensional incompressible Navier-Stokes equations,

$$u_x + v_y + w_z = 0, \tag{7.15.11}$$
$$u_t + uu_x + vu_y + wu_z = -p_x + \nu(u_{xx} + u_{yy} + u_{zz}), \tag{7.15.12}$$
$$v_t + uv_x + vv_y + wv_z = -p_y + \nu(v_{xx} + v_{yy} + v_{zz}), \tag{7.15.13}$$
$$w_t + uw_x + vw_y + ww_z = -p_z + \nu(w_{xx} + w_{yy} + w_{zz}), \tag{7.15.14}$$

have the exact solution given by

$$u(x,y,t) = -a\left[e^{ax}\sin(ay+dz) + e^{az}\sin(ax+dy)\right]e^{-d^2\nu t}, \tag{7.15.15}$$
$$v(x,y,t) = -a\left[e^{ay}\sin(az+dx) + e^{ax}\sin(ay+dz)\right]e^{-d^2\nu t}, \tag{7.15.16}$$
$$w(x,y,t) = -a\left[e^{az}\sin(ax+dy) + e^{ay}\sin(az+dx)\right]e^{-d^2\nu t}, \tag{7.15.17}$$
$$p(x,y,t) = -\frac{a^2}{2}\left[e^{2ax} + e^{2ay} + e^{2az} + 2\sin(az+dx)\cos(ay+dz)\right.$$
$$\left. + 2\sin(ay+dz)\cos(ax+dy) + 2\sin(ax+dy)\cos(az+dx)\right]e^{-2d^2\nu t}, \tag{7.15.18}$$

where a and d are arbitrary constants [35]. I like this solution very much because it is, again, a very simple solution for such a complicated nonlinear system of equations. I just want to point out that there is a typo in the original paper [35].

7.15.3   Couette Flow

Consider a constant-density flow between two infinite plates that are h apart with the upper plate moving at speed U relative to the lower. Also, we assume that the upper plate is held at temperature T_1 and the lower plate at T_0. I like this setting because it brings a dramatic simplification to the incompressible Navier-Stokes equations. In this situation, naturally, x-derivative of any quantity is zero as well as the y-component of the velocity is zero everywhere. Also the pressure is constant everywhere since the flow does not accelerate in any direction. Then, the continuity equation is trivially satisfied (all terms drop out), and we have the momentum and energy equations as follows,

$$0 = \mu u_{yy}, \tag{7.15.19}$$
$$0 = \kappa T_{yy} + \mu(u_y)^2. \tag{7.15.20}$$

where μ is the viscosity coefficient and κ is the heat conductivity. This is very nice. We can easily obtain the exact solutions: integrate the momentum equation to obtain the velocity, and then integrate the energy equation to obtain the temperature as follows,

$$u(y) = U\frac{y}{h}, \tag{7.15.21}$$

$$T(y) = T_0 + (T_1 - T_0)\frac{y}{h} - \frac{\mu U^2}{2\kappa}\left[\left(\frac{y}{h}\right)^2 - \frac{y}{h}\right]. \tag{7.15.22}$$

This simple flow is called the Couette flow. It can be a nice and simple test case for a two-dimensional Navier-Stokes code in a rectangular domain with a periodic boundary condition in x-direction, i.e., no in-flow and out-flow. Note that the velocity is a linear function of y. Most practical incompressible Navier-Stokes codes should be able to produce the exact velocity profile since they are typically designed to be exact for linear solutions.

7.15.4 Unsteady Couette Flow

The Couette flow can be made unsteady by oscillating the upper plate with the following time-dependent speed,

$$u(h) = U\cos(\omega t), \tag{7.15.23}$$

where ω is a constant which defines the frequency. In this case, the momentum equation is given by

$$u_t = \mu u_{yy}. \tag{7.15.24}$$

This problem was solved in Section 6.1.5 by the separation of variable with complex variables. The solution is

$$u(y,t) = Q\left[\sinh(ky)\cos(ky)\sinh(kd)\cos(kd) + \cosh(ky)\sin(ky)\cosh(kd)\sin(kd)\right]\cos(\omega t)$$
$$+Q\left[\sinh(ky)\cos(ky)\cosh(kd)\sin(kd) - \cosh(ky)\sin(ky)\sinh(kd)\cos(kd)\right]\sin(\omega t), \tag{7.15.25}$$

where

$$Q = \frac{U}{\cosh^2(kd) - \cos^2(kd)}, \quad k = \sqrt{\frac{\omega}{2\mu}}. \tag{7.15.26}$$

I like this one because the solution does not decay in time and oscillates forever, preserving the same amplitude. It would be interesting to see how long a numerical scheme can simulate this solution accurately, of course with a periodic boundary condition in x-direction.

7.15.5 Couette-Poiseuille Flow

I like flows with constant pressure gradient because it is often possible to obtain the exact solution. For example, I like the Couette-Poiseuille flow, which is the Couette flow with an added constant pressure gradient, $dp/dx = $ constant. The momentum equation and the energy are then given by

$$0 = -p_x + \mu u_{yy}, \tag{7.15.27}$$
$$0 = kT_{yy} + \mu(u_y)^2. \tag{7.15.28}$$

We can integrate these equations straightforwardly and obtain

$$u(y) = U\frac{y}{h} + \frac{h^2}{2\mu}\left(-\frac{dp}{dx}\right)\left[\frac{y}{h} - \left(\frac{y}{h}\right)^2\right], \tag{7.15.29}$$

$$T(y) = T_0 + (T_1 - T_0)\frac{y}{h} - \frac{\mu^3}{12k}\left(\frac{dp}{dx}\right)^{-2}\left[V(y) - V(0) - \{V(h) - V(0)\}\frac{y}{h}\right], \tag{7.15.30}$$

where

$$V(y) = (u_y)^4 = \left[\frac{U}{h} + \frac{h}{2\mu}\left(-\frac{dp}{dx}\right)\left(1 - 2\frac{y}{h}\right)\right]^4. \tag{7.15.31}$$

It is more complicated than the solution of the Couette flow, but the flow itself is very simple. If your incompressible Navier-Stokes code is accurate enough to preserve quadratic solutions, it should be able to produce the exact velocity profile.

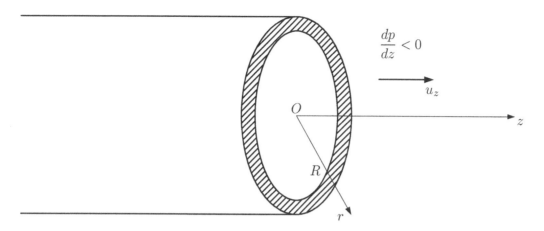

Figure 7.15.3: Hagen-Poiseuille flow.

7.15.6 Hagen-Poiseuille Flow

The Hagen-Poiseuille flow is an incompressible flow through a pipe of radius R driven by a constant pressure gradient in the z-direction in cylindrical coordinates. The only non-zero velocity component is u_z and all variables are functions of r only, i.e., $u_z = u_z(r)$. Then, the continuity equation is satisfied identically, and the momentum equation is given by

$$0 = \frac{dp}{dz} + \mu \frac{1}{r} \frac{d}{dr}\left(r \frac{du_z}{dr}\right). \tag{7.15.32}$$

Here, I like the fact that dp/dz is constant because it makes it very easy to integrate this equation. Simply integrating twice and applying the boundary conditions, i.e., $u_z(R) = 0$ and $u_z(0)$ is finite, we obtain

$$u_z(r) = \frac{1}{4\mu}\left(-\frac{dp}{dz}\right)(R^2 - r^2). \tag{7.15.33}$$

The velocity distribution is therefore parabolic, just like that of the Couette-Poiseuille flow. Oh, I forgot to work out a temperature distribution. Did you say that you wanted to derive it by yourself? OK, I'll leave it to you.

7.15.7 Axially Moving Co-centric Cylinders

Consider an incompressible flow between axially moving cylinders at constant speeds (see Figure 7.15.4). For this flow, clearly, the velocity components (u_r, u_θ, u_z) are such that $u_z = u_z(r)$ and $u_r = u_\theta = 0$. So, the continuity equation is satisfied identically, and the momentum and energy equations become

$$0 = \frac{1}{r} \frac{d}{dr}\left(r \frac{du_z}{dr}\right), \tag{7.15.34}$$

$$0 = \frac{1}{r} \frac{d}{dr}\left(kr \frac{dT}{dr}\right) + \mu \left(\frac{du_z}{dr}\right)^2. \tag{7.15.35}$$

Assume that the inner cylinder moves at $u_z(r_0) = U_0$ with a constant temperature T_0 and the outer cylinder moves at $u_z(r_1) = U_1$ with a constant temperature T_1. Using these as the boundary conditions, we integrate the momentum and energy equations and obtain the following set of exact solutions:

$$u_z(r) = U_1 \frac{\ln(r/r_0)}{\ln(r_1/r_0)} + U_0 \frac{\ln(r_1/r)}{\ln(r_1/r_0)}, \tag{7.15.36}$$

$$T(r) = -\frac{\mu}{2k}\left[\frac{U_1 - U_0}{\ln(r_1/r_0)}\right]^2 \ln(r/r_1)\ln(r/r_0) + \frac{T_1 - T_0}{\ln(r_1/r_0)}\ln r + \frac{T_0 \ln r_1 - T_1 \ln r_0}{\ln(r_1/r_0)}. \tag{7.15.37}$$

I like this solution because it is straightforward to generate a computational grid in the domain (i.e., a structured grid can be easily created between the two boundaries). Incidentally, there are lots of logarithms in the solutions, but many of them are identical. So, for example, $\ln(r_1/r_0)$ needs to be evaluated just once and we use this value elsewhere.

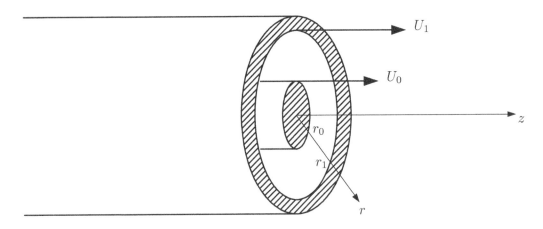

Figure 7.15.4: Axially moving co-centric cylinders.

7.15.8 Rotating Co-centric Cylinders

Consider an incompressible flow between rotating co-centric cylinders: the inner cylinder has radius r_0, angular velocity ω_0, and temperature T_0, while the outer cylinder has r_1, ω_1, and T_1 respectively. Note that the only non-zero velocity component is u_θ, and all variables are functions of r only. Then, the continuity equation is trivially satisfied, and the momentum equations and the energy equation are simplified as follows,

$$0 = -\frac{dp}{dr} + \frac{\rho u_\theta^2}{r}, \tag{7.15.38}$$

$$0 = \frac{d^2 u_\theta}{dr^2} + \frac{d}{dr}\left(\frac{u_\theta}{r}\right), \tag{7.15.39}$$

$$0 = \frac{k}{r}\frac{d}{dr}\left(r\frac{dT}{dr}\right) + \mu\left(\frac{du_\theta}{dr} - \frac{u_\theta}{r}\right)^2. \tag{7.15.40}$$

These are so simple that we can integrate them analytically to get the exact solutions:

$$u_\theta = r_0\omega_0\frac{r_1/r - r/r_1}{r_1/r_0 - r_0/r_1} + r_1\omega_1\frac{r/r_0 - r_0/r}{r_1/r_0 - r_0/r_1}, \tag{7.15.41}$$

$$\frac{T - T_0}{T_1 - T_0} = \frac{\mu r_0^2\omega_0^2}{k(T_1 - T_0)}\frac{r_1^4(1 - \omega_1/\omega_0)^2}{r_1^4 - r_0^4}\left(1 - \frac{r_0^2}{r^2}\right)\left[1 - \frac{\ln(r/r_0)}{\ln(r_1/r_0)}\right] + \frac{\ln(r/r_0)}{\ln(r_1/r_0)}. \tag{7.15.42}$$

Oh, I forgot to calculate the pressure. Well, you can get the pressure easily (if you wish) by integrating the momentum equation (7.15.38). I think that it can be a good exercise. By the way, unlike the previous cases, in this flow the pressure varies across the shear layer. This is because the domain is curved: the pressure gradient is needed to balance the centrifugal force, or equivalently, to create the centripetal force.

7.15.9 Flat Plate Boundary Layer

The flat plate boundary layer solution is a solution to the boundary layer equations,

$$\frac{\partial u}{\partial x} + \frac{\partial v}{\partial y} = 0, \tag{7.15.43}$$

$$u\frac{\partial u}{\partial x} + v\frac{\partial u}{\partial y} = \nu\frac{\partial^2 u}{\partial y^2}, \tag{7.15.44}$$

for a flow over a semi-infinite flat plate that extends $x \to \infty$ (Figure 7.15.6). First, to satisfy the continuity equation, we introduce the stream function ψ,

$$\psi = \frac{U_\infty x}{\sqrt{Re_x}}f(\eta), \tag{7.15.45}$$

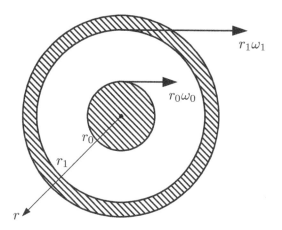

Figure 7.15.5: Rotating co-centric cylinders.

where U_∞ is the free stream velocity, and

$$\eta = \frac{y}{x}\sqrt{Re_x}, \quad Re_x = \frac{U_\infty x}{\nu}, \tag{7.15.46}$$

so that the velocity components are given by

$$\frac{u}{U_\infty} = f', \tag{7.15.47}$$

$$\frac{v}{U_\infty} = \frac{1}{2\sqrt{Re_x}}(\eta f' - f). \tag{7.15.48}$$

Therefore, the velocity component u depends only on η, i.e., self-similar. This is very nice. The solution is valid at any location on the flat plate (possibly far enough from the leading edge). On the other hand, v depends both on η and x. This may not sound very nice, but it makes sense. The boundary layer grows rapidly from the leading edge, but it will not grow so much in the downstream ($x \to \infty$) simply because there are no reasons for that. In order to compute the velocity components, we have to find $f(\eta)$ by solving the momentum equation which is written now in terms of f,

$$f''' + \frac{1}{2}ff'' = 0, \tag{7.15.49}$$

with the boundary conditions,

$$u(x,0) = v(x,0) = 0 \quad \longrightarrow \quad f(0) = f'(0) = 0, \tag{7.15.50}$$
$$u(x,\infty) = U_\infty \quad \longrightarrow \quad f'(\infty) = 1. \tag{7.15.51}$$

To solve this numerically, we rewrite this as a system of ordinary differential equations (ODE):

$$\frac{d\mathbf{F}}{d\eta} = \mathbf{G}(\mathbf{F}), \tag{7.15.52}$$

where

$$\mathbf{F} = \begin{bmatrix} f \\ f_1 \\ f_2 \end{bmatrix}, \quad \mathbf{G}(\mathbf{F}) \begin{bmatrix} f_1 \\ f_2 \\ -\dfrac{1}{2}ff_2 \end{bmatrix}, \tag{7.15.53}$$

with

$$f(0) = f_1(0) = 0, \tag{7.15.54}$$
$$f_1(\infty) = 1. \tag{7.15.55}$$

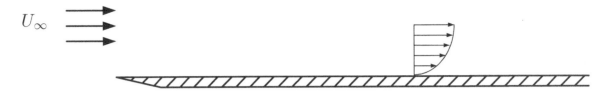

Figure 7.15.6: Flow over a flat plate.

Note that $f_1(\infty) = 1$ is not really at $\eta \to \infty$ because f_1 becomes 1 at some large finite value of η, say $\eta = 10$. To solve this system, some people employ the shooting method where they integrate Equation (7.15.52) from $\eta = 0$ to $\eta = \infty$ with an initial guess for $f_2(0)$, see if $f_1(\infty) = 1$ is satisfied after the integration, and if it is not, update $f_2(0)$ and repeat the calculation until they get $f_1(\infty) = 1$. This is not very simple to me; it doesn't make me feel like that I like it.

In fact, I know how to solve it in just two steps (no iterations!). As pointed out in Ref.[120], it is easily verified that if $g(\eta)$ is the solution of Equation (7.15.49), i.e., if it satisfies $g''' + \frac{1}{2}gg'' = 0$, with $g(0) = g'(0) = 0$ and $g''(0) = C_{g_2}$ for an arbitrary constant C_{g_2}, then

$$f = \alpha g(\alpha \eta), \quad \alpha = \frac{1}{\sqrt{g'(\infty)}}, \tag{7.15.56}$$

is the solution of Equation (7.15.49) that satisfies the desired boundary condition $f'(\infty) = 1$. So, this means that the correct initial condition for f_2 is given by

$$f_2(0) = f''(0) = \alpha^3 g''(0) = \alpha^3 C_{g_2}. \tag{7.15.57}$$

That is, integrating with this initial condition, you get what you want in one shot. So, here is a two step method. The following computes the velocity (u, v) at a given location $(x, y) = (x_p, y_p)$, $\eta = \eta_p$.

Computation of Exact Solution:
To really call this 'exact', you have to solve the ODE by a very high-order method with extremely small steps (e.g., the 4th-order Runge-Kutta method with $\Delta \eta = 10^{-5}$). Note also that we set $C_{g_2} = 1$.

1. Numerically solve Equation (7.15.52) with $f(0) = f_1(0) = 0$ and $f_2(0) = 1$, say, up to $\eta = 20$; call this solution $g(\eta)$ and set $\alpha = 1/\sqrt{g'(20)}$.

2. Numerically solve Equation (7.15.52) with $f(0) = f_1(0) = 0$ and $f_2(0) = \alpha^3$, up to the desired location $\eta = \eta_p$.

3. Compute the velocity there by

$$\frac{u}{U_\infty} = f_1(\eta_p), \tag{7.15.58}$$

$$\frac{v}{U_\infty} = \frac{1}{2\sqrt{Re_{x_p}}}(\eta f_1(\eta_p) - f(\eta_p)). \tag{7.15.59}$$

Honestly speaking, I like to skip the first step. So, I have computed (by using the 4th-order Runge-Kutta method with $\Delta \eta = 5.0 \times 10^{-6}$ and integrating the ODE up to $\eta = 100$) the value of $g'(\eta)$:

$$g'(100) = 2.085409176437910, \tag{7.15.60}$$

and thus

$$f_2(0) = \alpha^3 = \left(1/\sqrt{g'(100)}\right)^3 = 0.3320573362151946. \tag{7.15.61}$$

So, I can skip the first step from now on, and just perform the second step using this value. This is a very nice and simple one step method. I like it so much that I coded the algorithm and have made the code available at http://www.cfdbooks.com. I hope that you like it.

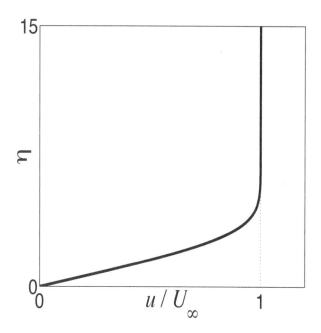

Figure 7.15.7: Velocity profile over a flat plate.

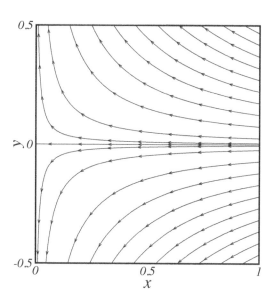

Figure 7.15.8: Viscous Stagnation Flow in 3D.

7.15.10   Viscous Stagnation Flow

For Stokes' equations, we cannot define the velocity potential in general, but we have the stream function for axisymmetric flows. Here is a simple and interesting one:

$$\psi = -\frac{1}{2}x^2 y^2, \tag{7.15.62}$$

in (x, y)-plane with the x-axis as the axis of symmetry. Streamlines are plotted in Figure 7.15.8. Yes, it represents a stagnation flow. And it is a viscous stagnation flow. Consider the velocity components,

$$u = \frac{1}{y}\frac{\partial \psi}{\partial y} = -x^2, \quad v = -\frac{1}{y}\frac{\partial \psi}{\partial x} = xy, \tag{7.15.63}$$

which shows that the velocity vanishes at the wall ($x = 0$). It thus satisfies the no-slip condition in contrast to the inviscid stagnation flow where $v \neq 0$ at $x = 0$. So, it is a viscous flow. For the viscous stagnation flow, it can be easily shown that the vorticity is not zero:

$$\omega = \frac{\partial v}{\partial x} - \frac{\partial u}{\partial y} = \frac{y}{2}, \tag{7.15.64}$$

except along the line of symmetry ($y = 0$). This is the reason that the velocity potential cannot be defined. I like the fact that the velocity components are quadratic function of space. I would expect that a third-order accurate scheme will produce exact results for the viscous stagnation flow. How about the pressure? You can find the pressure by integrating the momentum equations (4.28.2) and (4.28.3):

$$p = p_w - 2\mu x, \tag{7.15.65}$$

where p_w denotes the pressure at the wall. So, the pressure is a linear function of space. This is interesting. Recall that in the inviscid stagnation flow, the velocity is linear and the pressure is quadratic.

   If you compare the solution above with the inviscid stagnation flow solution in Section 7.12.1, you find that the viscous solution does not match the inviscid one as $x \to \infty$. In order to obtain the viscous solution that matches the inviscid one, it is necessary to solve ordinary differential equations. See Ref.[120, 169] for details.

   By the way, if you asked "Where is the stagnation point?", that would be a good question. I like it. If the stagnation point is defined as a point where the velocity is zero, then the entire wall will be the stagnation point in the viscous stagnation flow. But if the stagnation point is defined as a point from which the streamlines separate apart (the direction of surface streamlines cannot be defined), then the stagnation point will be at the origin $(x, y) = (0, 0)$, which matches the stagnation point in the inviscid stagnation flow. I like that. In fluid dynamics, I think we actually almost always take the latter definition; otherwise the word "stagnation point" will not be very useful in practical applications.

7.15.11   Stokes' Flow over a Sphere

Stokes' flow over a sphere is an exact solution to the steady Stokes equations (see Section 4.25), which may be expressed by

$$\operatorname{div} \mathbf{v} = 0, \tag{7.15.66}$$

$$\operatorname{div} \operatorname{grad} \boldsymbol{\omega} = 0, \tag{7.15.67}$$

where $\boldsymbol{\omega} = \operatorname{curl} \mathbf{v}$. We may introduce spherical coordinates (r, ϕ, θ) with axisymmetry with respect to ϕ (i.e., $\partial_\phi = 0$), and write the continuity equation as

$$\frac{1}{r^2} \frac{\partial (r^2 u_r)}{\partial r} + \frac{1}{r \sin \theta} \frac{\partial (u_\theta \sin \theta)}{\partial \theta} = 0. \tag{7.15.68}$$

Then, we can define the stream function as

$$u_r = \frac{1}{r^2 \sin \theta} \frac{\partial \psi}{\partial \theta}, \quad u_\theta = -\frac{1}{r \sin \theta} \frac{\partial \psi}{\partial r}, \tag{7.15.69}$$

to automatically satisfy the continuity equation. In terms of the stream function, the other equation (7.15.67) becomes

$$\left(\frac{\partial^2}{\partial r^2} + \frac{1}{r^2} \frac{\partial^2}{\partial \theta^2} - \frac{1}{r^2 \tan \theta} \frac{\partial}{\partial \theta} \right) \psi = 0. \tag{7.15.70}$$

Now consider a uniform flow with the speed U (in the direction of z) over a sphere of radius R at the origin, i.e.,

$$\frac{\partial \psi}{\partial r} = \frac{\partial \psi}{\partial \theta} = 0, \quad \text{on the sphere, } r = R,$$

$$\psi \to \frac{1}{2} U r^2 \sin^2 \theta + \text{constant}, \quad \text{as } r \to \infty. \tag{7.15.71}$$

Stokes discovered the following exact solution,

$$\psi = \frac{1}{4} U R^2 \sin^2 \theta \left(\frac{R}{r} - \frac{3r}{R} + \frac{2r^2}{R^2} \right), \tag{7.15.72}$$

so that the velocity components are

$$u_r = U \cos \theta \left(1 + \frac{R^3}{2r^3} - \frac{3R}{2r} \right), \tag{7.15.73}$$

$$u_\theta = -U \sin \theta \left(1 - \frac{R^3}{4r^3} - \frac{3R}{4r} \right), \tag{7.15.74}$$

the pressure is found from the momentum equation, $\operatorname{grad} p = \mu \operatorname{div} \operatorname{grad} \mathbf{v}$, to be

$$p(r, \theta) - p_\infty = -\frac{3\mu R U}{2r^2} \cos \theta, \tag{7.15.75}$$

(see Figures 7.15.9 and 7.15.10), the viscous stress on the sphere is given by

$$\tau_{r\theta}(r, \theta) = \mu \left(\frac{1}{r} \frac{\partial u_r}{\partial \theta} + \frac{\partial u_\theta}{\partial r} - \frac{u_\theta}{r} - \frac{3R^3}{2r^3} \frac{\mu U \sin \theta}{r} \right), \tag{7.15.76}$$

and therefore the drag force F is found by combining the viscous drag and the pressure drag,

$$F = -\int_0^\pi \tau_{r\theta}(R, \theta) \sin \theta (2\pi R^2 \sin \theta d\theta) - \int_0^\pi p(R, \theta) \cos \theta (2\pi R^2 \sin \theta d\theta)$$

$$= 4\pi \mu U R + 4\pi \mu U R$$

$$= 6\pi \mu U R. \tag{7.15.77}$$

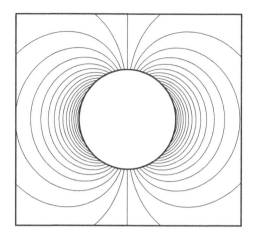

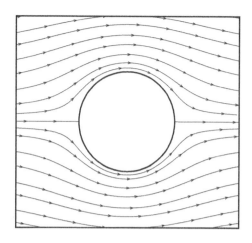

Figure 7.15.9: Pressure $p - p_\infty$ of the Stokes' flow over a sphere.

Figure 7.15.10: Streamlines of the Stokes' flow over a sphere.

I like this solution. The streamline pattern is similar to the inviscid flow over a cylinder, but here the fluid is at rest on the cylinder while the inviscid one is not (slip boundary). In fact, the pressure contours are very different from those for the inviscid flow. It is very interesting. By the way, this solution is exact only for the Stokes equations, not for the Navier-Stokes equations.

7.16   Compressible Navier-Stokes Equations

7.16.1   Compressible Couette Flow

An exact solution is available for the compressible Navier-Stoke equations for the Couette flow (Figure 7.15.2) [84]. With the same assumptions as made in Section 7.15.3, the continuity equation is again trivially satisfied, and the momentum equation and the energy equation are simplified as follows,

$$0 = \frac{d}{dy}\left(\mu \frac{du}{dy}\right), \tag{7.16.1}$$

$$0 = \frac{d}{dy}\left(\kappa \frac{dT}{dy}\right) + \mu \left(\frac{du}{dy}\right)^2. \tag{7.16.2}$$

Note that the pressure is constant everywhere (no acceleration) but the density and the temperature change in the y-direction. We also assume

$$\mu = \mu(T), \quad \kappa = \kappa(T), \tag{7.16.3}$$

i.e., the viscosity μ and the heat conductivity κ are functions of T only. Moreover, we assume that the Prandtl number Pr is constant,

$$Pr = \frac{c_p \mu}{\kappa} = \text{constant}. \tag{7.16.4}$$

Now, first, we integrate the momentum equation (7.16.1) once to get,

$$\mu \frac{du}{dy} = \mu \frac{du}{dy}\bigg|_{y=0} = \tau_w, \tag{7.16.5}$$

where τ_w denotes the shear stress at the bottom wall. Then, integrating once more, we obtain

$$u(y) = \tau_w \int_0^y \frac{dy}{\mu(T)}. \tag{7.16.6}$$

We will find that this is a very nice relation when we use it later to eliminate the same integral arising in an energy integral. Also, remember that we will integrate Equation (7.16.5) (in a slightly different way) to finally obtain the velocity distribution.

We now consider integrating the energy equation. First, we note that

$$\frac{d}{dy}\left(\mu u \frac{du}{dy}\right) = u \frac{d}{dy}\left(\mu \frac{du}{dy}\right) + \mu\left(\frac{du}{dy}\right)^2 = \mu\left(\frac{du}{dy}\right)^2, \tag{7.16.7}$$

where the momentum equation has been used in the last step. Using this, we can write the energy equation (7.16.2) as

$$0 = \frac{d}{dy}\left(\kappa \frac{dT}{dy} + \mu u \frac{du}{dy}\right), \tag{7.16.8}$$

so that it can be easily integrated once,

$$\kappa \frac{dT}{dy} + \mu u \frac{du}{dy} = \kappa \frac{dT}{dy}\bigg|_{y=0} + \mu u \frac{du}{dy}\bigg|_{y=0} = -q_w, \tag{7.16.9}$$

where q_w is the heat flux at the bottom wall (into the wall when $q_w < 0$) and the no-slip condition $u(0) = 0$ has been applied. This can be integrated once more by introducing the Prandtl number which is constant,

$$\mu \frac{d}{dy}\left(\frac{c_p T}{Pr} + \frac{u^2}{2}\right) = -q_w, \tag{7.16.10}$$

thus,

$$\frac{c_p(T - T_w)}{Pr} + \frac{u^2}{2} = -q_w \int_0^y \frac{dy}{\mu(T)}. \tag{7.16.11}$$

It is nice here that the integral on the right hand side can be eliminated by Equation (7.16.6):

$$\frac{c_p(T - T_w)}{Pr} + \frac{u^2}{2} = -\frac{q_w u}{\tau_w}. \tag{7.16.12}$$

It is very nice that the integral is gone. Just for convenience, we eliminate also T_w from above as follows: evaluate it at $y = d$ (the upper plate),

$$\frac{c_p(T_\infty - T_w)}{Pr} + \frac{U_\infty^2}{2} = -\frac{q_w U_\infty}{\tau_w}, \tag{7.16.13}$$

where the subscript ∞ denotes the state at the upper plate, then subtract Equation (7.16.12) from this to get

$$\frac{c_p(T - T_\infty)}{Pr} + \frac{u^2 - U_\infty^2}{2} = -\frac{q_w(u - U_\infty)}{\tau_w}, \tag{7.16.14}$$

which can be arranged into the following form,

$$\frac{T}{T_\infty} = 1 + \frac{q_w}{\tau_w U_\infty}(\gamma - 1)Pr M_\infty^2\left(1 - \frac{u}{U_\infty}\right) + \frac{\gamma - 1}{2}Pr M_\infty^2\left(1 - \frac{u^2}{U_\infty^2}\right), \tag{7.16.15}$$

where we have used $c_p = \gamma R/(\gamma - 1)$. Finally, in order to obtain the velocity distribution, we assume a power low

$$\frac{\mu}{\mu_\infty} = \left(\frac{T}{T_\infty}\right)^\omega, \tag{7.16.16}$$

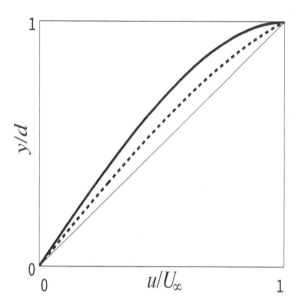

Figure 7.16.1: Velocity profile for the compressible Couette flow with $Pr = 0.73$, $\gamma = 1.4$; $M_\infty = 20$ (thick solid line), $M_\infty = 3$ (dashed line), $M_\infty = 0$, i.e., incompressible flow (thin solid line). Note that this is exact only when the viscosity obeys the power law (7.16.16) with $\omega = 1$.

where $\omega = 0.76$ is a good approximation for air, and insert this into the momentum equation (7.16.5) and integrate to get

$$
\begin{aligned}
\frac{\tau_w y}{\mu_\infty} &= \int_0^u \frac{\mu}{\mu_\infty} du = \int_0^u \left(\frac{T}{T_\infty} \right)^\omega du \\
&= \int_0^u \left[1 + \frac{q_w}{\tau_w U_\infty}(\gamma - 1)PrM_\infty^2 \left(1 - \frac{u}{U_\infty} \right) + \frac{\gamma - 1}{2}PrM_\infty^2 \left(1 - \frac{u^2}{U_\infty^2} \right) \right]^\omega du.
\end{aligned}
\tag{7.16.17}
$$

This can be integrated analytically (using the beta function), but the result will be very complicated. For code verification purpose, it suffices to take $\omega = 1$ and $q_w = 0$. Then, finally, we obtain the following implicit expression for the velocity distribution,

$$
\frac{y}{d} \left[1 + \frac{\gamma - 1}{3}PrM_\infty^2 \right] = \frac{u}{U_\infty} \left[1 + \frac{\gamma - 1}{2}PrM_\infty^2 \left(1 - \frac{1}{3}\frac{u^2}{U_\infty^2} \right) \right],
\tag{7.16.18}
$$

where τ_w has been determined by the boundary condition $u(d) = U_\infty$ and eliminated. This can be solved numerically. The velocity profiles in Figure 7.16.1 were obtained by the fixed-point iteration: compute V^{k+1} ($k = 1, 2, 3, \ldots$), at a desired location y, by

$$
V^{k+1} = \frac{\eta \left[1 + \frac{\gamma - 1}{3}PrM_\infty^2 \right]}{1 + \frac{\gamma - 1}{2}PrM_\infty^2 \left[1 - \frac{1}{3}(V^k)^2 \right]},
\tag{7.16.19}
$$

until $|V^{k+1} - V^k| < 10^{-15}$, where $V^k = u^k/U_\infty$, $\eta = y/d$, and the initial guess V^1 may be set any value between 0 and 1. Temperature can be calculated by Equation (7.16.15) for a given velocity profile, and density can then be calculated by the equation of state. Of course, I like this solution. It is very nice that the domain is very simple but the solution is not trivial. If you want a solution for a more general case, e.g., $q_w \neq 0$ or $\omega \neq 1$, you can derive it analytically. Good luck.

7.16.2 Viscous Shock Structure

Another useful exact solution to the compressible Navier-Stokes equations is the viscous shock structure solution. It is a viscous version of the steady shock wave solution in Section 7.13.8, i.e., a smoothly varying solution rather than a discontinuous solution. Consider the steady one-dimensional compressible Navier-Stokes equations:

$$\frac{d}{dx} \begin{bmatrix} \rho u \\ \rho u^2 + p - \tau \\ \rho u H - \tau u + q \end{bmatrix} = 0, \tag{7.16.20}$$

where

$$\tau = \frac{4}{3} \mu \frac{du}{dx}, \quad q = -\frac{\gamma \mu}{Pr(\gamma - 1)} \frac{d(p/\rho)}{dx}. \tag{7.16.21}$$

Integrating the Navier-Stokes equations (7.16.20), we obtain

$$\begin{bmatrix} \rho u \\ \rho u^2 + p - \tau \\ \rho u H - \tau u + q \end{bmatrix} = \text{constant.} \tag{7.16.22}$$

Then, we apply this equation inside and outside the viscous shock to get

$$\rho u = (\rho u)_o, \tag{7.16.23}$$

$$\rho u^2 + p - \tau = (\rho u^2 + p)_o, \tag{7.16.24}$$

$$\rho u H - \tau u + q = (\rho u H)_o, \tag{7.16.25}$$

where the subscript o denotes the outside state and the inside state has no subscript. Note that $\tau = q = 0$ outside the viscous shock where the solution is constant. Note also that the terms on the right hand side are all constant; we define them as

$$m = (\rho u)_o, \quad C_2 = (\rho u^2 + p)_o, \quad C_3 = (\rho u H)_o, \tag{7.16.26}$$

so that Equation (7.16.23) becomes $\rho u = m$ and can be substituted into other equations. We therefore obtain from Equations (7.16.24) and (7.16.25)

$$\tau = mu + p - C_2, \tag{7.16.27}$$

$$q = \frac{1}{2} mu^2 - \frac{m}{\gamma - 1} \frac{p}{\rho} - C_2 u + C_3. \tag{7.16.28}$$

By Equations (7.16.21), we obtain

$$\frac{du}{dx} = \frac{1}{K_u} \left(mu + \frac{m}{u} \frac{p}{\rho} - C_2 \right), \tag{7.16.29}$$

$$\frac{d(p/\rho)}{dx} = \frac{1}{K_T} \left(\frac{1}{2} mu^2 - \frac{m}{\gamma - 1} \frac{p}{\rho} - C_2 u + C_3 \right), \tag{7.16.30}$$

where

$$K_u = \frac{4}{3} \mu, \quad K_T = -\frac{\gamma \mu}{Pr(\gamma - 1)}. \tag{7.16.31}$$

If we assume that all variables have been nondimensionalized as in Section 4.14.3, then $p/\rho = T/\gamma$, and therefore these equations can be written as

$$\frac{du}{dx} = \frac{1}{C_u} \left(mu + \frac{mT}{\gamma u} - C_2 \right), \tag{7.16.32}$$

$$\frac{dT}{dx} = \frac{\gamma}{C_T} \left(\frac{1}{2} mu^2 - \frac{m}{\gamma - 1} \frac{T}{\gamma} - C_2 u + C_3 \right), \tag{7.16.33}$$

where

$$C_u = \frac{4}{3} \mu \frac{M_\infty}{Re_\infty}, \quad C_T = -\frac{\gamma \mu}{Pr(\gamma - 1)} \frac{M_\infty}{Re_\infty}. \tag{7.16.34}$$

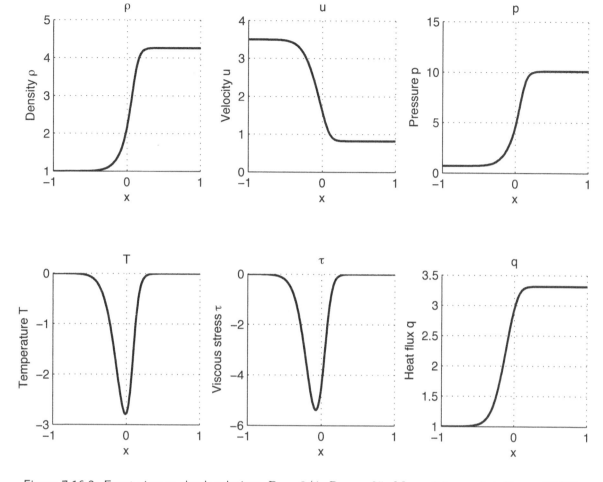

Figure 7.16.2: Exact viscous shock solution: $Pr = 3/4$, $Re_\infty = 25$, $M_\infty = 3.5$, $\gamma = 1.4$, $T_\infty = 400(K)$.

This is a system of nonlinear ordinary differential equations for u and T (the viscosity μ depends on T). It can be integrated numerically in a straightforward manner. In what follows, we employ this nondimensionalized system.

To integrate the system, we first choose the Mach number on the left side of the shock M_L, and set up the left and right states as follows:

$$
\begin{aligned}
&\rho_L = 1, &&\rho_R = \frac{(\gamma+1)M_L^2}{(\gamma-1)M_L^2 + 2}, \\
&u_L = M_L, &&u_R = \rho_L M_L/\rho_R, \\
&p_L = 1/\gamma, &&p_R = \frac{1}{\gamma}\left[1 + \frac{2\gamma}{\gamma+1}(M_L^2 - 1)\right], \\
&T_L = \gamma p_L/\rho_L, &&T_R = \gamma p_R/\rho_R.
\end{aligned}
\tag{7.16.35}
$$

Note again that all variables are assumed to have been nondimensionalized as in Section 4.14.3. The constants are computed by the left (or the right) state:

$$
m = \rho_L u_L, \quad C_2 = m u_L + p_L, \quad C_3 = u_L\left(\frac{\gamma p_L}{\gamma-1} + \frac{1}{2}m u_L\right).
\tag{7.16.36}
$$

We are now ready to integrate the system. Suppose we wish to compute the solution at every node in a one-dimensional grid defined by the set of coordinates $\{x_k\} = x_1, x_2, x_3, \ldots, x_N$ ($x_1 < x_2 < \ldots < x_N$), where N is the number of nodes in the grid. We consider integrating the system from the right state. First, we numerically integrate the system from $x = x_N$ to $x = x_{N-1}$ with the initial values: $u(x_N) = u_R + \epsilon$ where ϵ is a small number (e.g., $\epsilon = 1.0E\text{-}10$)

and $T(x_N) = T_R$ with a very small step Δx compared with $x_N - x_{N-1}$. Then, we do the same between two adjacent nodes, k and $k-1$, where $k = N-1, \ldots, 3, 2$, with the solutions at $x = x_k$ as the initial values. For example, we may employ the classical fourth-order Runge-Kutta scheme with $\Delta x = \frac{x_k - x_{k-1}}{1000}$ between the two nodes, k and $k-1$. Note that the last step in the integration between two nodes needs to be adjusted to finish exactly at $x = x_{k-1}$. Figure 7.16.2 shows the exact solution computed as described for $N = 200$. A sample program, which was used to generate the solution in Figure 7.16.2, is available at http://www.cfdbooks.com. I hope you like it.

I like the viscous shock structure solution very much for two reasons. First, the solution is not really physical. The flow inside the shock wave is unlikely to be in equilibrium, and therefore the Navier-Stokes system is not a valid model. However, it is still useful for verifying the accuracy of a Navier-Stokes solver. Second, the small number ϵ is essential to obtaining a meaningful solution. In fact, if $\epsilon = 0$, the computation will fail. Nothing will happen because the right hand side of the system (7.16.32) and (7.16.33) are exactly zero. We will not be able to reach the correct left state of the shock. This is very interesting. The location of the shock depends on the magnitude of ϵ. You might want to experiment with varying ϵ to see how it affects the shock location (or the existence). This solution has been used successfully for accuracy verification by many researchers [106, 110, 175].

7.16.3 Manufactured Solutions

We can manufacture exact solutions for the compressible Navier-Stokes equations. In fact, many people do. See Refs.[83, 91, 133] for examples. In particular, I like the idea of using exact solutions of the incompressible Navier-Stokes Equations as manufactured solutions for the compressible Navier-Stokes Equations: substitute an incompressible solution into the compressible equations and generate an appropriate source term. This way, we can have an exact solution which looks like a physical flow. For example, in Ref.[118], the Couette-Poiseuille solution described in Section 7.15.5 was used for verifying the accuracy of a high-order compressible Navier-Stokes code. You want to try it now? Then, pick any incompressible flow solution from Section 7.15, and generate an appropriate source term for the compressible Navier-Stokes equations (possibly by using a math software or by computing the source term numerically).

Bibliography

[1] S. Abarbanel and D. Gottlieb. Optimal Time Splitting for Two and Three Dimensional Navier-Stokes Equations with Mixed Derivatives. *ICASE Report 80-6* (1980) (see pp. 150, 152).

[2] I. Altas, J. Erhel, and M. M. Gupta. High Accuracy Solution of Three-Dimensional Biharmonic Equations. *Numerical Algorithms* 29 (2002), pp. 1–19 (see p. 55).

[3] J. S. L. Alzaeili and K. Mazaheri. Bulk Viscosity Damping for Accelerating Convergence of Compressible Viscous Flow Solvers. In: Proceedings of ECCOMAS CFD 2006. 2006 (see p. 133).

[4] M. Arora and P. L. Roe. On Postshock Oscillations due to Shock Capturing Schemes in Unsteady Flow. *J. Comput. Phys.* 130 (1997), pp. 1–24 (see p. 259).

[5] K. A. Bagrinovskii and S. K. Godunov. Difference Schemes for Multidimensional Problems. *Dokl. Akad. Nauk SSSR (NS)* 115 (1957), pp. 431–433 (see p. 37).

[6] B. S. Baldwin and T. J. Barth. *A One-Equation Turbulence Transport Model for High Reynolds Number Wall-Bounded Flows.* NASA TM-102847. 1990 (see p. 177).

[7] B. S. Baldwin and T. J. Barth. A One-Equation Turbulence Transport Model for High Reynolds Number Wall-Bounded Flows. In: AIAA Paper 91-0610. 1991 (see p. 177).

[8] B. S. Baldwin and H. Lomax. *Thin-Layer Approximation and Algebraic Model for Separated Turbulent Flows.* AIAA Paper 78–257. 1978 (see p. 177).

[9] A. F. Barannyk and I. I. Yuryk. Construction of Exact Solutions of Diffusion Equation. In: Proceedings of Institute of Mathematics of NAS of Ukraine. Vol. 50, Part 1. Institute of Mathematics, 2004, pp. 29–33 (see p. 190).

[10] T. J. Barth. Numerical Methods for Gasdynamic Systems on Unstructured Meshes. In: An Introduction to Recent Developments in Theory and Numerics for Conservation Laws. Ed. by D. Kroner, M. Ohlberger, and M. Rohde. Springer, 1997, pp. 195–285 (see pp. 32, 72).

[11] G. K. Batchelor. An Introduction to Fluid Dynamics. Cambridge University Press, 1967 (see p. 136).

[12] G. K. Batchelor. The Theory of Homogeneous Turbulence. Cambridge University Press, 1970 (see p. 173).

[13] E. R. Benton and G. W. Platzman. A Table of Solutions of the One-Dimensional Burgers Equation. *Quart. Appl. Mafh.* 30 (1972), pp. 195–212 (see p. 188).

[14] S. Blanes and P. C. Moan. Practical Symplectic Partitioned Runge-Kutta and Runge-Kutta-Nyström Methods. *J. Comp. Appl. Math.* 142 (2002), pp. 313–330 (see p. 38).

[15] A. Bonfiglioli. Fluctuation Splitting Schemes for the Compressible and Incompressible Euler and Navier-Stokes Equations. *Int. J. Comput. Fluid Dyn.* 14 (2000), pp. 21–39 (see pp. 80, 100).

[16] J. Boussinesq. Essai Sur Lar Théorie Des Eaux Courantes. *Mem. Présentés Acad. Sci.* 23 (1877), p. 46 (see p. 174).

[17] W. L. Briggs, V. E. Henson, and S. F. McCormick. A Multigrid Tutorial. Second. SIAM, 2000 (see p. 52).

[18] R. G. Campbell. Foundations of Fluid Flow Theory. Addison-Wesley Publishing Company, 1973 (see pp. 110, 167).

[19] D. Caraeni and L. Fuchs. Compact Third-Order Multidimensional Upwind Scheme for Navier-Stokes Simulations. *Appl. Math.* 15 (2002), pp. 373–401 (see p. 80).

[20] C. Cattaneo. A Form of Heat-Conduction Equations which Eliminates the Paradox of Instantaneous Propagation. *Ct. R. Acad. Sci., Paris* 247 (1958), pp. 431–433 (see p. 48).

[21] A. J. Chorin. A Numerical Method for Solving Incompressible Viscous Flow Problems. *J. Comput. Phys.* 2 (1967), pp. 12–26 (see pp. 36, 107, 133, 162).

[22] A. J. Chorin. Numerical Solution of the Navier-Stokes Equations. *Mathematics of Computation* 22.104 (1968), pp. 745–762 (see p. 263).

[23] C.-S. Chou and C.-W. Shu. High Order Residual Distribution Conservative Finite Difference WENO Schemes for Steady State Problems on Non-Smooth Meshes. *J. Comput. Phys.* 214 (2006), pp. 698–724 (see p. 221).

[24] B. Cockburn and C.-W. Shu. Runge-Kutta Discontinuous Galerkin Methods for Convection-Dominated Problems. *J. Sci.Comput.* 16.3 (2001), pp. 173–261 (see p. 51).

[25] B. Cockburn and C.-W. Shu. The Local Discontinuous Galerkin Method for Time-Dependent Convection-Diffusion Systems. *SIAM J. Numer. Anal.* 35 (1998), pp. 2440–2463 (see p. 48).

[26] W. J. Coirier and K. G. Powell. An Accuracy Assessment of Cartesian-Mesh Approaches for the Euler Equations. *J. Comput. Phys.* 117 (1995), pp. 121–131 (see p. 248).

[27] A. W. Cook and W. H. Cabot. Hyperviscosity for Shock-Turbulence Interactions. *J. Comput. Phys.* 203 (2005), pp. 379–385 (see p. 133).

[28] C. Corre, G. Hanss, and A. Lerat. A Residual-Based Compact Schemes for the Unsteady Compressible Navier-Stokes Equations. *Comput. Fluids* 34 (2005), pp. 561–580 (see p. 242).

[29] G.-H. Cottet and P. D. Koumoutsakos. Vortex Methods: Theory and Practice. Cambridge University Press, 2000 (see pp. 126, 164).

[30] A. Csik, M. Ricchiuto, and H. Deconinck. A Conservative Formulation of the Multidimensional Upwind Residual Distribution Schemes for General Nonlinear Conservation Laws. *J. Comput. Phys.* 179 (2002), pp. 286–312 (see p. 80).

[31] H. Deconinck, P. L. Roe, and R. Struijs. A Multi-Dimensional Generalization of Roe's Flux Difference Splitter for the Euler Equations. *Comput. Fluids* 22 (1993), pp. 215–222 (see pp. 75, 89).

[32] B. Diskin and J. T. Thomas. Convergence of defect-correction and multigrid iterations for inviscid flows. In: 20th AIAA Computational Fluid Dynamics Conference. AIAA Paper 2011-3235. Hawaii, 2011 (see p. 224).

[33] Weinan E and C.-W Shu. A Numerical Resolution Study of High Order Essentially Non-oscillatory Schemes Applied to Incompressible Flow. *J. Comput. Phys.* 110 (1994), pp. 39–46 (see p. 263).

[34] L. Eca et al. On the construction of Manufactured Solutions for One and Two-Equation Eddy-Viscosity Models. *Int. J. Numer. Meth. Fluids.* 54 (2007), pp. 119–154 (see p. 198).

[35] C. R. Ethier and D. A. Steinman. Exact Fully 3D Navier-Stokes Solutions for Benchmarking. *Int. J. Numer. Meth. Fluids.* 19 (1994), pp. 369–375 (see p. 264).

[36] C. A. J. Fletcher. Computational Techniques for Fluid Dynamics. Vol. 2. Springer, 1991 (see p. 30).

[37] C. A. J. Fletcher. Generating Exact Solutions of the Two-Dimensional Burgers' Equations. *Int. J. Numer. Meth. Fluids.* 3 (1983), pp. 213–216 (see pp. 188, 221).

[38] L. E. Fraenkel. On Corner Eddies in Plane Inviscid Shear Flow. *J. Fluid Mech.* 11 (1961), pp. 400–406 (see p. 235).

[39] E. Godlewski and P.-A. Raviart. Numerical Approximation of Hyperbolic Systems of Conservation Laws. Springer, 1996 (see p. 32).

[40] S. K. Godunov. An interesting Class of Quasilinear Systems. *Dokl. Akad. Nauk. SSSR* 139 (1961), pp. 521–523 (see p. 72).

[41] A. R. Gourlay and J. LL. Morris. On the Comparison of Multistep Formulations of the Optimized Lax-Wendroff Method for Nonlinear Hyperbolic Systems in Two Space Variables. *J. Comput. Phys.* 5 (1970), pp. 229–243 (see p. 38).

[42] H. Grad. Reducible Problems in Magneto-Fluid Dynamic Steady Flows. *Rev. Mod. Phys.* 32.4 (1960), pp. 830–847 (see p. 248).

[43] B. Grossman. *The Computation of Inviscid Rotational Gasdynamic Flows Using an Alternate Velocity Decomposition*. AIAA Paper 83-1900. 1983 (see p. 121).

[44] B. Grossman and S. K. Choi. *The Computation of Rotational Conical Flows*. AIAA Paper 84-0258. 1984 (see p. 121).

[45] H. Guillard and C. Viozat. On the Behaviour of Upwind Schemes in the Low Mach Number Limit. *J. Comput. Phys.* 28 (1999), pp. 63–86 (see p. 36).

[46] Y. H. and C. L. Merkle. The Applicaiton of Preconditioning in Viscous Flows. *J. Comput. Phys.* 105 (1993), pp. 207–223 (see p. 36).

[47] M. E. Habbard and P. Garcia Navarro. Flux Difference Splitting and the Balancing of Source Terms and Flux Gradients. *J. Comput. Phys.* 165 (2000), pp. 89–125 (see p. 198).

[48] N. D. Halsey. Potential Flow Analysis of Multielement Airfoils Using Conformal Mapping. *AIAA J.* 17 (1979), pp. 1281–1288 (see p. 235).

[49] A. Haselbacher and J. Blazek. Accurate and Efficient Discretization of Navier-Stokes Equations on Mixed Grids. *AIAA J.* 38.11 (2000), pp. 2094–2102 (see p. 145).

[50] C. Hirsch. Numerical Computation of Internal and External Flows. Vol. 1. A Wiley - Interscience Publications, 1990 (see p. 41).

[51] C. Hirsch. Numerical Computation of Internal and External Flows. Vol. 2. A Wiley - Interscience Publications, 1990 (see pp. 41, 67, 120, 122, 150).

[52] C. W. Hirt, A. A. Amsden, and J. L. Cook. An Arbitrary Lagrangian-Eulerian Computing Method for All Flow Speeds. *J. Comput. Phys.* 14.3 (1974), pp. 227–257 (see p. 28).

[53] W. F. Hughes and J. A. Brighton. Schaum's Outline Series, Fluid Dynamics. McGraw-Hill, 1995 (see p. 122).

[54] I. Imai. Fluid Dynamics. Vol. 1. Shokabo, 1973 (see pp. 54, 168–170, 183).

[55] R. I. Issa. Rise of Total Pressure in Frictional Flow. *AIAA J.* 33.4 (1995), pp. 772–774 (see p. 135).

[56] D. C. Ives. A Modern Look at Conformal Mapping Including Multiple Connected Region. *AIAA J.* 14 (1976), pp. 1006–1011 (see p. 235).

[57] Jr. J. D. Anderson. Computational Fluid Dynamics: The Basic with Applications. McGraw-Hill, 1995 (see pp. 30, 116).

[58] A. Jameson. Numerical Solution of Nonlinear Partial Differential Equations of Mixed Type. In: Third Symposium on Numerical Solution of Partial Differential Equations. 1975 (see p. 55).

[59] A. Jameson. Time Dependent Calculations Using Multigrid, with Applications to Unsteady Flows Past Airfoils and Wings. In: AIAA Paper 91-1596. 1991 (see p. 36).

[60] A. Jeffrey and T. Taniuti. Nonlinear Wave Propagation. Academic Press, 1964 (see p. 41).

[61] B.-N. Jiang. The Least-Squares Finite Element Method. Springer, 1998 (see p. 53).

[62] S. Jin and C. D. Levermore. Numerical Schemes for Hyperbolic Conservation Laws with Stiff Relaxation Terms. *J. Comput. Phys.* 126 (1996), pp. 449–467 (see p. 55).

[63] D. A. Johnson and L. S. King. A Mathematically Simple Turbulence Closure Model for Attached and Separated Turbulent Boundary Layers. *AIAA J.* 23.11 (1985), pp. 1684–1692 (see p. 177).

[64] I. W. Johnson, A. J. Wathen, and M. J. Baines. Moving Finite Element Methods for Evolutionary Problems II: Applications. *J. Comput. Phys.* 79 (1988), pp. 270–297 (see p. 219).

[65] Daniel D. Joseph. Potential flow of viscous fluids: Historical notes. *Int. J. Multiphas. Flow* 32 (2006), pp. 285–310 (see p. 121).

[66] R. Kannan, Y. Sun, and Z. J. Wang. A Study of Viscous Flux Formulations for an Implicit P-Multigrid Spectral Volume Navier Stokes Solver. In: 46th AIAA Aerospace Sciences Meeting. AIAA Paper 2008-783. 2008 (see p. 48).

[67] K. Karamcheti. Principles of Ideal-Fluid Aerodynamics. Reprinted. Krieger Publishing Company, 1980 (see pp. 19, 167).

[68] J. Katz and A. Plotkin. Low-Speed Aerodynamics. Second. Cambridge University Press, 2001 (see pp. 53, 184).

[69] S. Kawai, S. K. Shankar, and S. K. Lele. Assesment of Localized Artificial Diffusivity Scheme for Large-Eddy Simulation of Compressible Turbulent Flows. *J. Comput. Phys.* 229 (2010), pp. 1739–1762 (see p. 133).

[70] Soshi Kawai. Private Communication. 2013 (see p. 133).

[71] C. Kim. "Multi-Dimensional Upwind Leapfrog Schemes and Their Applications". PhD thesis. Ann Arbor, Michigan: University of Michigan, 1997 (see p. 213).

[72] B. L. Kleb. "Optimizing Runge-Kutta Schemes for Viscous Flow". PhD thesis. Ann Arbor, Michigan: University of Michigan, 2004 (see p. 150).

[73] A. N. Kolmogorov. The Local Structure of Turbulence in Incompressible Viscous Fluid for Very Large Reynolds Numbers. *Doklady Akademiya Nauk SSSR* 30 (1941), pp. 299–303 (see p. 173).

[74] A. M. Kuethe and C.-Y. Chow. Foundations of Aerodynamics. Fifth. John Wiley & Son Ltd., 1998 (see p. 123).

[75] D. Kwak et al. A Three-Dimensional Incompressible Navier-Stokes Solver Using Primitive Variables. *AIAA J.* 24 (1986), pp. 390–396 (see pp. 107, 162).

[76] L. D. Landau and E. M. Lifshitz. Fluid Mechanics. Second. Butterworth-Heinemann, 1987 (see p. 61).

[77] Dohyung Lee. "Local Preconditioning of the Euler and Navier-Stokes Equations". PhD thesis. Ann Arbor, Michigan: University of Michigan, 1996 (see pp. 36, 150).

[78] S. Lee, S. K. Lele, and P. Moin. Direct Numerical Simulation of Isotropic Turbulence Interacting with a Weak Shock Wave. *J. Fluid Mech.* 251 (1993), pp. 533–562 (see p. 133).

[79] W.-T. Lee. "Local Preconditioning of the Euler Equations". PhD thesis. Ann Arbor, Michigan: University of Michigan, 1991 (see p. 39).

[80] B. van Leer, W.-T. Lee, and P. L. Roe. Characteristic Time-Stepping or Local Preconditioning of the Euler Equations. In: 10th AIAA Computational Fluid Dynamics Conference. AIAA Paper 91-1552. Hawaii, 1991 (see p. 36).

[81] R. J. LeVeque. Finite Volume Methods for Hyperbolic Problems. Cambridge University Press, 2002 (see pp. 42, 56, 57, 61, 127).

[82] R. L. LeVeque. Balancing source terms and flux gradients in high-resolution Godunov methods - The quasi-steady wave-propagation algorithm. *J. Comput. Phys.* 146 (1998), pp. 346–365 (see p. 198).

[83] W. Liao et al. Textbook-Efficiency Multigrid Solver for Three-Dimensional Unsteady Compressible Navier-Stokes Equations. *J. Comput. Phys.* 227 (2008), pp. 7160–7177 (see pp. 198, 277).

[84] H. W. Liepmann and A. Roshko. Elements of Gasdynamics. Dover, 2002 (see pp. 122, 123, 244, 252, 254, 272).

[85] K. Liffman and A. Siora. Magnetosonic Jet Flow. *Mon. Not. R. Astron. Soc.* 290 (1997), pp. 629–635 (see p. 248).

[86] T. Linde and P. Roe. On a Mistaken Notion of of "Proper Upwinding". *J. Comput. Phys.* 142 (1998), pp. 611–614 (see p. 259).

[87] R. B. Lowrie. "Compact Higher-Order Numerical Methods for Hyperbolic Conservation Laws". PhD thesis. Ann Arbor, Michigan: University of Michigan, 1996 (see p. 240).

[88] R. B. Lowrie and J. E. Morel. Methods for Hyperbolic Systems with Stiff Relaxation. *Int. J. Numer. Meth. Fluids.* 40 (2002), pp. 413–423 (see p. 55).

[89] D. J. Mavriplis and S. Pirzadeh. Large-Scale Parallel Unstructured Mesh Computations for 3D High-Lift Analysis. *J. Aircraft* 36.6 (1999), pp. 987–998 (see p. 47).

[90] K. Mazaheri and P. L. Roe. Bulk Viscosity Damping for Accelerating Convergence of low Mach number Euler Solvers. *Int. J. Numer. Meth. Fluids.* 41 (2003), pp. 633–652 (see p. 133).

[91] J. M. McDonough. *A Class of Model Problems for Testing Compressible Navier-Stokes Solvers.* AIAA Paper 88-3646. 1988 (see p. 277).

[92] F. R. Menter. Two-Equation Eddy-Viscosity Turbulence Models for Engineering Applications. *AIAA J.* 32.8 (1994), pp. 1598–1605 (see p. 177).

[93] L. M. Mesaros. "Multi-Dimensional Fluctuation Splitting Schemes for The Euler Equations on Unstructured Grids". PhD thesis. Ann Arbor, Michigan: University of Michigan, 1995 (see pp. 53, 54).

[94] L. M. Mesaros and P. L. Roe. Multidimensional Fluctuation Splitting Schemes Based on Decomposition Methods. In: 12th AIAA Computational Fluid Dynamics Conference. AIAA Paper 95-1699. San Diego, 1995 (see pp. 36, 118).

[95] K. W. Morton and P. L. Roe. Vorticity-Preserving Lax-Wendroff-Type Schemes for the System Wave Equation. *SIAM J. Sci. Comput.* 23 (2001), pp. 170–192 (see p. 46).

[96] E. M. Murman and P. M. Stremel. *A Vortex Wake Capturing Method for Potential Flow Calculations.* AIAA Paper 82-0947. 1982 (see p. 121).

[97]   R. E. Neel, A. G. Godfrey, and W. D. McGrory. Low-Speed, Time-Accurate Validation of GASP Version 4. In: 43rd AIAA Aerospace Sciences Meeting and Exhibit. AIAA Paper 2005-686. Reno, Nevada, 2005 (see pp. 107, 162).

[98]   H. Nishikawa. A First-Order System Approach for Diffusion Equation. I: Second-Order Residual-Distribution Schemes. *J. Comput. Phys.* 227 (2007), pp. 315–352 (see pp. 36, 49, 50).

[99]   H. Nishikawa. A First-Order System Approach for Diffusion Equation. II: Unification of Advection and Diffusion. *J. Comput. Phys.* 229 (2010), pp. 3989–4016 (see pp. 33, 36, 50).

[100] H. Nishikawa. Adaptive Quadrature Fluctuation-Splitting Schemes for the Euler Equations. *Int. J. Numer. Meth. Fluids.* 57 (2008), pp. 1–12 (see p. 118).

[101] H. Nishikawa. Beyond Interface Gradient: A General Principle for Constructing Diffusion Schemes. In: 40th AIAA Fluid Dynamics Conference and Exhibit. AIAA Paper 2010-5093. Chicago, 2010 (see pp. 47, 207).

[102] H. Nishikawa. Divergence Formulation of Source Term. *J. Comput. Phys.* 231 (2012), pp. 6393–6400 (see p. 29).

[103] H. Nishikawa. First, Second, and Third Order Finite-Volume Schemes for Advection-Diffusion. In: 21st AIAA Computational Fluid Dynamics Conference. AIAA Paper 2011-2568. San Diego, California, 2013 (see pp. 36, 50).

[104] H. Nishikawa. First, Second, and Third Order Finite-Volume Schemes for Diffusion. In: 51st AIAA Aerospace Sciences Meeting. AIAA Paper 2011-1125. Grapevine, Texas, 2013 (see pp. 36, 49, 50).

[105] H. Nishikawa. Multigrid Third-Order Least-Squares Solution of Cauchy-Riemann Equations on Unstructured Triangular Grids. *Int. J. Numer. Meth. Fluids.* 53 (2007), pp. 443–454 (see p. 53).

[106] H. Nishikawa. New-Generation Hyperbolic Navier-Stokes Schemes: $O(1/h)$ Speed-Up and Accurate Viscous/Heat Fluxes. In: 20th AIAA Computational Fluid Dynamics Conference. AIAA Paper 2011-3043. Hawaii, 2011 (see pp. 7, 33, 277).

[107] H. Nishikawa. "On Grids and Solutions from Residual Minimization". PhD thesis. Ann Arbor, Michigan: University of Michigan, Aug. 2001 (see pp. 53, 208).

[108] H. Nishikawa. Robust and Accurate Viscous Discretization via Upwind Scheme - I: Basic Principle. *Comput. Fluids* 49.1 (2011), pp. 62–86 (see pp. 47, 207).

[109] H. Nishikawa. *The Roe-Averaged density.* http://www.cfdnotes.com/cfdnotes_roe_averaged_density.html. March, 2011 (see p. 93).

[110] H. Nishikawa. Two Ways to Extend Diffusion Schemes to Navier-Stokes Schemes: Gradients or Upwinding. In: 20th AIAA Computational Fluid Dynamics Conference. AIAA Paper 2011-3044. Hawaii, 2011 (see p. 277).

[111] H. Nishikawa, B. Diskin, and J. L. Thomas. Critical Study of Agglomerated Multigrid Methods for Diffusion. *AIAA J.* 48.4 (2010), pp. 839–847 (see pp. 47, 145).

[112] H. Nishikawa and K. Kitamura. Very Simple, Carbuncle-Free, Boundary-Layer Resolving, Rotated-Hybrid Riemann Solvers. *J. Comput. Phys.* 227 (2007), pp. 2560–2581 (see p. 4).

[113] H. Nishikawa, M. Rad, and P. Roe. A Third-Order Fluctuation-Splitting Scheme That Preserves Potential Flow. In: 15th AIAA Computational Fluid Dynamics Conference. AIAA Paper 01-2595. Anaheim, 2001 (see pp. 36, 118, 235).

[114] H. Nishikawa, M. Rad, and P. L. Roe. Grids and Solutions from Residual Minimisation. In: Computational Fluid Dynamics 2000. Ed. by N. Satofuka. Springer-Verlag, 2000, pp. 119–124 (see p. 218).

[115] H. Nishikawa and P. L. Roe. On High-Order Fluctuation-Splitting Schemes for Navier-Stokes Equations. In: Computational Fluid Dynamics 2004. Ed. by C. Groth and D. W. Zingg. Springer-Verlag, 2004, pp. 799–804 (see pp. 48, 51).

[116] H. Nishikawa and B. van Leer. Optimal Multigrid Convergence by Elliptic/Hyperbolic Splitting. *J. Comput. Phys.* 190 (2003), pp. 52–63 (see p. 36).

[117] H. Nishikawa et al. A General Theory of Local Preconditioning and Its Application to the 2D Ideal MHD Equations. In: 16th AIAA Computational Fluid Dynamics Conference. AIAA Paper 2003-3704. Orlando, 2003 (see pp. 33, 34, 36, 41, 53).

[118] T. A. Oliver. "Multigrid solution for high-order discontinuous Galerkin discretizations of the compressible Navier-Stokes equations". MA thesis. Massachusetts Institute of Technology, Aug. 2004 (see p. 277).

[119] K. Ou, P. Vincent, and A. Jameson. High-Order Methods for Diffusion Equation with Energy Stable Flux Reconstruction Scheme. In: 49th AIAA Aerospace Sciences Meeting. AIAA Paper 2011-46. Orlando, 2011 (see p. 205).

[120] C. Pozrikidis. Introduction to Theoretical and Computational Fluid Dynamics. Oxford University Press, 1997 (see pp. 26, 167, 269, 270).

[121] L. Prandtl. Uber die ausgebildete Turbulenz. *Z.A.M.M.* 5 (1925), pp. 136–139 (see p. 177).

[122] L. Quartapelle and M. Napolitano. Force and Moment in Incompressible Flows. *AIAA J.* 21.6 (1983), pp. 911–913 (see p. 136).

[123] M. Rad. "Residual Distribution Approach to the Euler Equations that Preserves Potential Flow". PhD thesis. Ann Arbor, Michigan: University of Michigan, Aug. 2001 (see pp. 53, 235).

[124] J. D. Ramshaw and V. A. Mousseau. Accelerated Artificial Compressibility Method for Steady-State Incompressible Flow Calculations. *Comput. Fluids* 18 (1990), pp. 361–367 (see p. 133).

[125] M. Ricchiuto et al. On Uniformly High-Order Accurate Residual Distribution Schemes for Advection-Diffusion. *J. Comput. Appl. Math.* 215 (2008), pp. 547–556 (see p. 51).

[126] F. Ringleb. Exakte Losungen der Differentialgleichungen einer adiabatischen Gasstromung. *Z.A.M.M.* 20.4 (1940), pp. 185–98 (see p. 248).

[127] P. J. Roache. Verification and Validation in Computational Science and Engineering. Hermosa, 1998 (see pp. 198, 201).

[128] P. L. Roe. Approximate Riemann Solvers, Parameter Vectors, and Difference Schemes. *J. Comput. Phys.* 43 (1981), pp. 357–372 (see pp. 6, 68, 74, 80, 88, 92, 93).

[129] P. L. Roe and M. Arora. Characteristic-Based Schemes for Dispersive Waves I. The Method of Characteristics for Smooth Solutions. *Numer. Meth. Part. D. E.* 9 (1993), pp. 459–505 (see p. 49).

[130] P. L. Roe and H. Nishikawa. Adaptive Grid Generation by Minimising Residuals. *Int. J. Numer. Meth. Fluids.* 40 (2002), pp. 121–136 (see p. 244).

[131] S. E. Rogers and D. Kwak. Upwind Differencing Scheme for the Time-Accurate Incompressible Navier-Stokes Equations. *AIAA J.* 28.2 (1990), pp. 253–262 (see p. 107).

[132] S. E. Rogers, D. Kwak, and C. Kiris. Steady and Unsteady Solutions of the Incompressible Navier-Stokes Equations. *AIAA J.* 29.4 (1991), pp. 603–610 (see pp. 107, 162).

[133] C. J. Roy et al. Verification of Euler/Navier-Stokes Codes Using the Method of Manufactured Solutions. *Int. J. Numer. Meth. Fluids.* 44 (2004), pp. 599–620 (see pp. 198, 277).

[134] J. Roy, M. Hafez, and J. J. Chattot. Explicit Methods for the Solution of the Generalized Cauchy-Riemann Equations and Simulation of Inviscid Rotational Flows. *Comput. Fluids* 31 (2002), pp. 769–786 (see pp. 54, 121).

[135] C. L. Rumsey and J. L. Thomas. *Application of FUN3D and CFL3D to the Third Workshop on CFD Uncertainty Analysis.* NASA TM-2008-215537. 2008 (see p. 198).

[136] H. Schlichting and K. Gersten. Boundary Layer Theory. Eighth. Springer, 2000 (see p. 132).

[137] A. Sescu, A. A. Afjeh, and C. Sescu. Numerical Solution to Nonlinear Tricomi Equation Using WENO Schemes. In: Eighth Mississippi State - UAB Conference on Differential Equations and Computational Simulations. 2010, pp. 235–244 (see p. 55).

[138] A. H. Shapiro. The Dynamics and Thermodynamics of Compressible Fluid Flow. Vol. 1. The Ronald Press Company, 1953 (see p. 191).

[139] A. H. Shapiro. The Dynamics and Thermodynamics of Compressible Fluid Flow. Vol. 2. The Ronald Press Company, 1954 (see p. 249).

[140] A. Sherif and M. Hafez. Computation of Three-Dimensional Transonic Flows Using Two Stream Functions. *Int. J. Numer. Meth. Fluids.* 8 (1988), pp. 17–29 (see p. 167).

[141] T. W. H. Sheu, S. K. Wang, and R. K. Lin. An Implicit Scheme for Solving the Convection-Diffusion-Reaction Equation in Two Dimensions. *J. Comput. Phys.* 164 (2000), pp. 123–142 (see p. 213).

[142] C.-W. Shu. "Essentially Non-Oscillatory and Weighted Essentially Non-Oscillatory Schemes for Hyperbolic Conservation Laws". In: *Lecture Notes in Mathematics*. Ed. by A. Quarteroni. Vol. 1697. Springer, 1998, pp. 325–432 (see p. 242).

[143]   A. M. O. Smith and T. Cebeci. Numerical Solution of the Turbulent Boundary-Layer Equations. Douglas Aircraft Division Report DAC 33735, 1967 (see p. 177).

[144]   G. A. Sod. A Survey of Several Finite Difference Methods for Systems of Nonlinear Hyperbolic Conservation Laws. *J. Comput. Phys.* 27 (1978), pp. 1–31 (see p. 258).

[145]   P. Solin and K. Segeth. Description of Multi-Dimensional Finite Volume Solver EULER. *Appl. Math.* 47.2 (2002), pp. 169–185–315 (see p. 97).

[146]   P. R. Spalart and S. R. Allmaras. *A One-Equation Turbulence Model for Aerodynamic Flows.* AIAA Paper 92-0439. 1992 (see p. 177).

[147]   P. R. Spalart et al. *Comments on the Feasibility of LES for Wings and on the Hybrid RANS/LES Approach.* Advances in DNS/LES, Proceedings of the First AFOSR International Conference on DNS/LES. 1997 (see p. 174).

[148]   Michael Spivak. Calculus on Manifolds: A Modern Approach To Classical Theorems Of Advanced Calculus. Addison-Wesley Publishing Company, 1965 (see p. 27).

[149]   J. L. Steger and R. F. Warming. Flux Vector Splitting of the Inviscid Gas-Dynamic Equations with Applications to Finite Difference Methods. *J. Comput. Phys.* 40 (1981), pp. 263–293 (see p. 101).

[150]   G. Strang. On the Construction and Comparison of Different Splitting Schemes. *SIAM J. Numer. Anal.* 5 (1968), pp. 506–517 (see p. 38).

[151]   M. Strelets. Detached Eddy Simulation of Massively Separated Flows. In: 2001 (see p. 174).

[152]   A. Suddhoo and I. M. Hall. Test Case for the Plane Potential Flow Past Multi-Element Aerofoils. *Aeronaut. J.* (Dec. 1985), pp. 403–414 (see p. 235).

[153]   H. Sun, D. L. Darmofal, and R. Haimes. Flux Vector Splitting of the Inviscid Gas-Dynamic Equations with Applications to Finite Difference Methods. *J. Comput. Phys.* 231 (2012), pp. 541–557 (see p. 211).

[154]   Y. Sun, Z. J. Wang, and Y. Liu. Spectral (finite) Volume Method for Conservation Laws on Unstructured Grids VI: Extension to Viscous Flow. *J. Comput. Phys.* 215 (2006), pp. 41–58 (see p. 219).

[155]   J. C. Tannehill, D. A. Anderson, and R. H. Pletcher. Computational Fluid Mechanics and Heat Transfer. Second. Taylor & Francis, 1997 (see pp. 30, 35, 120, 122, 150, 163, 211, 219, 244).

[156]   H. Tennekes and J. L. Lumley. A First Course in Turbulence. The MIT Press, 1972 (see p. 173).

[157]   J. L. Thomas, B. Diskin, and H. Nishikawa. A Critical Study of Agglomerated Multigrid Methods for Diffusion on Highly-Stretched Grids. *Comput. Fluids* 41.1 (2011), pp. 82–93 (see p. 145).

[158]   G. T. Tomaich. "A Genuinely Multi-Dimensional Upwinding Algorithm for for the Navier-Stokes Equations on Unstructured Grids Using A Compact, Highly-Parallelizable Spatial Discretization". PhD thesis. Ann Arbor, Michigan: University of Michigan, 1995 (see p. 170).

[159]   E. F. Toro. Riemann Solvers and Numerical Methods for Fluid Dynamics: A Practical Introduction. Second. Springer, 1999 (see pp. 42, 116, 134, 255).

[160]   U. Trottenberg, C. W. Oosterlee, and A. Schüller. Multigrid. Academic Press, 2000 (see p. 55).

[161]   E. Turkel. Preconditioning Methods for Solving the Incompressible and Low-Speed Compressible Equations. *J. Comput. Phys.* 72 (1987), pp. 277–298 (see p. 36).

[162]   E. Turkel. Preconditioning Techniques in Computational Fluid Dynamics. *Annu. Rev. Fluid Mech.* 31 (1999), pp. 385–416 (see p. 224).

[163]   E. Turkel. Review of Preconditioning Methods for Fluid Dynamics. *ICASE Report 92-47* (1992) (see p. 36).

[164]   E. van der Weide et al. A Parallel, Implicit, Multi-Dimensional Upwind, Residual Distribution Method for the Navier-Stokes Equations on Unstructured Grids. *Comput. Mech.* 23 (1999), pp. 199–208 (see p. 80).

[165]   B. van Leer. Computational Fluid Dynamics: Science or Toolbox? In: 15th AIAA Computational Fluid Dynamics Conference. AIAA Paper 2001-2520. Anaheim, 2001 (see p. 56).

[166]   B. van Leer. Upwind and High-Resolution Methods for Compressible Flow: From Doner Cell to Residual-Distribution Schemes. *Commun. Comput. Phys.* 1.2 (2006), pp. 192–206 (see p. 65).

[167]   S. Venkateswaran and C. L. Merkle. Analysis of time-derivative preconditioning for the Navier-Stokes equations. In: Fifth International Symposium on Computational Fluid Dynamics. 1995 (see p. 150).

[168] J. M. Weiss and W. A. Smith. Preconditioning Applied to Variable and Constant Density Flows. *AIAA J.* 33.11 (1995), pp. 2050–2057 (see p. 36).

[169] F. M. White. Viscous Fluid Flow. Second. McGraw-Hill, 1991 (see pp. 132, 134, 270).

[170] G. Whitham. Linear and Nonlinear Waves. Wiley-Interscience, 1974 (see p. 57).

[171] D. C. Wilcox. Turbulence Modeling for CFD. DCW Industries, Inc., 1998 (see pp. 173, 177).

[172] B. R. Williams. An Exact Test Case for the Plane Potential Flow about Two Adjacent Lifting Aerofoils. *Royal Aeronautical Establishment, R&M No. 3713* (1971) (see p. 235).

[173] W. A. Wood and W. L. Kleb. 2-D/Axisymmetric Formulation of Multi-Dimensional Upwind Scheme. In: 15th AIAA Computational Fluid Dynamics Conference. AIAA Paper 2001-2630. Anaheim, 2001 (see pp. 97, 98, 142).

[174] Y. Xing and C.-W. Shu. High Order Well-Balanced Finite Volume WENO schemes and Discontinuous Galerkin Methods for a Class of Hyperbolic Systems with Source Terms. *J. Comput. Phys.* 214 (2006), pp. 567–598 (see p. 198).

[175] K. Xu. A Gas-Kinetic BGK Scheme for the Navier-Stokes Equations and Its Connection with Artificial Dissipation and Godunov Method. *J. Comput. Phys.* 171 (2001), pp. 289–335 (see p. 277).

[176] Yan Xu and Chi-Wang Shu. Local Discontinuous Galerkin Methods for High-Order Time-Dependent Partial Differential Equations. *Commun. Comput. Phys.* 7.1 (2010), pp. 1–46 (see p. 48).

[177] H. C. Yee, M. Vinokur, and M. J. Djomehri. Entropy Splitting and Numerical Dissipation. *J. Comput. Phys.* 162 (2000), pp. 33–81 (see p. 242).

[178] H. Yoshida. Construction of Higher Order Symplectic Integrators. *Phys. Lett. A* 150 (1990), pp. 262–268 (see p. 38).

[179] S. T. Yu. Convenient Method to Convert Two-Dimensional CFD Codes into Axisymmetric Ones. *J. Propul. Power* 9.3 (1993), pp. 493–495 (see p. 97).

[180] E. C. Zachmanoglou and D. W. Thoe. Introduction to Partial Differential Equations with Applications. Dover, 1986 (see p. 35).

[181] X. Zhang and C.-W. Shu. Positivity-Preserving High Order Finite Difference WENO Schemes for Compressible Euler Equations. *J. Comput. Phys.* 231 (2012), pp. 2245–2258 (see p. 28).

[182] O. C. Zienkiewicz and R. L. Taylor. The Finite Element Method. Volume 1. McGraw-Hill Company, 1994 (see p. 48).

Extra Contents

What is Katate Masatsuka?

It is a pen name. It was introduced when a series of high-school mathematics books called "Seishun High-School Mathematics" was published in Japan in 2007. "Katatema" is the Japanese word for spare-time or part-time. "Satsuka", which is actually pronounced like 'Sakka', is the Japanese word for writer, author, or artist. Therefore, Katate Masatsuka means a spare-time author. He writes in his spare time.

A pen name is used because these books are written in an eccentric style. The Seishun Mathematics series is a collection of unique stories rather than of formulas and theorems. In "I Do Like CFD", Katate Masatsuka talks about the topics that he likes and explains why he likes them. The intent is to give a kind of impression by which readers can remember each topic easily. For example, students would easily remember what the Cayley-Hamilton theorem is useful for by a story of dating with a girl without a girl along with a mathematical demonstration. In CFD, students would more easily remember casual conversations with a professor more than mere technical explanations, like what a professor personally likes about a particular equation or a numerical method and why he/she likes it.

There is another focus he has in his mind when he writes these types of books. That is to include contents that can hardly be found in other books. The high-school mathematics books contain explanations and interpretations that can hardly be found in other textbooks. The CFD book contains eigen-structures of three-dimensional equations, derivatives of eigenvalues, detailed instructions of computing exact solutions, etc., that can hardly be found in other books. It is, of course, not difficult to figure them out, but it takes time and effort, and people often won't or can't do. Also, there are things that are known in a local community but not widely known perhaps because they may not have been published (e.g., 3D dissipation matrix without tangent vectors). He believes that it is important to share such useful information with everyone for promoting the development of CFD research and education. His intention is, however, not to provide ready-to-use formulas or codes, but rather to provide examples, which he believes are essential elements in learning. As he often says, once you understand what is going on, you will want to develop your own formulas or codes in a more efficient or elegant way.

His books are not just a collection of stuff taken from other books or research papers. Many are known, but others are his original ideas. For example, the simple wave solution for the ideal MHD equations is from his original derivation. He wrote it when he saw a wrong simple wave solution used for verification in a rather prestigious journal paper. He tends to write a note every time he has learned something, apparently in order not to forget what he learned.

As of now (September 2013), he has two tasks in mind. One is to revise the high-school mathematics books and the other is to finish "I do like CFD, VOL.2". To accomplish these tasks, he needs spare time.

"So, who is Katate Masatsuka?"
The answer can be found somewhere in this book.
But it can be anyone by definition.
Do you have spare time?

Katate Masatsuka and Professor Emeritus Haruo Oguro

Professor Oguro was Katate Masatsuka's undergraduate advisor at Tokai University in Japan, who graduated from the University of Tokyo with a major in physics in 1948, then from California Institute of Technology with a major in aeronautics in 1957, taught at universities in the U.S. afterwards, and later joined Tokai Univeristy. He is one of the most influential mentors Katate ever had. He is the reason that Katate got interested in fluid dynamics, that Katate survived graduate school in the U.S., and that Katate was able to write this book. Lots of the contents of this book are from Professor Oguro's course notes and homework problems. The extensive use of gradient, divergence and curl notations largely reflects Professor Oguro's style. The derivation of the exact solution to the Couette flow in Section 7.16.1 was a homework problem (for arbitrary ω) in Professor Oguro's class. Even after Katate graduated Tokai University and started to attend a graduate school in the U.S., he constantly exchanged letters with Professor Oguro for several years. At some point, however, he stopped hearing from Professor Oguro. Later, in 2003, he learned that Professor Oguro had passed away. He remembers Professor Oguro every time he consults the course notes and homework that he still keeps. Then here is a book, "I do like CFD, VOL.1". It is a book as largely influenced by Professor Oguro that will remind Katate of the fond memory of Professor Oguro, and will let him share Professor Oguro's influence with others who like fluid dynamics.

Dr. FUN3D

FUN3D is a suite of CFD simulation and design tools developed at NASA Langley Research Center. Katate Masatsuka, as soon as he joined the FUN3D team in 2007, created a character called "Dr. FUN3D" as shown above. It is printed on the mouse pad that he uses to work on FUN3D. It is one of his spare-time activities to create characters made of letters, typically from names. Many examples can be found in his online gallery: `http://www.ossanworld.com/art.html`.

Dear Readers

I'm so glad that my book has found a place to sit on your desk or book shelf. I hope that it will serve your purpose and you will find the book useful. If any comments/suggestions should occur to you, please feel free to send them to me. I'll be very happy to receive feedback from you.

I believe that it doesn't matter who the author really is. All that matters is the value of the contents. I myself find this book very useful. I have three copies myself and often consult them when I work on CFD. Nevertheless, as I read it myself, I feel that it is not yet comprehensive enough. I hope that I can continue to revise it and make future updates. Your purchase is a great encouragement.

Thank you very much,
Katate Masatsuka